ENEMIES OF HOPE

Enemies of Hope

A Critique of Contemporary Pessimism

IRRATIONALISM, ANTI-HUMANISM AND
COUNTER-ENLIGHTENMENT

First published 1997 by
MACMILLAN PRESS LTD
Houndmills, Basingstoke, Hampshire RG21 6XS
and London
Companies and representatives
throughout the world

ISBN 0–333–61109–8

A catalogue record for this book is available
from the British Library.

This book is printed on paper suitable for recycling and
made from fully managed and sustained forest sources.

10   9   8   7   6   5   4   3
06   05   04   03   02   01   00   99   98

Printed and bound in Great Britain by
Antony Rowe Ltd, Chippenham, Wiltshire

Published in the United States of America by
ST. MARTIN'S PRESS, INC.,
Scholarly and Reference Division
175 Fifth Avenue, New York, N.Y. 10010

ISBN 0–312–17326–1

To Chris and Kelly Verity – friends of hope – in gratitude for
forty-seven years of wonderful friendship

Contents

Acknowledgements

The author and publisher would like to thank the following who have kindly given permission for the use of copyright material:

Little, Brown and Company, for extracts from Dudley Young, *Origins of the Sacred*, 1991.

Faber and Faber Ltd for extracts from the *Collected Poems* of T.S. Eliot.

John Murray Ltd for extracts from *The Crooked Timber of Humanity*, edited by Henry Hardy, 1991.

It is a special pleasure to acknowledge the continuing kindness and support of Charmian Hearne and Tim Farmiloe of Macmillan, without whom this book, and its predecessors, would not have seen the light of day. I can only repeat what I have said on the occasion of a previous publication: it is difficult to imagine how I could have been better treated. Many, many thanks.

I would also like to thank Ruth Willats for her excellent editorial labours on the manuscript.

Preface

This book could have just as well been called *Enemies of the Enlightenment*. It has a filial relation to other volumes I have published over the last few years – *Not Saussure* (Macmillan, 1988, 1995), *In Defence of Realism* (Edward Arnold, 1988), *The Explicit Animal* (Macmillan, 1993) and *Newton's Sleep* (Macmillan, 1995) – taking up and developing themes and arguments figuring in those books. In one respect, *Enemies of Hope* is an explanation of why I felt so strongly about the often rather recherché issues dealt with by its predecessors: it exposes some of the passions behind their more detailed or technical discussions.

Enemies of Hope has no pretence to being a work of primary scholarship. In my discussion of both the Enlightenment and subsequent anti-Enlightenment figures it will be obvious that I have been heavily dependent upon secondary sources. No one with a life of finite duration could hope to be an expert on the entire *œuvre* of figures as prodigiously productive and diverse as Freud, Marx, Durkheim, Helmholtz, the contemporary structuralist and post-structuralist thinkers, or to have an intimate first-hand acquaintance with contemporary ethological, anthropological, archaeological, etc. writing. The only areas where I can lay claim to any kind of expertise are in twentieth-century American and European philosophical thought, the writings of post-Saussurean theorists, and modern biology and medicine. My greatest single debt is to the brilliant and passionate essays of Isaiah Berlin, whose masterpieces of scholarship, read or re-read when this book was well underway, changed its character dramatically, determining the content of the Prologue, helping me to see the underlying theme of the succeeding chapters and shaping the agenda of the final discussion. A not inaccurate, if rather inelegant alternative title for the entire book, could be *Yes-But to Berlin's 'Yes-But to the Enlightenment'*.

Reliance upon secondary sources is not as disabling as may at first sight be imagined, especially if the secondary sources are chosen with discrimination and one depends upon writers such as Isaiah Berlin and Dudley Young rather than, say, Paul Johnson or James Burke. For this book is predominantly concerned with how the writings of contemporary thinkers have entered into common currency and determined the received ideas and the 'self-image' (to use Alasdair MacIntyre's term) of the age. *Enemies of Hope* should be read as an argument, even a polemic, rather than an original contribution to the history of ideas.

In the pages that follow I take issue with a web of notions, attitudes and prejudices that have been important in Counter-Enlightenment thought over the last two centuries or so, and which seem at present to be again in the ascendant. Except in the first chapter, my focus is on twentieth-century Counter-Enlightenment thought, which is characterised above all by scepticism, or even hostility, towards the idea of a human being as a conscious, rational agent and of human society as susceptible to progressive improvement as a result of the efforts of conscious, rational agents. Out of this negative assumption, at the heart of the web, arise many disparate filaments: mockery at the hope of progress; rejection of technology; scornful dismissal of the claims of scientific truth and of the distinction between science and magic; contempt for the democratic process; a profound suspicion of rationalism, rooted in a debased view of the true nature of human beings; and a fixed and cynical belief that objectivity is a mere mask for vested interests and that the sense of justice and fair play is mere rhetorical mist.

I have written this book out of irritation with ideas that are often wrong and – since anti-rationalists implicitly claim a validity and scope for their own positions that even the most universalising rationalist would not dare entertain – pragmatically self-refuting. I believe that such ideas not only reflect the contemporary *Zeitgeist*, but actively – and adversely – influence it. Historians of ideas, properly exaggerating the importance of their special area of interest, would like to believe that large-scale philosophical beliefs are the motor of history. This is probably naive; but there is no doubt that ideas have a shaping force: they change history as well as being secreted and absorbed by it. Just as I do not accept that the horrors of the mass slaughter, or the persecutions, imprisonment and enslavement, of individuals in modern times, were incubated in classrooms where a post-Enlightenment secularised conception of humankind has been taught, I likewise do not blame Darwin for the brutalised social practices legitimised by the writings of the Social-Darwinists, or Herder for *volkisch* ideology that the Nazis invoked to justify their *Judenpolitik*. Nevertheless, I feel that the contemporary attack on Enlightenment values carries great dangers. At the very least, it is part of a process by which contemporary humanity is talking itself into a terminal state of despair, self-disgust and impotence. The self-congratulatory pessimism of many humanist intellectuals who have 'seen through' the pretensions of mankind, and dismiss science and, equally, the hopes of those who want to bring about a better future, is likely to be self-fulfilling. If we do not believe in the possibility of a better future – brought about by progressive and patient social reform, and by the application of ever-more effective technology – that future will surely not come about.

We have a responsibility to those who otherwise will be without hope to keep alive the hope of progress. To deny that hope, and to mock the agents and the means by which it might be fulfilled, is, as Peter Medawar said, the last word in meanness of spirit. Such meanness of spirit should not be allowed the last word.

Prologue
Isaiah Berlin's 'Yes-But to the Enlightenment'

INTRODUCTION

Most historians would concur with Lucien Goldmann's assertion that 'Eighteenth-century France is the country of the Enlightenment in its most developed and most thorough form'.[1] If this is the case, then it would appear, as Berlin has noted, that opposition to its central ideas 'is as old as the movement itself'.[2] This is less surprising when it is appreciated that many of these ideas grew out of a climate of opinion and from intellectual attitudes that had been established at least a century before the emergence of the archetypal Enlightenment thinkers – those *philosophes* associated with the *Encyclopédie*.

The refusal to accept custom and culture as the basis for belief, and the cultivation of scepticism and systematic doubt, is of course traceable back to Descartes' *Discourse on Method*, published in 1637.[3] Francis Bacon, who died in 1626, was an even earlier and possibly yet more powerful influence. His 'utilitarian empiricism', his dream of a kingdom of man founded on the sciences and his critique of the 'idols of the mind' that stand in the way of the acquisition of objective knowledge and human progress, was an inspiration to the *philosophes*.[4] Major elements of the latter's world-picture were also derived from the epistemological and political writings of John Locke, whose most important and influential works – *Essay Concerning Human Understanding* and *Two Treatises of Civil Government* – had been published in 1690. And then there was Isaac Newton, whose achievements and personal example were of the utmost importance in firing the hopes of the *philosophes* and showing them what might be achieved by a critical mind using the right method. Had he not, after all, revealed the inner principle underlying the clockwork of the universe and shown how nature – and hence man and society (which, the Encyclopaedists importantly believed, were parts of nature) – was in principle amenable to progressively greater understanding? Newton had established his European reputation by the end of the seventeenth century.

In short, Enlightenment thought did not emerge abruptly in the eighteenth century: its coming was well prepared. It is less surprising,

1

therefore, that the voices of dissent were heard so soon after the Enlightenment had come to consciousness of itself. Hamman (1730–88),
traditionally regarded as the first major Counter-Enlightenment figure, was born only seven years after d'Holbach (1723–89), the Enlightenment's most systematic spokesman. Herder (1744–1803),
Hamman's friend and disciple, launched his major critique of the
universalistic pretensions of Enlightenment thought – and his defence
of cultural pluralism and affirmation of the inner reality of different
value-systems – in 1769, over 20 years before the universalist dreams
of the Encyclopaedists received their clearest and most vulnerable
expression in Condorcet's *Sketch for a Historical Picture of the Progress
of the Human Mind* (1795).

It is arguable that, in the person of Rousseau, Enlightenment and
Counter-Enlightenment were born twins. Inasmuch as Rousseau was
fired by outrage at the irrational hierarchies based on custom and
tradition that dominated the civilised world ('Man is born free but
we see him everywhere in chains') he was a representative Enlightenment figure. But when he argued that these inequities, and the iniquities that flowed from them, were intrinsic to a contractual civilisation
that had supplanted the family-based society of the Noble Savage,
his voice was assuming one of the characteristic registers of the Counter-
Enlightenment. When, moreover, he severely restricted the role of reason
in public affairs, preferring to appeal to faith and feeling, he left an
ambiguous message. His assertion that right action could not be derived from right thinking, that reason was not enough to make us
behave reasonably, anticipated a strain of thought that was to become evident in thinkers as diverse as, and with as disparate agen-
das as, Coleridge, Kierkegaard and Freud. Rousseau's notion of the
Social Contract based upon the voluntary subjection of the individual
to the General Will is consistent with Enlightenment thought in its
recognition of the equal validity of all as contributors to the body of
the Sovereign and/Counter-Enlightenment in its identification of the
will – of unexamined, unreflective, impulses rather than reasoned views
– as the element the individual can and should contribute. The ambiguity of Rousseau's ideas is reflected in the way they influenced both
the liberal Kant and the conservative Hegel. In the muddled genius
of his soul, Rationalism and Romanticism – antecedents of, and reactions against, Enlightenment – engaged in bitter civil war.[5]

Enlightenment thought, then, has, almost throughout its history, been
inseparable from a Counter-Enlightenment shadow. The terms in which
their opposition has been expressed have changed many times in their
250-year dispute. The evolution of the quarrel may in part reflect the
fact that, in some respects and in some parts of world, the Enlightenment programme has been achieved: autocratic hierarchies based on

tradition – monarchies, aristocracies – have been displaced or rendered less powerful, or at least have become less secure; social gradi-
ents are less steep; rational contract has displaced custom; ignorance
of Nature (which d'Holbach identified as the mainspring of human
unhappiness) has been greatly amended by scientific investigation,
with consequent huge advances in human capacity to enrich life and
palliate suffering; and the power of the priesthood to determine what
people think and the choices they make in life has in many places
been greatly curtailed. The critique of the Enlightenment and its dream
of a progressively liberated and materially improved humanity is, in
Europe at least, now rarely conducted on behalf of a deposed priesthood
or an exiled monarchy reasserting the divine right of kings. Most of
the broad aims and ideals of the *philosophes* – if not their Utopian
dreams – are, theoretically at least, shared by the governments of many
countries.

The contemporary critique, however, has deep roots which draw
nourishment from the folk memories of two catastrophes that took
place soon after Enlightenment thought reached its ascendancy: the
politico-military catastrophe of the Terror and the Napoleonic Wars;
and the civil catastrophe of the First Industrial Revolution, where life
and labour were brutalised beyond anything that had been seen before. Both of these are still seen by some as inevitable consequences
of Enlightenment thought. And for many contemporary commentators, the lamentable record of slaughter and oppression in the twentieth century – total war and totalitarian regimes – simply combines
the horrors of both catastrophes in the industrialisation of terror. 'The
melancholy commonplace that no century has seen so much remorseless and continued slaughter of human beings by one another as our
own',[6] is seen by some as a remote consequence of the Enlightenment, or even as an expression of Enlightenment values.

This is a desperate and terrible libel upon what Berlin has described
as 'one of the best and most hopeful episodes in the life of mankind'.
Moreover, as I shall argue in due course, there are more persuasive
connections between Counter-Enlightenment (and anti-Enlightenment
thought more widely construed) and the most lamentable episodes in
twentieth-century history. First, however, it will be useful to set out
the essential elements of Enlightenment thought.

ELEMENTS OF ENLIGHTENMENT THOUGHT

At the heart of the philosophy of the Enlightenment was a belief in
the power of reason informed by empirical experience, with a corresponding rejection of unthinking prejudice and argument from authority.

The authority of reason had been dramatically reasserted by Descartes:
'We ought never to allow ourselves to be persuaded of the truth of
anything unless on the evidence of our reason.'[7] This, as Ernest Gellner
says, is a declaration of war:

> The battle-lines are now clear: individual reason versus collective
> culture. Truth can be secured only by stepping outside prejudice
> and accumulated custom, and refashioning one's world. It can only
> be achieved by means of proudly independent, solitary Reason. We
> pursue it rationally and we do it alone.
>
> (Gellner, p. 8)

Descartes' *Discourse on Method* had been the inspiration of the intellectual giants of the Age of Reason – the mathematicians and meta-
physicians of the seventeenth century, whose great exemplars were
Spinoza and Leibniz.

But there was a counter-current of thought active even at the height
of The Age of Reason. Its most celebrated figure was, of course, Pascal,
for whom the heart had its reasons that reason knows not, and who
could not forgive Descartes for reducing the role of faith and marginalising God. But, from the point of view of the Enlightenment, Pierre
Bayle was a more important critic of rationalism. Like Pascal, he argued that reason alone was not enough. Granted, it was a powerful
weapon in the battle against uncritical acceptance of traditional ideas;
but it was insufficient to generate, or even to support, positive beliefs:

> The same theory which serves as a weapon against error, may some-
> times do a disservice to the cause of truth. One thing you will always
> find if you look long enough, and that is antinomy, contradiction.[8]

The untrammelled application of reason would always result in
unalleviated scepticism. Bayle concluded from this that man must rely
on faith to justify his belief that there is a real external world and his
equally unprovable belief that God is not committed to deceiving man
as to his true nature and that of the world. He would have agreed
with Hamann that 'Reason is not given to you that may become wise,
but that you may know your folly and ignorance',[9] and that, like 'the
Attic philosopher, Hume', we need faith when we eat an egg or drink
a glass of water:

> faith is not a work of reason and therefore cannot succumb to any
> attack by reason; because believing happens as little by means of
> reasons as tasting and seeing.
>
> (Hamann, p. 182)

For Voltaire and the *philosophes* the emptiness created by the pure use of reason and methodological doubt could be filled by empirical experience.[10] Locke was the main philosophical sponsor of the Enlightenment belief in the primacy of the evidence of the senses. His assertion that the sole source of positive knowledge came from the senses connected in the mind of the *philosophes* with the inspiring example of Newton, whose great achievements were the result of the sharpest observations combined with the most elegant mathematics, and the Baconian philosophy of science which emphasised induction.[11] Observation and experiment provided the positive content of knowledge that could be organised into powerful general principles by reason, typically in its mathematical form. Both were necessary: as Kant, a late apostle of the Enlightenment, famously proclaimed in his *Critique of Pure Reason*, 'If ideas without experiments are empty, experiments without ideas are blind.'

Goldmann has summarised the Enlightenment philosophy of science, which also underpinned their social philosophy:

> The majority of the thinkers of the French Enlightenment occupied a third position, intermediate between rationalism and empiricism. They ... denied the existence of innate ideas ... holding that individual consciousness is invariably based on experience. None the less they generally acknowledged, expressly or by implication, the active role of reason in collecting the knowledge which has been acquired through perception and preserved in the memory, organising it in the form of thought and science, and directing action, under the influence of feeling, towards the greatest happiness of the individual.
> (Goldmann, p. 19)

Reason, in its twin role as a critique clearing away the unfounded customary and traditional assumptions that lie in the path to true knowledge and as a means of organising empirical data into laws which can preferably be expressed in mathematical form, united with observation in advancing human understanding.

This was not merely an epistemological position; it was also an ethical one. Reason and empirical science would be essential elements in the fight against the inequities and iniquities that have added immeasurably to human suffering and blocked the way to progress. The refusal to accept the validity of anything for which one could see neither the reason nor the evidence was also a rejection of the authority of those who have hitherto claimed unique possession of the truth. Bayle had already argued for tolerance on the grounds that no one could be in secure possession of religious truth. In future, religious belief was to be a matter of individual conscience and to use force in such matters was utterly repulsive:

The ethic basic to the Enlightenment is tolerance.
see pp. 21 & 165
but what happens when tolerance is not reciprocal?

That is the conclusion we are bound to come to, the conclusion, I mean, that any particular dogma, whatever it may be, whether it is advanced on the authority of the Scriptures, or whatever else may be its origin, is to be regarded as false if it clashes with the clear and definite conclusions of natural understanding, and that more particularly in the domain of ethics.[12]

Bayle's outrage at a book that praised Louis XIV for having used his power to make his country wholly Catholic again, instigating a campaign of persecution culminating in the brutalities that followed the revocation of the Edict of Nantes, was exemplary:

It is now a long time since those who arrogate to themselves the name of Catholic *par excellence*, have been perpetrating deeds that excite such horror in every human heart that any decent person must regard it as an insult to be called a Catholic. After the evils you have wrought in that most Christian kingdom of yours, it is evident that to speak of the Catholic religion and the religion of the unrighteous is one and the same thing.

(quoted in Hazard, p. 128)

Bayle bequeathed this call for religious tolerance, and his conviction that intolerance was profoundly immoral, to the Enlightenment thinkers. Voltaire's reading of Locke and his experience of living in the relatively tolerant England of the 1720s confirmed religious tolerance as the correlative of his critique of authority.

Descartes had attacked custom and tradition but only by using customary and traditional methods; it could be argued that his rationalism, even in the form of methodological doubt, was an inverted form of scholasticism. Locke's critique of authority, which questioned even the epistemological legitimacy of innate ideas, was more radical. It created a heightened sense of possibility. Kant's characterisation of the Enlightenment captures this sense:

Enlightenment is man's emergence from his self-imposed minority. This minority is the inability to use one's own understanding without the guidance of another. It is self-imposed if the cause lies not in a lack of understanding, but in the courage and determination to rely on one's own understanding and not another's guidance. Thus the motto of the Enlightenment is *'Sapere aude!'* Have the courage to use your own understanding.[13]

These, then, are the elements of Enlightenment thought: critical individualism hostile to an authority – both in matters of knowledge

De puis Voltaire l'histoire s'est enfermée dans une difficulté qui déchire toute littérature engagée, et que Voltaire n'a pas connue: pas de liberté pour les ennemis de la liberté: l'homme ne peut plus donner de leçon de tolérance à personne. Roland Barthes

and of power – based upon custom and tradition; tolerance of differences of belief in areas where there could be no certainty; and a faith in reason and unprejudiced observation, more specifically in its most dramatic and regulated expression in science. The cooperative, universalistic enterprise of natural science was the paradigm of how progress might be made. Like science, Enlightenment thought was fundamentally democratic: it presupposed the equality of all men and their equal right to freedom and self-development. It identified two main barriers to human happiness: ignorance (of nature, of the right way to govern societies so as to ensure the greatest happiness of the greatest number); and the obstacles placed in the way of progress by those whose vested interests were threatened by the advancement of the poor. These reactionary elements naturally preferred to maintain the mass of the population in a state of ignorance; for custom, tradition and authority, which underpinned a society that was hierarchical, unjust and intolerant, had served them well.

Using the faculty of reason supported by the growth of empirical knowledge, seasoned by irony and wit, driven by scepticism, irreverence and anger, eschewing dogma, mystical inner light, and the voice of supernatural authority, the members of the poignantly self-styled 'Party of Humanity', worked for social reform, embarking on what they believed would be an indefinite improvement in the lot of mankind. The barriers to advancement of human happiness would be pulled down one by one. Science would increase our understanding of nature, and, since humans were part of nature, our understanding of ourselves. Science would help men to know their true needs and how to fulfil them; the methods that had led to a deeper understanding of nature could also be applied to society and mankind considered as a piece of nature – to improved methods of government, and to developing ever-more successful cooperative enterprises and institutions. Scientific knowledge and a scientific approach, combined with a programme of universal education to disseminate new knowledge, would ensure that men could more effectively pursue the goal of eliminating unmet material needs, and even of achieving freedom, and possibly even of happiness. For many of the *philosophes*, Nature itself would replace God as the place of transcendence. When the people are freed, the will of the people will coincide with the will of Nature and of Reason and, in its new form of Nature, the Supreme Being will be worshipped at the Feast of Reason.

We have already alluded to the 'autocritique of the Enlightenment' within Rousseau's thinking and it is important to acknowledge that there were serious disagreements between individuals and also between schools of thought, as Berlin has pointed out:

Locke believed in intuitive truths in religion and ethics, while Hume
did not; Holbach was an atheist, like most of his friends and was
castigated for this by Voltaire. Turgot . . . believed in inevitable
progress; Mendelssohn did not, but defended the doctrine of the
immortality of the soul, which Condorcet rejected. Voltaire believed
that books had a dominant influence on social behaviour, whereas
Montesquieu believed that it was climate, soil and other environmental factors that created unalterable differences in national character
and social and political institutions. Helvetius thought that education and legislation could by themselves wholly alter, and indeed
perfect, the character of both individuals and communities; and was
duly attacked for this by Diderot. Rousseau spoke of reason and
feeling but, unlike Hume and Diderot, suspected the arts and detested the sciences, laid stress on education of the will, denounced
intellectuals and experts and, in direct opposition to Helvetius and
Condorcet, held out small hopes for humanity's future. Hume and
Adam Smith regarded the sense of obligation as an empirically
examinable sentiment, while Kant founded his moral philosophy
on the sharpest possible denial of this thesis; Jefferson and Paine
considered the existence of natural rights to be self-evident, while
Bentham thought this nonsense on stilts, and called the Declaration
of the Rights of Man and Citizen bawling on paper.

(Berlin, *The Crooked Timber of Humanity*, pp. 106–7)

Peter Gay, who is similarly deeply conscious of the differences between the men of the Enlightenment, and who resists the temptation
to treat the Enlightenment as a compact body of doctrine, also refuses
to fall into 'despairing nominalism' and he suggests that the *philosophes*
should be thought of as 'a family of intellectuals united by a single
style of thinking' (Gay, *The Enlightenment*, p. x). We may characterise
this style as a faith in reason and science, a burning hatred of injustice and inequity and of the institutions that maintain the status
quo, and a belief in the possibility of progress once the shackles of
prejudice and of oppressive social institutions have been shaken off.
In so far as there was a common aim, this could be characterised as
the liberation of men from fear and servitude and the restoration to
individuals of the right to sovereignty over their own lives. For the
philosophes the mission of man:

which gives meaning to his life, lies in the effort to acquire the
widest possible range of autonomous and critical knowledge in order
to apply it technologically in nature and, through moral and political action, to society. Furthermore, in acquiring his knowledge, man
must not let his thought be influenced by any authority or any

prejudice; he must let the contents of his judgements be determined only by his own critical reason.

(Goldmann, p. 2)

CRITIQUE OF THE ENLIGHTENMENT (1) EPISTEMOLOGY

Criticism of the Enlightenment has ranged in tone from temperate, reasoned dissent to screaming invective. Some of the more measured criticisms have come from those who felt that Enlightenment epistemology did not provide a basis for the very knowledge, not to say values, upon which the *philosophes* pinned their hopes for progress. The *philosophes* failed, they argued, to see that perception, or the evidence of the senses, and reason are not together sufficient to give an individual the knowledge necessary to address the important questions of life. At the most fundamental level, reason is not sufficient to organise sense impressions in order to create a coherent world. The world of the empiricist is a blizzard of fugitive impressions which is hardly likely to give rise to the notion of a stable reality 'out there'; psychological atomism and senationalism, gave no grounds, as Coleridge pointed out, for distinguishing between ordinary experience and raging delirium. Locke's descendants, such as Hume and Hartley, made the *tabula rasa* mind too passive, too mechanically composed and driven by the impingements of an external material reality whose ontological status was, as Berkeley and Hume himself had shown, problematic. It was not possible to reason oneself from a world of impressions and ideas (in Hume's sense) to a physical world of material objects, 'out there', set off from and yet accessible to the experiencing subject.

The most famous response to this epistemological defect in Enlightenment thought was Kant's *Critique of Pure Reason*. Kant argued that experience was itself structured and synthesised by the apparatus of the human sensibility and understanding; experience is possible only within certain formal boundaries – the so-called 'forms of sensible intuition'. It is through the workings of the understanding that sense experience is ordered into experience of an objective world. We ourselves introduce the order and regularity in objects, which we call nature. The understanding is itself the lawgiver of nature.

There is a sharp irony in Kant's position as the demolisher of the epistemological basis for Enlightenment thought. For he was, as we have noted, very much a child of the Enlightenment and a supporter of its aims and principles; and yet *The Critique of Pure Reason* has inspired a rich and wide tradition of idealistic thinkers for whom the nature of the external world refers us back to the nature of the human mind – and beyond this to the Universal Mind reflected in individual minds; inspired thinkers, in short, whose beliefs and arguments and passions pointed in the opposite direction to the Newtonian world of the Enlightenment. One has only to list Fichte, Hegel, Schopenhauer (not to speak of creative artists such as Kleist) to make the point.

Kant's postulation of a deep interrelationship between knowledge and its objects – so that the latter were a kind of internal accusative of the former – not only undermined the Lockean epistemology of the Enlightenment, in which experience resulted from the impingement of objects upon a mind understood as something entirely outside of them. His postulated divorce between human agency (and the moral law within) and material reality (the unknowable world of noumena) also sounded the death knell of the Enlightenment view of man as a material piece of material nature, and consequently undermined the hope of understanding him through the means employed by the natural sciences. At a less fundamental level, his thought opened the way to a tradition of thought in which knowledge was embedded in a dialectical relationship with the situation of the knower and his/her actions, and above all with values. By denying that knowledge was part of a whole, as opposed to something outside the whole, the Enlightenment had separated knowledge from value. Kant reunited them and thus began the process of undermining the notion of timeless, disinterested, objective knowledge. That was not what he intended, but was certainly how he was interpreted – and used.

CRITIQUE OF THE ENLIGHTENMENT (2)
THE PARADOX OF NATURAL VALUES

There was, anyway, a major difficulty inherent in the Enlightenment notion of *natural* values. For there was nothing in nature itself – or the laws, derived by reason reflecting on experience, that mirrored nature – to explain why anything in nature should be valued, even less why it should be cherished. The material world does not have the means to value itself and we, who are supposedly part of material nature, have nothing in us to explain how we come to value anything – how we take pleasure in some things, and dislike others, never mind how we come to have those 'higher values' necessary to underpin our concern for others. Not all the Enlightenment thinkers would have subscribed to La Mettrie's view that there was no distance between us and the natural world:

> The soul, then, is an empty symbol of which one has no conception, and which a sound mind could employ only in order to denote

that which thinks in us. Given the least principle of movement, animate bodies will possess all they need in order to move, sense, think, repeat, and behave, in a word, all they want of the physical; and of the mental, too, which depends thereon.[14]

But even without such extreme materialism, which denied any distance between mankind and the material world, it is difficult to see where the realm of distinctively human values could be accommodated in a world-picture dominated by the paradigm of physical science and in which there was explicit commitment to the idea of man as part and parcel of nature.

Even reason was problematic: it had a double status as both a natural phenomenon ('the brain secretes thought as the liver secretes bile', as Cabanis, La Mettrie's disciple explained) and as that by means of which nature at large and human nature in particular will be understood. In this context, Kant's Critiques may be seen as a way out of the tension he perceived in Hume's rationalistic philosophy, and which saw human nature as both rational and natural, as being able to understand and reason about nature whilst, at the same time, being itself a part of nature, a nature which enables us to *judge*, as well as to breathe and feel. That tension was most manifest, of course, in Hume's commitment to science, whilst he failed to provide epistemological justification for the mechanistic world-picture of the physicists (including the 'matter' of which it was largely composed) on the basis of his own account of experience and the sources of justified belief. Kant met this difficulty by arguing that we are not simply effects of the natural world that impinges upon us, but, in a sense, create that world out of faculties that are not too remote from the faculty of reason we apply to it. By this means we are granted sufficient transcendence from nature to be able to judge it, act upon it and import distinctively human values into it. It was a solution that gave birth to many more questions than it answered and to deeply anti-Enlightenment world-pictures that would have made Kant shudder.

There are, then, fundamental problems in Enlightenment thought: the epistemological one of accounting for our ordered world-picture; and the axiological one of accounting for our *values*. This last becomes particularly urgent when we try to account not merely for appetites, but for those impulses that inform the civic virtues necessary to ensure the health of society – the very goodness upon which the Enlightenment thinkers' hopes for a better future depended so crucially.

Goldmann has set out the Enlightenment difficulty over the origin of our values very clearly:

Between the age of traditional Christianity and the beginning of

dialectical philosophy there grew up the great individualist traditions which have continued to develop to this day: rationalism,
empiricism and the Enlightenment. These traditions dispensed with
all trans-individual concepts of God, community, totality and being. In doing so, they completely separated the two forms of individual consciousness, knowledge of facts and judgement of values.
Science had become 'morally neutral' in the seventeenth century,
and the problem of the Enlightenment was to find some other objective basis for value-judgements.

(Goldmann, pp. 25–6)

The great challenge was to find grounds for the assumption, which
lay at the heart of Enlightenment optimism, that, when both are properly
understood, private and public interest coincide. This challenge is the
equivalent, on the axiological plane, of the epistemological challenge
to find the basis upon which the delirium of individual impressions
and ideas could disclose a stable, coherent, intersubjective world to
which many millions of consciousnesses co-subscribe.

The belief that private and public interest, when seen aright, would
coincide belongs to a very long tradition of thought, reaching back to
Socrates. It was most poignantly expressed by Condorcet in his final
testament:

Is not a mistaken sense of interest the most common cause of actions contrary to the general welfare? Is not the violence of our
passions often the result either of habit that we have adopted through
miscalculation, or of our ignorance of how to restrain them, tame
them, deflect them, rule them?
(*Marquis de Condorcet, Sketch for a Historical Picture of the Progress
of the Human Mind*)

This strain in Enlightenment thought, its tentacular roots, its complex
foliage, has been lucidly encapsulated by Berlin:

They believed in varying measure that men were, by nature, rational and sociable; or at least understood their own and others'
best interests when they were not being bamboozled by knaves or
misled by fools; that, if only they were taught to see them, they
would follow the rules of conduct discoverable by the use of the
ordinary human understanding; that there existed laws which govern
nature, both animate and inanimate, and that these laws, whether
empirically discoverable or not, were equally evident whether one
looked within oneself or at the world outside. They believed that
the discovery of such laws, and knowledge of them, if spread widely

enough, would of itself tend to promote a stable harmony both between individuals and associations, and within the individual himself.

(The Crooked Timber of Humanity, pp. 107–8)

There is, of course, no guarantee that there will be a perfect fit between the individual and the common interest, or that where there is, individuals can be persuaded to reflect long enough in order to see this fit and modify their behaviour such as to choose only those self-interested actions that will support the public welfare. This optimistic assumption seems even more questionable in the context of Enlightenment belief in the 'natural nature' of man. To quote Berlin again:

They [the Enlightenment thinkers] thought that education and legislation founded upon 'the precepts of nature' could right almost every wrong; that nature was but reason in action, and its workings were in principle deducible from a set of ultimate truths like the theorems of geometry, and latterly of physics, chemistry and biology. . . . The more empirically-minded among them were sure that a science of human nature could be developed no less than a science of inanimate things, and that ethical and political questions, provided that they were genuine, could in principle be answered with no less certainty than those than those of mathematics and astronomy. A life founded upon these answers would be free, secure, happy, virtuous and wise.

(p. 108)

Since the eighteenth century, the reputation of nature (and consequently of the human animal as a part of nature) for beneficence has taken something of a nose-dive. Darwin revealed nature as a battleground in which the welfare of individual organisms was subordinated to that of the species and the welfare of one species was likely to be consumed in a process whereby species in general were honed to ensure progressively better adaptation to a hostile environment. And even before *The Origin of Species*, the findings of geologists such as Lyell had captured the public imagination with a vision of nature as a slaughterhouse. Nature 'red in tooth and claw' cared for nothing, not for individuals nor even for species, as grieving Tennyson observed in *In Memoriam*:

'So careful of the type?' but no.
From scarped cliff and quarried stone
She cries, 'A thousand types are gone:
I care for nothing, all shall go.'[15]

For twentieth-century neo-Darwinian thought, the interest of the individual soma is subordinated to the fundamental imperative to ensure the survival of the genome, a *forma informans* that outlives all mere instances and types. This objective is best served by ruthless and unremitting competition between individuals, resulting in the survival of increasingly better adapted vehicles for carrying 'the selfish gene'. The true nature of nature is better reflected in the 'arms race' between predator and prey, described vividly by Richard Dawkins in *The Blind Watchmaker*,[16] than in the operation of geometrical or other reason. If man is a successful piece of nature, if he survives and flourishes, he is more likely to be a predator than a benefactor; and the chances of public and private interest coinciding are therefore slim indeed.

Strictly naturalistic accounts of humanity do not hold up. As I have argued at length elsewhere[17] – and I shall return to these arguments in the present book – humanity cannot be reduced to animality: man 'the Explicit Animal' cannot be understood purely as an animal, or, indeed, as a piece of nature. Although, as already noted, few Enlightenment thinkers fully subscribed to La Mettrie's total assimilation of human beings into the clockwork of the physical universe, naturalistic accounts of humanity are nevertheless central to the world-picture of the *philosophes*. They underpin something that is deeply unsatisfactory in Enlightenment thought – its too-ready presupposition that there is the basis of a universal secular value-system to be found in nature, a source of values that would be sufficient to hold any human society together and to ensure that it would utilise the fruits of scientific investigation to bring about the progressive improvement of collective life.

Critics of the Enlightenment have tended to focus on one or the other facet of this assumption: either the belief that human beings are so formed that their natural impulses, so long as they are uncorrupted by the mystifications and menace of those who would suborn them to their own evil purposes, will serve the common good; or the belief that these impulses, and the values they serve, are universal. These have both been the subject of ferocious criticism and they, and their critiques, warrant separate examination.

CRITIQUE OF THE ENLIGHTENMENT (3)
THE UNTRUSTWORTHINESS OF MANKIND

The most savage critic of Enlightenment thought was Joseph de Maistre, whom Berlin has dubbed 'the Voltaire of reaction'.[18] His thought has been brilliantly summarised in Berlin's essay 'Joseph Maistre and the

Origins of Fascism', from which I have already quoted. Maistre's fundamental belief was that man – and nature – are intrinsically evil:

> In place of the ideals of progress, liberty and human perfectibility, he preached salvation by faith and tradition. He dwelt on the incurably bad and corrupt nature of man, and consequently the unavoidable need for authority, hierarchy, obedience and subjection. . . . he defended the importance of mystery and darkness – and above all of unreason – as the basis of social and political life.
>
> (Berlin, pp. 108–9)

Men are not only evil; they are also anti-rational. Successful societies take account of that and are not themselves founded upon reason: they are organisms that have grown over time, and their growth has been regulated by traditions that have little to do with reason. It follows that any attempt to reconstitute, reform or revolutionise society on the basis of rational principles is highly dangerous. Societies are intrinsically unstable, and tinkering will lead to chaos and disorder: 'men can only be saved by being hemmed in by the terror of authority' (p. 118), an authority underwritten by the executioner who must be ever-active in his terrible ministrations. Reason is too weak, a mere 'flickering light', to be the foundation of civic society. The bloodbath of September 1792 was an inevitable, not an accidental, consequence of the arrogant progressive dreams of the Enlightenment thinkers.

We shall return later to the Counter-Enlightenment idea of society as an organism and the organic mysticism which has challenged the notion that society can be regulated and reformed by reason. For the present we shall focus on the anti-rational and evil elements that are supposed to inhere in individual human beings. Man is, Maistre concedes, part of nature (though those two abstractions – 'Man' and 'Nature' – are, he believes, highly suspect). Nature, however, is not a beneficent mother but a charnel house:

> The whole earth, perpetually steeped in blood, is nothing but a vast altar upon which all that is living must be sacrificed without end, without measure, without pause, until the consummation of things, until evil is extinct, until the death of death.
>
> (de Maistre, quoted Berlin, p. 111)

This is reflected in the human propensity for war: 'men's desire to immolate themselves is as fundamental as their desire for self-preservation or happiness' (ibid., p. 121).

It is easy to dismiss Maistre as a poisonous madman whose thoughts – seemingly informed by *schadenfreude*, by a slavering delight over

the suffering of humanity and the miscarriages of progressive attempts
to improve the lot of mankind – are symptoms rather than arguments.
(It is difficult, for example, to be sure whether Maistre would see
Auschwitz as the dreaded consequence of the dreams of the *philosophes*
or their desired outcome – the universalisation of death that results
in the death of death.) But disregarding Maistre would, as Berlin convincingly argues, be a mistake. Maistre is the fountainhead of a major stream of anti-Enlightenment thought – and, indeed, of European
thought – in the century and a half after his death. His belief in the
intrinsic untrustworthiness of the mass of the people does not merely
look backwards to the doctrine of Original Sin but forwards to Fascism and the huge body of proto- and para-Fascist thought that has
figured so large in nineteenth- and twentieth-century discourse. The
combination of militarism, authoritarianism, irrationalism, blood-and-
soil nationalism, pessimism and a virulent hatred of what he called
'la secte' – Jansenists, Freemasons, Jews, democrats, idealists, middle-
class professionals, secular reformers – make his works a comprehensive anticipation of much that was to emerge from the darker recesses
of the European mind. At least two elements of Maistre's thought are
worth more than a shrug of contempt.

The first is the political conclusions Maistre draws from his belief
in the intrinsic evil and irrationality of humankind – a universalist
assumption, incidentally, that is at odds with his own attack on the
Enlightenment universalism (to which we shall return shortly). The
plausibility of the Enlightenment programmes for society depends upon
certain optimistic, even sentimental, assumptions about the nature of
human beings.[19] De Maistre is sure that these assumptions are profoundly
erroneous. Most Enlightenment thinkers believed that man is born free,
and born good; and that only injustice and social maladministration
placed him in chains: the social revolution would transform human nature. According to Maistre they had forgotten Original Sin and so
had failed to appreciate that underneath everything in mankind was
an impulse to do evil that was at least as strong as the impulse to do
good, a longing to destroy even stronger than the desire to build.

This, in essence, is what has been expressed by a thousand anti-
Enlightenment thinkers since: man is not to be trusted on his own. If
God does not exist (or his existence is denied or forgotten) then, as
Dostoyevsky claimed, everything is permitted; or there will be no
constraints – from fear of eternal punishment – upon individuals' pursuit
of their own desires. Discipline, authority and repression of individual
thought is therefore essential if the state is not going to decline into
lawless anarchy. The power of the state must be reinforced by the
mystery of ritual violence, even if the citizens abjure religious belief.
W.H. Auden expressed this in his *Vespers*:

For without a cement of blood (it must be human, it must be innocent) no secular wall will safely stand.

There is no doubt that it is an act of faith to assume that men are intrinsically good and well-disposed to their fellow men. Or even, as Helvetius believed (see note 19) that they are neither intrinsically good nor intrinsically bad, but can be made to be either, according to their education – where education is understood in the widest sense. The empirical evidence for the Essential Goodness of Mankind is slight, but no more slight than is available to support the Doctrine of Original Sin. The belief that positive values can be unpacked from human nature once humans are liberated from oppression, freed to be good, is perhaps over-optimistic. And even if individuals are intrinsically good, their congenital propensity to goodness may not be given full expression or may be twisted into bitterness, calculation and hatred, because they do not act out their lives in a good world. The appropriate conclusion a political philosopher should draw from this, however, is that, if society is not to collapse into an endless brutal, Hobbesian war of all against all, its institutions must be better than its individuals; the former must embody values that reflect human beings at their best. And this is scarcely an argument for autocracies of any sort, whether they are secular tyrannies or theocracies. Indeed, to conclude from the presumed original sinfulness of mankind that power in society should be deposited in an unaccountable elite, in leaders of great wisdom and strong will, seems at best naive in the extreme and at worst self-contradictory. 'Power tends to corrupt and absolute power corrupts absolutely' – Lord Acton's famous law surely holds if, as Maistre believes, those who have power are, like the rest of the human race, corrupt in the first place. If the Doctrine of Original Sin tells us that no individual is to be trusted, it would seem to follow that power is better widely distributed than concentrated in a few untrustworthy hands. The totalitarian states of the twentieth century have shown how the seizure of power by a small cadre of elite officers of the state removes the last barriers to untrammelled oppression, tyranny and outrage. There can be no exceptions to this – as Acton again emphasised – on behalf of the presumptive transcendent authority of those who are the repositories of God's will or (in the case of recent secular totalitarian states) of the forces of history; for no one can tell who these are – who really have God, History, Reason or the Good on their side – and there can be no way of protecting society against those who arrogate this status to themselves without legitimation. Moreover, there is no obvious distinction between accountability to invisible auditors – God, the forces of history, the remote future – and simple unaccountability. In short, the more pessimistic one's view of human nature, the more one should concur with the Enlightenment demand that society be based upon reason and evidence, and the more passionately one should advocate an accountable, contractually based community of individuals who are equal in the sight of the law. One does not best protect human beings against their ferocious conspecifics by giving some of them unlimited power.

CRITIQUE OF THE ENLIGHTENMENT (4) THE DEFENCE OF LOCAL CULTURE AGAINST UNIVERSAL REASON

Central to Enlightenment thought was a faith in the applicability of general methods to advancing our understanding of the universal properties of man and nature. For Maistre, the belief in universal human characteristics was a characteristic fallacy of the Encyclopaedist thought. He declared that there was no such thing as *man*; there were only men – who belonged to different cultures; one can be a Persian or a Frenchman, but not a man as such. The universalist solutions of the Enlightenment thinkers, the universalist declarations of the legislators of the French Revolution (and of subsequent left-wing revolutions), are thus meaningless. There is no 'species being' (to appropriate Marx's phrase), no human essence. Likewise, there is no such thing as Nature *per se*: there are plants, animals, inanimate beings, but they have no common essence.

Maistre's protestations against essentialism are, of course, undermined by his pessimistic vision of the world; for the latter presupposes that both nature and humanity have an unchanging, universal and unchangeable essence – if only as the unchanging, crucial elements in the eternal bloodbath. Nevertheless, his anti-essentialist claims warrant consideration because they connect illuminatingly with the views of other thinkers who need to be taken more seriously; and it is Berlin, again, who has made these connections and seen their implications for intellectual history and history *per se* most clearly.

The anti-universalist theme in Counter-Enlightenment thought may be thought of as having two elements: ethnocentricity or pluralism; and the refusal to accept the applicability in human life and the human psyche of the most explicitly universal tool of the Enlightenment, namely reason. The first element questions the extension of Enlightenment thought, aspirations and values; the latter their depth. The former points to an apparent antithesis or tension between universal reason and local culture; and the latter opens the way to a more profound irrationalism.

In *Reason and Culture*, Gellner has shown how, since Descartes, reason and culture have been seen in an adversarial relationship. The assumption of the sovereignty of Reason requires liberation from 'the

curse of custom and example' (p. 1). For Descartes, the truly rational
thinker was a solitary individual, using methodological doubt to set
aside all the beliefs and superstitions he had drunk in with his mother's milk. Individualism and rationalism were thus closely linked:

> that which is collective and customary is non-rational, and the over-
> coming of unreason and of collective custom are one and the self-
> same process.... Error is to be found in culture; and culture is a
> kind of systematic, communally induced error. It is of the essence
> of error that it is communally induced and historically accumulated.
> It is through community and history that we sink into error, and it
> is through solitary design and plan that we escape it.... Truth is
> acquired in a planned orderly manner by an individual, not slowly
> gathered up by a herd. Complete intellectual autarchy is, it would
> seem, feasible. It had better be, for it is our salvation.
>
> (Gellner, p. 3)

It was this apparently anti-cultural, anti-pluralist strand in Enlightenment thought that attracted the earliest and most enduring protests.
These protests have been eloquently summarised by Berlin at many
places in his *œuvre*, in particular in *Against the Current* and *The Crooked
Timber of Humanity*.

Berlin places Vico, Herder and Hamann at the origin of this stream
Counter-Enlightenment thought. Vico's writing, of course, antecedes the
explicit universalism of the *philosophes* and cannot be seen as opposing
it, but his observation that certain things in the past – notably the great
achievements of the Greeks in architecture, sculpture, philosophy and
literature – had never been surpassed is regarded by Berlin as an early
and decisive rebuke to the claim that history is a story of progress towards universally agreed goals, a gradual progression from (cultural
and local) endarkenment to (rational and universal) Enlightenment:

> For Vico there is no true progress in the arts; the genius of one age
> cannot be compared with that of another.... Each culture creates
> masterpieces that belong to it and it alone, and when it is over one
> can admire its triumphs or deplore its vices: but they are no more;
> nothing can restore them to us. If this is so, it follows that the very
> notion of a perfect society, in which the excellences of all cultures
> will harmoniously coalesce, does not make sense. One virtue may
> turn out to be incompatible with another.... There is both loss and
> gain in the passing from one stage of civilisation to another but,
> whatever the gain, what is lost is lost for ever and will not be restored
> in some earthly paradise.
>
> (*The Crooked Timber of Humanity*, p. 67)

From this, we may conclude several things:

1. Progress is not absolute, to be measured by reference to a single
   ultimate goal.
2. The past cannot be dismissed, as it was by some of the *philosophes*,
   as a period of relative endarkenment.
3. As mankind develops, there are losses as well as gains. Each stage
   of development has its own value – and its own values.

For Herder, whose writings were in conscious opposition to the
universalism of the *philosophes*, 'the general progressive amelioration
of the world' is a 'fiction'. Each stage of development has its own
value:

> The youth is not happier than the innocent, contented child; nor is
> the peaceful old man unhappier than the energetic man in his prime.
> (quoted in Berlin, *The Crooked Timber of Humanity*, p. 83)

This is not only a chronological, or diachronic, truth but also a geographical or synchronic one:

> Each nation has its centre of happiness within itself, just as every
> sphere has its centre of gravity.
>
> (ibid.)

It follows, therefore, that there can be no single prescription for human progress, because the goal of progress will differ from culture to
culture. Indeed, a universalist science of society may create a new
form of oppression to replace the despotism of traditional tyrants such
as kings, nobles and the grandees of the Church. The universalism of
a reason-based science of social progress implicitly assumes that the
differences between the cultures of different times and places are
unimportant compared with what they have in common. Reason, which
sees cultural specifics as various allotropes of the Great Sleep from
which it alone can enable men to awake to truth and progress, may
well overlook the very things that men value most, things that are
expressed in dress, song, rituals, institutions, language, diet, pastimes
– things that are not wholly amenable to reasonable explanation.
Cultures, and the values they implicitly express, are incommensurate
with one another: they are heterotopic, imaginative universes. They
cannot, therefore, be subsumed under a Utopian super-culture based
upon reason and the desire to alleviate human suffering using technologies unfettered by the irrelevancies of irrational human wishes
and institutions.

Recognising this, Berlin argues – speaking not only on behalf of writers he is explicating, but also it seems in his own voice – undermines a fundamental assumption in the Utopian ideals of the *philosophes*: the belief in a harmony of all excellences in an ideal world. People from different cultures differ not only in the means they choose to achieve the same (universal) ends; they may actually have different ends. No single kind of state – captured in an Enlightenment blueprint – could realise the fundamentally different and often deeply incompatible value-systems of men and women from different corners of the world. The failure to appreciate this, Berlin argues, has resulted in the characteristic outrages of the twentieth century: despotisms driven by the rhetoric of Enlightenment and reason crushing all dissenters who wished to retain the flavour of their own lives in irrational religions, unenlightened dress, an unprogressive love of wine, women and song, and so on. It is this, the argument runs, that has led to the persecution of those who wish to retain their own style of life as 'reactionaries', to state terrorism and rational terror, to the 'slave camps under the flag of freedom, massacres justified by philanthropy' and to other hideous paradoxes and ironies that Camus observed as symptomatising the sicknesses of recent history.[20]

This is, of course, deeply unfair. Toleration of difference, in particular of religious differences, was one of the key characteristics of Enlightenment thought and what Peter Gay has poignantly titled 'the politics of decency'. Indeed, a rich variety of opinions was seen as a necessary safeguard against fanaticisms that took ideas too seriously and prompted people to kill for them. As Voltaire observed,

> If there were only one religion in England, one would have to fear despotism; if there were two, they would cut each others' throats; but they have thirty and they live happily and in peace.
>
> (*Lettres philosophiques*, quoted in Gay, p. 400)

Toleration of diversity was a necessary precondition for the successful pursuit of knowledge – a view central to John Stuart Mill's *On Liberty*, published 150 years later. And whereas that toleration did not extend to toleration of wilful unreason, of the abusive mystifications by which those in power retained their positions, or of those whose intolerances prompted them to persecute others, it was as generous as it could be without yielding to a Tolstoyan 'resist not evil' quietism. It was expressed in Montesquieu's setting out the rights of accused persons, Lessing's advocacy of tolerance of Jews, Rousseau's defending the rights of children and Voltaire's efforts on behalf of the victims of the miscarriage of justice.

And there is a profound generosity in the Enlightenment belief that, fundamentally, human beings, whatever their superficial differences, have a common core of shared humanity. So committed were the *philosophes* to the idea of a universal humanity underlying cultural differences that Fénélon, by expressing these sentiments (in *Dialogue des Morts*), earned his place in the Enlightenment Pantheon despite his being a prelate of the Church: 'Each individual owes incomparably more to the human race, the great fatherland, than to the country in which he is born.'

It must be acknowledged that this generous sentiment – reminiscent of the Roman playwright Terence's 'Nothing human is alien to me' – has some ambivalence. It is generous inasmuch as it implies a recognition of the equal humanity of all human beings and, as a consequence, their right to be treated in practice as equals; it is less attractive if it amounts to the assertion that all human beings are really Parisians underneath: that humankind is Pariskind; that humans would be Parisians, or seem like Parisians, or behave like Parisians, or share Parisian values, if only they were given the opportunity to realise their true nature or express their full potential. Enlightenment universalism is less impressively generous; in other words, if it implies that completely to realise one's humanity, it is necessary to be a Parisian *philosophe*, so that anyone who fails to achieve this condition is not-quite-human; and the Noble Savage, or the German farmer, is a Parisian *manqué*, and so only potentially, rather than actually, fully human.

This latter is how the universalism of the *philosophes* has been interpreted by many, especially those, such as Herder, who resented what they saw as the French cultural dominance in the eighteenth century. The dangers of an apparent failure to accept that much of human essence is inescapably culture-bound, ethnocentric – and the connection between this oppressive universalism and faith in reason and the methods of science – is set out by Berlin in *The Crooked Timber of Humanity*:

> It is an accepted truth that the central view of the French *philosophes* is that (in the words of . . . Clifford Geertz) man is 'of a piece with nature': there is a human nature 'as invariant as . . . Newton's universe'. . . . Manners, fashions, tastes may differ, but the same passions that move men everywhere, at all times, cause the same behaviour. Only 'the constant, the general, the universal' is real, and therefore only this is 'truly human'. Only that is true which any rational observer, at any time, in any place, can, in principle, discover. Rational methods . . . can solve social and individual problems, as they have triumphantly solved those of physics and astronomy . . . once knowledge of man's true nature is attained, men's real needs will be clear: the only remaining tasks are to discover how they may be satisfied, and to act upon this knowledge.
>
> (p. 70)

Presented in this fashion, universalism can be seen less as the pure generosity of intellectuals who wish to acknowledge the equal reality of all human beings and the equality of their rights to exist and to pursue happiness, than as itself an ethnocentric superstition whose ethnocentricity is invisible because the ethnic group in question belongs to the economically dominant group – namely the Parisian variant of the European tribe.

The suggestion that universalism is simply a Parisian dialect of a post-Renaissance world-picture is, however, implausible because the rationalist and universalist tradition behind it goes back a long way – at least to Plato. A more telling ethnocentric attack on reason came from Herder's great friend Hamann, who, Berlin says, lit the fuse that set off the great Romantic revolt; he was 'the forgotten source of a movement that in the end engulfed the whole of European culture'.[21] For Hamann, reason might be a universal, but its concrete operations and prescriptions were culture-bound, if only because reason was expressed through, and hence embedded in, language. The entirety of the human ability to think rests upon language:

> All idle talk about reason is mere wind; language is its organon and criterion.
>
> My reason is invisible without language. . . . Togetherness is the true principle of reason and language, by means of which our sensations and representations are modified.
>
> With me the question is not so much: What is reason? but rather: what is language? and here I presume to be the basis of all paralogisms and antinomies which one blames on the former; therefore it happens that one takes words for concepts and concepts for things themselves.[22]

Language has an ontological priority over reason:

> it needs no deduction to prove the genealogical priority of language and its heraldry over the seven holy functions of logical propositions and inferences. Not only does the whole ability to think rest upon language . . . but language is also the central point of reason's misunderstanding of itself.[23]

At the very least, it is inextricably embedded in language:

> Even if I were as eloquent as Demosthenes, I should not have to do more than thrice repeat a single phrase: Reason is language, Logos.[24]

> I am quite at one with Herder that all our reason and philosophy
> amount to tradition. . . . For me it is not a matter of physics or the-
> ology, but language, the mother of reason and revelation, their alpha
> and omega.[25]

It follows from the embeddedness of reason in language that it is
also embedded in the specifics of communities; for 'community is the
true principle of reason and language' (ibid.). It is language – irreducibly
local, historical – that is the mark of the human. Human beings, reason, human reason, are inextricably caught up in the particulars of individual languages, which are, in turn, rooted in the specifics of individual
cultures. Reason is inseparable from language and consequently culture-bound and, in this respect, quite different from mathematics:

> Lastly, it is to be understood that if mathematics can claim for it-
> self the precedence of nobility on account of its general and necess-
> ary reliability, human reason must rank below the infallible and
> unerring instinct of the insects.[26]

Reason is deceived as to its own nature if it fails to see how its concrete operations lie in the stench and perfume of connotation-rich
discourse rather than the odourless matrix of purely denotative equations. And it is deceived as to the nature of humanity and human
values if it believes that the latter are objective, timeless and universal and are to be revealed by the operation of an objective, timeless
and universal reason. Reason is used by particular peoples, at particular times; it is universal only in so far as it is empty of concrete
content (like abstract mathematical operations) and, conversely, has
rich content only in so far as it is 'contaminated' by cultural particulars. In summary, reason that sees itself as opposed to culture, or
freed from it, will fail to understand where it gets its content from. It
will certainly find it difficult to translate into practical action. Even if
abstract logic is universal, transcending all ethnic particulars, reasonableness and good sense are rooted in the experience, needs and values of individual groups of men and women.[27] Mathematical, Cartesian
reason with its pretensions to universality is repulsive, an emptying
emptiness that creates stunted minds, hearts, people.

Hamann's assertion of the ethnicity of reason has had a major revival in the late twentieth century and will be discussed in future
chapters. Suffice it here to note that the arguments associated with
post-modernism have led to the sceptical view that reason is simply
the rhetorical figure of the discourse of the dominant group. (This
overlaps with the view that truth is relative to interpretive communities.)
The question as to whether universalising reason is a liberating or

oppressive force has been raised with particular energy in the debate between the two main strands of feminism – equity and gender feminism. A brief digression on this seems to be in order – to illustrate the problems of anti-Enlightenment anti-universalism.

For *equity* feminists, who may be seen as daughters of the Enlightenment, the heirs of Mary Wolstonecraft, reason would seem to lie at the root of their case for equal treatment. They would applaud Condorcet:

> Among the causes of the progress of the human mind that are of the utmost importance to the general happiness, we must number the complete annihilation of the prejudices that have brought about an inequality of rights between the sexes, an inequality fatal even even to the party in which favor it works. It is vain for us to look for a justification of this principle in any differences of physical organisation, intellect, or moral sensibility between men and women. This inequality has its origin solely in an abuse of strength, and all later sophistical attempts that have been made to excuse it are vain.[28]

For the more radical *gender* feminists, on the other hand, difference is as important as equity, and the assertion of separate value-systems of more importance than equal treatment according to supposedly universal, genderless values. Gender feminists have rejected the rights and privileges of men as standards or norms to which they aspire and, instead, seek political power to change those standards and norms. Reason is not a natural universal that can be accepted as given; on the contrary, it is an instrument of male oppression and for this reason has been subjected to a ferocious 'clitique'. Reason and objectivity are norms created by a patriarchy, the central instruments of a phallogocentric culture that has enjoyed hegemony 'at least since Plato'. To appeal to either is to collude with the oppressor.

Nussbaum has summarised this mode of thought very clearly in her critique of certain strands of radical feminist philosophy:

> We are told that systems of reasoning are systems of domination, and that to adopt the traditional one is thus to be co-opted. Our liberation as women, it is said, requires throwing over the old demands for objectivity and cutivating new modes of reasoning, which are not clearly specified, but which are frequently taken to involve immersion in a particular historical and social context, and to be closely allied with cultural relativism.[29]

Nussbaum has noted how, for radical feminists, the idea that detachment from context is the road to objectivity and hence to truth 'is bound up with a traditionally male denigration of intimacy, emotion and the body'. Even – or especially – logic, which may be seen as a minimal requirement of reason, is viewed in this unfavourable light. Some radical feminists, for example, have asserted that *modus ponens* – 'If p, then q; p; therefore q', the very cornerstone of logic – is not a necessary and inescapable (if rather trivial) truth. On the contrary, it, too, is merely an element in male-dominated Western thought. Ruth Ginzberg (quoted by Nussbaum) argues that *modus ponens* is a male patriarchal creation oppressive of women, on the grounds that 'it is a male-invented way of defining who counts as a rational being, and that women very often (more often, it is suggested, than men) fail to recognise *modus ponens* as a valid form of inference.'[30]

Of course, the very arguments invoked by gender feminists against logic and reason themselves appeal to reason and utilise (though very badly) a form of logic not dissimilar to the one that they reject. Without the constraint of logic, any conclusion could be drawn from their initial premises; for example, the following would be an acceptable argument:

1. *Modus ponens* is an obstacle to the advancement of women.
2. We should oppose any obstacle to the advancement of women.
3. Therefore we should embrace *modus ponens* in order to ensure the advancement of women.

More directly relevant to the present discussion of universalism is the way in which gender feminism inadvertently exposes the contradiction at the heart of affirmations of ethnicity and the denial of the validity of any universalising principle. For gender feminism wants to deny universalism at the same time as it is strikingly universalist in the scope of its claims and assumptions; it simply cuts the cake a different way – and with mouth-blockingly large slices. Gender feminism may be ethnocentric inasmuch as it denies that men have anything fundamentally in common with women; but it is universalistic inasmuch as it assumes that all women have things in common with one another – that they share the same hopes, wishes, needs and problems, and that there are universal solutions for these. Those women who point out that this is not the case – and that a child prostitute in a Bombay cage may have little in common with a feminist Professor of Law at an East Coast American university – and that they sometimes find radical gender feminism even more oppressive than the constraints of a male-dominated world – are dismissed with scorn. Gender feminists argue that, though there is superficial evidence to the contrary, underneath, all women share the same values and aspirations, so long as they are free of false consciousness inculcated by living in a world dominated by men. In a curious way, this echoes the belief of the rationalists from Socrates onwards and expressed in its canonical form in the passage we have already quoted from Condorcet's testament:

Is not a mistaken sense of interest the most common cause of actions contrary to the general welfare? Is not the violence of our passions often the result either of habit that we have adopted through miscalculation, or of our ignorance of how to restrain them, tame them, deflect them, rule them?

However, gender feminism seems to be more arrogantly universalising than Condorcet or any of the *philosophes*.

Whatever we think about the anti-rationalist posturing of some radical feminists – and few women really think that the path to salvation lies through irrationality, illogicality, multi-valued logics and the logic of 'p and not-p' – the example of gender feminism opens up wider questions about those who are allergic to the notion of a common human nature. If we deny universal solutions to problems and even question the universality of ultimate ends (for example, happiness) how shall we meet the challenge of improving the lot of humankind – assuming, as most people do, that it is ripe for improvement and that we have responsibilities beyond looking after our own immediate interests? The anti-universalistic, ethnocentric argument tells us that there will be no general solution for all people, for people have incommensurate value-systems. What it does not tell us is the size and scope and boundaries of the groups for whom a common solution might be found. There is no reason why separatism should stop at respecting the differences marked by gender boundaries; or the boundaries of a state. A nation-state is an historical accident and will consequently number individuals of many different ethnic origins among its citizens. These may, despite commonality of language and even of choice of beer, have profoundly different affiliations, dreams of happiness and value-systems. It may be necessary, within a state, to respect other differences. Perhaps the women will be so different from the men that their values may be considered not only incommensurable but also incompatible. Similar distances may divide the old from the young, the rich from the poor, the professional musician from the secondhand car salesman.

One of the great challenges for a government that sincerely wishes to ensure justice and prosperity for all is to mediate between the needs and values of individuals with profoundly different beliefs. Now the ethnocentric critique of universalism fails to suggest a natural point at which the respecting of boundaries should end. Since (as Wittgenstein

pointed out in the *Tractatus*) 'the world of the happy man is different
from the world of the unhappy man', the terminus of the division of
the state according to ethnocentric thought is probably the single individual. Since each individual is different and irreplaceable, each
inhabits and embodies an imaginative universe incommensurate (and
certainly incommensurable) with that of any other, then to legislate
for more than one individual is already to exert a kind of universalist
tyranny. In short, the natural terminus of anti-universalistic thought
is an end-stage anarchism, with the dissolution of the nation-state
into a colloidal suspension of individual wills pointing in random
directions.[31]

Ultimately, a radical anti-universalism is doomed to end in inco-
herence. But these are not the only grounds for anxiety at this rejection
of what is essentially generous rather than arrogant in Enlightenment
thought. The most powerful and achieved expressions of anti-
universalistic thought in the centuries since Herder and Hamann, the
assertion of *Volkerpsychologie* against the universalising tendencies of
the *philosophes*, have not had very attractive results: the rise of a bloody
and aggressive nationalism – both primary, and reactive, in response
to the chauvinism of other nations; and equally bloody ethnic conflicts within states whose boundaries do not coincide with the distribution of particular ethnic groups. Both may be characterised by an
intimate relationship between the affirmation of one's own humanity
(and that of the group to which one feels one belongs) and the denial
of that of others: I/we are human; you are different from us; therefore you are not human. Often this has been a reaction to an earlier
denial of full humanity – or the sense that it has been denied. Berlin
traces German nationalist thought, for example, to the feeling that it
had 'remained on the edges of the great renaissance of western Europe' (*The Crooked Timber of Humanity*, p. 245) and was peripheral to
the dominant discourses of eighteenth-century thought. More generally, nationalism ('an inflamed condition of national consciousness')
seems to be caused by wounds, to be rooted in some form of collective humiliation. Whatever the cause, the ultimate result has usually
been bloodshed. This is particularly evident in the case of sub-na-
tional self-assertion in ethnic conflicts within a national boundary,
inside a nation-state. There the conflict, the revenge-cycle of atrocity
and counter-atrocity, can have no natural term, except partition, or
dissolution into a multiplicity of independent neighbouring nation-
states fuelled by mutual loathing.

This is well illustrated by the current epidemic of ethnic conflicts.
From the point of view of the present argument, there are two op-
posing interpretations, illustrated perhaps by the wars in the former
Yugoslavia. One is to see these wars as the work of demagogues –

Milosevic, Tudjman, Karadzic – who, lacking any policy for improving the civic life of the peoples over whom they wish to gain or retain power, appeal to a wounded collective consciousness, and a feeling of folk solidarity rooted in an imagining of an idealised past, to rally the masses behind them against their rivals and opponents. The hostilities between the Slovenes, Croats, Serbs, Bosnian Muslims, Albanians, etc. could, on the other hand, be seen as the unintended consequence of a socialist, an 'enlightened' universalist, denial of the reality or the importance of the ethnic or expressive, as opposed to utilitarian, dimension in human life. As Berlin has pointed out, socialist thought has always 'regarded nationalist or regional loyalties as irrational resistance by lower forms of development, which history would render obsolete' (*The Crooked Timber of Humanity*, p. 253).

It might be argued, then, that the present, seemingly unresolvable, civil wars in the former Yugoslavia may be in part an explosive expression of ethnic feelings and loyalties that have been denied expression under 40 or more years of socialism. But to interpret the wars in this way, and see them as a critique of the universalism of socialism, is to overlook what may be the most important factor – the time and the manner in which socialism was imposed; the fact that it was the expression of a dictatorship forged in conditions of war and underwritten by the threat of an even more alien imposition. The stability of Yugoslavia under an anti-ethnic regime was achieved by a dictatorship whose strength was at least in part due to a fear instilled by the threat of a worse dictatorship: if Tito fell, the Soviets would invade. Moreover, the conditions under which socialism was forged in Yugoslavia were those of a forced forgetting of ethnic conflict, of atrocities committed during the Second World War. The evil committed by the Croats in the 1940s, for example, was on such a scale as to be unforgotten despite 40 years of official amnesia. The destruction of Yugoslavia, and the blood-boltered reassertion of ethnic loyalties, is less a pure resistance to universalist reason than a refusal to forget religious, economic and social grievances. This, and the need for those such as Milosevic to discover a new basis for legitimising their power after the party to which they owed their elevated positions had collapsed.

The yet more hideous ethnic conflict in Rwanda is even less a revolt against universalising reason that has failed to respect ethnic differences. It has been pointed out that there is no fundamental difference between the Hutus and the Tutsis: they speak the same language, have most customs in common. However, such differences as there were between the two – mainly that of wealth, the Tutsis being better off than the Hutus – were exaggerated and exploited by the colonisers, the Belgians, who found it easier to retain control over the country by suborning one class to join them in suppression of the other. Although the radio broadcasts that orchestrated the massacres of between 500 000 and 1 000 000 Tutsis were overtly racist in content, there was no truly ethnic basis for this.

Contemporary events in former Yugoslavia and Rwanda do not, therefore, support the case for a necessary head-on collision between ethnically rooted values and universalising reason. An individual's right to worship her own gods, to dress, sing and eat as she wishes, must always be respected; but this should not be assumed necessarily to feed into a blood-and-soil nationalism that links love of one's own group to hatred of another's, to a re-definition of rightful citizenship on genetic or racial grounds, or to a denial of those universal rights and values that are affirmed in the implicit equity of a universalising reason to which all have access. Reason is loose-textured enough to permit humanity a wide variety of self-expression: universalising reason will not of itself legislate for Coca Cola against wine and beer; for rational dress against national dress; for state-sponsored military parades against maypole dancing; for the Politburo against the Tynwald. Moreover, the ethnocentric denial of a universal humanity – not only by the wounded who are affirming their own humanity against hated economic, cultural or military oppressors but also by those who would liberate the world – may not be entirely benign. It can rapidly modulate into a denial that some people are human.[32]

These dangers of *anti*-Enlightenment thought have been well illustrated by the right-wing tyrannies of the twentieth century. The essential generosity of Enlightenment thought and the catatrophes that may follow when its values are rejected have been noted by Berlin:

> According to [some eighteenth-century thinkers] ... all men possessed certain unchanging characteristics in common, called human nature. ... The most important common characteristic was considered to be the possession of a faculty called reason, which enabled its possessor to perceive the truth, both theoretical and practical. The truth, it was assumed, was equally visible to all rational minds everywhere ... In the twentieth century this claim to universality, whether of reason or any other principle, is no longer taken for granted ...
>
> This is most obvious in the case of Fascism. The Fascists and the National Socialists did not expect inferior classes, or races, or individuals to understand or sympathise with their own goals; their inferiority was innate, ineradicable, since it was due to blood, or race, or some other irremoveable characteristic; any attempt on the part of such creatures to pretend to equality with their masters, or

even to comprehension of their ideals, was regarded as arrogant
and presumptuous. . . . The business of slaves is to obey.

(*The Crooked Timber of Humanity*, pp. 175, 176)

Fascism, Berlin argues, is only an extreme expression of this attitude,
which infects all nationalism (or assertive ethnocentrism) to some degree.
The assumption that others cannot share my ideals, that they cannot
understand them, is no less dangerous than the assumption that they
must do so and that their failure in this regard is evidence of their
primitiveness or their maleficence.

Hegel, then, seems to have been right in his observation that the
philosophy of the Enlightenment, which wanted to deliver man from
the irrational, was unifying. 'Reason unites mankind while the irrational
destroys unity' (Camus, *The Rebel*, p. 101). The assertion that man is
irrational, that the shallow light of reason cannot penetrate to man's
deepest essence, is not politically neutral. Irrationalism is the favoured
weapon of demagogues. As Camus points out,

When Mussolini extolled 'the elementary forces of the individual',
he announced the exaltation of the dark powers of blood and instinct, the biological justification of all the worst things produced
by the instinct of domination.

(ibid., p. 148)

The cultural relativism that was placed in opposition to Enlightenment universalism and which asserted that human beings not only
had adopted different means to the same ends but actually had different ultimate ends, had a curious ally in the idealistic tradition of
which Kant – a supporter of the *Aufklärung*, as we have already noted
– was the largely unwitting progenitor. Kant had been forced by the
conclusions drawn by Hume from empiricism to argue that reason,
far from being part of nature, was a product of the mind that ordered nature. He extended this to values. They were not inherent in
the natural world; rather they were created in the minds and hearts
of men. This belief is reflected in Coleridge's great poem *Dejection:
An Ode*:

I may not hope from outward forms to win
The passion and the life, whose fountains are within.

O Lady! we receive but what we give,
And in our life alone does Nature live:
Ours is her wedding-garment, ours her shroud!
And would we aught behold of higher worth,

Than that inanimate cold world allowed
To the poor loveless ever-anxious crowd,
  Ah! from the soul itself must issue forth
A light, a glory, a fair luminous cloud
    Enveloping the Earth –
And from the soul itself must there be sent
  A sweet and potent voice, of its own birth,
Of all sweet sounds the life and element!

The beauty of nature, the inspiration, the values, we derive from it, are not inherent in nature itself but are the result of 'the shaping spirit of Imagination'.

This is an extreme expression of the view that values are culture-relative – extreme, since it assumes that values are created in the individual mind (and can be lost if that mind loses its way). The pluralist views that we have been examining hitherto see values as the creation of human collectivities; the individual's values reflect those of the collective to which he or she belongs. There can, therefore, be no universal values derived from nature, only the values of the group. Values are fashioned by the group rather as works of art are fashioned by an artist: they are human creations not natural impositions. Both individualistic and collectivistic views are expressed in Herder's assertion that:

There is not a man, a country, a people, a national history, a state which resemble each other; hence truth, goodness and beauty differ from one another.
          (quoted in Berlin, *The Crooked Timber of Humanity*, p. 84)

In the twentieth century, the individualistic version of the idea that values are human creations is exemplified in existentialism; while collectivist views are found in a wide range of thinkers, including the apologists of the aestheticising politics of Fascism (Pareto, Gentile, Heidegger, Maurras, Barres) and the post-modernists, whose views I shall address in a future chapter.

The belief that values are not natural – that there are no eternal values dictated by the nature of which we are a part – requires further examination. Surely there are *some* universals that can be addressed by a better understanding of our material nature? Surely human nature and human wants have some boundaries? I shall discuss this – and attempt to identify some core or culturally invariant values – in the final chapter, when I consider the idea of progress.

CRITIQUE OF THE ENLIGHTENMENT (5)
THE POVERTY OF REASON

A less radical attack on Enlightenment thought focuses not on the universalism of values that is seen to be implicit in the faith in reason, but upon the *power* – or rather the impotence – of reason. This attack has two elements: disbelief in the ability of reason to regulate human behaviour and in the possibility of a rational reorganisation of society along the lines laid down by universal moral and intellectual ideals and principles; and belief that a life guided by, or based upon, reason is spiritually impoverished, that reason is a prison-house of the human spirit.

The criticism of reason, on the grounds of its inadequacy, goes back to the Enlightenment itself. Rousseau, who clearly saw the difference between the concrete, self-centred, private person – one instance of which was lavishly described in his *Confessions* – and the abstract 'citizen' serving the general interest, asserted that right action could not be derived from right thinking. And Hume, an archetypal Enlightenment figure and lionised by the *philosophes*, asserted that happiness and virtue could not be achieved by the exercise of reason alone:

> We speak not strictly and philosophically when we talk of the combat of passion and of reason. Reason is, and ought only to be the slave of the passions, and can never pretend to any other office than to serve and obey them.[33]

The suspicion provoked by individuals' claims to being driven by reason, the imputation of mere 'rationalisation', has been a major element in anti-Enlightenment thought, with Schopenhauer, Nietzsche and Freud being only the most prominent of German-speaking examples. The dangers in assuming the universal applicability of reason to human affairs are spelt out by Camus:

> Law can reign, in so far as it is the law of universal reason. But it never is and it loses its justification if man is not naturally good. A day comes when ideology conflicts with power. Then there is no more legitimate power. Thus the law evolves to the point of becoming confused with the legislator and with a new form of absolutism. Where to turn to then? The law has gone completely off its course; and, losing its precision, it becomes more and more inaccurate to the point of making everything a crime. The law always reigns supreme but it no longer has any fixed limits. Saint-Just had foreseen that this form of tyranny might be exercised in the name of a

silent people. 'Ingenious crime will be exalted into a kind of religion and criminals will be in the sacred hierarchy'.

(Camus, *The Rebel*, p. 101)

The deeper critique of universalising reason, deeper than the argument that it is a weak force in regulating human behaviour, is that a
life based upon reason does not meet the true needs, answer to the
fundamental hungers, of human beings. Knowledge of the kind that
the Enlightenment saw as being necessary for the salvation of mankind was, it was argued, empty of personal meaning, and consequently
offered no nutrition to the spirit. To give onself up to that kind of
knowledge was, as it were, to give oneself up to knowledge that had
form but no content; it did not satisfy the whole man. This was particularly true of scientific knowledge which was explicitly value-free,
describing how things were without pausing to consider why they
were as they were or how they should be, and expressed in a technology that increased our power to get things done without adding
to our understanding of why we should want to do anything in the
first place. A life based upon value-free science and technology was,
ultimately, meaningless.

Those who see this as a valid criticism of the Enlightenment, however, misconstrue the *philosophes*: the knowledge (and the reason) which
serves human life is not intended to stand for, or stand in for, human
life itself. We may think of the Enlightenment as being concerned
primarily with the means to human prosperity and freedom, but not
as being itself prescriptive of the end or ends of life, in particular, of
the *content* of life in the Utopias towards which it directed its hopes.
The science of light that will enable human beings to see better, the
science of heat that will enable them to spend more, rather than less,
of their time warm rather than freezing to death, cannot pretend to
prove that it is better to see than to be blind or to be warm rather
than cold. Such value-judgements antedate science. Nor does the science
of light and heat legislate or override the meaning of being warm
(rather than cold) or the pleasure of seeing (rather than not seeing).
The enjoyment of consciousness, the content of consciousness – the
experience which underpins and goes beyond knowledge – is not
cancelled by advances in factual scientific knowledge. No scientific
advance can diminish the sensation – and the mystery – of warm
sunlight on one's arm. It simply creates the context in which this
experience can be enjoyed. Likewise, there is nothing intrinsically hostile
in science to visionary delight in the world; nothing, anyway, as hostile
to such delight as are hunger or cold unalleviated by science-based
technology.[34]

Those who criticise Enlightenment thought for its dependence on, and faith in, science because the latter is supposedly lacking in values are confused at several levels. First of all, the value-free nature of science – itself questioned by many writers, notably those who would sociologise scientific knowledge – is evident not in its choice of topics but in its methods; in particular its resistance to confusing the subjective and the objective, for example, personal ambitions and discovered fact. My researching into the physics of heat is, of course, driven by all sorts of values – the agreed human preference for being warm over being cold; my own desire to further my understanding; my wish for glory and/or academic tenure, etc. The facts uncovered by my researches, however, have to be relatively uninfluenced by these values: my personality cannot enter into them; nor (*pace* the sociologists of scientific knowledge) can a putative collective personality of the scientific community or of mankind itself. Physics does not even concern itself with the *sensations* of heat: nothing in physics would lead one to suppose that 'heat' is associated with subjective sensations, or that one range of temperatures is preferable to another, being associated with pleasant rather than unpleasant sensations. Inside science, the landscape is comparatively value-free – that is what distinguishes science-based technology from magic and wishful thinking – though, of course, science itself serves the preferences, the values, of those who use it. Which is the second point: the value-freedom of scientific knowledge (which does not, of course, extend to the choice of what to investigate, what to make known) does not mean that a society that uses science-based technology to improve its collective lot is itself without values.

This should be sufficiently evident, but is not fully understood even by the more intelligent critics of the Enlightenment. For example, Goldmann's characterisation of the view of the *philosophes* fails to distinguish between means and ends:

> For them, the mission of man, which gives meaning to his life, lies in the effort to acquire the widest possible range of autonomous and critical knowledge in order to apply it technologically in nature and, through moral and political action, to society. Furthermore, in acquiring his knowledge, man must not let his thought be influenced by any authority or prejudice; he must let the content of his judgements be determined only by his own critical reason.
>
> (Goldmann, op cit., p. 2)

The inaccuracy lies in the characterisation of the acquisition and application of technological knowledge as 'the mission of man' and the claim that the Enlightenment thinkers believed that this mission gave meaning to human life. This is evident nonsense. No one can regard the universal mission of man simply to be to serve mankind: this is a

kind of tautology and an eternal deferral of the meaning of life.[35] If
this generation lives only to serve the needs of the next, if its purpose is consumed in the Kingdom of Means, then the Kingdom of
Means itself lacks meaning because the Kingdom of Ends is forever
deferred and, anyway, houses no meaning.

If we clear up this misunderstanding about the Enlightenment, then
much of the force of the critique that was derived from Romanticism
and from the dialectical thinkers influenced by Hegel and Fichte is
removed. 'Getting and spending' may be necessary to achieve a modi-
cum of material comfort for ourselves and our fellow humans; but
these are not ends in themselves and they do not stop us from pausing to admire a daffodil. Indeed, a farmer not destroyed by crippling
labour because of the availability of affluence and high technology is
more likely to pause to admire a daffodil than his predecessors, whose
common destiny was lifelong torture on the rack of gruelling physical labour.

Some of the harshest critics of the Enlightenment have focused on
its supposed overstatement of the role of reason in human affairs and
its corresponding overestimation of the power of reason to contain or
even direct human passions. Ironically, many such critics would have
agreed with the quintessentially Enlightenment figure David Hume
that reason not only is, but should be, the slave of the passions. Amongst
these have been Romantics for whom the will, rather than reason,
was the true, the authentic, human essence and the font and origin of
all that was good and worthwhile in men and women, the source in
particular of their best feelings and their richest creative powers. Other
anti-Enlightenment thinkers for whom the will had supremacy over
reason in the soul of man did not necessarily see the former as a
force for good; or not, at least, for progress. Romantic pessimists such
as Schopenhauer (who envisaged the will as a trans-human force) and
Baudelaire (who greatly admired Maistre) regarded it as not only
powerful, but also as essentially evil – as perverse as well as wilful.
They argued that all progressive thought was based upon a deep
misreading of human nature. There was also a profoundly influential
succession of thinkers, leading from Dostoyevsky through to the Existentialists, for whom the perverse and destructive nature of the will
was, paradoxically, healthy. It made it possible for the individual to
rebel against the collective, against the Utopian dream of the life based
upon reason, against a society, a social machine, so well organised
and so complex and so vast as to negate the individual's real existence – reducing his outward self to a cog in a machine and discount-
ing the uncombed creative disorder of his inner self as irrelevant.

Dostoyevsky's anonymous, imaginary author of the *Notes from the
Underground* expresses this with angry precision:

See here: reason is an excellent thing – I do not deny that for a moment; but reason is reason and no more, and satisfies only the reasoning faculty in man, whereas volition is a manifestation of all life. . . . It is true that, in this particular manifestation of it, human life is all too often a sorry failure; yet it nevertheless *is* life, and not the mere working out of a square root. For my own part, I naturally wish to satisfy *all* my faculties, and not my reasoning faculty alone (that is to say, a mere twentieth portion of my capacity for living).[36]

Man, Dostoyevsky's anti-hero asserts, always prefers to act in conformity with his wishes, with his whims, and these may not conform to the dictates of reason, or even of self-interest. He may choose to act perversely simply to assert his will, to affirm his own existence, to behave in such a way as to confound the very laws propounded by the social scientists that seem to suggest that he has no will of his own, that he is as the keyboard of a piano, played upon by the laws of nature, of which he is a part. And even if it could be shown that this perverse behaviour whose purpose is to defy the laws is itself predictable according to natural laws, he would still be defiant:

But if you were to tell me that all this could be set down in tables – I mean the chaos, and the confusion, and the curses, and all the rest of it – so that the possibility of computing everything might remain, and reason continue to rule the roost – well, in that case, I believe, man would *purposely* become a lunatic, in order to become devoid of reason and therefore able to insist upon himself.

(ibid., p. 615)

Reason does not capture fundamental human impulses:

Does reason never err in estimating what is advantageous? May it not be that man occasionally loves something besides prosperity? May it not be that he also loves *adversity*?.

(ibid., p. 618)

The critique fastens on the very heart of the Enlightenment dream:

Who was it first said that . . . man does evil only because he is blind to his own interests, but that if he were enlightened, if his eyes were opened to his real, his normal interests, he would at once cease to do evil, and become virtuous and noble for the reason that, now being enlightened and brought to understand what was best for him, he would discern his true advantage only in what is good (since it is a known thing that no man of set purpose acts against his own interests), and would therefore of necessity also *do* what is good? Oh, the simplicity of the youth who said this! Oh, the utter artlessness of the prattler!

(ibid., p. 607)

Thus Condorcet – Spinoza – Socrates: prattlers!

Dostoyevsky's critique is valid, in so far as it is directed against the excessively mathematicised social science, the Newtonismus, the scientism, of certain aspects of Enlightenment thought, in which clockwork human beings function perfectly in the larger clockwork of society and the latter reflects the greater clockwork of the material universe. However, as a critique of the Enlightenment as a whole, it is wide of the mark. Nevertheless, there is a case to answer, even if one would not wish, say, to halt a programme for saving the lives of babies on the whim of a man like Dostoyevsky's hero, a splenetic solitary full of petty spites and illwill, whose pleasures in life include the further humiliation and degradation of vulnerable and abused prostitutes. Dostoyevsky's absurd anti-hero is a powerful dramatisation of Kant's observation – and its implicit critique of Enlightenment optimism – that 'out of the crooked timber of humanity no straight thing was ever made'. It is a dramatisation rather than an argument and does not, perhaps, warrant reasoned disagreement. Just because there are individuals who wish to destroy rather than build, and because there is an element of such an individual in all of us, does it follow that we should not attempt to build? Should we pander to what is, collectively and individually, or should we proceed as if we were committed to making things better? The question, I assume, is a rhetorical one.

Or is it? Adorno and Horkheimer in their classic *Dialectic of the Enlightenment* suggest that by overlooking the irrational, and creating a programme for dominating nature, the Enlightenment has made the world into a prison, a closed universe in the grip of an instrumental reason which invalidates anything outside of itself, including the irrational forces within humanity itself. The direct result of this, they argue, is a 'return of the repressed' with bitterly ironical consequences for contemporary Europe:

In the most general sense of progressive thought, the Enlightenment has always aimed at liberating men from fear and establishing their sovereignty. Yet the fully enlightened earth radiates disaster triumphant.[37]

This terrible outcome (reminiscent of Camus's 'slave camps under the flag of freedom') is, Adorno and Horkheimer believe, the entirely predictable consequence of the barbaric naivety of Enlightenment thought. Barbaric, because it opened the way to a progressive instrumentalisation of humanity and nature, desacralising both, and killing reverence and wonder,[38] replacing them with an abstract, administered, bureaucratic world, a techno-ontology in which our fundamental orientation towards man and nature is trapped in a rigid and programmatic stance. This stance has ceased to be a choice and is now an inescapable destiny. And innocent – dangerously, fatally innocent – because it did not appreciate the archaic, destructive impulses within man that are given no outlet in a bureacratic civilisation. Reason itself cannot give birth to values and the aseptic world of science does not answer to our true, our most deeply human hungers. It is hardly surprising, then, that, instead of assisting our progress ~~from~~ savagery to humanitarianism, it has merely helped the human race to move on 'from the slingshot to the megaton bomb'.

Ironically, Adorno and Horkheimer's left-wing critique of the naivety and sentimentality of the Enlightenment – that it mistook the nature of human beings and that it overestimated the contribution of goodness and reason to their makeup – coincides in important respects with Maistre's more lurid right-wing critique. Maistre mocked the belief that men were born free and born good and that only injustice and social maladministration placed them in chains and made them bad, and the related assumption that the transformation of society would bring about a transformation of human behaviour. Man, he argued, was a rapacious animal and had the additional theological distinction of being stained with Original Sin: he would kill for the pleasure of killing, and not merely to survive. Underneath everything was the impulse to do evil, to smash rather than to build. Adorno and Horkheimer, writing under the twin influence of Freudian theory (especially as expressed in *Totem and Taboo* and *Civilisation and its Discontents*) and the recent horrors of European war, agreed. If anti-Semitism and the Holocaust proved anything, it was that men were more profoundly in the grip of 'the dark impulse' than beneficent reason. Reason is a small island in the human psyche and 'the rational island' is readily overwhelmed by a non-specific urge to destruction. Anti-Semitism expresses something permanent and general in mankind that is only accidentally directed against Jews:

> The psychoanalytic theory of morbid projection views it as consisting of the transference of socially taboo impulses from the subject to the object. Under the pressure of the super-ego, the ego projects the aggressive wishes which originate from the id (and are so intense

as to be dangerous even to the id), as evil intentions onto the outside world, and manages to work them out as abreactions on the
outside world; either in fantasy by identification with the supposed
evil, or in reality by supposed self-defense. The forbidden action
which is converted into aggression is generally homosexual in nature. Through fear of castration, obedience to the father is taken to
the extreme of an anticipation of castration in conscious emotional
approximation to the nature of a small girl, and actual hatred of
the father is suppressed. In paranoia, this hatred leads to a castration wish as a generalised urge to destruction. The sick individual
progresses to archaic non-differentiation of love and domination.

                                    (Adorno and Horkheimer, p. 192)

Anti-Semitism vividly demonstrates how 'paranoia is the dark side
of cognition' (ibid., p. 195).

We are a long way from the world of Condorcet and the belief that
'the violence of our passions' is 'often the result either of habit that
we have adopted through miscalculation, or of our ignorance of how
to restrain them, tame them, deflect them, rule them'. For the Frankfurt School practitioners of *kulturkritik*, 'the fully enlightened earth'
was bound 'to radiate disaster triumphant' because it overlooked the
fundamental irrationality and destructiveness – the anti-civic nature
– of the instincts at the heart of human consciousness.

As for the barbarity of the Enlightenment – rooted in its instrumental relationships, at once technological and bureaucratic, to a
desacralised universe and driven by unacknowledged archaic energies and impulses – this is reflected even in those areas of modern
life which have not been disfigured by an industrialised brutality (such
as world wars and mass persecutions). Inner and outer life are both
empty. 'The most intimate reactions of human beings'

> have been so thoroughly reified that the idea of anything specific
> to themselves now persists only as an utterly abstract notion: per-
> sonality scarcely signifies anything more than shining white teeth
> and freedom from body odour.
>
>   (ibid., p. 167)

Freedom to choose 'everywhere proves to be freedom to choose what
is always the same'. Works of art are made more widely and continuously available, but only at the cost of being degraded to the status of consumer goods, comparable to the washing powders and
automobiles that the advertising industry persuades people that they
need. 'The Culture Industry' exemplifies the Enlightenment as 'mass
deception'.

The Frankfurt School critique has been elaborated by many writers, notably Herbert Marcuse, for whom contemporary affluent democracies are 'one-dimensional societies', in which there is a 'closing of the political universe' (a lack of real political choice), a 'closing of the universe of discourse' (limits to what can be thought) symptomatised by neo-conservative, quietist, one-dimensional philosophical discourse which eschews the large questions, and a denial of the true needs of human beings under a tolerant 'repressive desublimation':

> The distinguishing feature of advanced industrial society is its effective suffocation of those needs which demand liberation – liberation also from that which is tolerable and rewarding and comfortable – while it sustains and absolves the destructive power and repressive function of the affluent society. Here, the social controls exact the overwhelming need for the production and consumption of waste; the need for stupefying work where it is no longer a real necessity; the need for modes of relaxation which soothe and prolong this stupefication; the need for maintaining such deceptive liberties as free competition at administered prices, a free press which censors itself, free choice between brands and gadgets.[39]

The local, explicit, outer tyranny of despots that the *philosophes* fought against has been replaced – at least in affluent, advanced industrial democracies – by a global, implicit, inner tyranny inculcating social conformity and repressing the capacity to lead free and whole lives. In short, desacralised, industrialised life is spiritually empty: even where Enlightenment values have led to progressive material improvement, this has not brought increased happiness or inner enrichment.

It is further argued that, despite a convergence towards uniformity in our lives, this has not created a true organic coherence between individuals within societies; on the contrary, the replacement of an order based upon custom, authority and a sense of the sacred by one based upon reason and regulated by secular contracts has atomised society. Tönnies, in *Community and Society* (1887), developed the notion of two types ('ideal' types in the Weberian sense) of society: the *Gemeinschaft* (communal societies) and the *Gesellschaft* (associative societies). The former were typified by peasant communities in which personal relationships were defined and regulated on the basis of traditional social rules and people had direct face-to-face relations with one another determined by `natural will'. In *Gesellschaften*, on the other hand, typified by modern, bureaucratic, cosmopolitan societies, natural will is supplanted by rational will, and human relationships are more impersonal and indirect, being constructed with a view to efficiency rather than sentiment. Such a foundation for human relationships threatens to weaken the traditional bonds of family, kinship and religion. As Gellner has summarised it:

[*Gesellschaft*] conveys the notion of an open society of anonymous individuals related by contract rather than status, engaged in a free market both of goods and of ideas, freely pursuing their own aims, and having a light but provisional commitment to cultural background, whether gastronomic, dialectical, sartorial or religious. In contrast with such liberalism, there was also a romantic mystique of a closed cultural community, whose members found fulfilment in its very idiosyncrasy and distinctiveness, and in the affectively suffused, highly personal even if hierarchical relationships which it sustained.

(Gellner, op. cit., p. 116)

The conclusion of the Frankfurt critique of the Enlightenment is that, even where its principles have been successfully applied, even where they have not been perverted into mass brutality, the material benefits have been bought at an unacceptable spiritual cost, with individuals living the empty and solitary lives of 'the lonely crowd'.

That the spiritual price of rational societies outweighs the material and other gains associated with them is not self-evident, though this has been assumed by so many for so long and repeated so often that it is almost beyond challenge. Nevertheless, I think that few critics of modernity would prefer untreatable cystitis to anomie, chronic malnutrition to alienation; and few would find being under the thrall of the priest, the local squire, an unaccountable government or an unchallengable workplace bully in an organic community better than living in an atomic society. Organic, materially impoverished societies look attractive, perhaps, only to those who do not have to live in them. Only to those who are not hungry, in pain, afraid of destitution in communities whose only source of welfare is the capricious charity of the well-heeled, do abstract forms of suffering seem more important than the concrete ones that have been palliated by the material advancement of mankind acting in accordance with reason and a sense of justice and equity. We shall return to this argument (which I have discussed at greater length in *Newton's Sleep*) in the final chapter.

CRITIQUE OF THE ENLIGHTENMENT (6)
REASON THE DEMON ARCHITECT

Adorno and Horkheimer bemoan the impact of rationalism on the quality of individual and collective social life. Others have questioned whether societies can be remodelled and governed on rational grounds. If human beings are not individually amenable to rationality, what of nations? There has been a distinguished and influential line of political thinkers – beginning, in the English-speaking world, with Burke and continuing up to the present day with figures such as Michael Oakeshott – for whom there are severe limits to the extent to which rational thought can be applied to the reform of society and to the degree to which rationally-based policies should and could interfere with the organisation of society and seek to influence the conduct of individuals. This organicist thought holds that the excessive application of reason, far from leading to a spiritually empty society, a highly organised community that does not answer to human needs, actually undermines organisation and has enormous destructive potential. The problem with rationalism, these thinkers argue, is that it wishes to bring about changes that can occur, if at all, only spontaneously, as a result of an evolutionary process that is nourished and guided by the deep, implicit wisdom embodied in tradition and ancient social institutions. Society cannot be made; it can only happen. Blueprint rationalism is not only too shallow to bring about beneficial change; it may be catastrophically destructive.

In his *Reflections on the Revolution in France*, Edmund Burke expressed his horror at the speed with which the Revolutionary Assembly wished to change the social order and the completeness with which it proposed to do it. In this, he believed, there was a great danger:

> When ancient opinions and rules of life are taken away the loss cannot possibly be estimated. From that moment we have no compass to govern us; nor can we know distinctly to what port we steer.[40]

This, written in 1790, before the Great Terror, and Danton and Robespierre were still unheard of, was prescient indeed. Maistre, writing in the knowledge of the bloody history of the Revolution said much the same, when he objected to the idea of re-constituting society on the basis of explicit principles:

> One of the greatest errors of a century that embraced them all was to believe that a political constitution could be written and created *a priori*, whereas reason and experience agree that a constitution is a divine work and that it is precisely the most fundamental and essential constitutional elements in a nation's laws that cannot be written.[41]

Burke's reason for the same conclusion was less nakedly theological. He argued that abstract reason – especially where, as in the French

Revolution, it was energised mainly by negative emotions such as anger, hatred, envy and malice – was not adequate to create a new state. Such reason, such general principles as the *philosophes* and their political disciples adhered to – disparaged by Burke as 'geometrical' – would be insufficient substitute for the wisdom implicit in the old social order; nor would it be able to replace the emotions of affection and loyalty that tied individuals to the traditional groups and ensured a social order that was stable, just and coherent. Timeless, universal rationalism could not replace historically-based and culturally-rooted organic growth as a principle of social change. At the very least, it was necessary to maintain the social order throughout change if civic life was not going to give way to anarchy and the old despotisms of the *ancien régime* were not simply going to be replaced by the new despotisms of the *parvenus*. The analogy between a state and an organism is a powerful one: change in the one and development in the other must be constrained by the need to 'keep the show on the road' while changes take place. The vital functions of the embryo developing *in utero* must be maintained while it grows, like a ship being refitted at sea; there must, therefore, be continuity as well as change. Change must therefore be gradual:

> Our political system is placed in a just correspondence and symmetry with the order of the world, and with the mode of existence decreed to a permanent body composed of transitory parts; wherein, by the disposition of a stupendous wisdom, moulding together the great mysterious incorporation of the human race, the whole, at one time, is never old, or middle-aged, or young, but in a condition of unchangeable constancy, moves on through the varied tenor of perpetual decay, fall, renovation, and progression. Thus, by preserving the method of nature in the conduct of the state, in what we improve we are never wholly new; in what we retain, we are never wholly obsolete ...
>
> Through the same plan of a conformity to nature in our artificial institutions, and by calling in the aid of her unerring and powerful instincts, to fortify the fallible and feeble contrivances of our reason, we have derived several other, and those no small benefits, from considering our liberties in the light of an inheritance.
>
> (Burke, p. 237)

Any reformer must take account of this; and it is clear, therefore, that changes cannot be based primarily upon reason, shorn of tradition, upon timeless and placeless universal principles, eschewing cultural particulars:

The old building stands well enough, though part Gothic, part Gre-
cian, and part Chinese, until an attempt is made to square it into
uniformity. Then, indeed, it may come down upon our heads altogether in much uniformity of ruin and great will be the fall thereof.

We cannot re-build on a rational plan, and by fiat, an organism that
did not grow on rational lines, and was not made by conscious human
activity, in the first place. Such radical, abrupt reconstruction will kill
rather than improve the organism. Reformers should also remember
the affiliations and loyalties and sentiments that make a community
viable. No abstract general principle of reason and justice and universal rights can replace the traditional sense of what is right and
proper. Abstract reason is insufficient to generate, or liberate, standards of private conduct and public life that are other than self-centred: we need the 'prejudice' and 'wisdom' of our ancestors. Communal
reason and will are found only in the deep traditions of a nation.

We must, therefore, distinguish between reform, which is to be encouraged – 'a state without the means of change is without the means
of its conservation' – and radical innovation, which is not. Our policies should be shaped by the facts on the ground and their historical
relations – they should be sensitive to the cultural quiddities – and
not by vague notions of reason tied into the speculations of theorists.

The limited applicability of reason to the development of society,
and the consequent obligation of prudent reformers to rein in their
ambitions of transforming society in accordance with rational and
universal principles, has become a widely received idea. The catastrophes of totalitarianism have been seen as being at least in part due
to the impossibility of creative interference in the shaping of society.
A loathing of blueprint rationalism has been extended to a distaste
even for the welfare state. A recent impressive restatement of the case
against deliberate reconstruction of society on the basis of comprehensive planning is to be found in the writings of Michael Oakeshott.

Oakeshott argued[42] that belief in reason as an all-purpose tool applicable to all human activities whatever their nature is mere superstition, a scientism that is, paradoxically, indistinguishable from magic
thinking. Reason, being external to its objects, can never acquire true
knowledge of them. Universal and abstract, it overrides, ignores or is
ignorant of, the specific concrete knowledge – handed down with
constant accretions and modifications from generation to generation
– essential to the pursuit of a particular activity and necessary to any
understanding of it. Because the rationalist approaches human activity from an abstract and universalist viewpoint, his interventions are
invariably misconceived and their consequences usually disastrous.

Oakeshott's political philosophy – anti-rationalist rather than

anti-rationality or irrationalist – unlike Burke's, is rooted in a theory
of knowledge. This was advanced in his earliest book,[43] where he identifies four distinct modes of experience: practice, science, history and
poetry. Practice is concerned with the conduct of life and is the basic
and primordial mode, being concerned with ensuring the 'habitable-
ness' of the world, with the fulfilment of our desires, purposes, needs,
passions and interests. It opens on to or opens up the world of values
– shared values. The scientific mode of experience, on the other hand,
creates a world of ideas which are independent of our needs and
preferences and ambitions, regarding things as they are in themselves,
independent of such personal categories, and which are universally
intelligible and communicable. It is a world of quantitative conceptions, as reflected in the fact that truly scientific explanations boil
down to statements of mathematical relationships, and it attains coherence only by excluding the incorrigibly subjective secondary qualities.
Self-evidently, it cannot capture the full character of experience. History
deals with the causal antecedents of specific events and does not invoke
general laws, or not explicitly, anyway. The fourth mode of experience
is poetry, which is not concerned with practical outcomes or moral
values: its essence is contemplation and delight, offering 'a momentary release, a brief enchantment'. It is 'a sort of truancy, a dream
within the dream of life, a wild flower planted among our wheat'.

It is easy to see why Oakeshott should be opposed to an ambitious
science of society that endeavoured to draw up a blueprint for social
change; for science was only one element in human knowledge, not
the last word on it. Oakeshott

> is opposed not to scientific enquiry or the scientific understand-
> ing ... but ... to what is variously described as 'naturalism',
> 'scientism' or 'positivism', that is, to the belief that the scientific is
> the only proper mode of human judgement ... and ... the confu-
> sion that results when issues of practical politics are regarded as
> scientific problems.
>
> (Greenleaf, *Oakeshott's Philosophical Politics*, p. 90)

Reasoned argument has a limited application to practical affairs;
moreover reason itself cannot be reduced to rules (technical knowledge). A universalist rationalism provides neither a complete diagnosis of nor an adequate remedy for any concrete social problem:
political activity is properly conceived only when regarded as the pursuit
of intimations inherent in an already existing tradition of political
behaviour. As Greenleaf puts it:

abstract ideas do not show how they are themselves to be applied to specific, concrete situations. The belief that all men have natural rights under the law of nature does not give one a useful guide as to whether to raise or lower old age pensions. Every concrete situation, every specific decision, is unprecedented, especially the situations brought about by rationalist revolutions.

(ibid., p. 90)

Although his epistemology is novel, Oakeshott's emphasis on the importance of tacit wisdom – embodied in institutions and the know-how of individuals – over explicit knowledge, *verstehen* over *erkennen*, places him firmly in a tradition that is both long and wide. And he therefore falls foul of the contradictions that beset the critics of rationalism: that they mobilise extremely abstract, universal principles and reasons to underpin their criticism of universalist, abstract, rationalism. More importantly, although, like Oakeshott, they claim to be anti-rationalist rather than anti-rationality or irrationalist, they do not give any indication as to how one should determine the limits of rational intervention or the attempt to improve human society on rational grounds. It may be true that abstract ideas do not show how they are themselves to be applied to specific, concrete situations; but it is equally true that those who would limit the application of reason to concrete situations do not know how far to limit it and what should be put in its place. Granted that 'the belief that all men have natural rights under the law of nature does not give one a useful guide as to whether to raise or lower old age pensions' but the decision about pensions will take place within a certain framework of assumptions about what is desirable which will be more helpful the more it is made explicit.

Unaware of the difficulties, even contradictions, in their own beliefs, conservative critics of the Enlightenment, then, have placed sharp limits on what can be safely changed or successfully created by fiat. The *philosophes*, they argue, deeply misunderstood the nature of society, an organism that has grown as the unwilled consequence of the behaviour, partly rational and partly irrational, partly willed, partly unthinking, of millions of men and women from time immemorial, an entity beyond the reach of scientific thought or, indeed, explicit rational calculation.

This conservative strain of political thought has seemed to have received massive endorsement from the failures of socialist societies, with their universal welfare planning and command economies. The catastrophes of socialism seem to illustrate that, even without the corruption, brutality and untrammelled barbarism that seems always to come in the wake of revolutions, societies explicitly founded upon

reason and ambitious Utopian planning will always run into problems – because society itself is too complex to be deliberately run.

This is the central message of those, such as Hayek, who base their
critique of socialism on the observation that social order cannot be
made: democratic society is 'a spontaneous order' that derives from
the undeliberately coordinated activities of individuals. The self-
organising qualities seen in the market, and in society as a whole,
emerge from actions undertaken within a framework that simply encourages voluntary cooperation. Any attempt to impose social order
will destroy the spontaneous order that is to the benefit of all. As
Dobunzinskis says, Hayek 'articulated what was only implicit in the
writings of Bernard Mandeville, Adam Smith and others, namely that
there are social arrangements that are the result of human action and
yet have not been deliberately designed'.[44]

Similar considerations inform Karl Popper's left-of-centre critique
of 'Utopian' or 'holistic' social engineering. The piecemeal technol-
ogist recognises, he says, 'that only a minority of social institutions
are consciously designed while the vast majority have just "grown"
as the undesigned result of human actions'.[45] There are, therefore,
limited opportunities for creating new social institutions or transforming
existing ones instantaneously by decree or over a short time period.
Unlike the Holistic or Utopian social engineer who aims at remodel-
ling the whole of society in accordance with a definite plan or blueprint, the wise piecemeal engineer

> knows, like Socrates, how little he knows. He knows that we can
> learn only from our mistakes. Accordingly, he will make his way,
> step by step, carefully comparing the results expected with the result
> achieved, and always be on the look-out for the unavoidable un-
> wanted consequences of any reform; and he will avoid undertak-
> ing reforms of a complexity and scope which make it impossible
> for him to disentangle causes and effects, and to know what he is
> really doing.
>
> (Popper, p. 67)

The holistic method is, in fact, impossible since the greater the holistic changes attempted, the greater are the unintended and largely
unexpected repercussions forcing hasty, ill-thought-out 'unplanned
planning' to deal with them. Worse still, the widening gulf between
Utopian plan and chaotic outcome will not prompt the megalomaniacal
planners to reflect and rethink. On the contrary, believing that they
are possessed of a totally reliable method for bringing about the Golden
Age, they will attribute failure to sabotage, to the work of reactionary forces. This will be especially true if the planners are convinced

that they are on the side not only of progress, but of history itself. As Camus said:

> To the extent to which Marx predicted the inevitable establishment of the classless city and to the extent to which he established the goodwill of history, every check to the advance towards freedom must be imputed to the ill-will of mankind.
>
> (*The Rebel*, p. 207)

An incompetent, arrogant and paranoid leadership will, if given enough power, soon create a dystopian Hell on Earth. The dreams of the *philosophes* are mocked in the nightmare of left-wing totalitarian societies:

> The principles which men give to themselves end by overwhelming their noblest intentions. By dint of argument, incessant struggle, polemics, excommunications, persecutions conducted and suffered, the universal city of free and fraternal man is slowly diverted and gives way to the only universe in which history and expediency can, in fact, be elevated to the position of supreme judges: the universe of the trial.
>
> (ibid., p. 207)

The critique of blueprint rationalism on the grounds that no one can calculate the overall effects of social intervention has recently derived striking support from developments in the application of mathematics to dynamical systems. Chaos theory has shown how the effect of small inputs into complex systems may be totally unexpected and quite out of proportion to the size of the input. These effects may also be wide-ranging and long-lasting. Uncontrolled instabilities – as well as surprising stabilities – emerge in unexpected ways.

The argument from complexity is strengthened by the point made repeatedly by Popper (as part of his critique of the scientism of holistic social engineers) that, unlike objects in the natural world, human beings react to the intentions of those who try to alter society or even to predict its future. Beliefs and expectations about interventions will themselves affect the result of interventions and the reaction prompted by those results will further modify them. The huge difficulties faced by technologists working with the material world – that, for example, the solution of one problem creates new problems also requiring solutions – are as nothing compared with those facing the engineers or would-be engineers of society.

All of this may be understood and accepted without any clearly conservative, even less anti-rationalist or anti-progressive, conclusion

being drawn. For unless we eschew all attempts to alter things for
the better and deny ourselves the right to intervene in any way and
forebear to predict the effects of non-spontaneous change, we will
always be committed to a level of action whose outcomes cannot be
precisely calculated. But irrational interventions and inertia also have
unexpected consequences. Popper did not, of course, intend to argue
against rational reforms; rather to argue for modesty and humility –
for piecemeal rather than holistic engineering and for a willingness
to recognise that every intervention is an experiment. We must expect the unexpected and be willing to learn from our mistakes.

This is a hard lesson for many to understand, despite the eloquence
with which Popper has expressed his views. In her posthumously published testament, Petra Kelly argued as follows:

> I often hear people arguing about the world's many evils and which
> should be the first confronted. This fragmentary approach is part
> of the problem, reflecting the linear, hierarchical nature of patriar-
> chal thinking that fails to grasp the complexity of living systems.
> What is needed is a perspective that integrates the many problems
> we face and approach them holistically.[46]

Kelly believed, it seems, that we shall not be able to put anything
right at all unless we put everything right at once. Have people learned
nothing from the experience of 200 years of millenarian ambitions?
Anyone who 'grasps the complexity of living systems' will know that
modest rather than global ambitions are more likely to be realised
and considerably safer.

Support for an intermediate position between the holistic engineers
and those who would oppose *all* attempts at improving social conditions comes from an unexpected source: the poet and conservative
practitioner of *kulturkritik* T.S. Eliot:

> We slip into the assumption that culture can be planned. Culture
> can never be wholly conscious – there is always more to it than we
> are conscious of; and it cannot be planned because it is also the
> unconscious background of our planning.[47]

> The other direction in which confusion of culture and politics may
> lead, is towards the ideal of a world state in which there will, in
> the end, be only one uniform world culture. I am not here criticis-
> ing any schemes for world organisation. Such schemes belong to
> the plane of engineering, of devising machinery. Machinery is necess-
> ary, and the more perfect the machine the better. But culture is
> something that must grow; you cannot build a tree, you can only

plant it, and care for it and wait for it to mature in its due time. And a political structure is partly construction, and partly growth; partly machinery, and the same machinery, if good, is equally good for all peoples; and partly growing with and from the nation's culture, and in that respect differing from that of other nations

(ibid., p. 119)

THE ENLIGHTENMENT AND THE NEW DARK AGES

The failure of the French Revolution was a great gift to the Counter-Enlightenment thinkers. Recent catastrophes arising out of socialist revolutions have seemed to prove beyond all doubt that the Enlightenment has not only failed to deliver on its promises but that the Enlightenment project was doomed. The two centuries of experience between the Fall of the Bastille and the Fall of the Berlin Wall have seemed to constitute an experimental refutation of the hopes and ambitions of the *philosophes*. For some thinkers, as we have seen, the notion of a rationally-based and just society progressing to ever-increasing prosperity, is in principle flawed. For them, the modulation of the Enlightenment programme into nightmare – from the great terrors of September 1792 to the sea of blood spilt in the greater terrors of the twentieth-century revolutions – is an inevitable, rather than an accidental consequence of the application, or attempted application, of the programmes of the *Encyclopédistes*. The Enlightenment thinkers fatally misunderstood the nature of individual men and of communities, of societies and the people of which they were composed. Others critics are less certain that Enlightenment hopes were deluded in principle but they none the less assert that they have not led in practice to significant progress in the happiness of mankind. The gravity of this charge varies.

Some draw attention to the fact that many today are no better off materially than their ancestors 200 hundred years before; indeed, for a few, life is worse. This is a quantitative claim and cannot be resolved by armchair debate; qualitative impressions are simply not enough. Even so, if one considers a nation such as Britain, there can be no doubt about the material advancement of its people. The length of life, the duration of healthy existence, the level of nutrition, education and safety have all been transformed out of all recognition. So, too, has the variety of goods and the wherewithal to consume them. This is true for every class, even the underclass created by the conservative governments of the last 17 years. Few starve or freeze to death, go barefoot, or suffer from unrelieved physical torments of punishing labour. The life of the average person in the United Kingdom is unbelievably comfortable and affluent compared with that of the pre-Industrial Revolution peasant.

This is so clear that the denial of progress has to shift its grounds. Some argue that improved affluence in the West (the North, the First World, etc.) has been bought at the cost of a deterioration of the health and happiness of those in other parts of the world. However, it is not clear that (a) the deterioration of Third World countries outweighs the gains in the First World; (b) that this deterioration is universal – after all, India, the Pacific Rim nations, etc. are becoming increasingly affluent; (c) that where there is deterioration, this is due to Western affluence, rather than the result of indigenous despotic regimes, civil and foreign wars, and a failure to adhere to those very principles which the Enlightenment held dear. If there is a growing distance between the poverty of the poorest and the wealth of the richest countries, this obviously cannot of itself be an argument against the Enlightenment principles of equity, justice, etc. At any rate, the global felicific calculus is difficult to carry out – especially since it would have to take into account the indirect adverse material effects of material progress, such as pollution and more general environmental depradation and degradation, and the difficulty of applying brakes to the escalation of consumption.

I have already noted the arguments of the Frankfurt School and others that, while there may have been overall material progress in the West, this has not brought happiness; indeed, because it has been brought at the cost of spiritual emptiness, it has subtracted from the sum of human happiness. The myth of the organic, poor-but-happy societies of the past is just that – a myth – but, as I have argued elsewhere, an adhesive one.[48] There is, again, little evidence that being deprived of cold, hunger, unrelieved pain associated with chronic ill health, of petty and major oppression by unaccountable, unelected political masters and the casual and unpunished brutality of anyone in power over one, has resulted in a net loss of happiness. In the absence of evidence to the contrary, one must assume that it hasn't; indeed, that the reverse is the case. It is also doubtful whether 're-pressive desublimation' and other spiritual ills that the Frankfurt School blame upon the tolerance and plenty that has resulted from the success of technologies in tune with the Enlightenment project are as bad as toothache, cold, unchallenged bullying, lack of rights, etc. that previous societies offered to their lower orders. Admittedly, Taylorised labour, based upon rationalistic principles, in an early mass-production plant was probably little better – possibly even worse – than the excruciating, endless labour of the peasant. Current factory life, however – in physically pleasant and safe conditions – may be repetitive and narrow, but is a definite improvement on hand-milking a cow with chapped udders at 4 a.m. on a winter's morning. And most people's work is considerably more varied and interesting than this. This is something that requires further discussion and I shall address it in the last chapter.

The most serious charge of all is that the secularisation and industrialisation of society, which is seen as an expression of the rationalism and technophilia of the Enlightenment, has led directly to the horrors of twentieth-century totalitarianism. There are two elements of this argument: the first is that there is something special about the horrors of the twentieth century – human beings are not merely no better now than in the past, they are actually worse; and the second is that these horrors are causally related to, or spiritual consequences of, the Enlightenment. I shall discuss the question of whether human beings in the twentieth century have been uniquely inhuman to their fellow creatures presently and also return to it in a later chapter. Let me first, however, briefly consider the question of whether the atrocities of this century can be laid at the door of Enlightenment thought.

The leaps of thought that this hypothesis requires – and we have already touched upon them in our discussion of the Frankfurt critique – are not entirely counter-intuitive. The *philosophes* made a decisive assault on the truth claims of religion – though, as Goldmann has convincingly argued, their arguments were powerful only because religion had already lost its ubiquitous presence and become a matter of argument: The Enlightenment critique of the *ancien régime* required an essentially secular society for its arguments to carry such weight. Even though not all of the *philosophes* were atheists (indeed, Rousseau saw atheism as being so dangerous that it justified the death penalty), they were so deeply opposed to the institutionalisation of religious belief and the abuses of power that flowed from it, that they emptied most of their remaining religious beliefs of specific content. Voltairean deism naturally leads on the one hand to the Newtonian clockwork universe with God as simply He 'Who winds up our sundials' as Lichtenberg said and, on the other, to the Religion of Humanity and the Supreme Being adored by cohorts of maidens at the Feast of Reason, 'the ancient god disembodied, and launched like a balloon into a heaven empty of all transcendent principles' (Camus, *The Rebel*, p. 92). The death of religion (though it had many resurrections subsequently) also removed a transcendental constraint on human behaviour. (Not that such constraints seemed to have had much influence in the past.) 'If God does not exist', as Dostoyevsky said, 'then all things are possible.' The *philosophes* replaced argument from authority with argument from evidence and reason: the authority of scientific truth displaced the authority of revealed religion with its institutions and functionaries.

This passionless, abstract, instrumental way of thinking – consonant
with the view that man was but a piece of clockwork in a clockwork
Newtonian universe – was, its critics argued, spiritually emptying.
Not only did it deny God and the transcendental basis for values; it
uprooted man from the natural world, put him out of touch with the
values and constraints that (supposedly) come from a deep relationship with nature. An instrumental reason, expressed in a relationship
of domination over nature (in contrast to the traditional farmer's dialogic
relationship) and reflected in the mass production of goods by means
of processing technologies (and the industrialisation of farming) alters
our relationship with other human beings: they, too, become a substrate
for the operation of instrumental reason, for the kind of scientistic
approach pioneered by the early mathematicians of society and the
dreamers of felific calculi.

All of these elements, it is claimed, came together in the industrialisation of death, the Taylorisation of slaughter, that made its spectacular début in First World War and received its hideous culmination
in Auschwitz. The scientism of the *philosophes* – and more, widely,
the materialist world-picture derived from physical science – has generated what Appleyard has described as a 'vision of man as a fragile,
cornered animal in a valueless mechanism'.[49] The dangers inherent in
this and, more generally, in the separation of scientific facts from human
values, were not at first appreciated:

> The innocence of the easy, progressive Enlightenment myth, was
> struck down in the trenches, finally died at Hiroshima and Auschwitz.
> Scientific reason was as capable of producing monsters as unrea-
> son. It is no good arguing that Auschwitz was unreasonable – in
> its way, it was, and secular society cannot be sure that it can offer
> any higher, purer rationality – nor is there any point in claiming
> that Hiroshima was a necessary evil designed to prevent more deaths
> from a protracted conventional war. That may have been true, but
> the atomic genie had been let out of the bottle and new, evil
> rationalisms would find other justifications for its use. Above all,
> nuclear weapons seemed to confirm our sense that there was some-
> thing unprecedentedly and uniquely corrupted about our age.
>
> (Appleyard, p. 122)

Value–free knowledge increasing our power, inescapably, rather than
accidentally, leads to immoral exercise of power: war, the industrialisation of death. According to many contemporary thinkers, therefore, industrialisation, secular rationalism and the technological vision
of nature as a machine and of man, within it, as *l'homme machine* converged inescapably in the extermination camps.

Heidegger, in a lecture in 1949, likened 'the manufacturing of corpses in gas chambers and concentration camps' to the mass production of agricultural goods.[50] This disgusting analogy has actually a rather more complex significance than might at first be appreciated: the source is a Fascist sympathiser who believed that Nazism had 'inner truth and greatness' because it addressed the 'the encounter between global technology and modern man'. For this Fascist supporter, who subscribed to an organic mysticism about blood and soil and the *volk*, the unimaginably terrible outrages of Auschwitz – upon which he conspicuously refused to comment after the war – were comparable merely to the agribusiness he loathed. This perhaps says more about Heidegger than about the nature or origin of the industrialised genocide of the death camps. But it opens on to a more general point. Fascism was, as we have already noted, actually opposed to all the values of the Enlightenment: its rationalism, its universality, its egalitarianism, its cosmopolitanism, its concern with the downtrodden and its commitment to a technologically-based society.

While it is possible, therefore, by leaps of thought, to connect some of the worst atrocities of the twentieth century with secular Enlightenment thought, it is easier to connect them with Counter-Enlightenment thought. The line leading from irrationalists such as Maistre, anti-universalist nationalists such as Herder and Fichte, and those who participated in 'the deep and radical revolt against the central tradition of western thought' that Berlin has called the 'Apotheosis of the Romantic Will' (in *The Crooked Timber of Humanity*), to the rise of Nazism and the concentration camps is short and straight. It is marked out clearly in the prose writings of Wagner, where a pessimistic-romantic Schopenhauerian philosophy of the will, a vision of a national art epic, a reactionary and atavistic nationalism and a virulent anti-Semitism which darkly hints at 'the final solution' live in nauseating harmony.[51]

It may be argued that the Enlightenment cannot be exonerated from blame for the Great Terrors of the left-wing totalitarian regimes – that the regimented dystopias of Marxist states are true fruits of the Utopian dreams of the *philosophes*. But even this is difficult to support. Fantasies of precisely regulated societies – Plato's *Republic* and Thomas More's *Utopia*, and any number of celestial and secular cities – antedate the Enlightenment. And many writers whose blueprints foreshadowed the oppressive regimes of the twentieth century were at odds with the ethos of the Enlightenment – even the relatively benign holistic social engineers. After all, it was Comte, not Voltaire, who wondered why, when we allow freedom of opinion in mathematics, we should allow it in morals and politics. It was Saint-Simon and Comte who took scientism, the unthinking application of the rhetoric of science to the management of societies for the good of their citizens, to their limit. And it was Marx, above all, who created the intellectual framework in which this could be developed to the full by taking the Hegelian doctrine of alienation and false consciousness into his own theory of history, class warfare and the revolution. It was he, not the *philosophes*, who furnished the theories that enabled the scientologists of the revolutions of the left to justify their assumption of limitless power and their murderous intolerance of those who disagreed with them. It was Marx, the *anti*-Enlightenment figure – who did not believe in respecting the views of autonomous individuals – who paved the way for the left-wing totalitarian states of the twentieth century and 'the slave camps under the flag of freedom'. As Camus pointed out, the means Marx advocated for bringing about the left-wing revolution are precisely the same as those which Maistre advocated: 'political realism, discipline, force'. (We shall return to Marx in chapter 5.)

In many cases the horrors of the twentieth century have nothing to do with either the Enlightenment or the Counter-Enlightenment. The recent genocide in Rwanda – in which, in just over 100 days, nearly one million men, women and children in a small country suffered a Stone Age death at the hands of their fellow countrymen – is hideously illustrative in this regard. This eye-witness report says it all:

> The death squads even tied up children and stuffed grass in their mouths before throwing them in the lake. And if they didn't drown they pulled them out again and buried them alive.
>
> (*Observer*, 1 May 1994)

This may put into context the claim that contemporary mass slaughter is to be understood as a phenomenon peculiar to industrialised, secular, scientific, rationalistic societies reflecting Enlightenment values. It also puts into perspective Bryan Appleyard's claim that 'nuclear weapons seemed to confirm our sense that there was something unprecedentedly and uniquely corrupted about our age' (p. 122). The preferred weapon of the Hutus was not the atomic bomb nor even the machine gun, but the machete.

Genocidal bloodbaths have been sickeningly common in history and are not peculiar to advanced, industrialised societies. Consider this comparatively modern example, one that cannot have been laid at the door of either science or the Enlightenment. A protest in 1894 by the villagers of the Sassum district of Armenia against illegal double taxation triggered a state-sanctioned wholesale massacre which claimed between 200 000 and 300 000 lives, including unarmed women and children living hundreds of miles away from where the protest had taken place. In 1915, the Ottoman rulers, fearing that Armenia in the east could secede and gain independence in the way that the Balkans in the west had done earlier, planned and executed an attack on the entire Armenian population. The result was that some 1 500 000 innocent civilians, about half of the Armenian population in Anatolia, perished. Only those who accepted Islam remained.

Wherever we look in history, we encounter similar outrages. Indeed, there have been events in the remote past whose scale has exceeded even that of the atrocities of the twentieth century. Consider, for example, the record of the Mongols. In 1219, following the attack on the city of Bukhara, the surrounding plain was 'like a tray of blood'. All 30 000 inhabitants were slaughtered. At the battle for Samarqand, prisoners were sent ahead with the intention that their bodies should mop up the rain of arrows from the besieged city. When the city fell, the entire Turkish garrison of 30 000 was put to the sword. The civilians were then divided into sections and treated accordingly: the women were set aside to be raped and then sent off as slaves; the clerics were all spared; while the entire population of craftsmen and artisans were transported to the Mongol homeland to be employed at Genghis's court.[52] This was nothing compared with what was to come. When the city of Merv fell in 1221, Persian chroniclers claimed that Genghis Khan's son, Tolui, supervised the slaughter of 700 000 citizens, sparing only 80 craftsman. Other cities in Transoxania were treated likewise: Balkh, one of the greatest cities of the age, filled with mosques, hospitals and palaces, was reduced to 'an empty arena of walls guarding a windswept plain' (Marshall, p. 56). The slaughter went on and on:

All the chroniclers seemed to vie with one another in trying to capture the sheer scale of disaster.... One writer claimed that when the town of Herat suffered Genghis's terrible retribution, having rebelled and challenged the Mongol rule, the Mongol general Elchidei is supposed to have exterminated no fewer than 1,600,000 people. Another chronicler puts the figure at 2,400,000. At the destruction of Nishapur, the death toll was supposed to have reached 1,747,000. The heads of their victims were stacked in three pyramids: a pile each for men, women and children.... At Bamian, Balkh and Merv the entire populations, even domestic animals, were slaughtered. A similar fate struck the towns of Azerbaijan.

And these were no mere murderous frenzies that got out of hand: they were part of a deliberate policy intended to create a state of terror. The cold calculations at the heart of the slaughter are revealed by the sparing of officers and nobles, artisans, craftsman, scribes, clerics, merchants and occasional administrators – anyone, in other words, who might prove useful. Peasants – the majority of human life in the thirteenth century – were treated somewhat differently. They were regarded without exception

> as having no greater status in life than a flock of sheep. When the order came to put the population to sword, they were herded to-gether like sheep – and dealt with as such. Being pasture-based horsemen, the Mongols viewed men and women who worked on their knees in the soil as the basest form of human life. Even a horse was far more valuable.
>
> (Marshall, p. 66)

In northern China, the genocidal campaign was, if anything, even more thorough. The ravages of Genghis Khan's campaign during the 1220s and the wars of conquest that followed under Ogedei had un-imaginably horrible effects. The nearest we can get to a measure of their scale – and the most poignant proof of the truth of the his-torian's claims about Mongol's deeds – comes from demographic data. A census taken by the Chin empire in 1195, before the Mongols be-gan their campaigns, recorded a population in northern China of just under 50 000 000; and when the Mongols took a census of their con-quered territory in 1235–6, they counted 9 000 000 – a discrepancy of 40 000 000 (ibid., p. 69).

The extent of the cruelties was matched by the nastiness of their details: when Inalchuq, the governor of Utrar who had defied Genghis Khan in a doomed but brave last stand, was captured, he was ex-ecuted by having molten silver poured into his eyes and ears. And the consumerist approach to human life is vividly illustrated by the fact that, when Genghis was buried, forty jewelled slave girls were buried with him.

The terrible set-piece slaughters – battles, sieges, etc. – took place against a background of continuous destruction and pillage that laid waste vast areas of civilised Asia. Cities, towns, villages were flooded (by the ingenious misuse of rivers), put to the torch or razed to the ground. The intervening agricultural land was destroyed. The devas-tation wrought directly and indirectly on the irrigation systems built up over millenia in the Persian plateau made land uncultivatable for centuries after the Mongols had left. Much of the territory they con-quered was rendered uninhabitable by the 'scorched earth' approach to conquest. It is probable that they did not really want the land: they wished only to subjugate the people by terror to ensure their own undisputed mastery of the steppe.

The horrors of pre-modern Mongol conquests differed from other wars of the time only in scale. In other respects, they were typical. The reduction of the civilian enemy to cannon fodder, the rape and enslavement and slaughter of women, and the disposal of children, and the genocidal attitude of conquerors towards the conquered was the norm, irrespective of whether the conquering hordes were Mongols or Crusaders. Massacre of the inhabitants was routine following a successful siege of a city: it made no difference whether the aggressors were mercenaries or religious idealists, Christians, Muslims or Heathens.

It may seem unnecessary to point out that the religious beliefs of the participants did not predispose them to be merciful. The cruel treatment of the pathetic young idealists who participated in the Children's Crusade in 1212 was no better than the behaviour of the Nazis towards the children caught up in their wars. The behaviour of the Crusaders themselves – crusading on behalf of the God of Mercy, Love and Forgiveness – when they were victorious was no different from that of their 'heathen' enemies: when in 1098 the city of Antioch fell, the entire Muslim population was massacred. And religion *per se*, without the pretext of war, was often sufficient to justify slaughter. The attitude of the Incas in matters of childcare may be judged from their regular sacrifice of children – often several hundred at a time – on ceremonial occasions or when defeats, famine and pestilence called for royal blood. This should cause those who imagine that cruelty to children is a twentieth-century invention to think again. The Aztecs exceeded even the Incas' cruelties. They believed that it was their divine duty to wage war in order to provide the sun with his nourishment of blood and hearts – that is, to obtain sacrificial victims. The estimated 80 000 individuals who had their hearts cut out while they were still alive in a single ritual at Tenochtitlan in the sixteenth century were captured spoils of war.[53]

And it is important not to forget, when we brood on the 'unprecedented' horrors of war in modern times, that the development of yet more horribly destructive weapons has not been the only change in the conduct of war in the twentieth century. There has also been an increased reluctance to use these weapons; and the emergence of the Geneva Convention on the conduct of war. It seems unlikely that Richard the Lionheart would have resisted the temptation to 'nuke' Saracen strongholds, have passed up the advantage of using poison gas or refused to use dumdum bullets on moral grounds. And he would no more have dreamt up the Geneva Convention than would Genghis Khan or the priests at Tenochtitlan.

The Geneva Convention is a remarkable tribute to the emergence of a moral sensibility in relation to war that, while it is not unique to the present age – after all Grotius' *De Iure Belli ac Pacis*, which established the notion of international law was published in 1625 – is uniquely developed in it. For the Romans, law and war had nothing to do with one another: *inter arma leges silent*. The massacre at My Lai would not have been regarded as an outrage, or even an error. And whereas the Marshall Plan for the reconstruction of post-Second World War Europe may have been motivated by fear of communism and a repetition of the consequences of post-First World War reparations and blockades, it was an advance in humanity on pretty well anything that had been seen in the past – unless it be that history is curiously silent on the post-war humanitarian relief and welfare missions undertaken by the Mongols, the Crusaders, the Aztecs and others. Until the end of the nineteenth century, the laws of war took the form of uncodified custom and practice, where the limitation of violence and bloodshed was based at least in part on self-interest and fear of reprisal. The Geneva Convention goes far beyond the laws of restraint hitherto seen in war. Hitherto, these laws, inasmuch as they were evident at all, applied only to conflict between ennobled gentlemen; they rarely embraced the peasant foot–soldier, even less his wife and children.

From the above we may draw two conclusions. The first is that 'dehumanising' science – and supposedly scientistic Enlightenment thought – has never been necessary to ensure that human beings plumb the depths of inhumanity. The second is that the assumption that the present century has witnessed a uniquely terrible treatment of human beings by their fellows is untrue. It is arguable that the 'Fundamental Rules of International Humanitarian Law Applicable to Armed Conflicts' (published in 1978 in *International Review of the Red Cross*), summarising the agreed international laws of war, says as much about the nature of the twentieth century as does *Mein Kampf* or the Gulag Archipelago.[54]

Anyone who attempts to defend the present century against its detractors must take account of the Holocaust. However, many historians, including Jewish historians who could not be accused of trying to diminish its horror, have attempted to put this abomination into historical perspective. Yehuda Bauer's *The Holocaust in Historical Perspective* (1982) attempts to 'demythologise the Holocaust', because, he feels, it stops us thinking about it: if the Holocaust was caused by humans and its horrors inflicted on other humans watched by yet other groups of humans, then it is as understandable as any other historical event:

What the Holocaust has shown . . . is that the range of possibilities of human behaviour is very wide indeed. We might have learned that from studying Genghis Khan or Nero or Caligula or the early Aztec kings, but we thought, perhaps, that the Enlightenment and the rise of democracy in some western countries in the nineteenth century might have changed all that.[55]

In summary, the two assumptions that support one of the most important elements of the contemporary critique of the Enlightenment – that the modern era has been marked by unprecedented cruelty; and that this is due to rationalistic, secular Enlightenment thought – do not stand up to close scrutiny. It may be that, as I have argued elsewhere,[56] the very sensibilities that are outraged by the horrors of our age have been sharpened by the Enlightenment, perhaps even created by it; that it is the rationality and compassion first expressed by the *philosophes*, the refusal to tolerate the intolerable, that has led us to hope for, to expect, to demand, higher standards of behaviour by human beings, singly or in groups, in peace and in war. If so, we can hardly blame the Enlightenment for the fact that these expectations have frequently been disappointed.

There is a deep inconsistency in those who use the hideous events of the twentieth century as a critique not only of the Enlightenment but also of the Enlightenment dream of progress and the view exemplified in Condorcet's last testament of the history of the human mind as being one of continuous progress towards enlightenment and truth. 'Look', they say, 'people are still hideously uncivilised.' And when the scale of past horrors is pointed out to them – so that what would be regarded as an outrage now would have passed as ordinary life then – they seem to imply that we should not judge the past by the same standards: the child-murdering Aztecs, the genocidal Mongols, didn't know any better. Could it be that the Enlightenment has, after all, raised our expectations and that this is accepted by its opponents who do, therefore, after all, believe in progress? That the passionate decency of the Party of Humanity has set standards that enable us to know better, even if we do not always act on our knowledge.

ENEMIES OF HOPE

All that I have said so far, which is by way of being merely preliminary to the business of this book, could be summarised as follows: every argument against the Enlightenment and its values can be answered by another argument of equal force. The putative dangers of Enlightenment thought are more than matched by the dangers inherent in anything the Counter-Enlightenment would offer in its place. Rejecting the notion that men are, or can be educated to be, good and should therefore be allowed to be as free as is compatible with the minimum order necessary to ensure the freedom of other men, and replacing it with the assumption that they are intrinsically and incurably evil and need to be ruled with a rod of iron has, in a variety of ways, assisted the cause of brutally self-serving autocratic regimes the world over. Asserting that man never knows his own best interests but is incorrigibly fallible, or – the secular equivalent of this view – imbued with 'false consciousness', has given despots of the right and the left the paternalistic mandate to override the wishes and rights of their subjects and to dismiss or crush disagreement. Denying reason a sovereign place in human affairs has been a gift to demagogues and tyrants, who are threatened by rational dissent. Rejecting the universalism of Enlightenment thought has proved to be all too compatible with an emergent aggressive ethnocentricity whose fruits have been nationalist wars and inter-tribal civil wars. In summary, there is scarcely an element of Counter-Enlightenment thought that has not bred monsters even greater than those supposedly spawned by the Enlightenment. As Mark Lilla has written, with reference to Berlin and the Counter-Enlightenment thinkers whose cause he has espoused:

> In reading [Berlin's] essays one has the uncomfortable impression of watching someone welcome harmless stray puppies into his home, only to discover years later they are all grown up, straining at their leashes, teeth bared.[57]

The profound ambivalence experienced by those who have thought most deeply about the dialectic between Enlightenment and Counter-Enlightenment is vividly reflected in two passages from Berlin. In the first, he notes sorrowfully that the Enlightenment dream of 'perfect and harmonious society, wholly free from conflict or injustice or oppression' was an ideal 'for which more human beings have, in our time, sacrificed themselves and others than, perhaps, for any other cause in human history'.[58] A second passage, however, reveals a totally different attitude towards the *philosophes*:

> The intellectual power, honesty, lucidity, courage, and distinterested love of the truth of the most gifted thinkers of the eighteenth century remain to this day without parallel. Their age is one of the best and most hopeful episodes in the life of mankind.
>
> (*The Age of Enlightenment*, p. 29)

The conviction informing *Enemies of Hope* is that it is possible to resolve this contradiction by separating what is good in Enlightenment thought, what is feasible in the Enlightenment dream, from what is bad, dangerous and unacceptable. In short, by rethinking the Enlightenment in the light of the experience of the last two centuries, in particular by taking account of what is valid in the Counter-Enlightenment critique. And this is what I propose to begin in the last chapter of this book. I believe that when this task is complete, it will be clear that there is much in Enlightenment thought, or in the intellectual example of the Enlightenment thinkers, much that is despised, misrepresented or forgotten, that is not merely valid and good but will be essential if mankind is going to negotiate the tricky times ahead. The Enlightenment thinkers were driven by a noble vision: a vision of men and women, gifted with critical reason, informed by accurate observation, working indefatigably for justice and the progressive improvement of the lot of all mankind. That vision was corrupted when it degenerated into a secular millenarianism. We need, therefore, to see how we can retain the Enlightenment dream of progress, fuelled by hatred of injustice, cruelty and hypocrisy, and regulated by the application of reason to a secular society assumed to be inhabited by largely rational human beings, without falling into millenarial errors and a fanaticism that ushers in oppression and slaughter.

Before this should be attempted, however, it is necessary to bring into focus the major theme of this book – namely, the predominant mode of the contemporary assault on the Enlightenment. There is much in present-day Counter-Enlightenment thought that is not new: for example, the attack on reason as a force within human affairs, either in the public or the private sphere; the denial of universal values; the critique of science; and the call for a renewal of religious values and a restoration of religious powers in contemporary society. All of these are familiar enough. What is new is the level at which these attacks are directed. We might summarise the contemporary Counter-Enlightenment as an all-out assault on the notion of a disinterested, rational, autonomous, conscious agent. This assault is characterised by dissolution of the idea of the deliberative self and denial of the central role of reasoned action in human life. As Goldmann said:

> all the leaders of the Enlightenment regarded the life of a society as a sort of sum, or product, of the thought and action of a large number of individuals, each of whom constitutes a free and independent point of departure.
>
> (p. 32)

The notion of a society consisting of individuals who are in any sense 'free and independent points of departure' precisely defines the target of much contemporary pessimistic and anti-progressive thought. In the late twentieth century, the enemies of the Enlightenment are enemies of what I believe to be the distinctive feature of humanity: a reflective consciousness which has a margin of autonomy.[59] They try to marginalise – even eliminate – consciousness itself; they deny the role of the rational self in public and private life – because reason is embedded in things that reason knows not and we are anyway driven by forces that lead us to unreason – forces which can be ignored only at the perilous cost of breaking out into murderous savagery; they deny the centred self; and, not suprisingly, they dismiss the notion that progress will occur as a result of conscious, rational, human action. The enemies of hope, in other words, either deny the existence of rational consciousness; or, if they accept that it exists, deny its central role in human affairs; or, if they accept that it exists and enjoys such a role in human affairs, believe its influence to be wholly, or largely, bad.

It is this more fundamental attack on the founding assumptions of Enlightenment thought that is the central concern of the present book. Only when this has been addressed at the level at which it has been pitched, only when the contemporary Enemies of Hope have been answered, can a defence of what is defensible in the Enlightenment seem anything other than naive.

If we are to believe – as I do, in opposition to many of the thinkers whose views have dominated intellectual life in Europe in the twentieth century – that the hope of progress is well founded, we must also believe in the central role in human affairs played by the conscious, responsible, individual human agent, and refuse to cede this role to unconscious social, historical or linguistic forces. Clearly, if we do not believe in the reality or the beneficence of the conscious autonomous, rational individual human being able to work together with other such individuals towards the common good, then there is no certain way forward for humanity. Maistre's universal bloodbath seems as likely an outcome as any other, and there is nothing we can do to influence how things turn out. Consequently, if there is a moral obligation incumbent upon intellectuals at present, it must be to oppose the prevalent *trahaison des clercs* – deeper even than the one that Benda deplored – of humanist academics who deny (or pretend to deny) the uniquely non–animal nature of humanity and who refuse to recognise the superiority of reason to irrationality, of science to magic, of accountability to unaccountable power, of hard-won factual knowledge to myth.

It is a melancholy thought that this book should be unnecessary. Those who pretend that their fellow men are parts of a machine or a text over which they have no control, that discourse is about power not truth, still experience bourgeois individualist indignation when they feel that their own rights have been trampled on or they have not in some regard been well served by others. Those who feel that, underneath, their fellow men are ravening animals for whom reason is an unacceptable prison-house still expect good behaviour in a bus queue and insist that those who queue should file rather than swarm when the bus arrives and expect equitable treatment at interviews for

jobs. Those who equate modern medicine with magic, the accountable doctor in the white coat with the unaccountable witch doctor
terrifying his patients, would not hesitate to seek the help of a Western-
trained surgeon for a ruptured appendix abscess, to take H_2 antago-
nists for a peptic ulcers or to sue for damages if they felt that they
had been badly treated. It may well be that, as one eminent legal
theorist claims, post-modernism

> questions causality, determinism, egalitarianism, humanism, liberal
> democracy, necessity, objectivity, rationality, and truth. . . . [It] makes
> any belief in the idea of progress or faith in the future seem ques-
> tionable.[60]

But post-modernists still expect a pay cheque every month, treat the
finance departments of their institutions as if they were staffed by
conscious agents, working within a framework of logic and causal
relations, responsible for making decisions. Post-modernists, that is,
have a very robust, unpost-modernist sense of their civic rights and
sometimes even a dim intuition of their civic duties. And, like the
rest of us, they differentiate between technology and mumbo-jumbo,
expecting to be flown between their international symposia in planes
designed on sound scientific principles and embodying the most complex and sophisticated technology rooted in the scientific discoveries
of the last 400 years and piloted by individuals with a conventional
sense of personal responsiblity.

In short, the irrationalists and anti-humanists whom I target in this
book vividly illustrate Trilling's observation of several decades ago,[61]
that 'it is characteristic of the intellectual life of our culture that it
fosters a form of assent which does not involve actual credence.' This
was written before the republic of letters became so congested with
advanced critics who call into question the very notions of truth,
meaning and reference and for whom all analysis of texts ends in
total uncertainty – *aporia* – but who none the less have sufficient command of 'the rhetoric of authority' to say what literature – and indeed the world-text – has been, is and must always be. As Frank
Lentricchia pointed out apropos of Paul de Man, there is 'a strange
discrepancy'

> between a frightfully sobering literary discourse and the actual prac-
> tice of a critic whose judgements, authoritative in tone and style,
> betray the theory.[62]

And

Even while, in *Blindness and Insight*, he was telling us that there was no truth, or if there was, that it could never be known, he spoke transcendentally of 'the foreknowledge we possess of the true nature of literature'.

(Lentricchia, pp. 283–4)

One of the great exponents of the art of 'fostering a form of assent that does not involve actual credence' is Michel Foucault, who did not actually believe himself. According to Clavel, his close associate, Foucault had 'vanquished Marx and the Enlightenment'.[63] Foucault, in particular, repudiated the idea of progress; for example, he demonstrated to his own satisfaction the baselessness of the common belief that Enlightenment ideas had led to a humanisation of the law and, specifically, of punishment. In *Surveiller et Punir*, he compared the punishment of Robert-François Damiens in 1757, an unsuccessful regicide, with the treatments recommended for imprisoned criminals by Leon Faucher in 1838. First Damiens' punishment:

According to the terms of his sentence, the flesh was to be torn with pincers from his breasts, arms, thighs and calves, and his right hand, which had wielded the knife, was to be cut off. His wounds would then be daubed with molten lead, boiling oil, burning pitch and a mixture of molten wax and sulphur. His body would be quartered and his limbs torn from his torso by four horses. The corpse was to be burned and the ashes scattered in the wind. In the event, Damiens' sufferings were even more Hellish than the sentence. The four horses were unable to perform their task. Even the harnessing of a further two horses to the chains on his legs failed to have the expected result. Damiens limbs were severed by his human executioners. Only then could the horses do their work.

Next the post-Enlightenment punitive code:

In 1838, Leon Faucher published an essay on prison reform and described the rules he had drawn up for the *Maison des Jeunes Détenus à Paris*. Article 17 read: 'The prisoners' day will begin at six in the morning in winter and at five in summer. They will work for nine hours a day in all seasons. Two hours a day will be devoted to instruction. Work and the day will end at nine in winter and at eight in summer'. [64]

These two examples, might, Foucault expects, lead us to believe that some things really do change and improve. But he disagrees profoundly. The difference between the two modes of punishment – the one unsurpassably barbaric, the other relatively humane – does not reflect a process of gradual and humanising reform. According to Macey, Foucault saw the difference as being only one of style:

> The transition from Damiens' execution to the publication of Faucher's timetable is one which sees the disappearance of the body from public view as the art of inflicting unbearable suffering is replaced by an economy of suspended rights.
>
> (ibid., p. 330)

For Foucault, the cases illustrate that 'the object of the reforms of the late eighteenth century was not to "punish less" but to "punish better", to insert the power to punish more deeply into the body social: "to constitute a new economy and a new technology for the power to punish"'.

As Ewald, one of Foucault's closest disciples, wrote (in a review that Foucault clearly approved of as it led to his being taken on as the latter's assistant), the effect of *Surveiller et Punir* is to shake one's belief 'in ethics of any kind' and 'to confirm the suspicion that truth itself is built upon police and judicial procedures' (Macey, p. 336). It is scarcely surprising, therefore, that the effect of this book was to paralyse those who wanted to work for reforms. In a roundtable debate between Foucault and a group of eminent historians, one of the historians pointed out that

> If ... one works with prison educators, one notes that the arrival of your book had an absolutely sterilising or, rather, anaesthetising effect on them in the sense that your logic had an implacability they could not get out of.
>
> (ibid., p. 404)

Another asked him a pertinent question:

> Is it horrible to recognise that there are degrees of horror? Does recognising that ... there can be a humanisation of its modes of existence mean defending prisons?
>
> (ibid., p. 405)

Foucault's readers will be relieved to know that his apparent immoralism – his wish to 'shake belief in any ethics' – and his refusal to recognise the reality of the difference between the judicial torture and murder of Damiens on the one hand and the prison timetable suggested by Faucher on the other were, of course, insincere, mere posing. His outrage at the police treatment of dissidents and journalists in France and his protest against the conditions in prisons that culminated in the formation of the *Groupe d'Information sur les Prisons* showed this. He also had a particular abhorrence of the death sentence. Outside his writings, in other words, he was on the side of the kind of reforms that the Enlightenment had initiated and he enjoyed expressing precisely the kind of moral indignation he mocked in his writing.[65]

Enemies of Hope, then, should be unnecessary because those who deny the possibility of rational autonomous agents still believe that they live in a world composed of such people; and those who pretend to loath the post-Enlightenment world still expect of others the critical attitude and the rules of evidence and argument that were dear to the *philosophes*. But, alas, this book is not unnecessary. The enemies of the Enlightenment, those who refuse to acknowledge the great achievements of rational, human consciousness – or pretend to do so – dominate the airwaves in the republic of letters. Their opposition to the idea of an autonomous thinking agent contributing actively to a process of change – the necessary condition for a deliberate progression towards a better future – is, of course, empty as well as unfounded. But it is an active, obstructive emptiness and needs to be dealt with so that the work of a reformed and chastened Enlightenment can be continued with the minimum of impediment.

Part I
Pathologising Culture

Introduction: Man as a Sick Animal

The most honest description of this first part of *Enemies of Hope* would be that it was a book review that got out of hand: I was asked for 800 words and wrote 30 000. The book under review was Dudley Young's *Origins of the Sacred: the Ecstasies of Love and War*.[1] There were many reasons why Young's book prompted me so to exceed my commission; foremost amongst these is that Young has compiled a veritable *vade mecum* of some of the most widely discussed notions and received ideas of our time. His book has proved an ideal springboard for examining these ideas.

There is a powerful meta-mythology of myths that is both a prominent feature of modernism and of modernism's reflections upon itself – upon how it differs from what has gone before and what it is trying to achieve. Myths, we are to understand, still speak to us – though we have considerably difficulty hearing what they have to say – because the primitive is still alive within in the depths of our being, and it profoundly affects our behaviour. This notion – that the Bush is active in the Salon – has proved, in its many different guises, perhaps the single most influential idea of the twentieth century. The strength of its influence may be owing to its originating from many seemingly disparate sources: the emergent discipline of anthropology that awakened a general interest in the life and art of exotic and 'primitive' societies; the birth of 'depth' psychology relating the psychopathology of everyday life to immemorial instincts and to memories in which the individual's nursery years and the pre-history of mankind were merged; the revolution in classical scholarship, anticipated by Nietzsche's *Birth of Tragedy* that switched the focus in Hellenic studies from the Apolline light of reason to the Dionysiac darkness that was its underside; and, finally, the revolution in art, driven by painters, composers and writers who sought to renew the 'tired', 'played out' overly self–conscious traditions of Western art by drawing on the unconscious genius of primitive cultures.

Origins of the Sacred is a notable contribution to meta-mythology. The familiarity of its terrain gives a sense of *déjà vu* and yet the ideas are expounded so clearly and connected with so much else that they seem new again. For Dudley Young's prodigious learning and gift of

71

exposition have enabled him to write a mighty work of synthesis, encompassing a vast body of knowledge drawn from anthropology, archaeology, cultural (indeed multi-cultural) history and pre-history, primatology, ethology, literature ancient and modern, and even 'hard' sciences such as neurophysiology and molecular biology. In the terrain of his book, ethics meets ethology, metaphysics joins hands with anthropology, genetic theology touches primatology and the armchair is translated to the field. It is a terrain – first staked out by thinkers as disparate as Frazer and Scheler – in which everyone, not least the reader, is playing an away match. Here Freud talks to Robert Ardrey, Aeschylus cohabits with Jane Goodall, D.H. Lawrence and T.S. Eliot consort with members of the Chagga tribe, and Derrida and Durkheim dance with Dionysus. The conversation ranges over the discovery of fire, the significance of brachiation and of ventral intercourse, Beethoven's last sonatas, the dialogue between the neocortex and the limbic system, quantum theory, and the breakout of higher primates from the forest to the savannah.[2]

Origins of the Sacred is as important for its exemplary – and consequently illuminating – flaws as for its achievements. It sets itself three impossible tasks: providing a coherent, convincing and adequately documented account of the biological and cultural journey leading from arboreal monkeys to us; identifying the motor behind that journey; and using these findings to suggest cures for the ills of contemporary humanity. The flaws of Young's book run deeper than this totally honourable triple defeat. Indeed, they run so deep that I was left in the end, as will be evident from the ensuing discussion, wondering how seriously Young takes his own ideas – or at least their therapeutic implications.

But in this important respect, too, Young is an exemplary exponent of *kulturkritik*: the uncertainty as to the sincerity of the prescriptions is a characteristic feeling awoken by prescriptive meta-mythology and I shall examine it in some detail. Suffice it to say for the present that *Origins of the Sacred* cries out for the author to examine and to resolve his own attitude towards the sacred, to the gods and to ritual, and that cry is unanswered. He does not really make up his mind whether an openness to the *pneuma* is access to truth or merely an essential condition of social hygiene, whether he sees religion in theological or sociological terms, as a source of transcendent truths or as an instrument for ensuring the health and stability of society. These and other important ambiguities – as, for example, in his attitude to science – make one question his sincerity and wonder whether, after all, he is only a tourist in the past rather than a genuine spiritual pilgrim. This is again a characteristic feeling evoked by the writings of the practitioners of *kulturkritik* and one that warrants addressing in its own right.

In common with many contemporary humanist intellectuals, Young sees himself as (to borrow Merquior's phrase) a 'soul doctor to a sick civilisation' and this gives *Origins of the Sacred* an additional importance; for, in his diagnosis (and his belief that there can be diagnoses) of society and in his prescriptions, the author typifies an entire class of intellectuals and his book is representative. Young's assumptions may be set out as follows:

1. Man has lost his way and humans are individually and collectively sick.
2. The sickness has several elements which may be seen variously as causes or effects – or as both, since these elements work together in a vicious spiral:
   (a) we have lost touch with our animal nature – our instincts, our deep feelings, etc.;
   (b) we have lost touch with our spiritual nature – our sense of wonder, our reverence before the great world and its manifestations;
   (c) we have lost touch with each other, with our sense of belonging to a community.
3. The root cause of this sickness is a transformation of human consciousness which
   (a) began about 400 years ago with the assertion of individuality in Renaissance thought;[3]
   (b) was reaffirmed in the Reformation;
   (c) became irreversible through the triumphs of the value-free enquiries of the emerging sciences; and
   (d) was universalised in a rationalistic individualistic society based not upon custom and tradition but upon contract and reason.
4. The sickness has meant that modern man, modern society, modern life are not only empty, but also in great danger. The feelings with which we have lost touch are still active within us and they break out from time to time in individual and collective violence and madness.

In summary, man is a sick animal whose inner being is adrift: he is cut off from both the ape and the angel within himself. To recover from his sickness, he needs to be put in touch with both.

Like other soul doctors, Young has his management plan. The treatment he advocates is based upon the belief (also common amongst soul doctors) that we can cure two illnesses with the one nostrum. Young believes that reconnecting ourselves with our animal and 'primitive' origins is also a way of putting us back in touch with our spiritual selves. This economy of effort is possible because our spirituality is derived from our animality: for Young, religious sentiments owed their origin to the ecstasies of pre-human primate life; and religious rituals, which were developed to control the behaviour, will help to regulate the buried animal life within us. His pursuit of this notion enables him to explore many areas of contemporary thought, and for us is a way of exploring an entire climate of opinion. We shall see that, also in common with that of many soul doctors, his thinking is riven with self-contradictions: his prescriptions are not only unlikely to be effective but are actually impossible to implement.

At the heart of Young's (exemplary) muddle is a self-contradictory attitude towards human consciousness. If man is a sick animal, his sickness resides at least in part in his overdeveloped consciousness. So soul doctors, such as Young, who live by means of an extremely abstract mode of human consciousness, tend to think that human consciousness in general is a bad thing and abstract consciousness exceptionally bad. And this observation – of the enmity between soul doctors who earn their living by deploying advanced human consciousness and consciousness itself – connects the concerns of this first Part of the book with the overall theme of *Enemies of Hope*.

1

From Apes To Plato

MYTHOHISTORY AND THE DIAGNOSIS OF MODERN MAN

The thesis of Dudley Young's *Origins of the Sacred: The Ecstasies of Love and War* is foreshadowed in the connection between the title and the sub-title. The primitive revelations of the sacred came, he says, through the ecstasies of love and war, and these, in turn, derived from the mutilating frenzy of the hunt. At the heart of religious experience is the blood sacrifice: the ecstasy of the tearing and eating of flesh is the fundamental act of making sacred ('sacro-ficere'), and the means by which man gains access to the divine, making the *pneuma*, the breath of god, come into his presence and the means perhaps by which he himself becomes divine.

In support of his thesis (which derives much *prima facie* credence from the ubiquity of animal and human sacrifice, not to speak of theophagy, in religious rituals), Young reaches far back into what he believes to be the animal origins of human culture – to the apes on the African savannah where, he claims, we shall 'find the true story of man's relationship with god'.

This story begins six million years ago when arboreal monkeys evolved into the chimpanzees from whom, he says, 'all man's higher activities arise' (p. xix). 'Some apes stood up and left the African forest to live on the savannah' (p. 62) and this had momentous consequences:

When we stood up and moved from the forests on to the savannah, we became both more violent and, necessarily, more preoccupied with nurturing the young. To move simultaneously towards violence and domesticity put an obvious strain on the mind, and what we know as the sex war was underway. Our desires to make both love and war were clarified on the dancing ground where divinity was discovered. As he danced for the buffalo, as his dance *became* the buffalo, primitive man discovered the ideas of representation that underlie sympathetic magic and the ideas of totemism and pollution.

(p. xix)

75

Alas, he also 'discovered the almost limitless anarchy that comes loose when sex and violence become truly human (which is to say god-haunted). This is the Fall...' In their ecstasy, the hunters turn on each other: the alpha becomes the prey of his conspecifics. The story of this discovery and of subsequent attempts to control the ecstasies of love and violence, ultimately to deny them and consequently to lose sight of them and at the same time to be in danger from their uncontrolled eruption, is the burden of the book and underpins its therapeutic message. Crucial to Young's case is his belief that 'violence is built into our brains'.

Emigration from the forests to the savannah was associated with the loss of the canine dagger, as the animal was no longer an obligate hunter. This, taken in conjunction with the loss of the refuge of the trees, meant that chimpanzees had to learn to stand and face predators with sticks in the hands. But it also meant that violence, which had been simply a necessary condition for survival in the case of obligate carnivores, was now available as a recreation. This was particularly likely because arboreal monkeys returned to the ground and bipedalism with specialisation of eyes and hands and a brain formidably enriched. This led to a new form of hunting: not, apparently, for survival – this is ensured by gathering nuts, berries, insects and foliage – but for fun. The delight in the murderous fun of the hunt was particularly likely to generalise outwards to other less innocent fun – murder of conspecifics and even cannibalism and genocide – because of the violent propensity of the chimpanzee brain.

Young adduces evidence for his belief that 'violence is built into our brains' from a 'triune' (three-in-one) theory of cerebral function put forward by Maclean in the 1970s. According to this theory, there are really three brains within the primate brain. The most primitive of these is a reptilian core which houses the instinctual programmes of sex and aggression and, crucially for Young's argument, the ritual behaviour by which they are defused. This 'cool' brain is surrounded by the limbic system which is less programmed, more intelligent and responsible for the emotions 'invented' (his word) by the mammals in the nurturing of their young. The balance between these two systems is destabilised by a newcomer: the neocortex, which expands spectacularly in the higher primates. According to Young's interpretation of the triune theory, as the neocortex grows, so does the imagination – for new forms of aggression. The chimp's cleverness leads him through boredom to becoming the most aggressive (as well as the most affectionate) of the non-human primates. This is expressed in his domestic relations but also, and more particularly, in his cultivation of hunting and the occasional practice of cannibalism and fratricide:

The primate codes that forbid intraspecific violence, defusing it through rituals of dominance and submission, are destabilised in the inventive complexity of the chimpanzee neocortex (as in our own)...

... our boundless capacity for destruction was 'created' when the burgeoning neocortex applied for permission to unlock the instinctual codes forbidding intraspecific violence.

(pp. 59, 60).

Combine this with the fact that our weapons were no longer built into our bodies (canine daggers) but optional extras (sticks and stones) and we have the scene set for the discovery of gratuitous violence and the ecstasies and the sacred terrors not only of the hunt, but also of murder, cannibalism and genocide. The hunting pack readily turns into the war pack: *organised* violence is added to the portfolio of animal behaviour. It is undertaken not in response to scarcity but for fun and for profit and aggrandisement – the desire to become bigger, greater, more abundant, to lose one's boundaries, a desire that Young identifies as a religious instinct. The victim's corpse, when it is that of a conspecific, makes the idea of death explicit: *pace* Yeats, it was not man but the savannah apes that discovered or invented death.

These ideas are set out in the first part of Young's book; the question of how 'the ecstasies the *pneuma* calls forth, and that overflowed so disastrously in the cannibal meal, were harnessed and regulated' (p. 127) is addressed in Part Two: 'The Lawlines of Culture'. One solution is love. Love, invented by the female animal as an alternative avenue to the sacred, was, according to Young, made possible by the forward migration of the vagina and the elongation of the penis in higher primates. This permitted ventral intercourse and the transformation of coitus 'from a ten-second jump and a shriek into something altogether more formidable' (p. 68):

When man was invited to look his woman in the eye and lie upon her much extended breast and be enfolded, not only was he being both delivered from the anxiety and distracted from the pleasure of hunting with the boys, but taken back to when he lay upon her breast as a suckling child, and she was all the world to him.

(p. 74)

The female's adoption of the ventral position while being penetrated turns the quick rut into a more protracted carnal ecstasy, personalised and individualised by a thou-making gaze.

It is in the Bonobo chimps that this countervailing force – the transformation of sex into love – is effected by the female in order to call back the male from the ecstasies of the hunt and the kill so that he will do his share in nourishing her young. This ensures that 'feckless males', who would otherwise 'be high on hunting and wanderlust and bring home no bacon', stay at home. Love – which is not to be identified with pair-bonding: it is much less widely distributed in the animal kingdom – thus joins war as the other main avenue to the sacred.[1]

The connection between making love and hunting – so that, for males, the one can be traded for the other – is that the penetrated body is, in a symbolic sense, sacrificed. One can meet the gods in the sacrificial tearing of one's own or another's body. This may be literally the case when, as not uncommonly happened in palaeolithic times, the act of giving birth ended in death.[2] Unfortunately, the invention of love didn't do the trick: 'empathic' female values did not prevail over abstract' male ones.

In 'The Lawlines of Culture', Young surveys the various, mainly mythological, strategies that have been deployed by primitive man for containing the sacred ecstasies. These are:

> The measures taken by palaeolithic man to live with his loss of innocence, the cultural moves he made to protect himself from further exposure to the sacred monster that had originally tempted him ecstatically into cannibalism and worse.
>
> (p. xx)

They were 'the ways in which magic, myth and ritual were used to allow us to talk to the gods without being swallowed by them'.

Young believes that the most successful strategy is expressed in Egyptian mythology, where the energies of mutilating violence are contained by 'the energies of lamentation'. The least efficacious is denial and repression which characterises secular and rationalistic ages – including our own. To this he devotes the entire final section of his book, where he observes the 'occlusions of the sacred' and traces the transition from myth to history. Many of the discontents, not the least the violence, of contemporary civilisation are attributed by him to these 'occlusions'.[3]

Young draws on familiar sources – Freud, Fraser, *et al.* – and yet his story seems even more vulnerable than the ones they tell. To support his crucial assertion that religious sensibilities have a pre-human origin, he claims that there is evidence that chimpanzees participate in religious rituals. He cites a single observation by the primatologist Jane Goodall of four males charging up and down a hillside, tearing off branches and hooting wildly, at the outbreak of a heavy thunderstorm. This, apparently, was a 'dance' and its significance was essentially religious: it was a response to the archetypal power inherent in the natural world – a power that would come to symbolise or to be symbolised in the gods. In responding in this way, the chimps would not only meet the storm as their adversary, but meet with it as their partner. The chimps' raindance was a way of making the gods visible and present and responsive to their needs. In a sense, too, the dance, participated in by some chimps and watched by others, underlined the collective power and unity of the troop. Thus engaged in their religious ritual, the individual chimps were aggrandised and enjoyed life more abundantly as they were possessed by and possessed the *pneuma*. Young, as we can see, unpacks an awful lot from a single observation.

How does this ritual relate to hunting? The ecstatic slaughter of the hunt and the ecstasy of the ritual dance converge in the climactic moment when, during the course of a ritual dance, one of the troop is slaughtered and eaten by the others. In explaining how this catastrophe happens and how it comes to be the defining event in the religious ritual, Young draws attention to certain aspects of the hunt that, he feels, make it assimilable to a sacrificial rite. Hunting involves stalking the animal and this partakes of both stealing and magical deception. In adopting the necessary disguises, the hunter enjoys an anticipatory merging with the prey. The war-whoops of the pack advertise and establish the flow of *pneuma* that binds the pack together. (Primitive hunting expeditions are often preceded by dances designed to invoke the spirit of the animal in question. In the dance, the hunters may wear the skins of the prey.) All of this makes the ecstatic religious dance amenable to transformation into a sacrificial rite, where the invocation of the gods and the slaughter of the prey are merged.

Ritualised religion, Young's story goes, begins with the dance in which the dancers invoke the hunted animal and become one with the natural forces – such as the *pneuma* – that are greater than they are. He then imagines a primal scene in which the alpha-male dares to leap into the magical space delineated by the dance, and so becomes momentarily deified, the incarnation of the *pneuma*; he becomes the dance itself; and at the same time he embodies the spirit of the hunted animal. This is highly dangerous; for the dancing ground is already ripe to become the theatre for ritualised aggression, a model of the hunting ground. It is scarcely surprising, then, that the subdominant males turn on the alpha-male and kill and eat him. It is even less surprising in the light of the ambivalence that imbues our attitude towards authority figures, especially those, who, like the alpha-male, have more than their fair share of the goods of the world – the spoils of the hunt, sexual partners, etc. The sub-dominant males are

physically able to kill him because the discovery of weapons narrows
the gap between their natural strength and his.

This is Young's version of Freud's meta-myth of the primal horde
commiting parricide. The alpha-male is killed and eaten, just as a real
prey would be. Following the murderous and ecstatic frenzy of the
dance, shame and hangover ensue. It is acknowledged that a new
order is needed to ensure that the murder is never repeated. After
the catastrophe on the dancing ground, the abiding preoccupation of
mankind will be how to contain the energies captured and released
in the sacrifical ritual.

Drawing upon a huge literature, Young surveys the strategies, the
'lawlines', used by primitive man from palaeolithic times to the cult
of Osiris, to regulate his exposure to the energies of *pneuma*, to harness the ecstasies that *pneuma* called forth, and that overflowed so
disastrously in the cannibal meal.

The first step is to separate the role of the alpha-male from that of
the shaman who, more than anyone, is responsible for invoking the
gods and bringing them into the dance. In future, no single individual
shall be both boss and shaman: lord of the hunt and lord of the dance.
In future the shaman, the individual who incarnates the *pneuma* on
the dancing ground, should not be the real, tangible, and so destruc-
tible, alpha-male who arouses so many ambivalent feelings but something more elusive – preferably one who, in his trances, has died and
so become discarnate. This strategy, though understandable, is deplored by Young. In destroying the unity of the alpha-shaman, and
so separating Church and State, the powers spiritual and temporal,
'we have destroyed and consumed *ourselves*, since in his prime alpha-shaman gathers and collects all our powers and all our luck. And
thus our world is truly shattered'. This amounts to an early version
of the much spoken of modern 'dissociation of sensibility': 'The splitting of these roles opened a rift in the human psyche that we have
been trying ever since to repair.' The split is between the doer and
the thinker, the soldier and the poet-priest, between the bow and the
lyre.

The shaman remains a shadowy figure. His life is a testimony to
the fact that 'supernatural power flows somewhere behind the world
of appearances, and is to be contacted through more or less ecstatic
ritual on the dancing ground'. He divides the world into two realms,
the sacred where *pneuma* presides, and the profane, or secular, realm
of everyday experience, and acts a mediator between them. He is fitted for his calling by an ability – which, incidentally, unfits him from
everyday life – to fall into ecstatic trances. In these he dies and visits
the other worlds and makes contact with the gods.

Young offers a wonderfully rich survey and rethinking of various

'Illuminations of the Sacred' afforded to the shaman and, through him, to his fellow tribesmen. He explores the concept of pollution (a crossing of boundaries, threatening to break down the border between categories, most notably self and not-self, and so represents chaos); the significance of the 'sacred swarms' (maggots, hives, throngs, crowds that symbolise changeless change, unmoving movement, and the marginal state that is ecstasy – between increase and loss, pleasure and pain, love and war, sanity and madness, cosmos and chaos – an original marginal state that is our introduction to the paradoxes of transcendence . . . the very mystery of Being itself); and the Trickster, who is like a child (and the earliest hominids) 'caught between two worlds'. The Trickster has lost the animal's instinctive certainties. But has not yet found the lawlines of culture: he is a creature of boundaries, the messenger between the gods and man, and supervises the transitions between the sacred and profane.

The second step is to introduce incest taboos. Since the slaugher of the alpha-male was at least in part provoked by jealousy at his expropriation of the females, exogamy is to be preferred: henceforth, breeding females will be assigned by rules rather than simply snatched by the powerful alpha-male. Man shifts 'from hominid harems to the arrangements almost universally adopted by primitive humans, matrilineal kinship networks built upon exogamy in accordance with the incest taboo'. Remorse will also prompt the pack to unite as a clan around some totem (who stands in for the devoured patriarch they wish to forget) and renounce their sexual claim upon his (now their) women. These women (their kin) will be offered as wives to another clan in exchange for theirs. From time to time, they will, in a primitive Eucharist, ceremonially devour an otherwise taboo totem animal, thus ritually recalling the infamous day of alpha slaughter.

This is the basis for Young's most important claim – it is hardly original, but his reassertion of it is based on a wonderfully imaginative rethinking of the major anthropological facts – that the most elementary form of religious life is the ritualisation of the hunt. Hunting is only partly about eating: the other part, bloodlust, is about the gods. In the hunt we are are out for treasure, to increase our substance. It does not arise from the pressure of need but, like the dance, from play. Hence the disdain for scavenging carrion: omnivore chimps will eat only meat they have themselves killed. There is a taboo against food unearned by the 'ecstatic exertions of killing' – against a free lunch. In the hunt, the pack becomes one body to subdue the body that opposes it. The process of hunting – and the distribution and eating of the spoils afterwards – is increasingly ritualised to make sure that the hunt does not degenerate into a free-for-all in which hunters fight with, and end up eating, one another in an ecstatic and

shameful recapitulation of the primal horde's murder and cannibal-
isation of the alpha-male who stepped into the ring and danced. The
pneuma is harnessed by the lawlines of the hunt which enable the
horde to ride right up to the threshold of lawlessness without crossing the line into chaos. The tearing of the living flesh is, according to
Young, *the* elementary form of religious life – the sacrifice that is a
sacro-ficere, a making sacred.

And there is another consequence: love, also, is made taboo; for it
is now recognised not only to be insufficiently weak to control the
bloodlust of the hunter, but to be positively dangerous. The power
females have discovered in love will have to be regulated. 'Kinship
institutionalises the power they [females] discover in love.' After the
catastrophe, love is outlawed by kinship: kinship regulates 'our ancient instincts for promiscuous sexuality . . . and also provides us with
a wealth of ties that bind us to one another in various simple and
complex ways'.[4]

All these strategies for containing the ritualistic violence of the hunt
may be quite inadequate:

> Given the overlapping of the energies that lead to love and war, it
> follows that any malfunction in the one will tend to be reflected in
> the other. If . . . erotic love was marginalized by kinship and then
> further sequestered by patriarchy, one would expect this to be re-
> flected in a coarsening of the energies deployed in hunting and
> war: the rites of lamentation that monitor and refine these energies
> will fall into neglect.
>
> (Young, p. 242)

Something more is needed to contain the energies. In the course of
his examination of the various attempts to release and to contain the
mysteries of sacrificial violence, Young looks at the literal containers
that humans have built to house the sacred. He identifies the five
primitive locations for the coming and going of divinity: the womb,
the Jewish Ark of the Covenant, the hearth, the sacred wood at Nemi,
and the dancing-ground of Dionysus. None of these devices is entirely successful; but at least they indicate mankind's ability to acknowledge its propensity towards, and need for, sacrificial violence.
Failure to do so will mean that we are more helplessly in its grip.

An alternative strategy is to project our violence and all our destructive potential outside of ourselves in a sacrificial animal that bears
it away: scapegoating. The divine pollution moves within the community and it can be expelled by being localised in a sacrificial animal
which is expelled – usually by being driven into the wilderness, stoned,
mutilated or dismembered. The rituals that surround scapegoating are

often very complex and refer back to ritual itself: as Young reflects, 'to have a ritual re-enactment of the origin of ritual is impressively subtle.' There is sometimes a self-consciously carnival element in the trial of the scapegoat. Scapegoating may also be deployed to resolve political crises – to arrest blood feuds, to halt the revenge cycle which may decimate primitive populations. 'Feuding clans were often reconciled by pooling their pollutions and sending them off on the back of a scapegoat.' This, however, requires quite a high level of agreement and presumably the mutual goodwill that is usually absent in such circumstances – not to speak of the parallel resolution of any secular basis for conflict.

By now, we have moved a long way down the winding road leading from apes to Plato. The *Bouphonia* or scapegoat ritual lasted until the fourth century BC. It evolved into the Dionysian ritual. Young traces the emergence of tragedy from the annual Passion (*pathos*) of Dionysus, the ritual play in which the *dromenon* ('the thing done', 'the thing enacted'), was the sacrificial slaughter of the god, either in human form or as a bull. Dionysus, the god of ecstasy and excess, expressed the unhousability of *pneuma*. The less stereotyped dramas that developed from the fixed *dromenon* of old still had as their theme the mark of Dionysus turned savagely mad, the ritual blasphemously mistaken, in which the violent energies overspill the ritual lawlines and mutilate the wrong victim. Greek tragedy is constructed around the original primitive plot: 'the restless refusal to abide within the lawlines of the sacred'.

A modification of the scapegoating solution was to harness the energies of lamentation: these are in principle capable of containing the mutilating violence. Young finds this expressed in the Osiris/Isis myths, which reveal 'intimacy and reconciliation between god and goddess, masculine and feminine . . . unique among the ancient civilisations'. His discussion of these myths is deeply moving, a pool of sad calm amidst the blood-boltered narrative that precedes and follows it. The annual mystery for neolithic culture was the lifecycle of the corn spirit born in green shoots, violently killed in the harvest and resurrected in the seed sprouting from the dead earth. This is most spectacularly expressed in the Nile valley, which intermittently floods the surrounding dry land. Young interprets the widespread practice of annual periodic human sacrifice – in which the 'Old Year' or the 'King', as the goddess's seed-bearer and harvest (her consort and son), would be killed, dismembered and offered up to her (as his wife and mother) in order to ensure the potency of 'New Year' – as a modification of the basic scapegoat story and the catastrophic feast of the alpha-shaman.[5]

Osiris was the lineal descendant of the neolithic fertility spirit, sacrificed, dismembered and buried annually to make the corn grow. What, according to Young, the Osiris/Isis myths brought was 'an adequate theology of lamentation' to control the sacrificial ecstasies. It is the lamentation of his wife and sister Isis that prevents the ever-dying Osiris from fading into oblivion. The moistening eye of the moon, the gift of Isis, makes dark things discernible. 'The one who holds him in her mind is Isis, who knows him as no other does; and when he is lost, as in the dry season, and when the moon disappears, it is she who has to find him' – she who has the sorrowing eye of tendance, which promises that the sufferings of violence can be borne and redeemed in the heart that commemorates them.

With his discussion of Isis and Osiris, Young concludes his search for ritual lawlines that are strong and subtle enough to govern the ecstasies that lead to love and war. He proceeds, then, to examine 'the occlusions of the sacred' that have accompanied the development of civilisation and the journey taking mankind further and further away from myth into history. He mourns the lost ecstasies, bemoans the loss of contact with the gods, and regrets both patriarchal man's abandonment of the epic endeavours to kill Death with his sword and with 'the feminine touch, then as now our chief stay against violence'. These losses are examined through three texts: the story of Adam and Eve and the Fall (Love); of Cain and Abel (War arising out of the failure of ritual to contain the envious rivalry of nearly matched figures); and the weird and wonderful *Epic of Gilgamesh* (the failure to kill death by 'glorious mutilations that will be forever remembered'). He then proceeds to examine crucial passages in the *Odyssey*, seeing this epic as representing Western man's decisive break with the primitive wisdom he wishes, through his book, to remember and perhaps reawaken.

In a brilliant reading of the Polyphemus episode, he exposes the tendentiousness of Homer's attempts to put Odysseus in the right and Polyphemus in the wrong, even though the latter, a harmless giant, is left blind and sheepless as a result of Odysseus' sojourn on his island. There is implicit a trumped-up charge of Polyphemus' cannibalism to justify Odysseus' piratical action and to make the reader desire his eventual triumph. However, the poem works against itself: the ten years' wandering, as a result of the invocation of Poseidon by Polyphemus (who is Poseidon's son), is, however inadvertently, made to seem justified and the need for cleansing evident. Odysseus has wronged Polyphemus and been appropriately punished: 'the blinding of Polyphemus is the pollution that keeps our hero from sleeping in his own bed throughout the poem' (p. 291). Moreover, Odysseus' strength – his *metis*, his craft, his artfulness – is built on deceit (central to the huntsman's powers), is ambiguous and a sign of the occlusion of the sacred by secular, merely human skills; 'it is a mark of man fallen into the necessity of politicking with his fellow man, and its dialectic with the open-handed ideals of godlike aristocracy is taken into the heart of Greek tragedy . . .' (p. 297).

Young's own *Odyssey* ends with the Minoan religion which, he believes, provided an answer, even more wonderful than the myths of Isis and Osiris, to the problems of sex and violence. The evidence he adduces for this claim is, to put it mildly, scanty. He sees in the ritual of bull-vaulting – where graceful leaping is associated with the risk of serious mutilation – a representation of the play of culture and nature and of the human wager with divinity. The Minoan culture was able to accept the feminine:

> Thus I would propose the Cretans as the first masters of the abyss, and would suggest that such mastery has everything to do with their men's ability to forego the macho reassurance of phallic jujus and be suitably impressed (not threatened) by the strikingly attractive women who dominate their artworks in those distinctively flounced skirts.
>
> (p. 302)

The Minoan vision was soon lost – displaced by the 'ascendant pugnacity of the Myceneans'. In the Mycenaean and classical Greek culture, the Cretan bull's mysteries were 'imported, distorted, dispersed, and shared out among Zeus, Poseidon, and Dionysus, the three Mycenaean bull gods'. The unacceptable bits went to Dionysus:

> The Cretan bull was severely lamed when parcelled into three by the Greeks; and in separating his wisdom from his violence much was lost; for much of wisdom can only be learned in the context of violence.
>
> (p. 305)

When this wisdom is lost 'as it is lost in the Homeric poems, the savage breast turns the mind to mindless mutilation on the high seas and the battlefield'.

THE PRESCRIPTION

Out of his magnificent, compendious, eccentric tour of primate, primitive, pre-classical and classical culture, Young draws lessons for the present time and suggests certain therapies for our age. His diagnosis is that modern man has lost touch with the sacred and, in particular, with the Dionysiac ecstasies associated with the mutilating frenzy of sacrifice. They, however, have not lost touch with him: his animal self, the violence built into his brain, will reassert itself in the organised violence of war and the casual violence of civil and domestic life. And his loss will also live on as a felt emptiness haunting the consumerist frenzies of contemporary life. The occlusions of the sacred, marked by the transition from a sense of a mythical to a merely historical past, lie at the heart of the discontents of contemporary civilisation.

His somewhat tendentious account of mythology and history points to his prescription for modern man. The cure will come from 'a return to the mythic voice to heal the divisions in his soul' (p. xviii). The great challenge of the twenty-first century, he tells us, is the recovery of innocence. 'A reacquaintance with our primitive selves', he claims, 'can make us *more* human rather than *less*' (p. xxix). Young mourns the passing of 'the shaman or the myth-making poet' – whom he identifies as the 'hero' of his book – in whom both secular power and priestly authority are invested. The present division of spiritual and secular authority, a remote consequence of the fission of alpha-shaman, symbolises the deep fissures in the soul of modern man. We must heal the fissure and bring the two authorities together.

Young wrote *Origins of the Sacred* for the sake of its prescriptions. The faith behind the huge effort and the massive achievement of the book is that it will be therapeutic to remember where we have come from, to retrieve our formative cultural experiences, to recover our sense of the sacred and of the animal powers in which that sense first originates. 'So much of how we think we should proceed today and tomorrow', he argues, 'depends on where we think we came from' (p. 121). Goethe is cited in one of a handful of epigraphs to the entire book:

> He who cannot give to himself
> An account of the last three thousand years
> Must remain unacquainted with the things of darkness,
> Able to live only from day to day.

So that we do not sleepwalk into the future, with consequences that are potentially even more disastrous than the catastrophes of the present century, we must wake up to, and so in a sense possess, or wake up out of, the past that, whether we acknowledge it or not, possesses us. Accordingly, 'we must put ourselves back to school with our forebears, to recall the myths that legitimize our existence and tell us how to live with godly power.' 'Reacquaintance with our primitive selves may make us *more* human rather than *less* . . . . We in the late

20th century may be sophisticated enough to benefit from regression to primordial states of mind.' Only then will we come to appreciate that 'pollutions both ecological and religious have much to do with the mismanagement of sex and violence.' And vice versa; for 'the worst of our pollutions, the nuclear bomb', 'the desire (disowned, of course) to nuke the planet is but the tumorous extension of the old sacrilegious dream of making war on the gods', to produce 'a phallus sufficiently potent in its erection to beat down the gates of heaven'. We have forgotten this.

The chronology of this forgetting is, as I have already mentioned, uncertain: it is variously reported as beginning in the mists of hominid pre-history when the role of alpha-shaman was separated into those of alpha and shaman; in the fourth century BC, with the repression of the true meaning of the Cretan *Bouphonia* – the ox-murder, the scapegoat ritual and Athenian harvest festival; and 'some 400 years ago' with 'the rift that opened in the Western soul ... when science and religion went their separate ways' (p. xvii). Regarding this last, Young tells us that 'evidence is increasingly accumulating to suggest that the occidental experiment in secular scientism over the last four centuries has involved the suppression of certain vital processes of a more or less religious nature, and that this is becoming intolerable' (p. xvii).

So much for the diagnosis and the therapeutic aim of the book. The question then arises: how is that aim to be achieved? At the very least, it would seem, we need to remember. But what is the work of memory? There is obviously more to it than the kind of knowing that reading a book – Young's mighty tome or any other – affords us. (The analogy here is the kind of reminiscence that a patient in psychoanalysis has to undergo: it is not sufficient that she should read the relevant Pelican telling her about the Electra complex.) How we move on from what we may or may not have learnt from Young – or from any of the long register of soul-doctors who peddle their insights about our primitive nature and the meta-mythology of myth – is unclear. Shall we abandon salaried labour for hunting and gathering and dancing and ancestor-worship? Shall we reject democratically elected governments for charismatic shamans? Shall we prostrate ourselves before an elected or emergent or reconstituted alpha-shaman in whose person are united the powers of the lords spiritual and temporal? How shall we unite together to do this? Will a small group of individuals who have read Young's book get together and slowly undermine the secular–scientific world order? Or is it simply a question of inner change without outward alteration? We still go to work on the tube and vote Labour and expect our governments to be accountable and continue to treat our spouses as equals, whilst inwardly we are moonlit with Osirian dreams and tasting the ecstasy, the mutilating frenzy of the flesh by introducing a smidgeon of S. & M. into our sexual relations with our partner?

Either solution raises serious questions. The outward, world-changing approach, even if it were remotely likely to be successful in a world where every individual is at a different stage of development, every country has different order and disorder and yet all are interlocked with one another, would be fraught with danger. The degree and kind of organisation required would put the logistic and strategic triumphs of the Second World War into the shade. And the pusillanimous attention to detail required for such a conscious reinventing of the world's culture – even one based on a kind of ur-culture derived from what was supposedly common to all mankind when it was emerging from its pre-hominid state – would be entirely at odds with the spirit of ecstasy that it would be designed to serve. It would, of course, for all sorts of reasons that it is hardly necessary to elaborate further, be impossible.

And even if it were possible, every step towards such a revitalised world culture would surely carry with it grave perils. The history of the present century has shown how the slightest nod in the direction of primitivism or *volkisch* organic mysticism can plunge societies into howling mass misery. And how could anyone with any knowledge of recent history hope for the return of alpha-shaman? Is this not the century that has produced Hitler, Mussolini and Ceaucescu in Europe, Bokassa, Amin and Mengistu in Africa, Pol Pot and Mao in Asia? Did they not have sway over every aspect of life and did they not create Hell on Earth? Any modern alpha-shaman would assume office in a world dominated by powerful technology (technology and the concentration of power that comes with it will not go away and will certainly not be foresworn by an all-powerful leader). If the experience of all the revolutions that have taken place over the last few centuries of rebellion and revolution has reinforced one message, it is about the corruptions of power.[6]

And as for the purely inward way, in which the individual who has regained his primitive vision, innocence and unity of being, loiters invisible among his fellow commuters as glass in water, this would surely not be possible: the inward change would demand correlative change in the outer world or it would prove unsustainable; the shearing force between inner and outer would otherwise make a kind of jet lag between world and soul the permanent condition of life. (This is something we shall return to when we discuss 'Religion and the Re-Enchantment of the World' in chapter 3.)

The question also arises whether any of the changes Young would like to bring about would be desirable. There seem to be three principal elements to his prescription: liberation from science and abstraction and its expression in technology and in our 'book-bound' culture saturated in bodiless abstractions; a change not so much in the role as in the presence of women and the female principle in the world at large; and a return to the gods and a restoration of the sense of the sacred enjoyed by the primitives so that we might better manage sex and violence. Do we really want to lose science, technology and the rich culture of the book? I think not. If the male principle were replaced by the female principle, what would be the consequence of this? Nobody has any idea. And what, finally, would be the result of restoring our primitive sense of the sacred? Are we sure that it would guarantee that we, whose lives are inconceivably different from those of our primitive forebears, would behave better? We cannot answer these questions with any certainty. But we can guess. And, as I have said, recent history has given us a lot of material to guess with.

Young's book, in short, raises more questions than it answers. But the questions it raises go beyond the book itself and its specific diagnoses and prescriptions. They open up to question the massive industry of *kulturkritik* and the intellectual sport of diagnosing the ills of society which has, in my opinion, become, towards the end of the century, one of the chief of our social ills. Young's book, in short, makes an excellent starting point for a critique of those humanist intellectuals who, armed with a soupçon of mytho-historiography, proceed to give expert opinions on modern life, and who seem to believe that they serve a useful job by diagnosing society, and indulging in a *kulturpessimismus* that affords them such comfortable and well-remunerated despair. Before I do this, however, I shall focus on the assumption, evident throughout Young's book and pandemic in modern thought, that we can somehow deepen our understanding of modern man by reconnecting his psyche with the putative soul of his animal forebears.

2

Falling into the Ha-Ha

I have set out the main lines of Young's account of the origin of religious experience, and of the intuition of the sacred. For Young, the blood sacrifice – the sacrificial tearing of one's own or another's body – lies at the heart of religious devotion; and the most elementary form of religious life is the ritualisation of the hunt. This shocking interpretation of human spirituality – an essential component of Young's theory of 'the origins of the distinctly human' – is not, of course, entirely original. Indeed, it is best understood as a brilliant synthesis of the well-known and less well-known ideas of thinkers such as Freud, Eliade, Fox, Campbell and those many other metaphysical anthropologists who have had such a marked influence on the way twentieth-century thinkers have thought about mankind. It is precisely because his views are typical rather than eccentrically original that Young's methods, assumptions and conclusions are worthy of sustained and respectful, if critical, attention. The critique of *Origins of the Sacred* that follows is therefore able to open on to a wider critique of the kind of 'cultural criticism' that has so dominated the thinking of humanist intellectuals in the present century.

For Young, trying to identify our origins is important because 'so much of how we think we should proceed today and tomorrow depends on where we think we came from'. 'Our distant past', he asserts, 'holds the key to so much in the present' (p. 39). In common with many thinkers – and, in particular, critics of civilisation – he believes that where we come from is the Animal Kingdom and that consequently many of the discontents of civilisation arise from the conflict between the demands, values and beliefs of civilised society and our animal instincts.[1] We are animals, yes, but we are sick animals, because life in human society denies expression to our animal nature and, worse, even prevents us from acknowledging that nature. We are possessed by instincts that we can neither express nor even admit to. The result is that, at best, we suffer in secret the consequences of inner division between our civic and our instinctual beings and that, at worst, we give distorted external expression to our denied instincts, so that others suffer – in, for example, the gratuitous violence and petty tyrannies of domestic life, and the oppressions, wars, etc. of public life. E.R. Dodds made very similar claims in his famous book on the irrational or Dionysiac element in Greek thought:

> To resist Dionysus is to repress the elemental in one's own nature; the punishment is the sudden collapse of the inward dykes when the elemental breaks through perforce and civilisation vanishes.[2]

This story of human beings as frustrated animals – or as frustrated primitives (who once enjoyed a closer relationship to our common animal nature) – seems to have received a huge boost from the recent advances in life-sciences: Darwinian evolutionary theory, ethology and molecular biology all seem to point to a closer kinship between mankind and animalkind than the comfortable are traditionally thought to be comfortable with. We share, Young reminds us, 99 per cent of our DNA with chimpanzees, which makes them more like our brothers than our cousins:

> we have strong grounds for taking our animality seriously: in each case we find 'hard' sciences proposing that our biological past may have a firmer grip upon our present/has been hitherto assumed. ⟨ than
>
> (p. 40)

The message of the biologists has inspired many creative writers: biological determinism is a dominant assumption amongst hugely influential twentieth-century writers as disparate as Gide, Hamsun, Lawrence, London, Hemingway, Golding, Céline and Nathalie Sarraute. Many twentieth-century novelists (and I have referred only to the most obvious examples) see the truth about character as being located in those immemorial primitive tropisms and animal instincts that really regulate behaviour and belief. They believe, as Lawrence did, that underneath personality there is something much more important and powerful and true than personality – the impersonal inheritance from the primitive and animal past. For such writers, civilisation is always assumed to be 'a thin veneer' over a seething, pullulating mass of uncivilised appetites, instincts and emotions. There is hardly a schoolchild who has not learned from a compulsory study of *Lord of the Flies* that, left to themselves, human beings will naturally revert to condition of primitives and wild beasts, creating a lawless society in which the strongest rule and the weakest are exploited or become the object of arbitrary cruelty or the unwilling victim of a blood sacrifice. The influence of the novelists has, in its turn, greatly amplified the influence of biologists. And biologism has also been mediated to a wider public by quasi-scientific writers such as Freud, who in their turn have had an incalculable impact on the literary imagination.

The story told in various ways by biological scientists, novelists and others depends upon, and reinforces, the assumption that there is no sharp break between humans and other animals. The enormous distance between animal origin and contemporary human destination that has accrued over millions of years since a distinctively human species emerged amounts to nothing. Man may remove himself from within the animal kingdom but the animal kingdom cannot be removed from within mankind. Howsoever the fundamental opposition between (animal) Nature and (human) Culture may dominate human experience, discourse, attitudes and behaviour, this opposition is simply, as Lévi-Strauss pointed out, the master-myth by which humans organise their pictures of the world and make the world over into their thing. In reality, our human nature remains rooted in our animal origins; and we are up to our leaves in our roots.

In contrast to the true story unearthed by biologists, anthropologists and others, the stories we tell ourselves about our origins exaggerate the sharpness of the distinction between man and the infra-human animals. Young wishes to correct this exaggeration:

> It is a structural necessity that all stories have to begin with a bump or a bang, *ex nihilo*, and this disconcerting fact is reflected in every myth of human origins, for all are constrained to say 'One day something happened, and man appeared'. Much of my argument thus far has tried to soften this bump by indicating that many things we usually think distinctly human can in fact be found among the chimpanzees.
>
> (p. 121)

My present purpose is precisely the opposite: to look critically at the business of 'softening the bump'. I shall argue that 'softening the bump' amounts to eliding an absolutely fundamental difference, jumping over a huge gap, and that identifying what is seemingly distinctively human with characteristics more widely distributed in the animal kingdom requires us to overlook, indeed traduce, humankind and to deny the central mystery of what we are. I shall argue that understanding humanity in terms of animality, and seeing men as beasts self-deceived by a surface rationality as to their true nature, prevents us from thinking about our own natures in a helpful and constructive way. It is a crucial factor in the emptiness of much contemporary social thought and cultural criticism.

ELIDING THE DISTANCE (1) MAN = ANIMAL

The supporting framework for this kind of thinking is, of course, provided by evolutionary thought which sees us as literally descended from the beasts. In disputing with the 'animalisation' of mankind, I shall not, of course, deny our beastly origins. (As a clinical physiologist and doctor trained in biological science, I am unlikely to be attracted by creationism.) Rather, I shall question the extent to which one can understand what we are now in terms of our supposed origins and I shall emphasise the *discontinuity* between ourselves and non-human animals. I shall argue that man, 'the explicit animal', is qualitatively different from any other form of life and that an observation of similarities between him and other animals should not blind one to the spectacular and obvious differences. I argue this not because Young's views, shocking though they are, shock me (I do not number doctrinal religious belief among my problems), but because I agree with Young that, if we are to think seriously about the present and future of man, we should know how to relate to the past; or, if this is too much to expect (and I think it is), we should at least know how *not* to relate to the past.

Some of the views I shall sketch here have been developed at length in *The Explicit Animal,* and the reader may wish to consult this book for further elaboration of my case against the comprehensive 'naturalisation' of humanity and the assimilation of human culture to animal nature.

Let us begin by examining the claim, central to Young's thesis, that 'the religious life was not a human invention, but was in fact discovered by the chimpanzees'. This is a startling assertion; for if anything is considered to be distinctively human, it is the propensity for religious belief. Man, surely, is the only theological animal. The evidence Young mobilises in support of his claim is scanty, to say the least. He relies, it will be recalled, almost exclusively on a handful of observations by a single scientist, Jane Goodall, of behaviour that she interprets as a 'raindance'. What Goodall saw was

> four males responding to the outbreak of a particularly heavy thunderstorm by charging up and down a hillside at Gombe, tearing off tree branches and hooting wildly. This continued through twenty minutes of thunder, lightning, and downpour, and was witnessed by a congregation of females and youngsters who had climbed into the trees nearby to watch.
>
> (p. 93)

Young sees this as the moment at which religion was born: 'the old joke that religion began with man shaking his fist at the heavens is thus not far wide of the mark.' The storm, he asserts, 'has provoked in [the chimp] the notion of an almost-present adversary who is, at least on this occasion, to be defied. It is a striking illustration both of the chimp's courage and of his powers of symbolisation.'

Goodall's observation has captured the imagination of many commentators in addition to Young, and they, too, have placed upon it a heavy weight of interpretation. Young suggests that it is also a theatrical performance. Mary Midgeley[3] believes that it is connected with 'the origin of human sport and dancing', adding that 'although this sort of thing is not just what goes on at the Bolshoi ballet . . . it does something to indicate on what bush, growing out of what soil, ballet is a flower' (ibid., p. 248). Religion, theatre, dance, sport: the Council of Trent, the queue at the box office, the cult of the tango, the European Cup – there seems to be no limit to what thinkers are able to unpack from a few brief moments of chimp whoopee. It is important to recall that the behaviour lasted for only a few minutes and that Goodall saw it take place only three times in ten years' close observation of chimpanzees.

Young supports his case with reference to another form of chimpanzee behaviour which he sees as proto-religious: the chorus of drumming and hooting sounds that takes place in the fading light. From this signal 'emerges both the experience of the troup [sic] as unified substance and the sky as ubiquitous animation – a preliminary sketch of divinity'(p. 96). But, of course, such mass vespers are not peculiar to higher primates: winter starlings fill the air with their roosting cries. Are they, too, asserting their collective solidarity before the Mighty One – or (to use Young's analysis) 'addressing something sufficiently abstract and amorphous to reflect and engender the idea of "the troop as a whole" – to wit the darkness descending' (p. 96). And why stop at starlings? Doubtless one could read into even more primitive organisms behaviour that could be arguably proto-religious, reaching further and further down into the animal kingdom until we had traced the origins of religion to the very first piece of DNA to employ the mediation of messenger (or 'angelic') RNA.

The interpretation of the chimpanzees' behaviour in the storm is clearly rank anthropomorphism – but of a very potent and telling sort. In reading it as a raindance ('with a pattern of chorus and principal dancers'), as a 'performance' 'constructed from symbolic materials with considerable scope for esthetic emendation' ('Almost operatic, we are in the foothills of art'), he is attributing to the chimpanzees a set of concepts, of abstract ideas, for which there is no independent evidence. His over-interpretation goes beyond the usual anthropomorphism of feelings – 'the wasp is grieving for her young' – to an anthropomorphism of abstract concepts. When he interprets the drumming 'as a preliminary sketch of divinity' he is supplying what is missing in the chimps' behaviour, without which it would not in any arguable sense be considered religious: a reference to an entity (Yahwe, God) with certain properties, to a certain picture of the world (the world is created by God, is encompassed by God, leads after death to other worlds, etc.), and, by implication, to certain consequent obligations placed upon the individual. In short, Young infers a set of beliefs which he imagines is implicit in this behaviour and makes them explicit; thus made explicit they are referred back to the behaviour. The riches he unpacks from it are precisely the riches he has put into it.

This is the essence of my critique of Young's vision of chimp proto-religiosity and his subsequent determination of what is right for us from what is true of them. The fallacy of misplaced explicitness is ubiquitous in ethology, even where gross Disneyfication of the animal world is avoided. Its characteristic manifestation is a tendency to unpack propositional discourse – this being the mark of explicitness – from non- or pre-linguistic behaviour. The more excusable examples are to be found in discussions of animal signalling systems. Honey bees are seen as dancing the equivalent of 'There is honey four hundred yards away to the south' and a 'grammar' is unpacked from 'their language'. Less excusable examples are derived from the notion that the world 'signals' to animals, transmitting 'information' to them: the flowers 'signal' 'danger' to a toad, as if an abstract category could be born out of the interaction between dumb fauna and even dumber flora. The root error is that of transferring epithets from our own thought processes to the natural world. Taken to its limit, this would justify regarding a falling stone as continuously stating Newton's law of universal gravitation, as a kind of protracted and highly numerate silent scream, as it fell.[4]

Man is, uniquely, an explicit animal. It is through human beings that the world becomes represented in abstract, general categories. We not only sense, or experience, the world but also have knowledge of it; and we know that we know; and we consciously enhance our knowledge and conceal it and boast about it and pretend to it. And we judge others' knowledge and our knowledge of their knowledge . . . and so on. This infinitely folded, endlessly unfolding, explicitness may be underlined by some specific comparisons between man and higher animals. Compare a chimpanzee learning not to visit a particular place in the jungle because it was once stung there by a bee with my attending evening classes devoted to a country I am going to visit next year in order to learn about the hazards to be avoided when on my visit. Compare the 'sexual signalling' of chimpanzees with the sexual signalling of an adolescent who learns lines of verse in order to impress his girlfriend and manipulates the conversation to create the occasion for his reciting them. Compare the feeding rituals of chimpanzees with the considerations that surround the choice of items on a menu when one is being taken out for a meal by a colleague. Compare the chimpanzee's moment of spontaneous 'dancing' with the years an individual may spend learning how to perfect the tango, entering tango contests, worrying about what to wear when participating, hurrying to catch a train to get to a competition on time, polishing the cups on the shelf, discussing 'the form' with fellow enthusiasts, reading articles about the politics of dance competitions, following the debates of the statutory body wanting to introduce rule changes, etc. And, finally, to return to religion, consider the distance between the chimpanzees' few minutes of the raindance (if that is what it was) and the mighty debates that have riven Western civilisation – and formed the pretext for massive, confessional wars – about the issues such as the *filioque* clause and the doctrine of transubstantiation. It is simply impossible to imagine an ape deliberately avoiding the most expensive meal in a restaurant in order not to embarrass his friends. Or (as Young does in his remarks about Beethoven and Wagner)[5] making incredibly general and complex remarks about the music of another ape who had live 150 years ago in order to support an even more complex point about the entire trend of ape history and culture over the last 150 years.

Religion, like so much else in human life, is a form of behaviour mediated through abstract knowledge and referential discourse. Again and again, in common with so many other cultural commentators, Young overlooks this unique aspect of human activity. And yet, he does have some inkling of it. This becomes clear in the pages he devotes to 'two kinds of knowing – practical and theoretical, instinctive and learned, intuitive and conceptual', a distinction, he reminds us, that is retained in the French *savoir* versus *connaître* and in the German *kennen* versus *wissen* (see pp. 69–71). Practical knowing, he tells us, is 'the way we knew the world before the neocortex began unbolting our instincts and pushing us towards the partiality of language, that tool whereby our experience can be not only mediated but replaced by symbols'. Young depores the displacement of *kennen* by *wissen*, and 'the general loss of instinctual and intuitive certainty about our world as the expanding neocortex unlocked our instincts', as 'one of the most important falls from grace into fumbling that our mythologies lament' (p. 71). However, he then spoils his case by suggesting that we *ken* things that really seem seem to be the province of explicit knowledge – such as the relationship between sexual intercourse and pregnancy several months later, or the parricidal orgy which gave birth to the incest taboo and exogamous marriage. This latter –

quite a complex story as we have seen – 'hums in the bloodstream, calling us to call it our own' (p. 107). If *kennen* has such scope, then, clearly, *wissen* hardly represents an advance; it has nothing left to do. Indeed, it seems to be a retreat; for if 'Nature's *kennen* knows its stuff rather better than man's *wissen*' (p. 71), then *wissen* can only be bad news; if *kennen* were unerring, it would be a shame to have buried it under the halting, error-ridden uncertainties of *wissen*. It is, of course, absurd to suggest that (outside the world of Winnie the Pooh)[6] 'hums in the blood' can be referential, propositional and have truth-values in the way that factual assertions can. It is equally absurd to think of *kennen* being as powerful as *wissen*, or practical know-how unreformed and unenhanced by *wissen* as being better than *wissen*-riddled human knowledge: the appallingly high infant mortality and premature death rate in all animal species (including kenny-canny primitive man) shows that it is not. If you want to achieve what most people want to achieve – at the very least, a secure supply of food and drink and living to a ripe old age – then *kennen* must be transformed by *wissen*. Young's lamentation is rather reminiscent of the story behind Kleist's *Puppet Theatre*, in which reflection, self-consciousness, signals a departure from animal or mechanical grace, which cannot be recovered until knowledge is perfected. Except that, in Young's case, knowledge is not to be perfected so much as laid aside, so that we can recover 'the way we knew – and know – the world before we began – and begin – to talk too much, read too many books, build too many machines, and abolish the family' (p. 70).

Young, in short, accepts that there is, after all, something fundamentally different about human consciousness. His *wissen* is not so different from my 'explicit knowledge'. However, he wishes to dismiss it as unimportant or, indeed, a backward step. It should be unnecessary to point out that explicit knowledge has a major advantage over implicit understanding: the former is open to questioning, to reform, to progressive modification in the light of external reality. Intuitive understanding, instinctive knowledge, may be more effective in some respects, in some contexts; but it may simply be unchallenged prejudice, error, misconception. The explicitness of *wissen* also permits its generality to be continually defined, refined and delimited. This is in sharp contrast to the *kennen* of animal instinct, exemplifed in the robin's extending the 'notion' of 'rival male' to include any surface in which a red patch is placed on a brown background and attacking it accordingly. The destructive religious, racial and tribal prejudices that have caused so much suffering throughout human history have a similar basis in unreformed *kennen*.[7] Of course, *wissen* is predicated on *kennen*: some things have to be taken on faith and assumed instinctively. (We have already quoted Hamann's observation that Hume 'needs faith if he is to eat an egg and drink a glass of water'.) It is even possible to talk, as the physiologist Walter Cannon did, of 'the wisdom of the body'. But this wisdom, which is evident even in unicellular organs, falls far short of the explicit knowledge whereby human beings steer themselves with comparative safety through the world.

The supplanting of instinct by conscious, rule-governed behaviour, by planning on the basis of knowledge, is no small thing: it is not only the major difference between humans and other animals but our distinctive mystery and glory. There may be a special appropriateness in quoting Jane Goodall here:

> [it] is only through real understanding of the ways in which chimpanzees and men show similarities of behaviour that we can reflect, with meaning, on the ways in which man and chimpanzees *differ*. And only then can we really begin to appreciate, in a biological and spiritual manner, the full extent of man's uniqueness.[8]

The fundamental difference between man and all the other animals, the extent of man's uniqueness, is so obvious, that it is difficult to see how it could have been overlooked. There are several explanations. First of all, consciousness (and in particular self-consciousness) is transparent, rather as language, in successful communication, is transparent. Consciousness, as Sartre would say, 'sacrifices' itself to its objects, to the things it is conscious of; explicitness to things that are made explicit.[9] Hume, too, spoke of the 'mind's great propensity to spread itself on external objects, and to conjoin with them any internal impressions they occasion'.[10] In, short, to lose itself in that of which it is mindful.[11] Secondly, there is a tendency to elide the differences between consciousness and unconsciousness: Freudians talk of 'unconscious thoughts' and conscious intentions are dismissed in favour of supposed 'unconscious impulses'. The historical unconscious (Marx et al.); the psychological unconscious (Freud et al.; not to speak of cognitive psychologists for whom conscious perception is rooted in unconscious processes); the social unconscious (Durkheim et al.); and the linguistic unconscious (post-Saussureans) – all these have marginalised consciousness, making it a minor affair and diminishing the distance between highly self-conscious human beings and instinctive animals. (A critique of this way of thinking is the main theme of Part II of this book.)

The assimilation of the explicit and conscious to the implicit and unconscious achieves its ultimate expression in the reduction of the former to neural structures. Young realises that his story of the birth of religion and its relation to the origin of totemism and exogamy, which revolves around the cannibalistic slaughter of the patriarch, raises one or two fundamental questions. When did the parricidal catastrophe take place? Did it happen only once? If so, how was it remembered? And how has it managed to influence the fundamental structures of kinship and culture in so many spatially and temporally and climatically diverse peoples? How did the word get round? Did the event happen at all? Are we talking history here, or myth?

Are we talking about catastrophic events that actually took place in hominid evolution and then became wired as archetypes in to the phyletic memories of mankind (Jung's Collective Unconscious) or are we telling 'just-so' stories to amuse and console our ignorance?

(p. 106)

He argues, against Lévi-Strauss, that (a) the events really did take place, and that (b) they did become hard-wired into the collective memory. He imagines that a dozen or so bands of hominids wandering the African savannah 'in the year dot' eat the cannibal meal. Half of these 'break up in psychic and political confusion, wandering away into madness, death and perhaps assimilation by the other five'. The remainder manage to establish the kinship structures necessary to prevent the recurrence of the parricidal meal and to ensure that its 'terrible beauty' is remembered on ceremonial occasions, so that any inclination to repeat it is nipped in the bud.

It would be something of an understatement to suggest that the story is speculative and constructed more out of a need to bring together Darwin, Frazer, Freud and Fox than it is occasioned by any adequate empirical data. For the present, let us focus on the question of the nature of the supposed memories that are 'hard-wired' into the Collective Unconscious. According to Young, hard-wiring takes place in the brain: the Collective Unconscious is the collection of individual brains, all of them hard-wired in the same way. The process of hard-wiring is the result of a mixture of repeated exposure to the relevant stimuli (it may take several gos to get the kinship structures and the terrible memory upon which they are based to stick) and Darwinian selection, which favours those bands of hominids who get it right. This process is necessary if events that took place only here and there and at great intervals of time should be accessible to all men and women at all time: hard-wiring permits the greater part of humanity to treat itself to a free (cannibalistic) lunch.

Young's account of the hard-wiring of phyletic memories cannibalises scientific discourse but, in accordance with his contempt for science (to be discussed in due course), draws neither nourishment nor wisdom from it. A handful of meals by a handful of tribes would

simply not be enough to effect a permanent and irreversible change
in the hominid brain. Millions of events would be required, and even
then there would be no guarantee of irreversible changes: other repeated events might be equally potent inducers of contrary, or interfering, phyletic memories wired into the brain. Young's position is
weaker even than that of other Lamarckianising Darwinians: *his* gi-
raffe gets its genetically transmissible long neck from a few stretches
of a very few giraffes. But, for our present purposes, there are more
important objections than Young's failure – not untypical amongst
humanist intellectuals – to achieve the level of understanding required
to pass GCSE biology.

The hard-wired phyletic memories are not really memories. No one
really remembers them – apart from anthropologists, palaeontologists
and cultural critics looking to support a thesis, and these remember
them not as first-person experiences energising and shaping impulses,
but as third-person facts. It might be argued that many of our ordinary memories are not remembered in the first person, that they are,
rather, embodied in habits. In short, that they are implicit in our patterns of behaviour. Once we go down that road, however, it is possible to read any memory into any habit, and Young's thesis, in particular,
becomes emptily circular. When memory is extended beyond individual experience and is, in addition, separated from experienced, first-
person content, then we are well-embarked on a journey into absurdity
that ends with ascribing memories to knicker elastic, DNA and stones.[12]

ELIDING THE DISTANCE (2) MODERN MAN = PRIMITIVE MAN

Eliding the difference between explicit and implicit knowledge – and
so between, for example, knowledge-based techniques and instinctual
behaviour – is thus a potent means of concealing the difference between Man, The Explicit Animal and infra-human animals. Arguing
that implicit memories laid down in early hominids are still active in
us now, is an example of this (not always conscious) ploy. But it also
exemplifes another conceptual legerdemain by which we are connected
through a series of bridges receding further and further into an ill-lit
past, with primitive man and worse: contemporary man is related to
early historical man; early history merges with pre-history; pre-historic man elides into hominids; and the latter into higher primates.
The agenda here is to deny that we have in any sense progressed
intellectually beyond 'primitive man'. This has two elements: denying that 'primitive' thought is as primitive (i.e. simple) as it looks;
and denying that 'advanced' or 'civilised' thought is as advanced or
as civilised as we think it to be.

Both elements are prominent in the structuralist anthropology of Lévi-Strauss. He argues that there is no fundamental difference between the 'pre-logical' mentality of Primitives and the 'logical' mentality of Modern Man. As Edmund Leach summarises it:

> Primitive people are no more mystical in their approach to reality than we are. The distinction is rather between a logic which is constructed out of observed contrasts in the sensory qualities of concrete objects – e.g. the difference between raw and cooked, wet and dry, male and female – and a logic which depends upon the formal constrasts of entirely abstract entities – e.g $+$ and $-$ or log x and x^e. The latter kind of logic ... is a different way of talking about the same thing.[13]

The 'logic of the concrete' is, according to Lévi-Strauss's interpretation, not at all primitive in the sense of being simple. On the contrary, it is mind-bogglingly complex. Sets of oppositions between concrete categories are opposed to one another to produce higher-order oppositions and structures. These latter reflect the fundamental, universal and permanent structure of the human mind and, in so doing, permit mankind – Primitive or Modern – to italicise the opposition of his Culture to Nature and to remake the world in the image of mankind, so that he makes it 'his thing'.

What Lévi-Strauss does not point out (perhaps he overlooks it) is that the difference between the Primitive (concrete) logic and the Modern (abstract) logic is that the former is implicit and the latter is explicit. This is reflected most tellingly in the difference between the mythologic of the Primitives he studies and the explicit logic he, the Modern, has unpacked out of them. Lévi-Strauss has a theory about the Primitive mind that owners of primitive minds themselves have not entertained and a conception of the universal that would have been beyond them. (Whether Lévi-Strauss's theory is correct or not is another matter.) The fiendish complexity of the Primitive mind was not accessible to those minds: it lay inside the structuralist's mind, not that of the Primitives from whom he unconvincingly denies himself to be any different. The complexity structural anthropology attributes to Primitive thought is borrowed from structuralism, for whom alone the structures of thought exist.

'Their thoughts are at least as complex as ours' is one way of narrowing the gap between Primitive and Modern thought. The complementary approach is to deny that there is any difference between the distinctive activity of the primitive mind – indulgence in myth and magic – and the distinctive activity of the modern mind – the pursuit of scientific knowledge both for its own sake and for its practical application in technology. This is a large topic and not to be dealt with in an aside. It will be addressed in chapter 3.

ELIDING THE DISTANCE (3) TENDENTIOUS DESCRIPTIONS OF MAN AND ANIMALS

A fourth way of 'softening the bump' is by generating tendentious descriptions of man and animals: the gap between man and animals is crossed by arranging for animalomorphic descriptions of man to meet anthropomorphic descriptions of animals. Tough-minded, naturalistic, materialistic descriptions of human behaviour and the impulses that drive it merge with sentimental, Disneyfied accounts of the feelings and motives of chimps and other primates. A few whoops and leaps by chimps are upgraded to a proto-religious ritual and the extraordinarily complex liturgies and devotions of human religion reduced to a primitive sacrificial essence. In this way, the gap between a gallop before a thunderstorm and the Council of Trent, or between a few minutes of random leaping and a season of the Ballet Rambert, is closed. This technique for eliding the gap is all the more powerful when the discourse about things on either side of the gap is almost continuously ambiguous as to whether it is describing, speculating or explaining; whether it is presenting facts, history-telling or mythologising; whether its account is synchronic or diachronic; whether the sequences of events it sets out are meant to be chronological or causal, logical or teleological; and where the main motor of the discussion is an association of ideas.

I don't think I am being unfair. Young's book is full of passages like this:

> The preacher [who reported that 'after intercourse all animals are sad'] may have been a gorilla. Like us he often prefers the ventral position and can spend up to fifteen minutes getting there. Like us too, he seems to be aware of others' suffering and to carry some melancholy sense that the world is out of joint, either that he has squandered his birthright or that the gods have plotted against him. Some males even choose not to breed, and mortality among unmated males is high. It may be that the gorilla's extraordinary sensitivity to failure derives from some recent awareness of ecological doom, which he finds confirmed in the arms of his lady gorilla. On the other hand, it could be that one day in the dim past he stumbled upon love, and its dreadful lessons turned him into a philosopher who despaired of evolution: which left the unphilosophical chimp and the quasi-philosophical human to continue the story.

(p. 74)

(Judging by this passage, 'quasi-philosophical' is about right!)

The process of assimilating human to animal nature is greatly facilitated, then, by two sorts of fallacy: animalomorphic descriptions of man and anthropomorphic descriptions of animals. The gaps in the fossil and cultural record also make it easier to leap across the space between man and animals. The circle that is completed by Disneyfying animals and biologising man is reinforced by two other projections.[14]

The first is projecting into primitive man, or even beyond him into hominids, our present selves and present preoccupations and then reading ourselves out of primitive man. For example, the interpretation of primitive rituals is shaped by what we know of the meaning of our own rituals and, in particular, by our knowledge of what they are like when experienced from within. 'Flintstoning the primitives' in this way is particularly easy because we know so little about them: the poverty of the data makes the past vulnerable to tendentious interpretation: there is little in the way of constraint of fact; and, in the absence of an adequate database, we are required to put in a lot of ourselves in order to make a coherent story. The remote past is thus like a Rorschach inkblot into which we can pour ourselves and our preoccuptions. And 'The Present Age', against which we evaluate the past in order to assess the present, provides another opportunity for projection. This, too, is a Rorschach inkblot, a meaningless sign into which we can read whatever concerns us, but for the opposite reason: we have too much information about it. As we shall discuss in chapter 4, 'our present age', 'society', 'the contemporary epoch', etc. are far too large for any mind to house. The image of the present will therefore inevitably be partial and so be able to reflect the preoccupations of the image-maker.

Both 'sophisticated present' and 'primitive past' – or the illness of the present and the therapeutic clues of the past – are therefore reflections of ourselves – the first because of the lack of data, the second because of the overwhelming, unmanageable abundance of it. (Any summarising, rounded off, sense of society, the world or the century as a whole, is delusory.) A new circularity will therefore be possible, tighter and more inescapable than either of the others: we shall read the past in the light of our preoccupations with the present and we shall read the present in the light of what we believe that the past has to tell us. These readings confirm pre-existing notions rather than expanding the horizons within which we think about our nature and its problems: just as Disneyfying animals deprives us of what they might have to offer us – a sense of their mysterious remote otherness – so Flintstoning the past by infusing it with the present and reading the present in the light of a Flintstoned past takes to extreme the familiarisation process by which we conceal the uncapturable otherness of others under general thoughts that seem to capture them. Inside this closed circuit, the sense of the past is subordinated to the diagnosis of the present (thinking about the past is usually a way of thinking against the present) and to the search for treatments for the supposed ills of the present. Descriptions, explanations and prescriptions are intertwined. (Comparisons between the present and the past are, of course, deeply suspect, in the way that comparisons between an individual one lives with and an individual one has only read about in a newspaper are suspect.)

Young's easy habit of connecting the issues of the present day with the soul of the remote past, the immediacy of his link between *Cosmopolitan* editorials and the dilemmas of palaeolithic man, illustrates the sterility of his closed-circuit thinking. And its absurdity. Take this, for example:

> Our relations with Mother Nature have gone wrong (the Ozone hole is gaping, CO_2 emissions are rising), and it will take more than a poem or two to get them right'. [You bet.] This is another reason for consulting primitive man, for he not only kept in touch with Mother Nature, but did so through the application of startlingly strong mythic medicine.

(p. xxiii)

This is deeply vulgar in the way that a Fred Flintstone cartoon is – but it is important because of its typicality. Morover, it undermines one of the other, deeper, purposes of Young's book – which is to open up a sense of the strangeness of the human world and the *distance* of our origins.

REMEMBERING THE DISTANCE: THE HA-HA

The overall result is to create a spurious appearance of continuity between human beings and infra-human animals; and the almost bottomless trench dug by human history between Culture and Nature is overlooked. With Young's help – and that of a thousand ethologists, literary biologists, anthropologists, metaphysical psychologists – we seem to be able to look straight through or past history and pre-history and find in the chimpanzees a reflection of ourselves, our needs, proclivities and preoccupations. The intellectual and other dangers of this fudge may be illustrated by a metaphor.

In the classical country house, the park or garden is separated from the countryside by a Ha-Ha. Essentially, it is a trench: on the side nearer to the house it is perpendicular and faced with stone; while the outer side is turfed and slopes gradually up to the original level of the ground. The most important feature of the Ha-Ha is its near-invisibility from within the house. It permits an unobstructed view of the countryside – hence its alternative name, *claire-voie*. From the house, the fields appear as a simple continuation of the park. This is, of course, deceptive. For the Ha-Ha presents a formidable barrier to sheep and cows and other animals that might wish to enter the gardens and graze on the prize roses or trample through the kitchen garden. Moreover, it a potential hazard to human beings, who might easily overlook it, as is indicated by its name: those who observed it said 'Aha!' and those who did not caused their friends to say 'Ha Ha' when they fell into it. According to most commentators, the Ha-Ha symbolises the boundary dividing human Culture from Nature, the domesticated house and garden from the wilder country beyond. It is possible also to see it as a rather precise metaphor of the invisible but very real divide between man, the explicit animal, and the rest of the living world. The non-human animals cannot cross this barrier; and those thinkers who do not observe it, who fall victim to the illusion of continuous ground between Nature and Culture, are destined to fall into it.

I commend this metaphor to anyone intending to trying to derive, or prescribe for, human nature on the basis of animal nature.[15] Such an approach overlooks the miracle, the mystery of explicitness. Pondering on such a metaphor might give pause to writers such as Young, who look to a return to animal knowledge and instinct, or to primitive wisdom, as the way out of what he perceives as 'our present dilemma'. (The scope of 'we' is always undefined.) I would also commend it to anyone else who might be tempted to denigrate human Culture as a deformed or unsatisfied or dangerously suppressed version of the animal Nature. If there is such a thing as The Human Predicament, and if it makes sense to offer any kind of treatment for it, I suspect that the prescription would take the form of more, not less, explicitness; of capitalising on the advances that we have made, rather than falling back into the instinctive world of primitive man, of hominids, or higher primates. The major challenge of the next millenium is not, as Young claims, 'the retrieval of innocence' and a return to our animal selves – in so far as we were human, we have never had such selves – but in dealing with and perfecting our knowledge. We cannot solve our problems by returning to, and perfecting, a nature we never had.[16]

3

The Wisdom of Myths and the Myths of Wisdom

INTRODUCTION

Origins of the Sacred is exemplary in both a positive and negative sense: on the one hand, it is a great work of scholarship and synthesis; and, on the other, it reflects, with admirable lucidity, a prevailing climate of opinion amongst contemporary humanist intellectuals. Young is in many respects a prisoner of the opinions current among 'culture critics', which is why a critique of the ideas advanced in his book naturally opens on to wider themes – on to a more general critique of contemporary cultural criticism.

His main thesis, which he shares with so many others that it must be accounted one of the great received ideas of our time, is that we have forgotten who we are; in particular, that we have lost sight of the sacred and animal powers that we once had and which, because we have forgotten them, now, in a sense, have us in their power. This thesis, which is developed in an arresting and highly original way, is buttressed by several assumptions which are more widely accepted and form the basis of the 'bitter line of hostility to civilisation' that Trilling spoke of as characterising so much modern literature. They are important elements in contemporary Counter-Enlightenment thought:

1. that there is an intrinsic wisdom in myths that goes deeper than the wisdom of our rationalistic and scientific thinking;
2. that science, and more particularly science-based technology, are not more advanced than so-called 'primitive' magic;
3. that we need to recover our sense of the sacred in this 'disenchanted' world;
4. and that there is something called 'wisdom', which we have lost, and which is more potent and effective than the 'cleverness' that is the only guide left to us.

An examination of these assumptions forms the agenda for this chapter.

106

THE WISDOM OF MYTHS

A cynic listening to Radio 4's 'Thought for the Day' or any modern preacher sermonising in a half-empty church or addressing a mass rally at the Hammersmith Odeon might conclude that the ability to find contemporary significance (or, indeed, any useful meaning) in Bible stories is a greater tribute to the ingenuity of the modern mind than to the inherent timelessness of those stories. In the twentieth century, thinkers have developed the habit of 'dipping into the myth kitty' (to borrow a wonderfully scornful phrase from Philip Larkin) in search of insights into what is variously known as 'the modern predicament', 'The Predicament of Man', 'the Struggle of the Modern', etc. (The singular – implying that there is a single such predicament, or Predicament – itself betrays a presupposition of colossal, indeed mythical, proportions.)

Dudley Young, in common with innumerable other commentators, laments that 'modern man no longer knows who he is':

> this is largely because he has forgotten where he comes from; and just as a neurotic individual can be strengthened by retrieving some repressed or disavowed aspect of his past, so can we all, both individually and collectively, be strengthened by remembering our sense of the sacred and the animal powers from which that sense arises.
>
> (p. xvi)

There could be no more luminous statement of the 'meta-mythology' – myths about myths – that has been a dominant influence in twentieth-century cultural discourse. Young's voice, proclaiming that recovering the mythical vision enjoyed by our primitive forebears is the road to self-knowledge, a means of renewing art and, perhaps, of healing the deep spiritual wounds that have led to so much private angst, public emptiness and collective bloodshed, is part of a vast chorus. The burgeoning and interacting disciplines of anthropology (armchair, functional and structural), of classical scholarship (which, in the writings of scholars as disparate as Nietzsche and Dodds, found Dionysiac darkness at the heart of Apolline light), and of psychoanalysis, have persuaded thinkers and chatterers alike that primitive forces are still active within us and that knowledge of them will instruct us as to our true nature and origins and show how we, and society, have gone off the rails. The influence of scholars and scientists has been, as we have already noted, amplified by that of creative writers: there was hardly a major literary figure in the first half of the twentieth century who did not at least dip into 'the myth kitty' and the work of many writers – Eliot, Yeats, Lawrence, to mention only a handful of those writing in English – has been myth-eaten to the core.

There is, it is claimed, a wisdom in myths and we have lost it, or thrown it away. 'The evidence is increasingly accumulating to suggest', Young tells us

> that the occidental experiment in secular scientism over the past four centuries has involved the suppression of certain vital processes of a more or less religious nature, and that this is becoming intolerable.
>
> (p. xvii)[1]

The voice of authority in our culture, Young says,

> is unquestionably the voice of science, and yet that voice is still unable (and often unwilling) to master those parts of speech . . . without which no utterance can be fully human, no author authoritative.
>
> (p. xvi)

As a result, we are impoverished; for 'the principal objective of mythic play was always the permission to live life more abundantly' (p. xxv). Because myths embody vital truths about ourselves, we are also self-ignorant, sleep-walkers: 'Man fishes for the logic of human desire with myth and ritual' (p. xxv):

> The myths (and rituals) that tell a man who he is and what he values compose his religion (from Latin *re-ligere* = to bind back): they bind him back through memory to the divine ancestors who call upon him to act and think in certain ways.
>
> (p. 18)

Man is a 'myth-making animal' and 'if you take away his gods he will replace them with devils: as Nietzsche put it, "Man would rather have a void for his purpose than be void of purpose"' (p. 36). In short, modern man 'must return to the mythic voice if he is to heal the divisions in his soul' (p. xviii).

This all sounds highly plausible, even if it owes some of its plausibility to its familiarity. The healing, renewing and inspiring effects of myths, their inherent therapeutic possibilities, are part of their immemorial meta-mythology: creation myths are invoked when individuals or societies are sick; healing through recitation of a cosmogony is a classical example of the use of myth as a magical incantation. The therapeutic of myth, however, is not all that easy to apply.

For a start, myths may not be believed. They may, as the Enlight-

enment thinkers judged them, simply be wrong; and believing something that one considers to be incorrect does not come naturally. Whilst
it may be easier to believe, for therapeutic reasons, that the wind is
the *pneuma*, the breath of God, the divine force that moved through,
energised and created all things, than to believe that the world is
poised on an elephant's back, it is not, for those of us who are used
to thinking that wind is only moving air, all that much easier.

There are several ways of avoiding the difficulty of requisitioning
belief for unscientific myths in a culture whose voice of authority is
science. The first is the *hermeneutic* approach, which tells us that myths
belong to meaning systems that are quite separate from those of secular
science: they are not to be judged by the latter, for they are, in a
sense, all of a piece, sealed off from other meaning systems, and equally
valid. The second is to say that myths are not be understood literally:
they are *expressive* rather than referential and, from the point of modern man, they are expressive of psychological truths. The third is to
assert that the world-picture dominated by secular science is itself
mythical: the seeming contrasts between science and myth, or between
science-based technology and magic, are overstated; what we have
are rival myths, rival magics. I shall deal with this third view later
and focus here on the first two.

We have already, implicitly, encountered the hermeneutic view in
the anti-universalist, Counter-Enlightenment arguments of Herder and
others. The relationship of this view to the defence of myth is set out
by Berlin:

> Myths are not, as enlightened thinkers believe, false statements about
> reality corrected by later rational criticism, nor is poetry mere em-
> bellishment of what could equally well be stated in ordinary prose.
> The myths and poetry of antiquity embody a vision of the world as
> authentic as that of Greek philosophy, or Roman law, of the poetry
> and culture of our own enlightened age – earlier, cruder, remoter
> from us, but with its own voice, as we hear it in the *Iliad* or the
> Twelve Tables, belonging uniquely to its own culture. Each culture
> expresses it own collective experience, each step on the ladder
> of human development has its own equally authentic means of
> expression.[2]

Unfortunately, this does not quite deliver what Young requires. If
myths belong to meaning-systems that are whole, entire unto themselves and incommensurate with contemporary meaning-systems, then
they cannot speak to us, cannot engage with our contemporary meanings, even to act as a critique of them. If myths are not mocked by
rationalistic thought, nor is rationalistic thought mocked by myths.

The belief that myths express important psychological (rather than factual) truths has been one of the most pervasive ideas of modern times. It has been sustained by translating the myths into modern terms, or into terms that make them relevant to the dilemmas, temptations, etc. of contemporary daily life. This is the 'Thought for the Day' strategy, whereby a preacher finds in myths expressions of familiar psychological, or more often, moral truths. Unfortunately, though this approach may rescue myths from the charge of simple error, it does so at the cost of downgrading them. When the kernel of contemporary good sense is delivered from the immemorial mythic husk, the wisdom of myths comes to look very like everyday common sense badly put – a banal signal coated in exotic noise.

A more impressive rescue operation is that mounted by the so-called depth psychologists. For them, as for Young, myths give expression to deep psychological truths that cannot otherwise be uncovered, even less spoken. Myths are the voice of the collective unconscious in an unrepressed state. It is only in mythic form that humanity could confess to itself that one of the founding acts of a distinctively human culture was the murder of the father by the primal horde and turning his corpse into the main dish in a collective meal. Once we understand this, we are able to make sense of many things – from the curious rituals of religion, such as the theophagous climax of the Mass, to our ambivalent attitude towards authority figures. Moreover, we gain insight into the forces that have moved us in the past and that continue to move us in the present. By remembering our remote past, to which myths give us unique access, we are made aware of the forces that potentially have us in their grip. Thus able to confront our inner darkness – the irrational, murderous instincts that are active within us – we are less likely to act out that darkness. If we recognise, for example, that 'the desire (disowned, of course) to nuke the planet is but the tumourous extension of the old sacrilegious dream of making war on the gods' (p. xvii), then we might be less likely to act on that desire. We need to remember, through acquaintance with myth, the violent child within us, if we are not going to follow, without knowing why, his path of destruction. Myth is the royal road to those healing insights that will liberate us from the violence immemorially implanted in our hearts.

There are many problems with the depth psychological rescue of myths, particularly as exemplified in the enormously popular meanderings of Freud and Jung. The first relates to the kind of memory required to gain access to myths and the process by which the faculty is activated. It is not enough, Freud et al. would tell us, to read the myths in a book or even to read an explanation of them, and their relevance, in a book. You cannot, for example, spare yourself weeks and years of therapy by reading the Pelican that tells you the story, linking your soul with the mythical past, conveyed in the myth. This may give you access to the facts and the concepts, but not to the wisdom. In order for the myth to be active in the surface of your life, it has to be excavated from your depths – depths that are not opened up by the mere act of reading. A special kind of remembering, of recognition, of acceptance, is required, an insight that links your own most private memories and most ancient feelings and instincts with the story, so that it shall live within you: the myth has to be energised by your sorrows so that its story shall become your story and you shall be nourished by its wisdom. You are required to *wake into* the myth through a special kind of anamnesis – a re-enactment, a re-living, a participative re-run. This requirement justifies the therapist's fees: he has to guide you into the underworld where Oedipus and you, the patient, where your unresolved desire for your mother and Oedipus' unfortunate choice of sexual partner, are one; where your mangled, dysfunctional feelings and the ancient myth, your fee-paying self and the collective memory, are inseparable. That, at least, is the standard depth-psychological story. And it has a momentary intuitive appeal: we cannot expect to be changed in our depths except by something that enters and is active in our depths. This immediate intuitive appeal does not, however, survive the reflection that the special kind of memory, with its exotic framework of magical thought, is rather out of place in the world in which the transactions between the therapist and patient take place – the secular, individualistic, rationalistic, contractual world in which the fee-paying patient meets the fee-paid therapist, for a fixed period of time at a pre-determined time and place. Nothing could be more remote from the world of myth and pre-history than the context of the therapeutic – or would-be therapeutic – transaction.

There are numerous other difficulties besetting the notion that the 'wisdom of myths' is to be located in their expressing permanent psychological truths about humankind and that 'reacquaintance with our primitive selves can make us *more* human rather than *less*' (p. xxiv). The argument that not to know the myths by which we live is to be in the grip of them (just as those who do not know their history are doomed to repeat it) and that, conversely, to know them in the sense of remembering them in a special way is to wake up out of them, as if out of a delirium and so to be liberated from them, does not withstand close scrutiny. Is it not just as likely that knowing our deepest feelings will activate them and place us more completely in their grip? The mere fact that those feelings are unreasonable or dysfunctional should cut no ice with a creature who, according to the psycho-mythologist, is intrinsically prone to be in the grip of

unreasonable feelings. The myth of the therapeutic value of releasing
the wisdom expressed or contained in myths, myths that have us in
their grip, is thus fraught with contradiction: it presupposes that reason and insight have those very powers that depth psychologists deny
that they have.

It might be argued that the bad – say, parricidal or matrophilic –
feelings expressed in myths are harmful only in so far as they are not
recognised as controlling forces in our lives and that acknowledging
the wisdom and psychological truth of myths is a way of bringing
those feelings to the surface where reason operates and can disarm
them. But that assumes that the feelings expressed in myths are absolutely wrong – that myths capture destructive rather than creative
forces in human life. This is not, however, consistent with a very positive
notion of the wisdom of myths: it reduces myths to mere access routes
to the nasty things in us that have to be dealt with so that we can
then proceed with civilised life. The notion of 'mythic play' as 'permission to live life more abundantly' (p. xxv) seems to have gone out
of the window. A deep acquaintance with myths becomes merely a
means of getting certain (primitive) things out of our system: a psychological dose of salts.

Unfortunately, there is no guarantee that such negative therapy,
aimed to liberate us from myths, or to use myths to liberate us from
the forces which myths (unconsciously) express, will produce predictable
or controllable consequences. The introduction of individuals to their
own supposedly amoral, Dionysiac, instinctual self may encourage
rather than suppress anti-civic feelings. After all, the general notion
that the fate of our character is fixed in the prehistory of our lives
and that of the human race – by, for example, the way we and the
primitive collective negotiated the transition from omni-libidinous
infancy to pre-citizenly toddlerhood – would seem to undermine rather
than reinforce the sense of personal responsibility. The wisdom of
myths may 'give us permission' to live life more psychopathically. So
much for the myth of 'myth as sacred physic'.

However, we may find it difficult to take myths seriously enough
for them to liberate us to live either more abundantly or more
psychopathically. For it is not easy to see how we can 'rescue' myths
– from their evident factual inaccuracy or inconsistency or groundlessness according to modern standards of evidence – without patronising them. To say that the Oedipus story is 'really' about the process
of negotiating the passage from the Mirror stage to the Symbolic Order,
or of dealing with the incestuous feelings infants entertain for opposite sex parents, is, implicitly, to presume that 'we' know more about
what the myth is about than those who accepted it literally did. Lacan
or Freud are assumed to transcend Sophocles or the belief-system his

plays drew upon. If Oedipus is a deeply true story, but only as an anticipation of something that had not been appreciated until psychoanalysis was invented, it is one whose deep truth we have access to and the Greeks did not. Their unconscious (and our unconscious) has reached consciousness in us. In short, the depth-psychologising of myths does not unequivocally rehabilitate them: they are upgraded to the house of deep truths; but they are downgraded to confused expressions of truths that are now set out more clearly in textbooks of psychology.

But even if myths could be rescued in this way, we need to ask what we should do if we wish to acquiesce to their superior wisdom. Do we bow down before all of them, as valid expressions of ancient wisdom? Are all myths of equal antiquity equally good? Or do we pick and choose between them? Presumably we must be selective, because (except in the system of Lévi-Strauss, of whom more presently) different myths say sharply different things. They differ not only at the level of surface, factual claims but also in their deep meanings and their fundamental prescriptions. If, however, we are to discriminate, what criteria shall we use to separate the sheep from the goats? How shall we know what myths are good for us? Shall we appeal to contemporary common sense? Or to our sense of what works or what would be right for us? Hardly, since acquiescence before myths is supposed to reform our common sense and revise our ideas of our true nature and true needs. The question of how we choose is unanswerable and it follows from its unanswerability that we should have an infinite hospitality to myths of any sort and swallow all available brands of 'strong mythic medicine' to purge the Augean stables of our modern souls.

And we are still left with the wider question of the dangers of giving credence to myths. The word 'myth' most fundamentally means 'story'. It tends to be used ambivalently – either of a story that is untrue but widely believed and so is highly influential; or, as in the sense we have been addressing already, one that is superficially or factually or literally untrue, but which is deeply, or expressively, or allegorically, or in some other way, profoundly true.

Once we recognise this ambiguity, we have grounds for uneasiness about prescribing Young's 'strong mythic medicine'. If myths are stories that are both factually untrue and yet are in tune with deep psychological truths, their influence in the social sphere might not be benign. Consider, for example, the urban myths; or the myths that underpin racial prejudice; or all those scapegoating myths that encourage dangerous and pernicious explanations of the misfortunes of daily life. Anti-Semitism was nourished on a rich diet of myths. As Margaret Anne Doody points out, 'The deployment of myth can serve to mould a public impervious to argument, and inoculated against discussion and dissent.'[3] This underlines the close relationship between the 'immemorial' knowledge and wisdom contained in myths and unaccountability. The unaccountability is two-fold: that of the myth and that of the myth-teller.

If the myth expresses the implicit knowledge, the unconscious, of the tribe, then it is not amenable to testing. The demand for evidence seems comically misplaced, almost a category mistake. What a myth demands is unquestioning acquiescence, not informed assent based upon critical evaluation: it says 'believe and then ye shall understand' and re-classifies faith, blind faith, as a moral requirement. A pure manifestation of folk wisdom, it is a form of spiritual nutrition, doing for the soul what the mother's milk it is taken in with does for the body. It is no more open to challenge, to question, to critical evaluation in the light of reflective experience than is that mother's milk.

The other, and equally dangerous, aspect of the unaccountability of myths is that of the myth-teller. Myths exemplify precisely that pernicious relationship between power and knowledge that Foucault, quite incorrectly, saw as characterising all knowledge. The myth-teller, who commands belief without being required to earn it, who can lay claim to knowledge without the obligation to declare the basis of his knowledge, is the archetype of groundless, and therefore unchallengable, authority. The yokel taking in tales of wonders and gobbets of wisdom through an open, toothless mouth widening from omicron to omega as the unchallenged claims extend their scope is the epitomic victim of manipulation and control, of arbitrary and untrammelled power.[4]

This asymmetrical power relation between myth-teller and myth-swallower is often enhanced by the mythology of myths – which is that they will not be understand by those who are not willing first to succumb to them; that their spiritual nutrition is unavailable to those 'shallow, clever' people who keep their wits about them and deploy their intelligence. First believe and then ye shall understand. Myths are often paradoxical and superficially stupid. This paradox, however, is not an oversight, a weakness, but an invitation – with or without menace – to the listener to lay down his arms and allow the myth to enter into him. The surface idiocy of the myth conveys an implicit command: leave your surfaces and enter the depths of the collective understanding where the stupidity will miraculously turn into wisdom. Those for whom myths remain simply stupid are excoriated for being incapable of escaping their own surfaces, for being superficial through and through, and so condemned to a 'shallow' rationalism, to mere common sense and reasonableness. The charismatic myth-teller addresses himself only to the pure or open of heart, to those who are willing to let go of their narrow understanding and let him and what he has to say enter into their hearts. And if he has to use unusual means to bring about this state of acquiescence – the famous paradoxes and silences and seemingly unprovoked slaps of the Zen master – then so be it: this is a reflection not of the weakness of the myths he is telling, of the groundlessness of the knowledge he is imparting, but of the intrinsic stupidity of humanity and the fact that, as Heracleitus said, 'The beast has to be driven to the pasture with blows.' The relationship between the transmitter and the receiver of the myth is unequal: the myth-teller's depths of understanding, his openness to the new, the different, to others' viewpoints, is not on trial. His resistance to the listener's objections is not taken as proof of his, the myth-teller's, shallowness and closed-offness. In this asymmetry, we see naked power and the possibility of infinite abuse. The silence of the Zen master is the acme of unaccountable wisdom.

The elevated, immune status of the myth-teller shows how dangerous is the attribution to myths of a deep wisdom accessible only to a few who may mediate it to the many. The shaman is the 'hero' of Young's book (p. xxiv). But he is not easily picked out of a crowd of others who could surely, in the twentieth century, be regarded as no one's hero and certainly not the prescription the soul doctors ordered: the charlatans, the demagogues, the tyrants who have trampled over the bodies and lives of so many in the last 100 years. The wisdom of myths has been, and still is, the lifeblood of corrupt prelates, genocidal dictators, leaders of totalitarian religious sects, and others whose dreams of power have been fulfilled through their uncanny ability to understand and manipulate the fears and hopes of ordinary people.[5]

But that is another story. In shifting our attention from the myth to the myth-teller, we also move from the wisdom of myths to the myths of wisdom which we shall explore in due course. For the present, let us look at the conclusion we must draw from our observation of myths: the best thing to do with myths is to be constantly on guard against them and to disbelieve them as hard as we can. Yes, we are deceived if we imagine that we are not always to some extent in the grip of myths, to some extent acting out unchallenged and groundless beliefs. Our duty, then, is not to so order our affairs that we succumb to more myths, or more completely to the ones we are already gripped by. Yes, much of the argument against the Channel Tunnel was a rationalisation of a hostility driven by the myth of Britain as an unviolated island at risk from violation, from pollution, by the uncontrolled influx of foreigners. But this is an argument for recognising the power of irrational thought in our lives, not for trying to extend or succumb more completely to that power. Just as we should recognise how much unreasonableness lies beneath our surfaces of reason in order that we should become less, not more, unreasonable.

'The best thing to do with myths is to recognise them for what they are and to disbelieve them as hard as we can.' Is that not a rather hard prescription? It is, until we think of the unbelievably oppressive customs and cruel practices that have been driven by myths. Take, for example, the pollution myths that have dominated and continued to shape so many societies. For Young, pollution myths seem on the whole to be good for the health of society:

> divine pollution moves within the community. We want to expel it. Since it must have been introduced by some carrier or agent of the divine, our task is to find this carrier and send it or him back to the chaotic realm from which it issued. This divination and expulsion is to be supervised and ritually undertaken by the shaman.
>
> (p. 230)

Such 'mythic healing' may indeed give permission to live life more abundantly:

> To discover that everything that disorders our lives may be systematically understood as divine pollution is potentially liberating.
>
> (p. 184)

This might sound all very innocent and jolly if one knew nothing about the history of the world and the terrible influence that irrational myths of pollution have played in it. Was not the master-myth of twentieth-century anti-Semitism the idea that Aryan blood would be polluted by Jewish blood unless the Jews were exterminated? Or does Young, who makes occasional knowing allusions to Auschwitz, think of Auschwitz as simply a term of rhetoric to be mobilised from time to time and not as the name of a terrible lesson from which every single one of us is obliged to learn? And did not myths of divine pollution lie behind the hideous cruelties inflicted by shamans – half-priests, half-physicians – on their wholly vulnerable patient-victims, who were usually distinguished less by their illnesses than by their powerlessness? And many women would certainly have strong and justified views on the role played by pollution myths in the degradations they have had to endure since time out of mind. And in many societies, for example in South Asia, myths of pollution have underwritten the frozen structures of oppression which have condemned whole sectors of the population to sub-human status and inhuman treatment. This is worth a few moments' further consideration.

The caste system classifies individuals according to the extent to which their occupations and habits involve them in polluting activities.

Fishermen who live by taking the lives of animals, road sweepers whose occupations bring them into contact with the emissions of the human body, are untouchable; as such they have until recently had severely curtailed civil rights. They still suffer grotesque limitations upon their life-chances and desperately low social esteem. Not all that much has changed since untouchables were herded into ghettos away from the Brahmins who might be polluted by contact with them or, indeed, were condemned to a nocturnal existence because even the sight of them might contaminate. The business of treating the lower castes as objects of disgust was quite burdensome for the higher castes as well:

> As the child learned to accept responsibility for its own bodily cleanliness, it was also taught the importance of avoiding the invisible pollution conferred by the touch of members of the lowest castes. The mother or grandmother would call him in and make him bathe and change his clothes if this should happen, until his repugnance for a low caste person's touch became as involuntary as his disgust for the smell and touch of faeces.[6]

The vast amount of time and effort that the higher castes spent avoiding direct or indirect contact with pollutants or with those who may have had direct or indirect contact with them was not only cruel – racism of the most naked kind – but also misdirected if its aim was to prevent illness and death. The irrational rituals of cleaning and pollution control left a heritage in which the rational methods of pollution control – for example, proper sewage systems – were neglected. Even now, infectious diseases – largely avoidable with simple public health measures – account for an appalling proportion of childhood mortality among the poorer classes in countries where pollution myths have had such a powerful influence.[7]

Young's generous openness to myths seems to blind itself to the cruelty of ritual and dogma, to the horrors of pogroms and scapegoating. It is difficult to imagine these things without perhaps getting down to cases. Let me end with a single example from my own experience as doctor in Nigeria: the case of a 14-year-old girl with a vesico-vaginal fistula. She had been married at the age of 13 as the third wife to a 60-year-old leper, considered a good catch because sufferers from leprosy do well out of begging. She became pregnant shortly after her wedding and, because her pelvis had not fully grown, she had an obstructed labour. The indigenous approach to midwifery in rural parts of Nigeria when I was working there in the 1970s tended to be governed by myth rather than reason, by incantation rather than by a sense of anatomical and physiological possibility. After over 80 hours of screaming hell, a surgeon was summoned to the bed of this 14-year-old mother. He was required to deliver bits of a dead baby using cranioclastic forceps. Thanks to his skill, not based on magic but on the principles of rational surgery, the mother survived. Even more astonishingly, she survived the standard post-natal ritual of purification. This involved being placed in a room heated to over 100 degrees Fahrenheit and drinking large volumes of a 'purifying' fluid which happened to be rich in various salts. The consequence of hyperthermia and severe post-partum anaemia, along with electrolyte disturbance, was acute peri-partal cardiac failure, a problem we saw very commonly. She was brought in to hospital, 'purified' but half-dead and, after standard Western treatment for cardiac failure and her anaemia, was returned to the bosom of her tribe. She was not collected by her husband, who was ashamed and angry at the death of his child.

Her shame, however, was only just beginning. As a result of the prolonged obstructed labour, the dead baby's head had opened up a fistula or passage between her bladder and her vagina. She was consequently permanently incontinent of urine which, inevitably, was also infected. Out of her vagina, therefore, there was a constant drip of smelly urine which, in a hot country, made her *persona non grata*. She was regarded with disgust and contempt: a stinking creature, making explicit the essentially polluted nature of all women. In the end, young as she was, she decided to flee to the only place that had helped her: the hospital where she had been treated for cardiac failure. In this hospital there happened to work a retired surgeon who had devoted the final phase of her professional life to repairing vesico-vaginal fistulas. This surgeon did not having a waiting list; rather, a waiting place. Individuals came to the hospital and stayed in a stinking barn called The Mothers' House where many other victims of primitive midwifery awaited their turn for surgery. After a three months' wait, during which she kept herself alive by begging, she received the skilful attention of the retired surgeon who saw the problem in rational terms: it was not a question of divine or secular pollution but of mechanical damage. The hole was repaired, the last urinary infection was treated with antibiotics (not purification rituals) and she returned to the bosom of her tribe.

Whether the story had a happy ending or not is uncertain. She had had to hitch-hike to and from the hospital; doubtless her smell would have chaperoned her on the way down. On the way back, sweet-smelling and quite pretty, she would be an obvious target for the polluting attentions of males. If she returned pregnant, the cycle of shame and expulsion would start all over again.

This story, a horrific tale of abuse, would hardly register in the annals of the kind of infamy that a mythological, as opposed to a rational, approach to the world would routinely foster. I began by referring to larger-scale outrages, including the standing outrage of the caste system which is still influential in Asia, despite its now being illegal. However, anyone inclined to believe in the wisdom of myths is invited to remember the little girl with the hole in her bladder whose myth-ridden society damaged her and then cast her out as a polluted monster. They might then ask the question: if the myths are so wise, how is it that those who subscribe to them are so stupid; how do the 'healing' myths cause so much damage? The larger resonance of the question might then dawn on them.[8]

MAGIC AND SCIENCE

The conscious desire to understand the world and so to increase our power to act on it and thereby palliate helplessness before natural and perhaps supernatural forces and in the face of the imperious needs – the hungers and pains – of our own bodies, is a prehistoric one. It arguably defines, or at least illuminates, what is unique about humankind. It therefore seems strange that now, of all times, in which we have witnessed unprecedented, almost miraculous successes of science both in interpreting the world and increasing our power to act upon it, there should be such widespread hostility amongst humanist intellectuals and others both to science and to the technology based upon it, if not to the material comforts they have brought. Young's hostility to science is far from being heterodox among humanist intellectuals.

'The voice of authority in our culture', Young says, 'is unquestionably the voice of science' (p. xvi). Along with many other exponents of *kulturkritik*, he deplores this:

> and yet that voice is still unable (and often unwilling) to master those parts of speech (many of which are locked up in old dictionaries) without which no utterance can be fully human, no author authoritative. The evidence is increasingly accumulating to suggest that the occidental experiment in secular scientism[9] over the past four centuries has involved the suppression of certain vital processes of a more or less religious nature, and that this is becoming intolerable.
>
> (pp. xvi–xvii)

These views are themselves expressed with considerable authority – as well they might; for Young is echoing the received ideas of his time and his sentiments are drawing on a massive tradition of techno-phobia and loathing of science reaching back to the Romantics two centuries ago. I won't deal with this here, as I have addressed it in some detail elsewhere.[10] I shall, however, examine Young's attempts to downgrade science's claims to truth by putting science on a par with magic. (This complements the other strategy, discussed in the previous section, of upgrading the truth-claims – or meaning-claims or wisdom-claims – of primitive, mythological beliefs.)

Let us examine Young's belief that science and magic are equally valid (and, indeed, that in some respects magic is superior to science because it incorporates spiritually nourishing knowledge and wisdom and understanding that science has forgotten). Young finds similari-ties between contemporary physical theory and the ancient ideas of the shamans:

> subatomic physics has decided that matter, the world-body, *res extensa*, is a fiction we can now do without. This is quite a good joke: when a physicist says that matter doesn't exist, and is *really* just an epiphany of the energy gods, he has almost come full circle, very close to the primitive animist who knows that everything is alive, and derives its liveliness from the winds of *pneuma* blowing through it.
>
> (p. xxiv)

It would be possible to devote an entire book to unpacking the errors in this short passage. It is tempting to do so, because, like so many of the other errors in *Origins of the Sacred*, they are exemplary – enjoy-ing the status, one might say, of 'received misunderstandings'. I shall confine myself to a handful of the most important.

There is little in common between primitive animism and subatomic physics – never mind all the supra-atomic physics that has been a not inconsiderable part of the physics of the last 400 years. First, there are fundamental differences in methods, grounds of belief and the social context of belief. The unchallengable assertions of the shaman are quite different from the provisional suggestions of the scientist, hypotheses rooted in his own and others' observations and arguments and open to ferocious criticism. (For nearly 40 years after he pub-lished *The General Theory of Relativity*, Einstein, despite his status as the greatest contemporary physicist and one of the greatest scientists of all time, could not persuade the scientific community of the validity of his view of quantum theory nor of the fruitfulness of his search for a unified field theory.) Secondly, subatomic physics does not deny the existence of matter; rather it redefines the conditions under which mass-energy takes the form of matter (as in the ordinary macroscopic world) and the conditions under which it takes the form of energy.

Energy has not displaced matter as the fundamental stuff of the universe. Thirdly, even if energy in the scientific sense had displaced matter, energy would be at least as remote from the *pneuma*, the wind, the movement of air, as matter would be. The scientific conception of energy has no relation to the 'sacred delight' of the *pneuma*; the physicists' 'energy' bears not the slightest resemblance to Young's 'energy gods'. And finally, subatomic physics is only a part of science and, although its concepts pervade those of many other scientific disciplines, they do not displace the latter. Subatomic physics cannot be regarded as standing for science any more than ethology, hydrodynamics or physiology is a valid metonym for all modes of this multiform activity. Much science works at a level where the ordinary concepts of matter and energy are in place. What many in the humanities seem to fail to appreciate is that, while the billiard ball world of the atom is no longer in place, the billiard ball world of the billiard ball most certainly is, and here the notions of contiguous causality and of locality are most definitely intact, as anyone who has played billiards will know. Even advanced, science-based technology still utilises the world of causal relations and locality.

Those who would relativise science – and put it in its place along with other 'useful fictions' – are very fond of subatomic physics because its major notions seem to undermine the very principles of robust non-magical common sense upon which science is itself based. The interaction between the observer and the observed in quantum mechanics, necessary to ensure the collapse of the wave-packet to a particle or wave form, is taken to imply that 'animism' is back – as if the macroscopic world of bridge-building and physiology were reghosted by the difficulties of interpreting the subatomic world of photons. It certainly does not license the return to animist mythology that Young advocates or give any point to his criticism of the scientist's failure to animise in the way that the Greeks did:

> even now when Einstein's monster [nuclear power] is on the loose, physicists choose, for their purposes, to ignore what the Greeks profoundly knew, that Poseidon gets *angry*.
>
> (p. 29)

Young's misunderstanding of the implications of the paradoxes of subatomic physics – based in part upon his failure to appreciate that science is not coterminous with subatomic physics and that the reality explored by science is not coterminous with the subatomic realm – is, as I have said, common amongst those who would relativise science. Equally common is the habit of proceeding from the observation that imagination plays a role in the generation of scientific

theories to the assumption that science is no different from any other
creative endeavour:

> modern science is not a revelation to end all revelations, merely
> the one that has prevailed over Western man in the past four cen-
> turies. It proceeds through the construction of hypotheses, induc-
> tive leaps that make connections hitherto undreamed of. Scientific
> breakthroughs, like poetic masterpieces, tend to arise in the middle
> of the night, and the genius appointed to transcribe them tends to
> have no idea where they came from . . .
>   In short, the scientist with his model weaves a fictional web not
> altogether unlike the poet's . . . and though the philosophers have
> known this for some time, the movement of modern physics back
> to idealism in quantum mechanics is making this clear to the inter-
> ested layman.
>
> (p. 27)

The intelligent (as opposed to the merely 'interested') layman would
appreciate that the role of imagination in science is crucial but severely constrained. This was made clear by Claude Bernard over a
century ago: 'outside of my laboratory, I let my imagination take wing;
but once I go into my laboratory, I put my imagination away.' It is
the second phase that distinguishes science from poetry: imagination
that distorts results in accordance with the experimenter's wishes leads
to fraud, to 'findings' that others cannot repeat, to failure and, in the
not too long run, personal humiliation. Science, unlike poetry, is a
deeply cooperative enterprise that goes beyond the implicit (and sometimes resisted) cooperation of intertextuality evident in poetic composition. Its theories and laws are advanced in an acid bath of criticism,
proof, refutation, counter-theory. Nothing could be more remote from
the context in which poetry is created and magical webs woven by
poets and/or magicians. Anyone who tried to subject Shelley's claim
that 'Adonais is not dead, he sleeps', to rigorous testing would be as
mad as anyone who tried to derive the technological specifications of
an aeroplane from 'The Ode to the West Wind'.
  Young also notes that theories come and go, go into and out of
fashion and 'old-fashioned theories are even less admired than old-
fashioned poems' (p. 27). In common with many others, he sees this
as evidence that science is not 'a technique whereby Nature is induced simply to reveal herself in facticity'. The claim that science
does not contain fiction is, he asserts, unfounded, merely central to
'the ideological strength of scientific materialism in the 19th century'
and we should, by now, know better.
  Here, his views come very close to those of the Strong Sociologists

of Scientific Knowledge (SSSK), who maintain that the authority of science comes not from the unique access it gives us to the truth about material reality, about Nature, but from the authority of its spokespersons. According to Young, '"models of reality" . . . retain their authority only as long as the relevant experts think they do' (p. 27). The implication is obvious: it is the sponsorship of accredited experts rather than the objective truth of the models that gives the latter their 'authority'.

Like many other ill-informed commentators on science, Young imagines that the succession of revolutions in science – which supposedly proves that science simply lurches from one fashion to another and that those fashions have nothing to do with objective reality – leaves nothing securely in place. As I have discussed elsewhere,[11] the SSSK brigade seem unaware of how much does remain even after the most drastic revolutions. There is, first of all, a huge and irreversible growing body of facts, inductive laws, principles and techniques. Science progresses by accumulation of reliable knowledge. Much pre-Galilean physics remains true and practically useful: what Archimedes discovered about levers and the forces exerted by fluids on solids immersed in them is still true. Newtonian celestial mechanics and optics were not obliterated by Einsteinian relativity and quantum theory. Einstein's relativity theory merely demonstrated that Newtonian theory was true if it was assumed that the speed of light was infinite – an assumption that was accurate for most (terrestrial, macroscopic) purposes. $E = mc^2$ has not wiped out, disproved or even superseded $F = ma$, though it has redefined the conditions under which such an equation may be invariant.

In short, the indubitable fact of progress in science, which may be expected to continue indefinitely, does not mean that science never touches on the truth, or that the authority of its theories come not from the insights they give us into nature but from the collective authority of its proponents which can be explained sociologically. Science, at no stage, has complete mastery of all truth – a theory of everything – but has in its possession an enormous collection of truths, different as to their scope, level and target area. (Indeed, as we shall discuss, it is part of the ethos of science and scientists not to claim to complete mastery of the truth: its essential humility – in contrast with the arrogance of magic – is bound up with its consciousness of ignorance and the provisional nature of some of its laws.) That is why the authority of individuals or groups carries the least weight in science.[12] 'Thou shall not a be-all nor an end-all be' could be the motto of any group of scientists.[13]

Those who assert that science is just another form of herd behaviour and that the authority of discourses has, like that of magic, a sociological not an epistemological basis run into a very big problem: that of explaining why science 'works' – in the sense of enabling us to predict and shape the natural world to an unbelievable degree – whereas magic most certainly does not. This should cause Young to pause more than he does, especially as, like many who relativise the epistemological status of science, he depends heavily upon the technology and findings of 'Western' science – not least to support his argument about the animal origin of certain aspects of human nature and to support his account of the putative journey of mankind from apes to Plato. Without his frequent use (and even more frequent abuse; see below) of science, he would not be able to say where we have come from, how we got to where we are now, what the diagnosis of our troubles is and what prescription he should advise. He uses evidence derived not only from 'soft' sciences, such as primatology, ethology and anthropology, but also from harder sciences such as the physics presupposed in the carbon-dating of artefacts to determine the historical order of the emergence of different cultures and the molecular biology subsumed in the DNA technology used to determine the evolutionary relationships between species. (He cites the fact that 99 per cent of DNA is common to chimps and humans as 'strong grounds for taking our animality seriously', p. 40.) Behind each fact he adduces lies a massive hinterland of scientific knowledge, of concepts, theories and general principles. The carbon dating of artefacts, for example, is built on the notion of 'isotopes' (and hence of elements, atomic weights and atomic structure), of the differential accumulation of isotopes of carbon by living things (and hence the very complex notions of plant metabolism) and the progressive loss of this selectivity after the plant is killed (and hence the notions of radioactive decay and the complex theory behind its measurement). Out of the technique of carbon-dating one could, that is, unpack pretty well all of modern mathematical physics.

To say that Young uses science for his own purposes actually understates the situation: he draws from science statements of such scope that would make any scientist blench; and under his uncritical gaze tentative hypotheses turn into iron laws. The combination of a contemptuous dismissal of science with a snobbish, opportunistic use of terms and notions and facts drawn from science and an uncritical use of what they believe to be the latest views of scientists is typical of many culture critics. They float too far above science to see it clearly and have only the dimmest idea, not only of its actual strengths but also of its self-professed areas of weakness. A couple of examples of Young's over-selling science are worth dwelling on.

In his Introduction, he talks about Wagner and the latter's supposed failure (according to Nietzsche) to keep in touch with the 'physiological presupposition' of earlier music. The word 'physiological' causes Young immediately to don a white coat: 'Our clinical adjective for such a breakage', he tells us, 'is "schizoid"' (p. xxxi). The 'our' refers, presumably, to the community of clinicians with whom he is inward. 'Schizoid' in this context is, of course, almost meaningless. It is, however, a robust adjective: it has survived the 1960s – when anyone who was the slightest bit fashion-conscious included being 'a bit schizoid' among their essential accessories; and it has survived the 1970s – when 'schizoid' was mobilised to describe mere indecisiveness (as in 'I'm a bit schizoid over the Everton versus Liverpool issue'). Notwithstanding its near-meaninglessness, 'schizoid', restored to its clinical authority, has the necessary ring about it to suggest a deeper, more scientific understanding of what Wagner's trouble ('foreshadowed' we are to understand, in late Beethoven) is about.

'Physiological' is also misused, being deployed in a way that would convey no meaning to a physiologist, for whom 'physiological' refers to the functions of the healthy human body that are the particular concern of a certain discipline – namely physiology. It is most commonly used of those processes and systems that serve to maintain the constancy of the internal milieu (vis-à-vis temperature, blood pressure, oxygen saturation, electrolye concentrations, etc.). In a looser sense, 'physiological' may be contrasted with 'pathological', as referring to the functions or properties of the healthy, as opposed to the ill, body. But this loose sense carries more weight than 'healthy' only if there is some justification for retaining the original scientific provenance. Young's use of the term does not have this justification.

There is, of course, no scientific basis for judging Beethoven's late music as lacking 'physiological presuppositions' or (even more problematically) 'foreshadowing' Wagner's lack of them. Young has a kind of train of thought: in the works of late, deaf Beethoven, music parted company with dance; dance is good for you; late Beethoven music is therefore less physiological than the music that went before. This groteque over-generalisation from an individual case might be a whisker less absurd if the individual case were at all convincing. However, although I am a great admirer of the 'Goldberg Variation' and the 'Art of Fugue', I am not persuaded that they are more dancable or walkable than the 'Ode to Joy' or many movements of the late Beethoven symphonies (one of which, the Seventh, was, of course famously described by Wagner as 'the apotheosis of the dance') Young deems unphysiological.

Young's use of 'physiological' epitomises the knowingness that oozes so disastrously through the works of so many cultural critics. For present purposes, it more importantly symptomatises his ambivalent relationship to science, which is at once a source of authority in arguments and at the same time to be rejected, as the source of much of our contemporary ills. This ambivalent attitude seems to license a cavalier attitude to the normal standards of proof. By taking over the rhetoric of science and at the same time despising its discipline, Young falls victim, as so many other have, to argument by association of ideas and simple dishonesty in the use of scientific evidence. Consider this passage of pseudo-science, used to justify Young's case for the reawakening of the 'primitive' within us:

> Recent research indicates that without dreams our thinking condenses and implodes, trying to say several things at once, and we 'dream awake' in the bad sense as schizophrenics do. Thus, dreaming in its proper context is a kind of licensed madness that ensures sanity. The suppression of such madness produces insanity. Primitive myth/ritual should be understood as the socially licensed attempt to clarify, interpret, and detoxify the nightly mad excursion into dream-time ... Although ultimately playful, such work is an utterly serious aspect of the human undertaking; and when a culture forgets this aspect of the human contract with the gods, and suppresses its dreams, as ours does, it too goes crazy.
>
> (p. 85)

The reference to 'recent research' without specification of what that research is and where it is published is intellectually suspect. Suppose, however, that Young is referring to the work that showed that selective deprivation of subjects of REM (or dream) sleep led to psychosis, this would not support his hypothesis. For his argument is that collective madness results not when nobody is allowed to dream, but when they are not permitted to acknowledge, act out, etc., the dreams they do have; and there is no research, recent or otherwise, to support this.[14] The argument is by association of ideas, or of handwaving citations, or of Shining Names. It generates some amusing mixed metaphors: 'As the culture sets, it hardens the arteries by insulating us from any vivid perception of morality' (p. 6). This is not the way of science.

The selective and uncritical use of scientific data, and their interpretation beyond the places where they may be safely used, is most spectacularly exemplified in the scientistic neophrenology that underpins much of Young's thesis about the origins of the sacred and the human propensity to violence. It will be recalled that he accepted as one of his building blocks the triune concept of brain and behaviour, according to which there are really three brains within the primate brain. The imbalance between the three brains, caused by the uncontrolled development of the neocortex, leads to big trouble:

The primate codes that forbid intraspecific violence, defusing it through rituals of dominance and submission, are destabilised in the inventive complexity of the chimpanzee neocortex (as in our own) . . . our boundless capacity for destruction was 'created' when the burgeoning neocortex applied for permission to unlock the instinctual codes forbidding intraspecific violence.

(p. 59)

This concept remains only a concept and very little evidence has supported it since it was advanced over 20 years ago. This, however, does not worry Young. His manner of using science and his respect for the huge body of neurophysiological literature over the last 20 years may be judged by the fact that he appeals to Koestler's quotation from Broca (d. 1860s) for evidence about the 'structural primitivity' of the limbic system versus the neocortex.

What 'primitivity' means non-tendentiously here, God only knows, but these are the kind of terms that would be useful to a neo-phrenologist. Young, like many neo-phrenologists, is also impressed by size: it translates into all sorts of other things. The fact that the neocortex is big means that it is necessarily 'more playful, imaginative, and inventive':

my story locates the devil in the neocortical expansion that unlocked our instinctual primitive codes and opened the door to what the moralists call 'free will', the philosophers 'rationality' or 'irrationality', the poets 'imagination', the psychologists 'consciousness', and so forth. What is important in all of them is to cleave to the paradox the snake so nicely symbolises: we become free enough to be enslaved by passion, rational enough to be utterly irrational, imaginative enough to invent heaven *and* hell, and conscious enough to lose touch with our unconscious.

(p. 122)

Young's use of the science he despises is rather reminiscent of the use that men who despise women make of them. Woman-hatred is often associated not with celibacy and isolation, but with the most frantic promiscuity that causes the maximum of damage. To put this another way, there seems/to a fatal law according to which those who attack science (good science) are prone to be indiscrimate in their support for certain scientific concepts which they half-understand and believe support their own intuitive views – to wit, bad science: anti-science and scientology are separated by a hair's breadth.

Young's pseudo-science may be intimidating when, as is so often the case with exponents of *kulturkritik*, it is combined with moral instruction. The best protection against it is to read Richard Feynman's moving commencement address to the students at Caltech in 1974, 'Cargo Cult Science'.[15] Feynman draws an analogy between the kind of para-science that many humanist intellectuals (not just Young – that is why his example is important) engage in and the cargo cult of some people in Papua New Guinea:

> I think the educational and psychological studies I mentioned are examples of what I would like to call cargo cult science. In the South Seas there is a cargo cult of people. During the war they saw airplanes land with lots of good materials, and they want to make the same things happen now. So they've arranged to make things like runways, to put fires along the sides of the runways, to make a wooden hut for a man to sit in, with two wooden pieces on his head like headphones and bars of bamboo sticking out like antennas – he's the controller – and they wait for the airplanes to land. They're doing everything right. The form is perfect. It looks exactly the way it looked before. But it doesn't work. No airplanes land. So I call these things cargo cult science, because they follow all the apparent precepts and forms of scientific investigation, but they're missing something essential, because the planes don't land.
>
> (p. 340)

Young is not entirely unaware that his attack on science in support of a primitive culture whose nature and unfolding has been revealed by science may lay him open to the charge of self-contradiction. He knows, like the rest of us, that science works not because it has the whole truth about everything, but because many of its claims are true, in the sense of amounting to reliable knowledge. However, his assertion that his book is 'to some extent an exercise in reconciliation, an attempt to enlist science in the cause of mythology and vice versa' (p. 5) is less than convincing, especially in view of his repeated attempts to knock science down to size. In places, he goes beyond merely suggesting that there has been no real gain in knowledge or understanding as a result of the last 2000 years of scientific activity: he suggests that science has actually taken away knowledge. Science has blotted out the memory of the myths 'that legitimize our existence and tell us how to live with godly power' (p. 22). Indeed, he singles out as 'the signal achievement of modern science' effectively 'to "deconsecrate" both space and time'. (How sacred pre-scientific and hence pre-technological space will look when it is wet and bug-infested to an ill-clothed and hungry man or woman is another matter to which we shall return.) 'Scientific theory', he tells us, 'corrupts the mind chiefly through its abstraction' (p. 29). This is a difficult doctrine for a man who has written, and would require us to read, a book of some 372 pages of abstractions of a very high order indeed. It also causes difficulty with his chronology, which I shall come to in a minute. His condemnation of science is scathing:

> Under the dispensation of science, our ears have been stopped up, our moral and religious faculties have atrophied, and our hands rendered incapable of prayer. Because of this stupefaction we have become, from time to time vulnerable to infection by lurid fantasies of salvation.
>
> (p. xxii)[16]

Here the emphasis is on the spiritual, the moral and imaginative, damage caused by science. And in this he joins hands with many other commentators who seem to take up where Galileo's inquisitors left off.

Prominent amongst these is Bryan Appleyard, whose naive book[17] has the virtues of clarity, passion and typicality. As the title of his book makes clear, science for Appleyard is the key to 'understanding the present', in particular, all that is wrong with the present age. Unlike Young, Appleyard does not even try to deny that science is more powerful than magic and does not try to equate the two:

> science has been more successful and effective than any other form of human knowledge, this has made it the primary determinant of our way of life and our attitudes to the world and other people.
>
> (p. xii)

This, however,

> is dangerous because science itself has no morality or faith and can tell us nothing about the meaning, purpose and significance of our own lives. . . . Hard scientific truth denies us a place in the world, an ultimate significance and a sense of the worth of our own actions, it subverts values by insisting upon the contingency of all that we do and are.
>
> (pp. xii–xiii)

Science has caused 'appalling spiritual damage' and there is worse to come because 'Science, quietly and inexplicitly, is talking us into abandoning ourselves, our true selves' (p. xviii).

How is this? Quite simply, science has separated truth and meaning:

Science was the lethally dispassionate search for truth in the world whatever its meaning might be; religion was the passionate search for meaning whatever the truth might be. Science can lay a claim to a meaning in the sense of establishing causality, and religion could claim truth in the sense of a transcendent order. But science's meaning does not answer the question Why? And religion's truth had no scientific relevance.

(p. 83)

The ascendancy of science has resulted in a picture of the world that is itself free of meaning, a spiritually empty universe that appalls the mind and a 'vision of man as a fragile, cornered animal in a value-less mechanism' (p. 112). Scientific knowledge severs truth and meaning because its whole ethos is to eradicate, so far as is possible, the scientist himself from his findings. As Frances Yates has pointed out, this is in sharp distinction to his magician predecessors:

The basic difference between the attitude of the magician to the world and the attitude of the scientist towards the world is that the former wants to draw the world into himself, whilst the scientist does just the opposite, he externalizes and impersonalizes the world by a movement of will in an entirely opposite direction.[18]

For Appleyard, the decisive argument against science is not the physical and environmental catastrophes it has brought about. On the contrary, he thinks the Green Movement, which aims to limit the damage caused by science-amplified production and appetite for consumption, misses the essential (spiritual) point. Even if science could be made green, it would be deeply harmful: the harm is a spiritual harm.[19]

However, Appleyard does accept that magic cannot compete with science when it comes to solving certain problems, for example, a life-threatening pneumonia in a child. In the case of pre-scientific cultures, the cure of the child 'humiliates local wisdom and it is difficult to believe that what remains can continue to be a unique culture in any meaningful sense':

the truth is that all the artefacts, clothes and rituals were diminished the moment penicilin was first administered. Their absolute meaning was made relative by the devastating effectiveness of science.

(p. 9)

Of course, we can protect the culture and local wisdom against this kind of humiliation: 'we can deny them the penicillin and let the child

die in the name of the long-term autonomy of his people's culture'
(p. 8). In short,

> We may wish to say prayers to make our crops grow, but we know
> that fertilizers work better; we may wish to protect isolated cul-
> tures from the inroads of technology, but we know that antibiotics
> will stop their children dying; we may wish to walk to America,
> but we know we must fly.
>
> (p. 69–70)

This is at least closer to the real issues than Young's vague (and
deeply irresponsible) hand-waving about 'mythic medicine' and the
power of sympathetic magic. Appleyard counts the cost of rejecting
science, instead of merely focusing on the supposed spiritual cost of
the scientific world-picture. A spiritual morality that accepts the avoidable death of children needs to have its arguments well sorted out.
And this, alas, is where Appleyard is no more impressive than Young.

First of all, he fails to consider the credentials of the scientific world-
picture. We are repeatedly told how Galileo, Newton, Darwin, etc.
have removed us from the centre of the cosmos. This is nonsense. If
the ideas of these thinkers are true – and Appleyard does not question their truth – then they have not removed us from the centre of
the cosmos: we were never there in the first place. They may have
destroyed the illusion that we were in the centre of the cosmos, but
losing such a fundamental illusion cannot, surely, be regarded as an
absolute, or unmitigated, spiritual loss. Secondly – and in this respect,
too, he is typical of many – he does not consider how comforting or
otherwise were the world-pictures that have been displaced by the
scientific one. Granted that the scientific world-picture diminishes man
in two ways: first, it makes the individual an insignificant, finite event
in an infinite and meaningless universe (Voltaire's 'man is an insect
on an atom of mud'); and secondly, the progress that is built into
science makes each generation only a step towards a receding future
that may despise its relative ignorance. But is this any worse than
most religious world-pictures? According to the Christian doctrine, a
man is not only a 'miserable worm', but a worm that is the plaything
of a capricious God and one, despite his extreme finitude, that is liable to punishment in an eternal afterlife in hell. At least science brings
with it hope – if not of eternal life, then of palliated mortality – which
is more than can be said for religion with its tales of gods with their
infinite and arbitrarily exercised power and the various modes of eternal
damnation.[20]

Nor does science eradicate significance, as so many, Appleyard and
Young amongst them, claim:

Wonder was further and more visibly expelled by the scientific loss of innocence which I have described. Lethal pesticides and atom bombs signalled that, far from being wondrous, our knowledge was little better than demonic ingenuity.

(p. 170)

We shall address the issue of the supposed 'disenchantment of the world' in later (pp. 141-4). Suffice it to say that, although certain modes of wonder and significance (those most approved by churchmen) have diminished, wonder and significance themselves remain alive: sexual desire, parents' love of their children, hope for the future – these are not things of the past. Moreover, there is nothing in science to justify our being less concerned about the mystery and wonder of human consciousness that lies at the base of everything, including science.[21] Appleyard may be right in claiming that science provides 'truth without significance' (though I am not sure that he is); but even if it did, this would not take away the significance of ordinary things. We still live in the brazier of meaning, just as much as our pre-scientific forebears did; but we are privileged that, thanks to science, we are less obliged to spend time in the brazier of unmeaning.[22]

Some of the meanings we have lost in our supposedly 'disenchanted' world are, in short, well lost. And the standard story that the old terrors have been replaced by newer, worse ones should be taken with a pinch of salt. Science, we are told, has made the world less homely. Before science we were at home in the world; and now we are disinherited children cast adrift in a boundless universe of which we are not even an insignificant part. The assertion that science has made us less at home in the world is, of course, rubbish. It is based, first of all, upon an exaggerated estimate of the extent to which we internalise large ideas. To Pascal's assertion that 'the eternal silence of the infinite spaces terrifies me', Paul Valéry, one of the most original thinkers of the twentieth century, retorted: 'the intermittent hubbub in the corner reassures me'.[23] I suppose we should be terrified by the world-picture of science, but we aren't because, like most people throughout history, we are more concerned about the intermittent hubbub in the corner than the eternal silence of the infinite spaces. Man is a metaphysical animal but only up to a point. And that goes for the critics of modernity as much as for the rest of us.

More importantly, the standard story overlooks the reality of our human nature. We are more likely to be at home in the world now than at any other time. Some have actually deplored this, complaining about the extent to which the world has been made over in the image of humanity, so that even the sky at night reflects the glow of sodium lights. And those of us who benefit from science-based and science-inspired technology are less, not more, likely to be invaded by the inhuman now than in pre-scientific times. Our sciences are ways of humanising the world. The world has always been cold; we are simply waking up to ways of making it a bit warmer.

By way of illustration, consider medicine, a practical, middle-of-the-road applied science – a human art made more human by being rooted in biological science. As an art that draws on many sciences – most obviously the biological sciences but through them on chemistry and physics – medicine reveals the essentially human nature of applied science and, since successful application depends upon theoretical advances, of science *per se*. Let us begin our consideration of medicine with some basic considerations, overlooked by those who despise science. Ordinary human meaning is limited in two ways: by death and by suffering. Death is an affair of the body and much suffering is also mediated by it. Human lives are pitched into an inhuman world of nature, and ordinary human happiness is predicated on the health of a body whose properties are remote from ordinary discourse and intuitive understanding. Scientific medicine attempts to mediate between human experience and this body that is at once our home and at the same time a piece of material nature and subordinated to its laws. The great advances in medicine have come from looking at the body from the point of view of a biologist and seeing it ultimately in physical and chemical terms. It is biologically-based medical science that separates the incompetent butchery and poisonous therapeutics of a century ago from the effective surgery and physic of today.

Consider surgery. Every aspect of surgery has been transformed and made less barbarous. Accurate diagnosis does not require opening up the patient, but can be accomplished using new imaging techniques, such as computerised tomography, that show the inside of the body. If this reveals a lesion requiring an operation, the offending object may in many cases be removed using minimally invasive methods deploying the latest technology for visualisation, incision and suturing. Post-operative care – pain control, fluid replacement, nutritional support, the prevention of infection and nursing care – has also benefited from basic biological science. The comparison between the relative civilities of present-day surgery and the brutalities of the past should nail once and for all the idea that high-tech equals nasty tech or, more generally, that there is something intrinsically inhuman about technology. A drip infusing pain relief, thirst relief and nutrition is infinitely less brutal than the natural situation of pain, dehydration and malnutrition.

Even so, it may still seem barbarous to some to approach human suffering through the conceptual framework of biological science, to see the individual in pain or facing death as a dysfunctioning organism exhibiting the consequences of neurochemical or biophysical deviation. And, if this were how a doctor approached a patient, it would indeed be barbarous. In so far as good medicine has an inescapably impersonal and hence seemingly barbarous element, this comes from the very nature of the human body: that, in relation to certain things, we are pieces of matter, subject to the laws of physics and chemistry, which fail in material ways. We cannot escape the fact that there are such things as diseases and that they are impersonal invasions of our lives, cutting across their meanings and purposes with meaninglessness and anti-purpose, bringing anti-meanings, defined privatively in terms of the things they take away from us. Severe pain, nausea, coma, death borrow any meanings they have from us, from the pleasures they deprive us of, the goals they frustrate.

In their critique of the 'spiritual emptiness' of science, Young, Appleyard and others ignore the totally devastating, emptying effect of the needs science sets out to solve: hunger, pain, nausea, cold. And so they miss the fact that, far from condemning men and women to solitude, science brings individuals closer together; for there is no more impenetrable solitude than that of unremitting pain. To be in pain is, as Valéry said, to be forced to pay supreme attention to something we did not choose to pay attention to; it is (to reverse what Barthes said of love) to be in the brazier of unmeaning. It is the assertion of the essentially inhuman nature of the laws under which the human body has to operate. Inasmuch as medicine has an impersonal aspect, it is dictated by the impersonal element in our bodies whose physico-chemical properties may, when they deviate from the norms associated with health, release upon us anti-meanings or limits to meaning in the form of coma and death.

Appleyard, Young and others are persuasive in their critique of the spiritual emptiness of science only inasmuch as they ignore – and so help us to forget – the inhuman experiences that has occasioned the scientific endeavour to control our bodies and their environment. They lack an urgent sense of the needs and hopes that make sense of science and technology. Perhaps they would have written more intelligent and honest books if they had been required to research and write them with unrelieved gout or cystitis or hunger.

Scientific medicine, in short, is an appropriate response to the needs of human beings caught up in the inhuman situation of being betrayed by the infirmities of their bodies. (The habit of condemning human beings for their inhumanity sometimes makes us forget that the inhuman does not always come from human beings.) Good medicine is, of course, not simply good science (just as good technology is not simply good science); nor is it simply just good technology. Medicine addresses impersonal disease processes that afflict human bodies, but it recognises that these processes are suffered by human beings: that what is of ultimate importance is not the dysfunctioning body but the distressed person. And so, in practice, medicine begins and ends with the person: it begins and ends with talk, and talk pervades the entirety of the medical encounter. In the beginning is the history: the patient's complaint, her anxieties, her suffering. 'This is what is happening to me, this is what matters to me, this is the meaning of what has happened to me.' In the beginning, that is to say, is the subject. Next there is the physical examination, to see what can be understood by an informed and sympathetic looking, touching, probing. This still involves the doctor as a human being, mobilising the tact necessary to negotiate this transition and questioning the patient as a subject: does it hurt here? etc. Next, there are investigations, the tests, that address the person as a body being examined: the X-ray that shows the body structures, the biochemical test that shows the body composition, etc. Those tests are objective – there is nothing subjective about my serum sodium levels – nevertheless, doing them needs to be justified to the patient and the results explained to her. And then there is a return to talk: 'This is what I think is happening, this is what it means for your life', etc. Talk, in short, pervades the entire interaction between doctor and patient; by this means, the doctor mediates between science that approaches the individual as a body with certain general properties – and so interprets her suffering as an instance of a generalised mode of dysfunction – and the humanity that recognises her as a person whose experience of disease is unique, impinging as it does on a unique life and a unique human consciousness.

Scientific medicine is the art and craft and human skill of applying general concepts to the diagnosis of types of human ills suffered by individuals who are both instances of general types and singulars. It takes account of the chimeric nature of the human body, which is both a piece of living matter in a material world and a conscious being in a human world; both our ownmost place and the potential source of the most alienating experiences, such as the dinning unmeaning of toothache (the brazier of unmeaning) or secondary deposits in the bones, which both reflect general processes in living matter and are unique experiences in individual persons. Failing to take account of the individuality of disease as well as its generality will leave the patient diminished: bodily whole but spiritually impoverished and prone to seek help from the more personal magic of alternative but ineffective therapies. Good medicine does not fail to take this into account and the higher competence of scientific medicine does not dictate that one should ignore the humanity of the patient.

Pre-scientific medicine also began and ended in talk. The only difference – and what made it less, rather than more, human – was that, not being based in biological science, it didn't pass through impersonal, objective science and so didn't work. The operations killed or maimed, the drugs poisoned. Magic may have had the advantage of offering a more numinous conception of the world and of disease, making disease more continuous with the soul and human institutions, but its results were disastrous. There is nothing more inhuman or deceptive than untried and ineffective remedies, than medicine that doesn't work.

Of course, there are still many diseases which contemporary scientific medicine cannot cure; and there are many forms of suffering that may present to the doctor that have nothing to do with the dysfunction of the body. Doctors employ science to establish the incurability of the disease or its lack of physical basis; but still mobilise their art to help. They see their role as being to alleviate distress where they can and to witness it, to interpret it, to sympathise, where they cannot. The goals of medicine remain what they always have been: cure some, improve many, comfort all. The primitive medicine of the magician often did precisely the opposite: kill some, worsen many, and terrify all.

The ethos of scientific medicine is captured in a recent report to the British Royal College of Surgeons on the future of surgery. After a discussion of the major advances expected in the practice of surgery arising out of new diagnostic and operative methods, the report adds:

> Not everything will change beyond recognition. The human body will remain susceptible to disease, injury and starvation. Afflicted human beings will remain frightened of such events striking themselves or their families. Whatever revolutions there may be in diagnostic techniques and methods of treatment, patients will still look for reassurance, comfort, understanding and dedication in those looking after them. Not even the most sophisticated computer will replace the reassuring voice, the squeeze of the hand, the sympathetic ear, the eye to eye contact that only humans can give to one another. The challenge for tomorrow's medical practitioner ... will be to cope with changes that inevitably occur in a manner that uses the advanced technology available in combination with an essential humanity that has been the hallmark of the medical profession since the time of Hippocrates.
>
> (Sir Miles Irving, personal communication of an unpublished report)

This is only the continuation of the immemorial challenge that has faced clinical science: to mediate between the transhuman vision of science and the human needs of patients: between two equally valid visions of the human body – one terrifyingly remote and objective arising out of pain and the fear of death, and the other reassuring and subjective – and to labour on behalf of the latter against the former. This is medicine at its best. (Which is not to say that it is always at its best, just as literary criticism is not always at its best: both are prone to corruption, abuses, injudiciousness, inhumanity.) Even at its best, medicine does not solve the human condition, but it returns us from the dark places of the body to daylit places where we can resume our normal lives and even start start thinking about the human condition: to normal Tuesdays.

In summary, then, medicine, which has in an extraordinarily short time been immensely empowered by biological science, challenges the prejudice that science is intrinsically dehumanising, and overturns the fatuous idea that science is inherently non-human, inhuman or even anti-human. Like the human hand, science, science-based technology, and even science-based medicine, may be used for inhuman, as well as for human and humane, purposes – to stroke rather than to caress; but, like the human hand, the inhuman and inhumane uses to which it may be put are not intrinsic to it.[24]

Anyone who would prefer the magician's version of medicine is welcome to it. Before they consign others to the tender mercies of the omniscient witch doctor, or even give themselves up to longing for the enchanted world of the pre-scientific era, they might do well to think of the little girl with the hole in her bladder, nearly destroyed by 'mythic' therapy and saved by dull conventional medicine and surgery. They should recall the mythic (or criminally incompetent) midwifery that put the hole there in the first place, the pollution myth-driven post-natal care that nearly killed her and the further pollution myths that turned her terrible misfortune into grounds for expulsion from the bosom of her warm organic community into the world at large. And they should compare this with the rational, science-based care and kindness that repaired the hole in her bladder and fitted her to be received back into her warm, organic community and, alas, to return to the bed of the all-powerful elder who had been chosen as her husband.

More generally, we need to recall the needs and hopes – the sufferings and woes served up by Mother Nature – that make sense of science and technology.[25] Science, Appleyard tells us, 'subverts values by insisting upon the contingency of all that we do and are' (pp. xii–xiii). This is simply untrue. As for values, they come from our needs and hopes, our longings, our fears and our dreams. We don't require science to look after them for us – any more than we need science to tell us that we live only once and that, during that time, there are things, like the life and death of our children, that matter infinitely to us. The conscious self that underpins so-called 'value-free' science is inescapably linked to value.[26]

This should seem quite obvious, but humanist intellectuals such as Young, Appleyard and a million others whinge that science doesn't tell us how to live or what to live for. And its silence, apparently, is infectious:

> modern liberal-democratic society has been created by the scientific method, insight and belief ... and ... both this society and science itself are inadequate as explanations and guides for the human life.
> (Appleyard, p. 228)

That science does not pretend to be an explanation of or a guide to human life is actually a saving grace, a virtue. It is connected with its essential humility which, in turn, is a source of its great strength (see note 18). Besides, it would be terrifying if science did provide complete explanations; if out of its general laws there came a universal, eternal prescription for life. Then, surely, we should have serious grounds for anxiety, for complaining of its moral, intellectual and even metaphysical tyranny.[27]

Anyway, the complaint that science has nothing to say about what we are and how we should live is nonsense. It has a lot to say about what we are (even if there are many people who don't like what it says): physiology, molecular biology, physics are not entirely irrelevant to understanding some things about ourselves. These are not, of course, the last word on our nature (as I have argued at length in my book *The Explicit Animal*). Science is far from comprehending the very things that are most distinctive about us – that we are conscious and that, if we can be thought of as matter, we are very peculiar forms of matter to whom things matter – but it does have a lot to say about how we should achieve many of our most important intermediate goals: avoiding pain, cold, starvation, postponing death and disability, etc. But, to repeat, it doesn't have the last word. Science doesn't tell us what our ultimate goals should be. But then neither do literature, painting, music, sculpture ... Nor are myths very helpful in this regard.

Appleyard's scathing attack on what he sees as the hubristic foolishness of science is the other side of Young's attribution of wisdom to magic and myths. Together, they stand for the numberless choir of Jeremiahs who find modern life spiritually and morally empty and who cannot link the word 'civilisation' with 'Western' without putting the former in lip-curling inverted commas. Neither seems unduly concerned about the major difference between magic and science: that the latter works and the former does not.

Let us take a poignant recent example: the eradication of African river blindness. For thousands of years, endemic onchocerciasis has meant catastrophic levels of blindness in certain areas of western and equatorial Africa. Until recently, there were villages where the sighted were a privileged minority in a country of the blind. The strongest mythic medicine, based upon the most elaborate mythological explanations – God blotting out the sun as a punishment, etc. – were unavailing. About a century ago, Western scientists found a non-mythological explanation: the blindness was due to a larva transmitted by a certain kind of fly that had a predilection for areas near the great rivers. And more recently, scientists, guided by an understanding of the respective vulnerabilities of the human body and the larva, have devised treatments. This, combined with non-mythologically driven public health measures – homely advice about hand-washing, etc., to break the chain of transmission – has started to eliminate the problem. Villages that were once colonies of the blind now have populations consisting almost entirely of sighted people. Whilst the earlier mythological explanations may have had more expressive force than the rational and factually accurate accounts of the scientists – though the notion of an angry god blotting out the sun is not very comforting, as well as being factually wrong – they did nothing for the health and happiness and the richness of the life-experience of the people.

Magic may be fun for the magician, but it is based upon error, leads to incompetence and is fundamentally fraudulent. The magical and scientific accounts of the problem are not, therefore, equally valid, and it is utterly heartless and irresponsible to suggest that they are.[28]

It is also rather dim in an important sense, which Appleyard and Young should acknowledge. Both, as we have seen, make much of the 'loss of wonder' that has accompanied the rise of science. This may account, perhaps, for their own failure in this regard – their insensibility to the deep mystery of science – the miraculous capacity of man, the explicit animal, to entertain notions that have such extraordinary reach and power. For all their talk of 'the recovery of wonder', Young and Appleyard seem exceptionally impervious to the wonder of scientific achievement. Their scornful, knowing, cynical attitude – utterly conventional among culture critics – shows a striking lack of that very astonishment, reverence and awe which they wish to restore to us. Anyone with the slightest hint of those 'primitive' virtues of wonder and piety would be too overwhelmed by the pace and magnitude of the achievements of science to be able simply to 'place' and dismiss it. Consider this:

About 400 million years ago, the first aquatic vertebrates evolved; at least 2 million years ago man's ancestors first chipped stones to make simple tools. Less than 10 000 years ago, in the neolithic revolution, animals and plants were first domesticated. If a film, greatly speeded up, were to be made of vertebrate evolution, to run for a total of 2 hours, tool-making man would appear only in the last minute. If another 2 hour film were made of the history of tool-making man, the domestication of animals and plants would be shown only during the last half-minute, and the period between the invention of the steam engine and the discovery of atomic energy would be only one second.

(Maynard Smith, quoted in Tallis, *The Explicit Animal*, p. 180)

This at least makes visible the two miracles that are most spectacularly expressed in the mighty achievements of science and technology: the great mystery of the partial intelligibility of the world (the greatest of all mysteries according to Einstein); and the almost equally great mystery of the peculiar status of man as a uniquely 'explicit animal', in the sense I have expounded in my book of the same title, able to elaborate that partial intelligibility. And there are many lesser mysteries; for example, how it is that, in the sphere of science, men have, almost uniquely in human affairs, been able to overcome their natural idleness, charlatanry, irrational opposition to one another's views and develop a collective, robust and effective view of the world. How did they manage to set aside the numberless and innumerate prejudices of common sense and produce (to use Gellner's description) 'a cumulative, trans-cultural science'? Young and Appleyard do not ask these questions because their conventional hostility to science makes them overlook its extraordinary achievements. They are unastonished at the mysterious fact that human abstractions can penetrate the opaqueness of the material world, permitting human beings to fly at 600 miles an hour, to send their voices tens of thousands of miles through space, to delineate the precise nature of the diseases that inflict their bodies, and so on.

Nor, finally, do they ask why science has achieved such an ascendancy over myths. Could it be that myths and magic declined because they did not achieve what people would like them to achieve? Could it be that science is more powerful than magic? And if the answer to these questions is yes, we have to ask what it says about the relative truth of science and myth and about those who urge us to abandon the former in favour of the latter.[29]

THE RE-ENCHANTMENT OF THE WORLD

The Disenchantment of the World

In his lamentations over 'the deconsecration of space and time', Young is far from being eccentric, or even original: he merely adds one voice to a vast chorus of lamentation over 'the disenchantment of the world', to use the phrase that Weber took from Schiller. For Weber, the phrase encapsulated the general trend of secular rationalisation. We have already noted Adorno and Horkheimer's strictures on the Enlightenment – how, by overlooking the irrational and creating a programme for dominating nature, it made the world into a prison, creating a closed universe in the grip of an instrumental reason which invalidates anything outside of itself, desacralising nature, human society and individual human beings. In the place of a magic, numinous world, in which anything was possible and meaning was unbounded, we have a mechanical nature composed of matter obedient to a handful of laws that have no human meaning and a rational society based upon contract in which the fulfilment of our wishes has left us 'bemused and blinking in the cluttered playroom' (p. xv). Modern man, Aldous Huxley complained, 'no longer regards nature as being in any sense divine, and feels perfectly free to behave towards her as an overweening conquerer and tyrant'.[30]

The journey to this pass had two stages. First of all, God was driven to the edge of the universe. This was, in fact, a temporary rehabilitation – better to be the all-wise, hidden clockmaker than the vengeful, interfering, petty-minded God invoked by the priests – and the work of pious scientists. As Diderot observed:

> It is not the hand of the metaphysician that has struck the sharpest blows against atheism. The sublime meditations of Malebranche and Descartes were less capable of shaking materialism than a single observation of Malpighi's. If this dangerous hypothesis is weakened in our day, the credit is due to experimental physics. It is in the works of Newton [and others] that we have found satisfactory proofs of the existence of a supremely intelligent Being. Thanks to the works of these great men, the world is no longer a god. Instead it is a machine, with its wheels ropes, pulleys, springs and weights.[31]

Believing in the all-wise clockmaker hidden in the clock was an act of faith, and a rather arduous one at that. This was less and less possible in a rational society. Soon, the Divine Clockmaker was no longer able to withstand the cold climate of contractual society. Science and capitalist economies have in common their rationalism. No God could survive their combined onslaught. Goldmann again:

By the 18th century, the area of bourgeois life occupied by rational thought and action had grown so far that the *nature* of the question had altered. It was no longer a question of the place to be assigned to reason in a life built on faith, but rather of what place there could be for faith within a world vision grounded on reason.

(p. 54)

For Weber, capitalist economies had the more potent deicidal effects. In an industrialised society, dominated by mass production, rational calculation is essential and this, inevitably, extends beyond production and consumption, beyond the necessary getting and spending. Rationalisation means 'that there are no mysterious incalculable forces that come into play, but that one can, in principle, master all things by calculation': every economic rationalisation of a barter economy has a weakening effect on the traditions which support the authority of a sacred law.[32]

In short, secularisation – not perhaps as complete as is usually assumed by cultural critics – was due more to industrialisation and economic changes (which until the last century or so largely took place independently of science) than to scientific discoveries about the nature of the world. Bryan Wilson (quoted in Wolpert, p. 29) emphasises how secularisation is less a matter of loss of belief, than 'a shift of power from elites with access to supernatural ordinances, to other bases of power, a shift dictated by economic forces'. Although technological advance made a major contribution, these, until recently, came, as Wolpert convincingly argues, more from 'the blacksmith's forge than from the Royal Society'. By the time science started to make its dramatic impact on the speed and direction of the evolution of technology (round about the time of Michael Faraday's great discoveries), the economically-driven process of secularisation was well underway. So much for Appleyard's claim that 'science, quietly and inexplicitly, is talking us into abandoning ourselves, our true selves'. If anything can be blamed for the secularisation of the Western world, it was Protestantism and its work ethic and doctrine of salvation that led to the worship of productivity, the admiration of material success as proof that one was one of the elect rather than the damned, and the rationalisation of economic activity. It is very difficult to assume a rational, disenchanted mind-set for six days a week and recover an enchanted world on the seventh.

It is important, however, not to exaggerate the scale of the change and, before we proceed to examine the calls for a re-enchantment of the world, we should question:

1. how enchanted it ever was, before the combined onslaught of bourgeois society, capitalism, craft-driven technological advance and,

latterly, science and science-based technology drove God into exile
and emptied the air of angels; and
2. how disenchanted it is now.

In relation to the first question, it is evident that Young imagines a
golden age before 'the rift that opened in the Western soul some 400
years ago when science and religion went their separate ways' (p. xvii)
when 'every grove and stream, the earth and every common sight'
did seem 'apparelled in celestial light'. In common with many cultural critics, he is satisfied with remarkably little evidence to establish this. He quotes *A Midsummer Night's Dream*:

> We have laughed to see the sails conceive,
> And grow big-bellied with the wanton wind.

and asks us to 'imagine how much more fun a day's sailing was when
we could experience the world in this way!' (p. 175). It is difficult to
refrain from sarcasm. How the Elizabethan sailors – malnourished,
overworked, brutally treated, usually ill and often in extreme discomfort
and frequently in danger – must have enjoyed their sailing and how
much more as they swapped metaphors! To place Young's Arcadian
fantasies in their true light, it might be as well to look at an historian's view of the 'enchanted world' as it looked a decade or two after
Shakespeare wrote *A Midsummer Night's Dream*:

> The insecurity and discomfort of life encouraged irresponsibility in
> the ruler. . . . The impact of war was at first less overwhelming than
> in the nicely balanced civilization of today. Bloodshed, rape, rob-
> bery, torture, and famine were less revolting to a people whose
> ordinary life was encompassed by them in milder forms. Robbery
> with violence was common enough in peace-time, torture was in-
> flicted at most criminal trials, horrible and prolonged executions
> were performed before great audiences; plague and famine effected
> their repeated and indiscriminate devastations.
>
> The outlook even of the educated was harsh. Under a veneer of
> courtesy, manners were primitive; drunkenness and cruelty were
> common in all classes, judges were more often severe than just,
> civil authority more often brutal than effective, and charity came
> limping far behind the needs of the people. Discomfort was too
> natural to provoke comment; winter's cold and summer's heat found
> European man lamentably unprepared, his houses too damp and
> draughty for the one, too airless for the other. Prince and beggar
> were alike inured to the stink of decaying offal in the streets, of
> foul drainage about the houses, to the sight of carrion birds pick-
> ing over public refuse dumps or rotting bodies swinging on the

gibbets. On the road from Dresden to Prague, a traveller counted 'above seven score gallowses and wheels, where thieves were hanged, and the carcasses of murderers broken limb after limb on the wheels'.[33]

So much for the not-yet-disenchanted world and the not-yet-deconsecrated space and time.

In relation to the second question, as to how disenchanted the world is now, we shall content ourselves for the present with noting that, according to Wolpert (p. 32), 75 per cent of people surveyed in the United Kingdom believed astrology to be scientific; and there are nearly 200 000 part-time and full-time astrologers in the United States. We shall return to this matter in due course.[34]

Religion: the Impossible Solution

One remedy for a disenchanted world is to restore its enchantment by tapping into immemorial sources of enchantment: the religions that have sustained mankind in the past, the belief-systems to which our pious ancestors subscribed. This is, of course, not as easy as it sounds. As Kierkegaard said, it is one thing to stand on one leg and prove the existence of God, quite another to fall on one's knees and worship Him. And it is one thing to prove that it would be a good idea to have religious beliefs and quite another to be overwhelmed or possessed by specific convictions. Religious beliefs cannot be requisitioned to order, either to solve the individual sense of emptiness and division or to provide the steadying influence that a potentially violent society requires. Bryan Appleyard, who toys with the idea of 'a return to orthodox religion' as a means of 'humbling' science, seems to realise this, though he does not fully appreciate the implications of the passage that he cites from Max Weber:

> The need of literary, academic, or café-society intellectuals to include religious feelings in the inventory of their sources of impressions and sensations, and among their topics of discussion, has never yet given rise to a new religion. Nor can a religious renascence be generated by the need for authors to compose books, or by the far more effective need for clever publishers to sell such books. No matter how much the appearance of a widespread religious interest may be simulated, no new religion has ever resulted from the needs of intellectuals nor from their chatter.[35]

The pragmatist/evolutionist argument that beliefs are true in so far as they serve our needs can hardly be translated to the realm of resurgent religiosity. Religious belief is not simply there to be taken off the shelf when required, irrespective of whether the occasion of need is a deep hunger of the soul or a more superficial desire for an intellectual accessory. It follows from the very historical processes, the sea-changes in the human spirit, that Young and Appleyard deplore, that religion is not there for the taking, cannot be assumed like a robe to cover our spiritual nakedness. Goldmann again:

> the development of the bourgeois has for the first time in history produced not merely a class that has generally lost its faith, but rather one whose practice and whose thought, whatever its former religious belief, are *fundamentally* irreligious in a critical area [economic activity] and totally alien to the categories of the sacred.
>
> (p. 55)

It is perhaps an oversimplification, and certainly an exaggeration, to suggest, as Goldmann does, that *homo economicus* has replaced the category of the right and wrong with that of the successful and the unsuccessful, but this does underline the distance that bourgeois society has travelled from a community dominated by the non- or counter-utilitarian category of the sacred. For this reason, it is simply not possible by will, by fiat, to restore the collective faith: the entire climate of assumption is alien to it. Religion cannot be restored as an occasional episode, a set-piece of reverence or frenzy, of worship or sacrifice, an aria of the sacred planted in or growing out of a recitative of the profane – if only because the religious episode presupposes the religious world-picture. The religious domain is empty if it is not a factor in every human feeling and common event. It is not possible to pass six days of the week in a rational, contractually based world, regulated by the usual sense of justice and fair play and then on Sunday worship a God who lives by anything other than fair play. As Goldmann says, once people started judging God as a person, He was doomed as a serious focus for human passion and belief, as the place where all our hungers, sorrows, appetites and fears could meet and be transfigured into wholeness of meaning.[36]

Religion cannot, in short, be embraced on rational grounds – just because it would be good for society (or for the hole in the ozone layer) if we were all a bit more pious and generally theophilic and prone to see every living thing as a theophany – precisely because religion has nothing to do with reason and, more often than not, is an outrage to practical reason. Religion can be embraced only because it is an engulfing truth that embraces oneself. 'God', Nijinsky said, 'is a fire in the head.' It cannot be revived as a prescription for the salvation of secular society, because it can truly survive only in a world glistening with superstition: its specific beliefs make sense only in a

wider reality that is experienced as a fabric of symbols – symbols
accepted rather than asserted.

Moreover, we may not actually approve of the moral and social
content of the religious beliefs that are available for us to take 'off
the shelf'. Religion may be not so much above and beyond us, ex-
ceeding the reach of our supposedly inlapsed imaginations, as beneath us:

> although the doctrines of Heaven and Hell, of Original Sin and the
> Last Judgement may have some nominal significance for us, they
> are no longer part either of our moral or intellectual reality. The
> fantasies which were once expressed in Christian demonology and
> in Christian visions of Hell have been progressively relegated to
> the thriving sub-cultures of satanism and science fiction, of horror
> comics and pornography. In the dissociated post-religious culture
> which has ... been brought into being, Christians and traditional
> humanists alike are often unable to bring themselves to believe that
> the very forms of fantasy they have been conditioned to revile once
> lay at the orthodox heart of the religious tradition which our cul-
> ture still tends to revere. Our modern cultural predicament has been
> most succinctly and poignantly expressed by the novelist John Updike:
> 'Alas we have become, in our Protestantism, more virtuous than
> the myths which taught us virtue; we judge them barbaric'.[37]

Young, who argues most strenuously for requisitioning religious
belief – 'modern man must return to the mythic voice if he is to heal
the divisions in his soul' (p. xviii); 'we must put ourselves back to
school with our forebears, to recall the myths that legitimize our existence and tell us how to live with godly power' (p. 22); etc – seems
particularly vulnerable on this point. If we no longer believe in the
immemorial myths, and if the scientific approach commands so much
authority (though very few people subscribe to the kind of metaphysical
scientism that Appleyard and Young seem to think is pandemic), and
if the social context in which religious belief is possible has now passed,
then there is no pharmacy available to dispense his prescriptions –
even if they were not dangerous or foolish or empty.

Young's book opens with a long and approving discussion of
Wordsworth's famous 'getting and spending' sonnet and he agrees
with the poet – as numerous others have done – that we would prefer to be pagans 'suckled in a creed outworn' than 'to live, as most of
us do, unmoved by beauty' (p. xv). The trouble is that, even if we
did prefer to be such pagans, the breasts from which we suckled our
creed would dry up at once if they discovered that we regarded the
creed as 'outworn'. In common with many writers who have called

for a return to earlier forms of belief, he wobbles between unconscious self-contradiction and downright insincerity.[38]

Fully to appreciate this conflict – through which Young's thesis tears
itself apart with at least some of the mutilating frenzy of the hunting
pack – it is necessary to step back a little and look at the interrelationship between attempts to determine the nature of religious experience and belief on the one hand and our view of the true nature of
humankind and its position in the universe on the other.

We may, roughly, adopt one of two sorts of attitude to religious
belief: we may see it as a means of access to the truth about sacred
realities that lie beyond us; or we may see it as an essentially human
institution, as a product of human society or the human psyche, as a
symptom of the endeavours of human beings to resolve their internal
tensions – for example, as a means by which solidarity is maintained
between members of a society, or as a compensatory device that makes
limitation, frustration and suffering, the gap between how one would
like things to be and how they are, tolerable. The former approach
emphasises the referents of religious belief – the real things (*realissimi*)
it is about; the latter focuses on its causes – the material or cultural
causes that prompt human beings to believe in such things. In the
latter case, trying to understand religious belief is, effectively, explaining
the aetiology of an illusion; in the former, the posture is one of gratitude for access to truth. The aetiological approach – which derives
from Enlightenment thought, from Biblical criticism, from Feuerbach
and Marx, as well as from Darwin and Freud – looks backwards and
inwards to the intra-human (historical, cultural, psychological) origins of belief; the referential approach looks forwards and outwards
to the objects that the belief is about and which have inspired it. These
different views of, and attitudes towards, religion are deeply incompatible. And they are connected with different notions of the place of
humankind in the universe:

1. The *aetiological* approach typically sees religious belief as a stage
in the development of the human species which it will outgrow or
has outgrown. It assumes that human beings are, in fact, metaphysically alone in the universe, acting according to meanings they create
and not according to a divine prescription. Religion may (or may not)
be a useful tool for living, but life itself does not have any divine or
even preordained purpose. The decline of religious belief is a 'mature' recognition of this, a sign (to expropriate Kant's analogy) that
man has shaken off his self-imposed minority.

2. The *realist* or *referential* approach to religion assumes that human
beings are not metaphysically alone in the universe, and that, far from
choosing their meanings and destiny, have a pre-fixed essence which
is enacted or unfolds during their life. Life itself has meaning as a

testing ground for the soul. The result of this test will determine the soul's destination. The decline of religious belief is seen to be a fall from fundamental truth, a metaphysical, even ontological, catastrophe for mankind.

It is worthwhile emphasising again how these two views of religion – the one historicising, psychologising, anthropologising or sociologising, the other granting belief a status that transcends humanity, indeed nature itself – are absolutely incompatible. Let us now look further at the question of the nature of human beings and the relationship of this to the nature of religious belief.

For much of our lives the mystery of who – or what – we are collectively does not trouble us individually. The great question of the nature of humankind is often occluded by the small question of who I, this individual, am. This smaller question is not positive or absolute, but comparative; it focuses on self-image and and is concerned with self-esteem; it is subsumed in our personal ambitions, our hopes and angers and jealousies, our projects, our responsibilities, our dreams of happiness; our preoccupation with what we are going to achieve, enjoy, become. So, although the question of what I, a human being, am *qua* human being should be the deepest and most preoccupying of all questions, it is usually pre-empted by the sense of what I, this human being, am. We are steeped in implicit answers, incorporated in our long-term and immediate relations to others, in our *curricula vitarum*, in our social role and status. Thus the answers to the little questions blot out the big question. Any question about our ultimate, collective destiny as human beings tends to be neutralised by the continually renewed alibi of a medium-term future, which, like a visual field, has no clear end, fraying only into receding dusk of old age and death: we think of ourselves in very specific terms that are at once incorrigibly personal and fragmentary. And much of the time, even the narrow questions about who and what we are do not trouble us: we simply get on with doing what we seem to have to do or be, with enacting an agenda that we in some small degree choose and to a great extent have thrust upon us, albeit often in part as a consequence of our earlier choices.

Every now and then, however, we 'surprise a hunger for more seriousness in ourselves' (to use Philip Larkin's phrase) and the question of who or what we are in a universal sense crystallises out of our smaller-scale, shifting, deliquescent or nascent senses of what and who. The question of ultimate ends, the Purpose to which all our purposes tend, has been the traditional context in which the Great Question as to our nature – our collective nature, our species being – has become explicit. The ontological question of what I am bears within it the germ of a teleological riddle – the question of what I am for –

what is the point of all this effort, this joy, this suffering – and even of my destination.

It shares the fragility of all our metaphysical questions, and has difficulty commanding sustained serious attention. It is threatened from without: it can scarcely survive the ring of the telephone, recalling us to our responsibilities and our customary small change answers. But it is also threatened by extinction from within, by the answer – implicit within its being asked as a question that has some point – that neutralises it. The presupposition of purpose and destination carries within it the idea – and answer – of a Being who has in His inscrutable keeping the purposes and destinations to which we tend. Our destination is Heaven or Hell – or some state more appropriate than our present condition to our worth and deserts – and our purpose is to come to an understanding of this and, even, to worship God. The point of our lives is to earn, in this Vale of Soulmaking, one or other of those destinations. The too ready availability of the answers inside the questions makes it possible to subscribe to the answers without feeling the questions; to take the answers off-the-shelf; and, for those who subscribe to organised religions to meet the questions (or answer-neutralised questions) by appointment on a regular basis (for example, weekly, at 11 a.m. on a Sunday).

In the last few centuries, for some of the reasons we have already discussed, there has been an astonishing secularisation of Western thought about the nature of man. This may be seen, as among other things, a switch from an emphasis on the definition and understanding of humankind in terms of its purpose or destination to one which focuses on man's (material) origins and his psychological and social health. This secularisation fits with the switch from the referential to the aetological interpretation of religious belief and is necessarily mirrored in a revolution in the way human beings, human lives, are understood. Humankind is no longer thought of as something made for a purpose, but as something that happened with as little, or as much, purpose, as anything else in the universe. Indeed, with a shift of emphasis from the why to the how, the interest in purposes has died away and, with it, not only the positive sense of purpose but also some of the need to ask questions about purpose. The soul, no longer something drawn upwards to God, is something that is shaped and driven from behind by its past, the past of its primitive forebears, the past of the animals from which it derives its material body and its instincts, the past of the particular piece of matter of which it is composed. There is, of course, nothing new about secular definitions of humankind: after all, Aristotle defined man as the 'political animal'. What is new is the assumption that the secular definition captures the whole of man.

There has been a turn, then, from a destination-based to an origin-based explanation of human beings. The search for the answer to the question 'What (ultimately) am I?' has switched from the theological to the psychological (our selves are the product of the early conflict between our prehistoric wishes and their finite satisfaction, between our infinite, anti-social desires and social constraints); to the anthropological (we are the late echo of early man); to the ethological (our souls are made of the tropisms that have come from our animal forebears); to the biological (we are rooted in evolutionary inheritance of our brains); and even to the physical (we are polyphasic systems in dynamic equilibrium acting out the biophysical consequences of the duplication of DNA and its disposable soma).

And this is how Young runs smack into a brick wall of self-contradiction: his religious interpretation of our destiny tries to root itself in a secular-scientific, cultural, material account of our origins – and, indeed, of the origins of our religious belief. A comparatively minor self-contradiction – and one that we have already noted – is that, despite his hostility to (secular) science, Young relies heavily on scientific evidence to support his argument about our spiritual needs. A much more serious self-contradiction is that his case itself advances an interpretation of religious belief and the notion of the sacred that lies within the very scientific framework he hopes that religion will overthrow and within the humanist framework that religion as he envisages it should transcend. He roots religious belief in a nexus of material, biological, sociological and psychological causes. His understanding of religion is precisely the one that would invalidate its referential truth and refer it back rather to its causes: his vision is aetiological rather than realist. The materials that, according to Young, lead to the formation of religious beliefs do not themselves seem to partake of the transcendent. We acquired our religious beliefs not from a revelation granted by God – they are not the consequence of God's self-epiphany – but from apes. Let us remind ourselves how Young saw this happening.

The purest element of the sacred is *pneuma*: wind or breath. The wind, Young tells us, is perhaps man's oldest divinity:

> In the beginning supernatural power is experienced by primitive man as energy that interrupts the normal flow of events – an obvious example is the thunderstorm. . . . By degrees, through the use of ritual and sympathetic magic, he seeks to harness this power so that it may animate and sustain the fabric of human orderliness we call culture. . . . Invisible, ubiquitous, unpredictable, formless, the breath of life may well convey our first intimations of divinity.
>
> (p. 311)

These early intimations of 'divine energy' were aroused in hominids and even chimps before they awoke in primitive men. The metaphysics of wind was the purest intimation of the divine. As we know, there were less innocent epiphanies: the *pneuma* was also invoked in the ecstatic frenzies of the hunt and the mutilation of living flesh, or, closer to home, in the hideously literal eucharist that took place when the prey and *pneuma* became merged after the alpha-shaman's fatal leap into the magical space of the dance. According to Young, the transformation of these elements into clearly developed religious experiences and beliefs depended upon the evolution of the brain: the birth of the gods was at least in part an affair of the neocortex.

If we are to make this aetiological account of religious belief (in which it is a mere epiphenomenon of neocortical activity) into, or compatible with, a referential account, we must assume that the neocortex is a kind of squarial tuning us into pre-existent transcendent entities. However, Young does not even attempt to rescue theological realism in this way. Indeed, he omits all discussion of the cognitive content of religion, of the truth (or falsehood) of its specific assertions about the existence of gods, their properties, propensities, etc. And this should be a matter of great concern to him because everything else he says about the origins of religious belief undermines its referential truth. Young's story of how the encounter with the wind, a desire for self-aggrandisement and life more abundant, sexuality, aggression and bloodlust are transformed into something that is recognisably religious belief, even if it were true or plausible, would give one little reason for believing that religion has anything to do with the truth about anything. No details are given about the process of transformation, but there seems to be little to inspire confidence in the validity of the cognitive, referential component of something that is so deeply rooted in pure affect, pure hunger, pure need. Quite the reverse: the misconceptions of apes and the bloodlust of hominids seem highly unlikely to give one access to eternal truths that are relevant to us now.

Young's advocacy of a deeper spirituality – encouraging us to 'suckle on pagan creeds outworn' – is, in summary, hardly strengthened by his suggestion that the content of spiritual belief is simply an elaborated delusion. His causal-material account of religion is scarcely a reasoned defence of religious belief. In particular, the relationship between man and god is seen exclusively from the viewpoint of man. Young shows us religious belief as the internal affair of a creature whose brain, whose emotions and whose social and physical conditions are evolving in such a way as to favour its emergence. The cognitive objects of religion scarcely exist; or do so only as the cognate or internal accusative of man's impulses, needs and desires. In short,

Young is in the position of trying to derive a spiritual destination consequence from a material cause analysis.

It is scarcely surprising, therefore, that, in common with many others who call for a re-enchantment of the world, he sees the return to religion as necessary only to meet needs internal to the human race. The scandal of impiety is understood in terms of the damage it does to our health rather than the offence it causes to the gods. Young urges us to retrain our sense of the sacred, not because we have a profound duty to give the gods their due so that we shall not upset them (and so risk punishment), but because it is good for the physiology of our soul and the hygiene of our spirit. This pragmatist position will be unacceptable to those who stubbornly adhere to the view that the best basis for holding a belief is that it is true, not that it is good for us to hold it. Few would be willing to try the difficult task of holding a belief because it would be good for us to do so, irrespective of whether this belief is actually believed to be true.

To summarise: the question of our origins tends to be a secular one, inasmuch as it tends to be understood in terms of material causes, biological ancestry and historical or cultural evolution, whereas that of our destination tends to be a theological one, insofar as that destination is interpreted spiritually. Young is in the contradictory position of trying to re-spiritualise us through emphasising the secular – and adaptive – origins of our beliefs in the sacred; of trying to restore to us a sense of spiritual destination on the basis of our material origin. He leaves undiscussed the crucial question of whether we are to take the religious beliefs he advocates cognitively – or merely affectively or politically. However useful it may be for social or even atmospheric hygiene (Young, it will be recalled, suggests that we should consult primitive man for 'a dose of the mythic magic that may help us to repair the hole in the ozone layer'), this possible usefulness cannot be sufficient to persuade the human collective to recover faith in such a complex and difficult set of beliefs as characterise the religions Young singles out for praise. We cannot believe in Osiris, or any other descendant of the *pneuma*, to order; we cannot have ecstatic visions of God on our own advice (any more than we can consciously and deliberately proceed with the project he identifies as the crucial one of the twenty-first century of making ourselves more innocent). Religion cannot be restored as a force for social coherence, or as a cure for a feeling of spiritual impoverishment, if its doctrines do not enjoy the status of being 'the truth'. Their being socially desirable is simply not enough.

Young seems to have little interest in the question of the reality of the gods. But it is no small detail, when we are religiously inspired, to know whether we are inspired because the gods enter into us; or whether our state of inspiration creates the illusion of the gods; whether our neocortex is created by God or God by the neocortex; whether God is an illusion or whether (comparatively) all else is. Young seems both to eat and to have his gods: he explains them physiologically but gives them moral-theological approbation.

Perhaps the deepest flaw in his account of religion – a flaw shared by many anthropologically-based accounts (notably that of Freud) – is his failure to explain how the inchoate sense of the Great Otherness experienced by chimpanzees in a thunderstorm was transformed into recognisable religions, into something so thickly encrusted with complex rituals and observances and, in particular, so freighted with specific abstract beliefs and claims about the world. It takes an awful lot of anthropomorphic reading of animals and animalomorphic reading of humans to get from a few excited chimps swinging in the trees to the dispute about the *filioque* clause that has riven the West since the early centuries after Christ, or the Council of Trent, or the 20 or 50 volumes of Karl Barth's *Church Dogmatics*. It is blatant anthropomorphism to suggest that the chimps' raindance has an implicit supernatural referent to the kind of god that humans have worshipped – and argued about and been puzzled by – over the millenia. A diffuse excitement and terror may inform the sense of a Great Other overarching, underpinning and governing the world; but this sense is not reducible to that excitement. Agreed, the cognitive element is not the whole of religious belief. But without the cognitive element, the world-picture, religious sentiment simply collapses into non-referential agitation or excitement.

Young's central message is that we must recover our sense of the sacred. At the same time, he explains the origin of the religious life as being due to the misinterpretation by chimpanzees of a storm as the sign of 'an almost-present adversary' (p. 93). The *pneuma*, which is what moves the invisible world-body, indeed pushes it around, is the invisible world-soul, which is wind, which is divinity, which is God. *Pneuma* arose literally out of the wind: 'The wind is perhaps man's oldest divinity' (p. 311). The profoundly blasphemous notion that our sense of the sacred has its deepest roots in the ecstasies and metereological confusions of chimps – in meta-wind – can hardly be a very firm basis for a return to religious values. Is he recommending that we should prostrate ourselves before the delusions of chimps? That we should pretend that we don't know that wind is moving air and, in this sense, is no more divine than still air or moving rocks or a man falling down stairs?

In summary, the failure of Young's religious prescriptions to address our present-day needs is three-fold (and in each of his folds he typifies much *kulturkritik* handwaving towards 'renewed spirituality').

First, he reduces religion from a set of truths to a set of symptoms and, connected with this, reduces its revival from access to a transcendental truth to meeting a human need: he does not seem to realise that you cannot have religion sociologically as an instrument for securing social cohesion and stability, and psychologically as a means of securing the inner balance of individuals, and, at the same time, eat it metaphysically. Confusing having and eating may be precisely what religion is about in Young's theophagy, but it is not the way to press a serious case. Secondly, he doesn't account for the transformation of religion from chimp excitement and bloodlust to the abstract forms it has taken (and so he falls into a bottomlessly deep Ha-Ha): by biologising, ethologising and anthropologising theology, he not only detonates any claim religion might have to the status of bearer of truth about a transcendental world but also overlooks the most extraordinary thing about man – that he makes his experience explicit in propositional form and raises complex and abstract institutions, rituals and power structures around it. And thirdly, as a result of this, he doesn't help us to work out what form a revived spirituality would take – what belief-content it would have, what words would be sculptured out of *pneuma* or apeish meta-wind – or what we should actually do now. Indeed, it is difficult to believe that he really wants us to do anything or to be sure how seriously he expects us to take the prescriptions he has based on his tendentious reconstructions of man's collective CV. Faced with nostrums that hint at a return to dancing and Dionysus, I am tempted to repeat F.R. Leavis's question (after hearing one of Edith Sitwell's poems read out in a suitably lugubrious tone of voice by one of his pupils): 'What should I do now? Ejaculate?' *Origins of the Sacred* cries out for the author to examine and to resolve his own attitude towards the sacred, to the gods and to ritual. Because that cry is unanswered, one suspects its author of being a tourist in the past rather than a spiritual pilgrim.[39]

Religion: the Dangerous Solution

For many writers, the re-enchantment of the world is urgently required, not only to make the world, life, modern society less dispiritingly empty of meaning but also to make it safe – safe from the runaway growth of consumption, from a science-based technology whose power seems to be limitless, and from the forces within individuals and society as a whole that, if unrecognised and untreated by spiritual therapy of one sort or another, could, with the help of a nuclear weapon or two, lead to universal apocalypse. This assumption that the desecularisation of the world will make the world more, not less, safe, cannot be allowed to pass unquestioned.

Young himself repeatedly reminds us of the dangers associated with the *pneuma*:

> The further one goes back, the more disturbing, even violent, sacredness becomes: that which sanctifies is always potentially polluting, the divine intrusion may become lethal if the dose is not properly regulated by art and ritual.
>
> (p. 310)

In this respect, he may not be as self-contradictory as might first appear. Yes, the mythic medicine that will reawaken our sense of the sacred carries with it the potential to reactivate forces that may in themselves be destructive and damaging – the blood-boltered ecstatic longing for self-aggrandisement through becoming one with the *pneuma*, the murderous frenzy that led to the sacrificial killing of alpha-shaman. However, Young would argue, those forces always exist within us. Just because they are repressed, it does not follow that they are not active within us. Liberating them from repression opens them up to regulation – by the 'lawlines of culture'. Unfortunately, Young's argument is itself highly dangerous. For it may not be safe to reawaken in society the sensibilities of the hunting pack – or to exhume those sensibilities from repression – if the weapons that lie to hand are not sticks and stones and bare hands, but nuclear weapons.

These are not merely theoretical dangers; for there seems to be something intrinsic to religion that makes it particularly apt to foment, elaborate and and organise conflict. This is captured in Diderot's famous fable about the origin of religion:

> A man had been betrayed by his children, his wife and his friends. Treacherous partners had destroyed his fortune and made him destitute. Filled with hatred and deep contempt for the human race, he left society and took refuge in a solitary cavern. There, pressing his fists into his eyes, and planning a revenge proportioned to his bitterness, he said, 'Monsters! What shall I do to punish their acts of injustice and make them as wretched as they deserve? Ah, were it but possible to imagine ... to put into their heads an illusion, which they would think more important than their own lives, on which they could never agree with each other!' ... At that moment he rushed out of the cavern crying 'God! God! ...' Countless echoes all around him repeated 'God! God!' The terrifying name was carried from pole to pole and everywhere it was heard with astonishment. Men at first fell down to worship, then they rose, asked questions, argued, became embittered, cursed one another, hated one another, and cut one another's throats. Thus was the deadly

wish of the hater of mankind fulfilled. For such has been the past history, and such will be the future, of a being who is as important as he is incomprehensible.[40]

The invention of the idea of God is seen as the supreme act of revenge upon mankind by one who had been treated evilly by men. For, in Diderot's view, religion is inseparable from conflict: in the space of a sentence, men pass from falling down to worship to getting up to cut each other's throats.

Anyone who would prescribe the reawakening of religious belief and a generalised desecularisation as a means of making the world safer needs to ask whether the relationship between piety and violence, between falling down to worship one's own god and rising up to cut the throat of one's neighbour who worships a different god, is inescapable. One would have to be very ill-informed indeed to be prepared to argue that the influence of religion in the past has been benign – either for individuals or for the planet as a whole. Few Jews would think that Christianity had done much to improve their lot or, indeed, to civilise the world. Hindus and Muslims, Shi'i and Sunni Muslims, Protestants and Catholics have used, and continue to use, their beliefs to justify despising, and making life worse for, each other. So we must ask: is the hatred inspired by religion – the hideous 'confessional wars' that have dominated so much of human history, the pogroms and persecutions, the even more widespread, if less organised, loathings and meannesses and unkindnesses – accidental or of its essence?

Clearly, once religious sentiments and beliefs are built into collective faiths, supported by fixed rituals and dogmas, and embodied in institutions that enjoy extensive secular power, then they are in prime position to collect, organise, foment and implement hatreds on the kind of scale we have seen throughout history. Our question, however, is Diderot's: 'Are the abuses and the essentials of religion inseparable?' Can we overcome the difficulty expressed by Voltaire – who was a deist and a believer in a good creator and Architect of the Universe whom he thought a friend of mankind – of knowing how to worship God rather than the deceit of men?

One way of separating the abuses from the essentials of religion might be to disestablish it completely, to strip it of its relations to temporal power structures, to remove the carapace of churches and even of ritual and dogma; to encourage a priestless religion of individuals whose beliefs may or may not coincide with one another – a religion of sensibility and intuitions, rather than of institutions and power, based upon an open-ended sense of transcendental otherness, of meanings that go beyond daily meanings. Such a total disestablishment of religious belief should be possible in a secular age; for, as Goldmann points out, whereas previously

> Previously unbelief had been an individual state and faith collective. Once scepticism had become a social phenomenon, faith tended to become an individual matter.
>
> (p. 57)

It is not, however, as easy as it at first appears.

First, even private religious experience may not always be entirely benign: a concern 'to attune one's inner being to the timeless and the transcendent'[41] more quickly blunts one's awareness of others' suffering than it reduces one's interest in one's own; it more efficiently distances one from one's neighbour's toothache than it liberates one from the aching of one's own teeth.[42] Secondly, it may not be possible to have truly religious experiences without religious beliefs and the latter may require, indeed demand, some kind of external acknowledgement, support, validation, legitimation. Very few people, I suspect, could pull off Young's feat of submitting to religious experiences while knowing their cognitive content to be based on the delusions of chimps. The sense that there are (to use Wordsworth's phrase) 'Unknown Modes of Being' needs, for all sorts of reasons, to root itself – or to feel that it is rooted – in some kind of seemingly objective truth about the world. It consequently demands external support; in particular the validating acknowledgement of others – if only to protect it from being downgraded to a mere series of episodes, a succession of bouts of transcendental epilepsy. The most characteristic expression of this desire for external validation is the wish to convert others or to see in others' faith a reflection of one's own experiences and hence to be oneself a convert. This is understandable: for what is at stake is the meaning of one's life, of the world, as seen from what is felt to be the truest perspective. The passage from feeling to cognitive content and from personal to collective feeling – in sum, the passage from personal feeling to collective belief – is thus direct and rational. The vision in the desert is preached in the marketplace and, suddenly, we have a sect crying its wares and competing with other sects, large and small, for a place in the sun, wanting to requisition its share of the collective consciousness. When what is at stake is the ultimate meaning of things, it is hardly surprising that the step from falling on one's knees to worship to standing up to cut one's neighbour's throat is a very short one.

It is not necessary to invoke corruption, or even desire for power, to explain how passionate convictions conceived in the desert can give birth to manipulative institutions. My longing that you should believe what I believe is at bottom a wish that you should acknowledge *me* – by accepting the reality of my vision, of my view of the world, the legitimacy of the meanings that seem to me to lie at the bottom of all meaning. My religious passion, my own private beliefs, can very quickly therefore become coercive. I need to fight to save my soul (though I will see it as a fight to save *your* soul) from the tragic error of having missed the true meaning of life.

Thus does solitary religious experience pass, even without the help of corruption and worldly ambition, from an episodic intuition that wishes to be granted the status of a body of higher truth to a power struggle which will give birth to institutions in which the solitary experiences are validated in collective rituals and dogma, and unbelievers explicitly or implicitly execrated. Then corruption and worldly ambition can do their work. The power required to protect the institution starts to be enjoyed for itself and, as the intuitions that gave birth to the vision fade through custom, the power becomes more important than anything else. The solitary vision scleroses into an established Church and the Church connives with other centres and sources of power to ensure its own survival and extend its realm. And because there is always more than one Church, the sect it represents will be a faction with a vested interest in destroying other sects – either peacefully by conversions or, if necessary, by other means.

No major religion has managed to keep blood off its hands, not even those religions such as Buddhism, which emphasise the primacy of personal experience (or impersonal experience) over rituals institutions and dogma, affirm the sanctity of all life and advocate renunciation of the will. Out of the will to transform individual experience into truth and the need to win acknowledgement for that truth comes the relation between religion and power, religion and institutions, religion and the maintenance of the social order, religion and corruption, religion and bloodshed.

That at least has been true of the religions that have antedated our supposedly secular times. But what of their revival in irreligious times, and the conscious attempt to re-enchant the world? A religious revival has a choice: either of being merely the scattered and episodic experiences of a few individuals, experiences which lack the status of objective truth and are in danger of being dismissed in a deconsecrated, rationalistic, industrial society as 'religiosity'; or of trying to become a mass movement that will attempt to dominate, if not the world, at least the meaning-systems and the values to which the world seems to subscribe. The latter path seems to be the more favoured and when it is taken, the result is typified not by pews more choked with pious souls in an Anglican church or Buddhist temples more thronged with contemplatives, but by the blood-boltered theocracies of Iran or the more localised, though hardly less profound, horrors of Rancho Apocalypse.

The necessary relationship between true religious belief (as opposed to mere spiritual tourism or strategic or unthinking conformity to the dominant faith) and aggressive fundamentalism is set out in Hamann's attack on Moses Mendelssohn's *Jerusalem: A Plea for the Toleration of the Jews*. Hamann loathed Mendelssohn for trying to separate what man owes God and what he owes Caesar, the public from the private. To do so, he argued, is to cut an individual into separate pieces, as if he were so much inanimate flesh:

> If religion is to be taken seriously at all it must penetrate every aspect of man's life; if it is true, it is the heart and soul of a man's being; a religion that is confined to its 'proper' sphere – like an official with limited powers, to be kept in his place, not allowed to interfere – that is mockery. Better deny religion altogether, like an atheist, than reduce it to a tame and harmless exercise within an artificially demarcated zone that it must not transgress.[43]

It is not sufficient, for a true believer, that others should believe what he believes. He also requires that his beliefs should penetrate the public sphere, should be all-pervading and, indeed, be woven in with the power structures of the society in which he lives. At the heart of a profound religious passion are the seeds of intolerance – of others' unbelief, of a society that does not recognise one's private vision and which, in failing to do so, undermines it – even, or especially, if it (merely) tolerates it.

There is thus a well-beaten path from the solitary prophet crying in the wilderness via the demagogue in the marketplace to the priest-king imposing a reign of terror on his spiritually less well-endowed subjects. Once what is owed to God and to Caesar are not kept apart and spiritual and temporal power are fused – as in an established religion or, more completely, in a theocratic or hierocratic state – then the conditions for the establishment of Hell on Earth are largely in place. The decade and a half since the Ayotollah Khomeini came to power in Iran is only the most prominent example.

The case of Iran illustrates another point about the dangers of mass respiritualisation of society. There is a fine line between a legislature and executive that is accountable to an invisible authority (God in religious states, the laws of history in communist ones) and straight-forward unaccountability, between power that is answerable to God and naked untrammelled power. According to Roy Mottahedeh, recent Iranian history has shown how the mullahs closed the gap between Khomeini's belief that there was no need for a constitution,

since power stems from God alone, and the actual result of the revolution, which was the promulgation of an Iranian constitution with
its declaration that the source of parliament's legislative authority was
'the will of the people'.[44] This enabled Khomeini to assume unprecedented power as *mujtahid* or interpreter of Islamic law. The consequences are, as we know, appalling, amply justifying Lord Acton's famous
rebuke to Bishop Creighton when the latter supported the move to declare the Pope infallible. Acton rejected Creighton's assumptions about
the special virtues of those in authority by divine appointment:

> That we are to judge Pope and and King unlike other men, with a
> favourable presumption that they did no wrong. If there is any pre-
> sumption, it is the other way against the holders of power, increasing
> as the power increases.... Power tends to corrupt and absolute
> power corrupts absolutely. Great men are always bad men, even
> when they exercise influence and not authority; still more when you
> superadd the tendency or certainty of corruption by authority. There
> is no worse heresy than that the office sanctifies the holder of it.

Dudley Young's lamentations over the separation of the roles of
alpha-warrior and shaman, of temporal and spiritual power – 'the
splitting of these roles', he complains, 'opened a rift in the human
psyche that we have been trying to repair ever since' (p. 115) – are
placed in an interesting light by a recent report of the all-party British Parliamentary Human Rights Group, which focused on the plight
of women in Iran 15 years after an alpha-shaman came to power.[45]
Failure to cover themselves from head to foot carries the death penalty, although the usual punishment is 84 lashes. This is applied even
if the exposure is accidental and involves just a stray hair slipping
from under a scarf. And an advanced state of pregnancy or extreme
old age are no protection against either the charge or the brutal penalty. The punishment for adultery is stoning to death for an unmarried woman; married women are buried up to their necks in stones
and may go free if they escape. Virgin women are raped before execution, the privilege of carrying out the rape going to the guard who
wins the lottery.[46] The latter's purity is protected by the issue of a
marriage certificate, sent by the religious judge to the family of the
executed woman, along with a box of sweets.[47]

The theocratic tyranny of Iran is replicated in many other places.
They give a clear indication of the consequences of a desecularisation
of society, and of the mobilisation of 'strong mythic medicine' and
the unification of the split psyche of mankind in the person of the
alpha-shaman. The re-enchanted world is unlikely to be filled with
little posies of flower people made yet more gentle by spiritual

experiences that loosen the hold upon them of the drive towards getting and spending. Enchantment will make the world more, not less, dangerous. Alpha-shaman sitting on a stockpile of nuclear weapons may, if he is as sincere in his religious beliefs as he is in his lust for power, be inclined to bring about the apocalypse that will fast-track the passage of mankind to its eternal reward.

Even where they have not themselves sponsored terror, the potentates of the established Churches have not always hastened to try to stop, or even to condemn, violence and injustice if it has been directed at those who do not share their own version of the Celestial City. Pope Pius X's silence when faced with incontrovertible evidence of the mass murder of Jews – perhaps due to cowardice, caution and indecision, but probably not (after all, the Jews had murdered Christ, hadn't they?) – was especially nauseating but not entirely uncharacteristic. It is echoed in the silence of the Catholic Church over the hundreds of thousands of Serbs killed by the Catholic Ustasha soldiers in the 1940s; of the Orthodox Church over the atrocities committed by the Serbs in the recent Balkan wars; and, most sickeningly, in the recent failure of the partisan local Catholic Church to condemn the recent slaughter of several hundreds of thousands of Tutsis by the Hutus in 1994. Even where they have spoken out against injustice, and the established churches have put their sentiments behind the poor and powerless, they have usually been careful to place their power at the disposal of the rich and powerful.

There is another reason for not looking to organised religion – or to a spirituality that would tend to promote or find its acknowledgement in organised religion – to secure the safe future of the world. Although it is unique in its insistence on clothing them from head to foot in its approved apparel, Islam is not the only religion that tries to imprison women's bodies. Many other religions demand their say over women's pelvises. Apart from the ultimate civil liberty outrage that this represents to the secular mind, it is also a great danger to the world.[48] In many underdeveloped areas, health care is provided predominantly or exclusively by the Roman Catholic Church and the Church's powerful position prevents effective access to reliable contraception.[49] The consequences for this are catastrophic – for the women who have to bear the unwanted children, for the societies in which they live, and for the future of the world.

Douwe Verkuyl has illustrated, through a series of terrible case histories, the results of the hostility of Church and Mosque to fertility control and to sex education for women. Anyone who imagines that the future security and happiness of the world will come from a revival of religious belief should read these case histories and reflect on them. Here are a couple:

Nthabiseng is a recently qualified nurse. With the economic structural adjustment programme she is happy to get a job in a remote
Catholic hospital, even though her husband is obliged to stay in
town to look for work. At Christmas her husband visits her; there
are no condoms at the mission. At the end of her 6 months' proba-
tion the nuns in charge of the hospital will make her have a pregnancy test. If she is positive, she will not get the job. Six weeks
before the urine test, she misses her period. A traditional doctor in
a nearby village tries to help. She dies of a perforated uterus.

Hajira is 12 years old. Her father, a staunch Muslim, thinks that
education is not important for girls. She has had her menarche, so
a marriage is arranged. She never had sex education and has no
idea about antenatal care and what is supposed to happen during
a delivery. Her mother cannot tell her: she died during her eleventh
delivery. After three days' labour at home she is encouraged with
hot irons on her back to push harder. In the end a dead baby is
born. Three months later she is able to walk more or less normally
and is rejected by her husband because the huge hole in her bladder
causes her to smell and leak. Her family does not want her back.[50]

If the second story has a familiar ring, echoing a tale earlier in this
chapter, it is because the events it describes are so common.

The most poignant measure of the terrible consequences of the pelvic persecution of women is maternal mortality rates. In Zaire, the
lifetime risk of death from pregancy is 1 in 25, whereas in Western
Europe, near the headquarters of the Roman Catholic Church, the risk
is 1 in 5000. This statistic highlights how the policies of the powerful
religions cause the most suffering where people are poorest. The explanation is partly that denial of a women's right to control her own
fertility has its most disastrous effects amongst the poor; but also because
the rich and those geographically nearest to the headquarters of the
Catholic Church obey the edicts of the Church with least conscien-
tiousness and, moreover, have the means to disobey. Fertility rates,
as has often been pointed out, are inversely related to the proximity
of the population to the Vatican.

The hideous completeness of this hypocrisy is captured in another of
Verkuyl's case histories, the only one in which a male takes centre-stage:

His name is Karol. He is a Polish gynaecologist and a staunch supporter of the Roman Catholic Church. In public he is a declared
opponent of the use of 'unnatural' contraception. His official income is not very high, but luckily the termination of many unwanted
pregnancies in wealthy women makes it possible for him to give
his two children a decent education.

The overall adverse effect of pelvic persecution is greater than the sum of the individual tragedies, great as the latter is. There is the added horror of the consequences of an uncontrolled increase in the population. 'Barring disasters', Verkuyl says, 'the natural annual increase of 5% would see the world population in December 2100 at 1049 billion. Not a cockroach would be left of nature'.[51] Of course, disasters have not been entirely barred. Of the 6 million live births in Central and Eastern Africa in 1993, Verkuyl estimates that at least 1 million were to HIV-positive mothers. About 300 000 of the babies will have died because of AIDS by 1995; a further 700 000 will be orphans a few years later.[52] So much for the class of 1993. The same will be true of the class of 1994 . . . the class of 1995 . . . and so on. These disasters – or 'natural' population control – will not, however, be large enough to prevent the population from rising, and from outstripping the growth of scarce resources. The fall of the death rates of children under 5 from 1 in 4 to a mere 1 in 5 has led to an ecological disaster in Rwanda – a classic case of the demographic entrapment to which Maurice King drew attention.[53] In 1993, 5.2 children were born per 100 of the population and each woman was expected to have 8.5 children during her lifetime at the present birth rates.[54] The complete collapse of the Rwandan ecosystem predicted 20 years ago could not be prevented: Rwanda has a very Catholic population and a powerful Catholic Church with a long tradition of strong opposition to family planning. The collapse took only a little longer than anticipated and the consequences when they came were, as we know, utterly hideous, with the death by murder of several hundred thousand, perhaps a million, people, in a few weeks. As Luc Bonneux commented:

> The Hutu extremists have been called possessed by devils by Western press. I hold that those who believe in devils should forget about studying the state of mind of the desparate Rwandese and concentrate on the devastating consequences of denying effective family planning to the poor of the world, who are smothering each other through the sheer growth of their numbers.

Postnatal population control, using a machete applied to the heads of innocent men, women and children, must surely be inferior to the use of the contraceptive pill or the condom.[55] In short, one would have to be crazily optimistic to believe that a resurgence of religious belief would be good for humanity, if only because the greatest threat to the future habitability of the planet comes from religious opposition to a rational approach to population control.

This resistance to rationality is rooted in a fundamental refusal to take the temporal world seriously – except in terms of gaining power

over it. There is simply no tradition amongst many religions of addressing the true problems of the world. Again and again, a practical
question is turned into a symbolic one and the practical solution is
lost sight of, or actually opposed. A small but telling example will
have to suffice. In 1942 the Dublin City Council put forward a plan
to provide schoolchildren with a hot lunch. This was at a time when
malnourishment and its consequences – TB, rickets, etc. – were rife.
The plan was opposed by the bishops on the grounds that it was
'interfering with the normal family life of the people'. This symbolic
concern for family solidarity was held to be more important than the
practical matter of preventing malnutrition. 'The leaders of the Church,
who watched every schoolgirl's knee with anxious eyes, had no concern for undernourishment and rickets'.[56] Protests were disregarded
because the then Taoiseach, Eamonn de Valera, was frightened to oppose
the bishops. In itself, the denial of hot food to malnourished children
is a comparatively small outrage – small compared with the horrors
of confessional wars, with religious persecutions, with instilling universal global terror at the prospect of eternal damnation – and yet it
is a revealing one. For it bridges the space between the major violations and the continuous minor injustices that a framework of religious belief, against which rational appeal is impotent, makes possible.
It shows how the enveloping fog of unreason and unreasonableness
is a seamless robe.

Time, perhaps, for a summarising pause. The rediscovery of the
sacred may not, after all, be the royal road to healing the world's
spiritual, let alone its material, ills if only because even private religious experience harbours within it a nascent public dimension. The
ecstatic moment of private vision longs for external legitimation to
liberate itself from its momentariness, from the subjectivity that makes
it frail and unstable – especially in a predominantly secular world.
There are two ways forward from the 'innocent' moment of transcendental wonder: either it can be donated to an existing institutionalised religion and the individual concludes that he or she has been
converted to, or returned to, such-and-such a religion; or, in the case
of more robust souls, the moment can be the founding experience of
a new sect, faction or denomination. Legitimating the moment of transcendental wonder will thus require a priesthood – either off-the-shelf
or brand new. It is, therefore, but a short step from there to those
uses and abuses of power we have already dwelt upon. There is, in
other words, a fundamentalist tendency in even the most innocent of
intimations of immortality, of unknown modes of being, of meanings
beyond ordinary meanings. Even from such private and innocent beginnings, religious intimations taken seriously inevitably demand that
the non-secular values they hint at should suborn everything else to

themselves: they do not want merely to be patronised by (repressive) tolerance. A revived sense of the sacred cannot accept merely being a colloid of intuitions suspended in a secular medium, or mere scintillations of consciousness winking on and off, like occultating stars in an infinite dark night. Religion cannot, therefore, be used to prop up, or to renew, a largely secular society: (a) because it creates its own world-picture, its own values, that swamp, subvert, overturn, those of secular society; and (b) because it cannot be believed to order, for some other purpose ('God is a fire in the head'). There is that in religious experience which insists on being translated into belief, ritual and institutions and that in religious conviction which feels that it should affect every aspect of the believer's life and of others around him and that anything short of external expression in a universal, fundamentalist society is inadequate.

Anyone who prescribes a reconsecration of the world as a means of saving the planet must consider these terrible dangers. That is why before I would even consider looking to religion for salvation, I would like to know what 'religion' includes. If its job specification includes the persecutions, the prejudice, the ignorance and the bitterness of past and present religions, I'd like to leave it out of the prescription. Although it is rarely true that religion on its own makes societies evil, it does seem to provide a focus, a forming up position, a higher alibi, for evil forces in them. The Coleridgean separation between 'belief' and 'faith', between private experience and institutionalised doctrine, between reverence and power, which some seem to hope will make religious conviction less potentially malignant, is not one that is likely to hold sway in Belfast, Amritsar, Mecca, Sarajevo or Rwanda. The relationship between what Rousseau called 'the religion of man as man ... without temples, without altars, without rites, and strictly limited to the *inner* worship of the Supreme God, and to the eternal obligations of morality' and 'the religion of the citizen' which is 'inscribed in a single country' and 'whose dogmas, rites and forms of worship are all prescribed by law' and which 'limits men's rights and duties to the territories in which its altars reign supreme' is complex and difficult, as Rousseau himself appreciated.[57]

Until this difficult and dangerous relationship is sorted out, it is arguable that, far from needing 'mythic medicine' and a 'reconsecration of space and time', humanity could do with a break for a while from organised (and disorganised) religious belief with its revelation-based certainties and oppressive institutions and the bumpy, blood-splashed road that leads from the one to the other. At any rate, until religion takes seriously the needs of human beings and addresses the real problems of the planet and accepts Milton's declaration that 'no ordinance, human or from heaven, can bind against the good of man'.

Any future attempt to re-establish social order on the basis of a collective spiritual institution rooted in transcendental belief and backed up by powerful organisations should be undertaken only when mankind has shaken off the destructive habits that have been inseparable from the institutionalised religions of the past and present and has learnt the difficult art of separating revelations of the divine from reasons for hating, controlling and persecuting other people. *Tantum malum religio potuit suadere* ('So many wrongs could religion induce to' (Lucretius)). It is, as Francis Bacon said ('Of Unity in Religion'),[58] too easy 'to make the cause of religion to descend to the cruel and execrable actions of murdering princes, butchery of people, and subversion of states and governments'. Too often does the Holy Ghost 'assume instead of the likeness of a dove . . . the shape of a vulture or raven'.

If religion mobilises what is deepest in us, it is not surprising that it organises, and so makes more powerful, inhuman nastiness as well as human goodness, so that the two work symbiotically towards a dubious end. Hatred of, and contempt for, others, love of power, sadomasochistic fantasies – all of these things are pandered to in and inflamed by religious institutions, doctrines and rituals. Too often, organised religion has proved to be the enemy of unthinking decency, spontaneous kindness – despite the Great Teacher's warning that 'Except ye be as little children, ye shall not enter the kingdom of Heaven' – for the sake of deeper emotions that do not compensate for the ordinary virtues that they outlaw. It is no wonder, therefore, that, if God existed in such a form that He could have emotions, human religion would most likely be one of His deepest sources of shame. (As Voltaire said, God's only excuse is that He does not exist.) Let us hope for humanity's sake and His that, if there are to be religions of the future, they will be different from those of the past. They will separate the cognitive, the affective, the moral, the political and the metaphysical and understand that these might not meet in the one place. Until such time, the answer to Diderot's question – 'Are the abuses and the essentials of religion inseparable?' – must, alas, be a resounding 'yes'.[59]

THE MYTH OF WISDOM

Where is the wisdom we have lost in knowledge?
Where is the knowledge we have lost in information?

(T.S. Eliot, Choruses from 'The Rock' 15–16)

We began by discussing the wisdom of myths and found that that wisdom was not easy to find. Of course, a certain amount of ingenious reintepretation could seemingly prove that the old myth-makers anticipated the discoveries of more recent depth psychologists or were in tune with what might pass for common sense in our present rationalistic world. But that is a poor yield and hardly compensates for the dangers incurred by a society that gives credence to myths. The balance between the pernicious and the therapeutic in myths is decisively in favour of the former – as recent history has all too clearly illustrated. 'The wisdom of myths' proves, we found, to be itself a myth and a far from benign one. The wisdom of myths is a subdivision of a larger, less well-defined 'folk wisdom'. This notion, when it is not patronising, is also not entirely safe. It encourages the belief that there are certain attitudes, assumptions and pieces of lore that are beyond challenge, and so discourages reflection on certain prejudices. And it is close to the idea of the *volk* as a repository of that which is truest in the collective consciousness. Fifty years after Auschwitz, there is no need to spell out what demagogues can do with this claim.

Questioning the wisdom of myths, then, leads naturally to questioning the reality or validity of 'folk wisdom'. And it seems equally natural to extend this process further and ask whether the notion of 'wisdom' is itself a myth – and a dangerous one at that – that has no place in a secular, rational society where individuals are legally or contractually answerable for their actions.

What do I mean by 'wisdom'? Or, rather, what do those who sigh for its return, mean by it? It is a faculty of insight and understanding that goes deeper and sees further than ordinary intelligence, good sense, competence and cleverness. It is a special ability to see what is really at stake in human affairs, to see what, on a given occasion, is really needed. The concept of wisdom is allied not only to that of depth (as opposed to shallowness) but also to that of authority – more specifically to transcendent authority. Its ultimate expression is the operation of God's will in creating the universe, in laying down its laws and in adjudicating in human affairs. The important criterion of wisdom is that it goes beyond, and is in the last more effective than, all those other things – cleverness, ingenuity, common sense, practical know-how, determination, energy, knowledge, information – we deploy in the pursuit of human ends. Wisdom, additionally, sees beyond those end themselves, to some deeper and fundamentally more meaningful end. Wisdom, in short, is even more likely to tell us to look beyond 'getting and spending' than it is to tell us how to 'get' more effectively and 'spend' more enjoyably.

In suggesting that 'wisdom' may be a myth, I am questioning whether in a world of increasing know-how, the wise man (it is usually men who are accounted 'wise') still has a distinct place and his distinctive contributions have anything to offer. There are two sources of his wisdom: *a priori* access to transcendental truths, which come ultimately from God; and *a posteriori*, as a result of accumulated experience. The first source is typified by priestly wisdom, and is afforded to certain individuals (witchdoctors, magicians, cult leaders, holy men, shamans, prophets etc.) who are either given or earn (through suffering, fasting, devotion, self-denial, even training) privileged insights into God's mind. This source of wisdom may be regarded as archaic and is largely dismissed in secular societies. The second source of wisdom is one's own life-experience; and wisdom is therefore granted to the elders, who have had more of their life than the young and, having reflected upon their experience, have learned from it. Outside of gerontocratic and gerontophilic societies, this second source of wisdom is not given much credence.[60] So wisdom has to survive by drawing on both – *a priori* and *a posteriori* – sources of authority, without too explicitly leaning on either. In the contemporary notion of the wise man are fused both a special perviousness to transcendent sources of understanding and a heavy burden of experience – in the idea of someone with a special aptitude for learning from experience and a lot of experience to learn from.[61] The ambivalence permits the notion of the wise man and his wisdom to pass relatively unchallenged through the inhospitable secular climate of opinion. Thus reinterpreted, wisdom seems a rather small addition to the notions of competence, know-how, etc., acquired through a relevant apprenticeship of experience. And yet people want to make much more of it: they want to restore the mythical status of 'wisdom' as some special knowledge and understanding that feeds into a special competence which would befit a leader, or someone to whose opinions and power one should defer.

The reasons for this are clear when one recalls the rhetoric that is used to attack the more conventional sources of competence. 'Information' and 'cleverness' are usually opposed to 'wisdom'; and the former are in abundant supply in this age, while the latter, it is complained, is in severe shortage. The technology which is regarded as the dominant cultural fact of our age is seen as proof of this. We are able to produce cleverer and cleverer gadgets to do our bidding and so improve the material conditions of our lives. And yet, as a result, because our cleverness has not been informed by wisdom, our gadgets are not bringing us happiness (think of the break-up of family life) and the planet is going every which way to Hell. As a result of modern technology, we are bombarded with increasing amounts of information, but we do not really further our understanding because we lack the ability to filter that knowledge and relate it to our lives. Our lives grow materially and cognitively richer while we become progressively emptier; we are offered more and more solutions to our problems and are given more and more advice to help us to implement them, but we are simply bewildered and disoriented. The ingenuity that enables us to create complex technology has not stopped us from engaging in useless and unprecedentedly destructive wars. It was cleverness that permitted us to imagine the atomic bomb and lack of wisdom that permitted us to make and drop it. In a century that has certainly been cleverer and, in many respects, more competent, than all previous centuries put together and yet has also been more destructive, intentionally and unintentionally, wisdom seems to have a lot going for it.

So much for the rhetoric in support of 'wisdom'. It does not, however, even suggest an answer to what form wisdom should take and how it would be effective in curing those ills that cleverness, competence, good sense have failed to address. For a start, the space made available to wisdom, if it is not simply going to fly in the face of good practice, is nowadays relatively small. To take a trivial example, what is the role of wisdom – which is not knowledge or competence or cleverness – when your car has broken down in the rain? If, in this situation, you were given the choice between wisdom without competence and competence without wisdom, which would you choose? If wisdom is 'a wisdom of the gaps' (cf. the god of the gaps), then it is very tightly squeezed indeed: much of the problem-solving necessary in the modern world is technological and the higher competence (or anti-competence) of wisdom simply has nothing to offer. It is significant to note, in this context, that in the places where wisdom ranks highly, the level of material prosperity is low. The wisdom of the East is closely associated with unforeseen flooding in the East, with famine in the East: where there is much wisdom, there is much suffering.

This association is open to several interpretations. The most charitable (and it is not very flattering to wisdom) is that the hope that wise men give out in an otherwise hopeless or unalleviated situation makes that situation more bearable. The groundless prognostications and the other authoritative pronouncements of the wise man – the fakir, the village elder – make disaster seem more predictable and, by magic thinking, more amenable to control. At the very least, it rescues a human, symbolic meaning from inhuman, random catastrophe. The less charitable interpretation is that material deprivation is in part a result, perhaps, of an over-reliance on wisdom, on the authority of those whose cognitive credit is due only to seniority, to caste, or rumour. In short, it is as if wisdom and practical competence have a certain amount of space to share out between them, so that as the one grows the other shrinks.

I mentioned the uselessness of wisdom in situations of practical need such as when the car breaks down in the rain. Let us move to more serious crises, to areas where pure technical competence is self-evidently insufficient. Medicine has been traditionally dominated by wise men, ever since the founding fathers – Aristotle, Galen et al. – laid down the fundamental misconceptions which it took physicians millenia to wake out of. Wisdom was the dominant attribute of successful physicians throughout the centuries during which doctors misled, poisoned, butchered, tortured and killed their patients. It is only a small exaggeration to state that, until the turn of the century, with the exception of certain, relatively simple surgical procedures, nothing doctors did for their patients contributed to their health or happiness; and frequently it damaged them irreversibly. This did not dent the confidence of doctors in the wisdom of their pronouncements. The dramatic success story of medicine – 'the youngest science', as it has been called – in the twentieth century has been one of the retreat of personal, individual wisdom in the face of science that is open to all. The idiosyncratic diagnostic methods and remedies of the charismatic physician, dictated by groundless fashion, personal taste and the whims of the moment, has been replaced by standard, objectively demonstrated diagnoses and agreed upon treatment protocols. These are not owned by any individual and an insistence upon clinical freedom – the immemorial right to reinvent medicine and to be eccentrically wrong – has given way to the humble recognition that painstakingly developed treatment methods evaluated and re-evaluated in carefully conducted trials are better than anything you might think up in your armchair. The advent of objective diagnostic methods put paid to the 'brilliant diagnostician' (whose reputation may have been founded as much upon force of personality as upon diagnostic accuracy); and the advent of the double-blind controlled trial put paid to the idea of the brilliant 'healer' (except in the case of surgery, where a uniquely personal competence is still crucial, though it has to function within closely defined protocols, guidelines and other constraints or clinical practice).[62]

The double-blind controlled trial is a remarkable example of humility: at its heart is the doctor's deep suspicion of his or her own ability to be objective where what is at stake is a judgement as to the outcome of a treatment. In short, the retreat of wisdom (or the doctor's and patient's belief in the former's special authority) has brought only benefits. Scientific competence is not, of course, enough: kindness, courtesy, sympathy, emotional sensitivity, common sense, a desire to be helpful, respect – all these are essential if the doctor is not going simply to be an agent for mobilising technology for the patient without finding out what the problem is and what the patient wants done about it. But none of these things – kindness, courtesy, etc. – justifies the term 'wisdom' or quite fills the space the word used to mark out.

There are those, inevitably, who long for the return of the 'wise' physician. They do not base this Arcadian longing on a precise knowledge of what 'wise' physicians actually achieved for their patients. Some of the more prominent, revered by their colleagues and pupils, were not very kind to their patients, often terrified them and frequently caused them considerable harm with their quack remedies. Their portraits, in the novels and memoirs of their patients, are not flattering. To read a biography of a chronic invalid such a Virginia Woolf, who suffered all her life from manic-depressive psychosis, is to see the Great Names in action and marvel at their confidence as they prescibed quack and baseless remedies and exerted their oppressive authority. The shift from authority-based towards evidence-based medicine, and the demise of the special authority of the Wise Physician, is one of the great triumphs of science-informed humanity.

Sometimes there is a call for doctors to reassume priestly authority:[63] this is chilling, not only because the hope of successfully suing Sarastro for malpractice seems a particularly forlorn one, but also because the values of doctors *qua* doctors and priests *qua* priests are utterly opposed. For a priest, death is a gateway; for a doctor, it is a terminus. The calls for the return of priestly wisdom are usually prompted by examples of unthinking use of high-tech investigations by a doctor who failed to take a simple history and consequently made a disastrous diagnostic error; or by equally depressing examples of the unthinking use of heroic treatments that have added suffering to those whose lives have been pointlessly prolonged. What these examples of stupidity call for is not some mysterious substance called 'wisdom' (with or without priestliness) possessed by a 'wise' physician, but less stupidity and less idleness. After all, everyone can see the stupidity of the error. The next step is to see how it came about. Clever idiots make a more powerful case for non-idiotic cleverness than for something called 'wisdom'.

We should be cautious about advocating the return of a commodity that was in such rich supply when the only effective medicine was brutal surgery and bonesetting – typified by a surgeon-butcher hacking off the leg of a screaming, puking wretch 'anaesthetised' by spirits – probably not preceded by a careful discussion with the patient of the pros and cons of the operation. The alibi of wisdom can be used to cover up a good deal of ordinary incompetence and to support the asymmetrical power relation between patient and doctor. Both of these are vividly illustrated by the case of Freud, the emblematic figure of medical wisdom, the doctor who saw 'deeper' into the suffering soul of humanity, the paradigmatic healer. The truth that has slowly leaked out about Freud is that he was a monster – a monster of genius, but a monster none the less. He was a strikingly incompetent diagnostician, incapable of learning from his own mistakes because he could not acknowledge them, preferring to rationalise them away; and he was a bully who forced his patients to provide him with confirmatory evidence of the groundless notions he had thought up at a time when he was addicted to cocaine.[64] (The fact that these fantasies are transcribed into brilliant and sober prose that apes the surface features of scientific discourse does not make them less fantastical.) The remarkable case of Sigmund F. illustrates dramatically how the attribution or assumption of wisdom permits abuse of power. The myth of wisdom supports power; the power supports the myth of wisdom. The attribution of wisdom asks us to suspend our critical sense, to move from equality to blind obedience, unquestioning belief. The fantastical notion of an individual who can tap directly into sources of knowledge that lie deeper and are more potent than the accumulated competence and know-how of the rest of mankind – like the priest who has one ear cocked for God's voice, or the witch doctor who knows your trouble straight away without proper history or examination – is an ancient one. (Freud's special source of knowledge lay in the ideas he had dreamed up himself and then found in your unconscious after a long search that gave him time to plant them there).[65] Those of us who have to put up with colleagues whose demeanour is one of wisdom may also have personal reasons for doubting the benefits that would follow from an epidemic of wisdom in the medical profession.[66]

The idea of the Wise Man as the political saviour – the healer of the state – has had as long a history as that of the Wise Healer. (The two, of course, would be united in Young's alpha-shaman.) The history of the twentieth century alone should be sufficient to discredit this idea for ever. If we list only the most prominent among the leaders who have been acredited with boundless wisdom, or have ascribed this faculty to themselves, the point is made: Hitler, Lenin, Stalin, Mussolini, Ceaucescu, Bokassa, Mao. The attribution of wisdom, or the arrogation of the title 'The Wise', is a crucial step in the transformation of the ordinary corruption that comes from ordinary power to the absolute corruption of absolute power.

Of course such leaders could not pull off the delusion were there not a corresponding desire among the led to believe in wisdom. The appetite for power is dangerous, not only on account of human cowardice and ruthlessness but also because of the human desire to prostrate oneself before authority. This was seen clearly by Maistre:

he underlined ... that the desire to immolate oneself, to suffer, to prostrate onself before authority, indeed before superior power, no matter whence it comes, and the desire to dominate, to exert authority, to pursue power for its own sake – that these were forces historically at least as strong as the desire for peace, prosperity, liberty, justice, happiness, equality.[67]

For Maistre, this is part of his case for a hierarchical society governed by beings of 'great wisdom and strong will'. But those of us who have lived in a century dominated by beings of 'great wisdom and strong will', and reddened by the blood they have shed, feel that the case against wisdom as a desirable characteristic in a leader (as opposed to honesty, competence, humility, etc.) is much stronger. The leader to whom the impalpable, transcendent quality of wisdom is attributed cannot be challenged, even when he leads his people to the kind of catastrophe that Mao visited at regular intervals upon the Chinese people. Wisdom is an inscrutable mode of competence that allows one to side-step the usual tests of competence. It is, therefore, ideally suited to providing alibis for an incompetent, arrogant, greedy, nepotistic, brutal scoundrel. The self-proclaimed wise man lacks humility; the publicly acclaimed wise man is on the road to ruin, but he will ruin the public before he ruins himself.

The natural homeland of wisdom is, of course, the realm of the sacred. Wisdom is what most importantly sets off the priest from his flock, the adept from the layman. The priest's special competence lies in the transcendental knowledge he enjoys, his special insights into God's will and ordinances. This should, in a predominantly secular world, be a small realm; for the distinctive wisdom of the nineteenth and twentieth centuries has been to have discovered how many problems are practical problems for which the solution is technical; and the priest has no special competence in the practical world. His wisdom refers to a place beyond the daily world in which we get and spend, or indeed struggle for survival. But this is not how things work out. We have already seen how the wisdom of the Church has blocked many practical solutions to urgent problems – such as how to ensure that one gives birth to a living child without dying oneself, and how to have only so many children as one can feed. The Church, whose only claim to consideration is wisdom, necessarily wishes to make everything a matter of wisdom. This is helped by the ambiguity of the figure of the priest: through his wisdom he represents a power, not himself, to others on earth; and although the power he represents is a power not his own, he becomes, by means of it, a power himself. In the attitude to the priest, the desire to prostrate oneself before God and the desire to grovel before another human being converge. As the representative of an unworldly wisdom on earth, he has an unworldly power that is not tested against any kind of practical competence. His advice is sought on many things outwith his competence and he is an authority on all things. The road to catastrophe and ruin is thus hard-metalled – whether it is the local horror of Rancho Apocalypse, brought about by a single charismatic unaccountable figure, or whether it is the more widespread horror of the current situation in Central Africa or the Philippines, which is in part the product of the concerted action of a mass of priests subscribing to a powerful Church.

The case for wisdom, or something like it, or something like the image of it – as of a knowledge and understanding that goes beyond competence, technical expertise, a *kennen* beyond *wissen*, implicit knowledge deeper than explicit knowledge – a moral impulse that goes beyond the individual, the selfish, a vision that looks beyond immediate ends to ultimate ones, beyond the immediate problems and their solutions to final aims and consequences – at first sight looks strong. The lovers of wisdom can point to the various pickles that clever specialists have got us into and remind us how the accumulation of knowledge (or, if they want to be really rude, 'mere information') has not led to spiritual satisfaction, etc. But this case – for more wisdom, for an extension of its influence – simply falls apart on close inspection. A cynic might say that progress will not consist in gradual elimination of traces of Original Sin but all traces of the Original Gullibility we enjoyed when we were infants and believed, on the authority of our parents, what we were told. A desire to believe in the myth of wisdom may be seen as a continuing symptom of this Original Gullibility. Of course, our attitude to wisdom may be ambivalent: we fear it, as well as needing its reassurance; for although wisdom was what our parents had when they were the keepers of the objective meanings in the world, this was a time when we were often in the wrong and always (compared with adults) incompetent and in need of regulation. Wisdom therefore reassures us – telling us that there are stable meanings in the world that make us safe – but it makes us uneasy, for it judges us as silly children and places us in an inferior and dependent position. Those wise eyes that reassure us also look at and through us.

The notion of wisdom as a special mode of knowledge, a transpractical competence, is indistinguishable from arrogance. At the religious level, the priests of the established Church cannot afford to admit to making mistakes – the humility that lies at the root of what remains of wisdom in a world of accountability – for the truths the Church promulgates are eternal as they are about an unchanging God. The charismatic leader cannot afford to admit to having made mistakes because he cannot share the fallibility of those from whom he is infinitely distanced by his privileged insight into the mind of God: his failure, if he fails, must be apocalyptic. He must lie to himself or pre-empt the necessity for this by denying himself the kind of examination and questioning that the rest of us expose ourselves to: his internal PR must be as good as his external PR. (The Shilling Lives of the Wise Men do not admit this, of course. We have stories of 'inner turmoil', 'agonies of self-doubt', etc. But in order for the life to be recorded, to qualify as a Shilling Life the doubts have to be overcome.) At a lower level, it is difficult to separate wisdom from the complacency which comes from awareness of possessing wisdom. Naturally, humility may be simulated – even the Pope who exerts power over a billion pelvises is a 'humble man' – but it is probably impossible to be 'wise' without being complacently aware of being wise.[68] True competence actually requires much more humility, if only because competence has to be maintained with great effort and it is tested again and again against performance. Competence is, therefore, not only more difficult to achieve than the mask of wisdom, it is more humble. It requires much more self-forgetting.

Wisdom, then, seems largely an anachronism, like a spiritual or cognitive phlogiston. It traditionally comes in, or is concentrated in, pearls and it seems as if these pearls can be linked together to form an amulet to ward off critical intelligence and to drive away the demons of actual expertise, common sense, genuine learning from experience, factual knowledge and ordinary sympathy and kindness. For all of these latter threaten – along with accountability – to leave wisdom without a place to call its own. And wisdom is a dangerous anachronism. The truth of Karl Popper's words about great men seem even more compelling now than when he wrote them in 1943:

> if our civilization is to survive, we must break with the habit of deference to great men. Great men make many mistakes; and . . . some of the greatest leaders of the past supported the perennial attack on freedom and reason. Their influence, too rarely challenged, continues to mislead those on whose defence of civilization depends, and to divide them.[69]

For 'great men' read 'wise men'. 'Where is the wisdom we have lost in knowledge?' the poet asks. If wisdom is a form of unearned authority, a legitimacy from nowhere, our reply must be an immediate and cheerful: 'Well lost!'[70]

CODA: KITSCH RE-ENCHANTMENT: MAGIC IN AN AGE OF SCIENCE

For many, the hope of a re-enchanted world does not focus on a return to the established religions – perhaps on account of suspicions and anxieties I have explored in this chapter or perhaps because its rituals and observances are too demanding or too dull, too reminiscent of the boredom and prisons of their childhood – but in dalliance with less respectable, more sensational forms of the sacred, less pin-striped manifestations of the transcendental mysteries. Such individuals prefer

> To communicate with Mars, converse with spirits,
> To report the behaviour of the sea monster,
> Describe the horoscope, haruspicate or scry,
> Observe disease in signatures, evoke
> Biography from the wrinkles of the palm
> And tragedy from fingers; release omens
> By sortilege, or tea leaves, riddle the inevitable
> With playing cards, fiddle with pentagrams . . .[71]

and so on. In short, to dabble in the occult, to sample the thousand available varieties of 'alternative therapies', and to seek out the paranormal.

It is easy to see why the occult, ESP, etc., are attractive, and why the attraction connects with deep needs as well as the shallower longing to be in on something only the cognoscenti have access to. John Jay Gould has expressed well the deeper origins of the fascination with the occult:

> I think it has to do with immortality doesn't it? . . . So much of our human life is our attempts to deal with this most terrible fact – which our large brains allow us to learn for non-adaptive reasons – namely that we must some day die and disappear and not be here any more. There's such a great desire to believe that mind is transcendent of the universe in the hope that, although we know our bodies disappear, there's something about us that will be immortal. If there is ESP, if mind can be transferred, if mind can live, if there is reincarnation, then we might have continuity. That's what it's all about fundamentally.[72]

Much of this is harmless, because powerless, though there are some individuals who, through their interest in the occult, scare themselves and others to death, and there are others who fall into the hands of unscrupulous and charismatic charlatans who may exploit them financially, sexually or in other ways. Nevertheless, even though it usually does not cause positive harm, preoccupation with magic, and in particular the paranormal, closes off areas of consciousness which might otherwise be developed.

First, such preoccupation is counter-educational. As Richard Feynman has pointed out in the essay mentioned earlier (see note 15), contemporary magic has to borrow an aura of authority by clothing itself in pseudo-scientific discourse. Science grew out of alchemy; and now contemporary alchemy – typified in various forms of 'alternative' medicine – imagines that, if it expropriates some of the language spoken by its child, it will take on some of its power. (The confusion of words and things is of the essence of magic and magic thinking.) Reflexology, for example, borrows the physiological concept of the 'reflex' and the notion of topographic representation in order to provide a theoretical basis for therapies that have, in fact, no basis in physiology nor any proven efficacy. Where scientific support is sought, the necessary 'observations' are usually obtained in the absence of any proper precautions to ensure that they are not subject to confounding factors. Often the names of famous scientists are invoked in order to add credibility – as if the rejection of argument from authority did not lie at the very heart of the scientific ethos. The revival of magic thinking frequently seeks justification in Ripley's Believe it or Not accounts of the mysteries turned up by scientists. Contemporary physics, in particular quantum mechanics, has been used in this way. For example, the rejection of causality and the upholding of the notion of non-locality at the micro-physical level are seen to confirm what magicians have always maintained about the interaction between thought and things. The fact that these breaches with classical physics and non-magic common sense can be observed only under special conditions in the laboratory and do not apply to the macrophysical world – the very place where the hopes and delusions of magic apply – is overlooked. So has the fact that universal physical laws cannot explain, support, validate or prove the existence of the rather local and special events in the magician's hut, the tea cup, the crystal ball or the seance. In short, the magician's use of science is an anti-education in the methods and principles and laws of science – and the painstaking care with which its conclusions have to be established and described:

> How poor are they that ha' not patience!
> Thou knowest we work by wit, and not by witchcraft,
> And wit depends on dilatory time.

(*Othello*, II.ii.360–3)

Secondly, preoccupation with magic traduces the true nature of the imagination, of enchantment and of piety. The fairies, spirits, mystic modes of communication, which our forebears conceived, grew out of their sense of the mystery of things. For them, they were of the first order. For us, they are second-order, hand-me-downs. Our imagination needs to confront the mysteries head on. And there is no shortage of such mysteries: they lie not in objects and events whose very existence is in doubt, but in the things about us that no one could doubt even if they tried. It is a curious misconception that an interest in, or a desire to believe in, paranormal phenomena, is a sign of a well-developed imagination. In fact, the reverse is true: those who need the excitement of extra-sensory perception are simply numb to the miracle of sensory perception; those who seeks visions of the mysterious are blind to the mystery of vision, etc. Such individuals are devoid of the sense of the enchantment of the everyday; they lack true piety, inasmuch as they prefer exotic localised mysteries to the mysteriousness of everyday things. For the truly awoken imagination, ordinary 'voyance' (about whose existence there can be no question) is more mysterious than clairvoyance, present sight more interesting than foresight, etc. It is perverse to hold that ghosts, which may or may not exist, are more interesting than radio waves, which certainly do; that SP is less interesting than ESP; or that the living dead (about whose existence there is considerable, dull dispute) are more interesting than the living. Radio waves are no more completely explained than ghosts; SP is no less an enigma than ESP; a body haunted by its consciousness, its mind, is infinitely more worthy as an object of wonder than a room possibly haunted by a departed spirit.[73] This – the mystery of the world and, above all, of the living people in it – is something that lovers know. In their passion for one another, they are privy to the true mystery of ordinary life: their mutual enchantment reveals how there is more true magic in the mind-transfer that takes place via the speech of one living person to another than in the voices of ghosts speaking to the communicants at seances. Lovers can see how incarnation is as mysterious as any apocryphal reincarnation. And the loving and delighted parents of a newborn child know that (as Paul Valéry said) every birth is a resurrection of the flesh.

Having said this, even lovers cannot always be entirely absolved of trying to find the mysterious in the doubtful rather than in the obvious and certain. Frequently, they marvel over the coincidence of their meeting and getting to know one another, and read into this evidence of predestination and the operation of hidden forces. The general passion for finding symbolic significances in coincidences is to be deplored. We cannot usually say whether or not they are against the odds and require occult explanation because the denominator – the number of unpaired incidences – is not given. This is partly because of idleness and lack of any scientific method and partly because it is impossible to count up incidences; they do not naturally divide themselves up into countables. Nevertheless, it is safe to say that during the writing of this sentence, many thousands of incidences (sounds, actions, thoughts, memories) unpaired with co's have taken place. The apparent excess of coincidences over and above the number that could be explained by chance is the result of recall bias – which biases recall in favour of the co- and the extraordinary co- against the ordinary incidence and the dull co-. The continuous background of non-co or even dull co- is overlooked. We have to ask ourselves whether individuals obsessed by coincidences are so bored by incidences that they have to find some way of making them interesting. The coincidence of a meeting – manipulated by chance or some putative force – is no more mysterious than a planned meeting that involves perception and locomotion, assisted and unassisted, communication assisted and unassisted – in short agency operating over space and time. Instead of saying, 'Fancy our meeting by chance!', we should say 'Fancy our meeting by design!' (I am reminded of Pascal's impatience with individuals who were fascinated with identical twins but not with humanity.) We should not seek signs, Musil says somewhere; the true state of man is when everything is a sign. To put this another way: true imagination is excited by the actual, rather than the merely fancied possible. Or, as Wilde remarked, 'It is only the unimaginative who ever invent anything.'

That, perhaps, is why a teacup in a circus tent is not the best of all possible theoscopes or the most appropriate means of re-enchanting the world.[74]

4

A Critique of Cultural Criticism

Sartre's case is exemplary but not unique. A sort of moralising masochism, inspired by the best principles, has paralysed a large number of European and Latin American intellectuals for more than thirty years. We have been educated in the double heritage of Christianity and the Enlightenment; both currents, religious and secular, in their highest development, were critical. Our models have been those men who... had the courage to tell and condemn the horrors and injustices of their own societies.... Their criticism and that of their heirs in the nineteenth century and the first half of the twentieth was creative. We have perverted criticism: we have put it at the service of our hatred of ourselves and of the world. We have not built anything with it, except prisons of concepts. Worst of all, with criticism, we have justified tyrannies. In Sartre, this intellectual sickness turned into an historical myopia: for him the sun of reality never shone. That sun is cruel but also, in some moments, it is a sun of plenitude and fortune.[1]

[There is] a particular theme of modern literature which appears so frequently and with so much authority it may be said to constitute one of the shaping and controlling ideas of our epoch. I can identify it by calling it the disenchantment of culture with culture itself – it seems to me that the characteristic element of modern literature, or at least of the most highly developed modern literature, is the bitter line of hostility to civilization which runs through it.[2]

Pessimism has become something of a fashion, a kind of intellectual pose to demonstrate one's moral seriousness. The terrible experiences of this century have taught us that one never pays the price for being unduly gloomy, whereas naive optimists have been the object of ridicule.[3]

INTRODUCTION: THE WISE MAN AS INTELLECTUAL

Although 'wisdom' seems a rather precarious quantity in a secular, rationalistic world of competence and accountability, the notion of 'the wise man', 'the sage' still carries some credence. The belief persists that there are individuals endowed with this elusive faculty, people with an understanding that goes beyond, and digs beneath, specific competences. What remains of religious belief in secular societies has supported a profusion of gurus – rather as a rotting tree provides sustenance to a rich variety of fungi. These individuals are less likely to be formal members of a priestly hierarchy (vicars of established Churches in secular societies are more often objects of scorn and pity or grudging respect than of reverence) than freelance sages related saprophytically to established religions.

Alongside these self-employed Knights of Transcendence, there are secular gurus whose views also command attention and respect, although these views are more likely to be lectured from podia than howled in the wilderness, given to the flock in paperbacks than intoned from high towers. I am talking of intellectuals. The credence given to the pronouncements of certain intellectuals – and, indeed, the very concept of an 'intellectual' – is a tribute to the remarkable persistence of the belief in the existence of a faculty of free-floating 'wisdom'.

There is something very positive, indeed admirable, in the notion of the intellectual, understood as an individual possessed of a sceptical intelligence, guided by reason and fact, regulated by a robust common sense and irreverent sense of reality, and fired with a passion for fair play; of an individual placing the advantages of his comparative freedom from want and his intelligence at the service of justice. The supreme example was, of course, Voltaire who had the added virtue of physical and moral courage: he was prepared to excoriate the powerful, though it earned him persecution, cudgelling and exile. Commitment to both rational thought and moral courage were fused in the notion of an unflinching commitment to the truth and to erasing the infamies – the vested interests, the preconceived ideas, superstitions, etc. – that concealed the truth and so blocked the way to human advancement.

In the notion of an abstract and indignant intelligence as an all-purpose tool that is able to see deeper and further into every situation which confronts it, there are, however, dangers – which Voltaire and the *philosophes* did not entirely escape – as his ultimate 'apotheosis' proves. These dangers seem to have become, in the case of many contemporary intellectuals, something between an occupational hazard and a job description. Chief amongst them are: (a) a delusion of omniscience; (b) a permanent state of moral superiority bordering on moral megalomania; and (c) a tendency to be hostile towards, and negative about, the society that feeds, clothes and waters oneself. This is nowhere more clearly seen than in France, amongst those numerous *maîtres à penser* who have assumed Voltaire's mantle (even though in many cases, they may execrate the memory of the man). But it is also seen elsewhere, perhaps because France defined the characteristics of the intellectual; and perhaps because some of these – the assumption of omniscience in particular – may be connected with deep propensities of the human mind. (We shall return to this at the end of the chapter.)

There are two paths by which an individual may become established as an all-purpose intellectual, a wise man whose field of understanding encompasses the entire modern world. The first, and longer, route is to acquire a reputation for scholarship, thought and innovation in a particular discipline (usually in the humanities rather than the sciences: the wider views of biochemists are rarely sought). Achievement in the native discipline is seen to signify a more general intelligence, an all-purpose profundity of mind, which can be brought to bear on issues outside the specific area of competence in which the reputation was gained. Thus historians, literary critics, anthropologists begin to command a hearing for their views on their own society, on politics, on morality, on the way forward for man, etc. Once this tradition has been established, the passage from scholar to seer gets shorter and shorter.

In recent decades, the abbreviation of the path between scholarship and sagehood has reached its logical conclusion in the birth of academic 'disciplines' whose primary function is to equip individuals with the ability to pronounce upon a wide range of issues relating to the nature of contemporary life and the way forward for society. These are 'higher order' or second-order 'multi-disciplines' that rarely generate new information of their own but rather synthesise or bring together information and ideas from other disciplines. The emergence of such disciplines in the 1960s had many causes. One was the publication of widely read books that brought to commentary on contemporary life scholarship from fields as remote as anthropology, modern literature, depth psychology, etc. The writings of Norman O. Brown and Herbert Marcuse were typical in this respect. Another cause was the structuralist shift from focus on content to a focus on underlying form and the boundaries of possible content. This began with phonology but spread to encompass many other subjects, in particular anthropology and 'psychiatry'. The superseding of linguistics by semiology under the leadership of the charismatic Barthes – of the study of language by the study of all sign systems – and the interpretation of society and consciousness as interacting sign systems, was a crucial event in breaking down the barriers between disciplines and, incidentally, in removing internal quality control on scholarship and argument. An all-purpose method – a science of forms that studied significations apart from their content – seemed to be available for permitting one, with remarkably little effort, to understand every manifestation of human life.

Although Barthes' 'euphoric dream of scientificity' (his own words) lasted only a decade, its example was enduring. He himself was appointed to a Chair of Literary Semiology at the Collège de France; and his contemporary, Michel Foucault, was appointed to the Chair of the Study of Systems of Thought at the same institution. Although the structuralism that established the reputations of stars such as Barthes and Foucault – notwithstanding that most of them either abandoned their structuralism or denied that they had ever been structuralists in the first place – was soon deeply unfashionable, it left an enduring impression on humanist intellectuals and wrought changes in the practice of humanities that are still apparent. For structuralism, with its emphasis on systems of the possible rather than examples of the actual, indicated ways in which it might be possible to avoid the initial passage through a particular discipline and become a free-floating intellect straightaway, in other words, to begin, rather than end, an academic career by being a Professor of Things-in-General. It was possible to be multi-disciplinary from the outset, be a free-ranging intellect, who is an authority on absolutely everything – for form is invariant across a wide variety of content.

This move from form to content was at once downwards – to the 'underlying' structures – and upwards – to an overview. A more homely upwards movement for some scholars (or quasi-scholars) was supported by the declaration that literary studies should be contextually sensitive, that scholars should examine how literature related to the real world, and recognise that literature was only a minute ('elite') part of a much larger mass of cultural manifestations. Literary studies should give way to Cultural Studies – which related the literary text to the larger Social Text and looked more widely at non-literary cultural phenomena, such as soap operas, football fanaticism, pop music, etc. This democratising move – driven in the UK by inspiring individuals such as Raymond Williams (who was concerned with the hegemonic power of the received culture), Richard Hoggart (who had discovered that the reading of literature was the last use to which the acquisition of literacy was put) and Stuart Hall – was another assault on the content-based expertise of traditional scholars, and enabled entrepreneurial young academics to leapfrog those who remained mired in particulars. The journey from post-doc in Cultural Studies to Professor of Things-in-General is much shorter than the journey from Lecturer in Seventeeth-Century Literature to Reader in Seventeeth-Century Literature.

Richard Rorty wrote of 'a kind of writing ... which is neither the evaluation of the relative merits of intellectual productions, nor intellectual history, nor moral philosophy, nor social prophecy, but all of these mingled together in a new genre'.[4] It was Derrida, however, who contributed most decisively to the final freeing of intellectuals from a specific area of peer-reviewed or otherwise tested expertise. Although he 'did time' in a particular area of scholarship – notably studying phenomenology – he did not make much of a reputation there and he transferred very early to the career of a Professor of Things-in-General. He arrived spectacularly with his *Of Grammatology*, which brought with it a method that emptied every text of specific content (except self-contradiction) and made everything a text, while his own texts retained a fullness, a density of signification, in virtue of being so badly written as to cover everything with a thick night of obscurity. The signified darkness – the site of endless quarrels among scholars – provided the very substance that his writing, his method, denied. He also memorably reinvented the intellectual as a kind of yeast in the dough of discourse, one who put all disciplines on guard against themselves, a free-floating mind alighting where it will and, like the false Government Inspector in Gogol's tale, leaving everything irreversibly altered.

The following passage from an interview with Derrida beautifully captures the all-purpose competence that characterises the new breed of intellectuals. Here Derrida is explaining the role he feels he can play in the development of contemporary architectural thought – despite never having built a house or had any training in the field:

> I thought at first that perhaps this was an analogy, a displaced discourse, and something more analogous than rigorous. And then ... I realised that on the contrary, the most efficient way to put Deconstruction to work was by going through art and architecture. As you know, Deconstruction is not simply a matter of discourse or a matter of displacing the semantic content of the discourse, its conceptual structure or whatever. Deconstruction goes *through* certain social and political structures, meeting with resistance and displacing institutions as it does. I think that these forms of art, and in any architecture, to deconstruct traditional sanctions – theoretical, philosophical, cultural – you would have to displace ... I would say 'solid' structures, not only in the sense of material structures, but 'solid' in the sense of cultural, pedagogical, political and economic structures. And all the concepts which are, let us say, the target

(if I may use this term) of Deconstruction, such as theology, the subordination of the sensible to the intelligible and so forth – these concepts are effectively displaced in order for them to become 'Deconstructive architecture'. That's why I am more interested in it, despite the fact that I am technically incompetent.[5]

Mere 'technical incompetence' – in this case in the field of architecture – does not stop him from having his say. For the guru has a general, free-floating competence that overrides the specific incompetences. This is the master-delusion inscribed in the notion of 'the intellectual' – one modern manifestation of 'the wise man'.

It might be thought that possession of an all-purpose method, neutral as to field, would not bring with it knowledge because it is addressed not to content but to form. However, the Method can yield the required dividends and turn omnicompetence into omniscience if it is suggested – as many cultural critics since the structuralists have suggested – that there is no such thing as content, only form. This has been roundly reaffirmed by Derrida's denial of presence. An understanding of form is the nearest we sublunary beings shall get to content. The all-purpose intellectual can therefore unpack the requisite omnniscience from his omnicompetence and can make seemingly empirical pronouncements of an enormous scope, perfectly exemplified in the kind of statements Barthes specialised in: 'the petit bourgeois is a man unable to imagine the Other' ('Myth Today'); 'replacing the feudal index, the bourgeois sign is a metonymic confusion' ('Index, Sign, Money', in *S/Z*); 'language . . . is neither reactionary nor progressive; it is quite simply fascist' ('Inaugural Lecture, Collège de France').

In a scathing comment on Ezra Pound, who seemed to have something wise and old-possumish to say about everything – from poetry to economics, Mencius to Amy Lowell – Gertrude Stein, called him 'a village explainer', adding, lethally, that this was 'all right if you were village, but, if you were not, not'. The Information Superhighway seems set to realise the dream of the Global Village (which links people, but not their experiences, together). And it is appropriate, therefore, that we should have had an epidemic of Global Village Explainers. Baudrillard and Lyotard are the best known, but they head up a vast chorus of intellectuals-as-seers. These second-generation *maîtres à penser* are the most developed representative of a whole class of humanist intellectuals who, without necessarily having made their reputation in a particular field of scholarship, are able to become Professors of Omniscience. They are able, like every barroom bore, to pronounce with charismatic authority upon everything under the sun, but they are also, unlike the barroom bore, able to command a hearing for their pronouncements. They address not a bored and sceptical wife or a solitary, switched-off drinking companion wondering what he has done with the car keys, but audiences of many hundreds of students and other acolytes hanging on their every word. The reason for the attentiveness of their audiences is obvious: those who absorb the wisdom of the Professor of Omniscience – those who believe that they might understand him – entertain the hope of becoming, themselves, in their own little spheres, Professors of Omniscience. To study with, or under, a Global Village Explainer is to learn how to become a Global Village Explainer oneself.[6]

HYSTERICAL HUMANISM AND THE DIAGNOSIS OF SOCIETY

Since the passage from junior academic to intellectual guru has been so greatly accelerated as a result of the discovery of techniques of omnicompetence and the emergence of subjects, such as 'Cultural Studies', that enable everyone who engages with them, however briefly, to be able to see deeper and further, to X-ray the very culture in which they live, it is hardly surprising that guruhood is no longer the privileged status of a few. Anyone who studies the works of a guru becomes himself or herself a guru-let of sorts.

Global Village explaining is not therefore confined to full-time academics. So numerous now are the cultural critics and society doctors that, although I do not believe that any age has characteristics that can be captured in a single adjective or a handful of adjectives, I am tempted to characterise the present as an age that has a predisposition to characterise itself – and in disparaging, even pathologicalising terms. A significant proportion of the intellectual and para-intellectual community (columnists et al.) sees its prime function to be that of scourge of the world in which it lives. This function used, of course, to be delegated to the priesthood, who took up arms against a secular world on behalf of the gods of which they, but not the world, were mindful. There is now a large unofficial priesthood, mostly without religious belief and certainly not invoking divine validation (only the legitimation of the Shining Names), whose views are addressed to an ill-defined congregation in a church-without-walls, and slipped out not in sermons but in books, articles, reviews and television programmes. The diagnostician of society tends to be scornful, angry, sorrowful. He may be paid for being a journalist, but he is in truth a minor prophet, a pocket Jeremaiah, a Cassandra of the Home Counties and/or the *arrondissements*. There has also been a 'trickle-down' effect: Global Village explaining has become such a normal part of journalistic commentary that it is hardly noticeable. When the television critic of the *Sun* newspaper talks about 'the madness of the age', it causes less stir than if he had used a semi-colon or written a paragraph composed of more than one completed sentence. We are all cultural diagnosticians now, though only a privileged few are full-time professionals.

And what they all – gurus and guru-lets – find when they diagnose modern society is that it is seriously sick. This was remarked by J.G. Merquior, whose brilliant critique of structuralist and post-structuralist thought ends as follows:

> That a deep cultural crisis is endemic to historical modernity seems to have been much more eagerly assumed than properly demonstrated, no doubt because, more often than not, those who generally do the assuming – humanist intellectuals – have every interest in being perceived as soul doctors to a sick civilisation. Yet is the medicine that necessary or the sickness that real? Perhaps we should be entertaining second thoughts about it all.[7]

Alas, Merquior died prematurely and time was not granted him to carry through such a programme of 'second thoughts', though his identification of *kulturkritik* – the refusal of modernity – as the thread linking so much of the thought or pseudo-thought of contemporary humanist intellectuals, and his penetrating comments about the genealogy and sociology of the merchants of the *kulturkrisis*, are a magnificent start.[8]

Foremost among Merquior's targets were the post-Saussurean thinkers and others for whom contemporary history and the writings of Nietzsche have created a pretext for diagnosing a fundamental sickness in modern civilisation:[9]

> Modern culture is deemed to be in crisis because its mental set is 'shown' to be fallacious; *kulturkritik* presupposes a *kulturkrisis*.

This doesn't have to be taken for granted, of course:

> it may well be that behind the cultural void and trash alleged by advanced nihilist thought there is nothing to be apprehended. *The crisis, then, would be less an object than a product of countercultural thought.* . . . Post-structuralist . . . theory should be prepared to accept that its One Big Thesis – the massive rottenness of our culture – is a creative fake. Hence the invariable phoniness of its thundering against Western (that is modern, enlightened) values: it must sound hollow, since it is all a construct rather than a record of a real situation.

Once this is conceded, then there are problems for the exponents of *kulturkritik*:

> In any case, how can such a school of thought, embattled as it is against all notions of objective truth, ever convince us that its picture of our cultural predicament is the *right* one?

It is in the spirit of Merquior's rational scepticism towards what he calls 'hysterical humanism' that the present enquiry is conducted.

Cultural critics present us with a variety of grave diagnoses, but do not usually offer any treatment (Young is an exception in this regard). Of course, a good doctor should make a diagnosis before proposing remedies. The question is how good, how accurate, how true, the diagnoses are. The first thing to notice about them is that they are mind-exceedingly large. Intellectual over-reaching is a necessary fault, almost a pathognomonic sign, of those who would wish to put up their plates and establish themselves as doctors to sick civilisations by pronouncing modern life, contemporary culture, the present age to be rotten and its denizens to be aspiritual, empty, imaginatively atrophied, moral pygmies or corrupt power-brokers. (It is interesting to note in passing that saying such rude things about entire epochs or societies rarely causes offence. Our contemporary prophets earn accolades for themselves and their 'uncomfortable' – ultimate term of praise – messages; they are certainly better paid and better treated than Jeremaiah or Cassandra ever were, with shorter and more regular hours of work, and enjoy more respectability. The standard fare for such prophets in the wilderness is not 'locusts and wild honey' but tenure, syndication, jet travel, hospitality suites and international cuisine.)

The question we need to ask is whether or not this, or indeed any other, age is diagnosable; whether general statements about 'contemporary man', 'the present epoch', 'the world today', have any validity whatsoever, or have the remotest chance of being true. I don't think they do. They are unsupported by anything like an adequate amount of data. A moment's thought should be sufficient to convince one of their vulnerability. A general statement about an individual – whether he is good or bad, whether he is or is not impatient, whether he can or cannot concentrate, is always unsatisfactory. It seems to fall short of the complexity of the facts, of the felt reality of his life – as observed in different situations, in relation to different people, different demands, different activities, different moods, different states of consciousness. (Think how an individual may fluctuate in the course of a day – and how uncapturable the experience of a day is, as soon as one genuinely sets oneself to recall it rather than merely to make some standard diary-entry comments about it.) How much more un-satisfactory must any general statement be about a small group of people – say, a group of friends. The gap between the simplicity of the assertions and the complexity of the facts will increase exponentially as one moves from small groups to larger groups – to the pupils of a school, to the inhabitants of a village, to the citizens of a town, to the people in the North of England, to the British race, to 'the' European culture and, finally, to the contemporary world, composed as it is of individuals of different sexes, ages, social classes, characters, states of nutrition, educational attainment, interests, upbringing, customs, life-chances, expectations, attitudes to the world and to others, habits, beliefs, joys, etc. And yet, general statements of the widest scope are made repeatedly and those who make them seem comfortable with them. Indeed, the greater their scope and the more complexity they gloss over, the more variety they conceal, the more easily they seem to evade critical scrutiny. A thoughtful person would be dubious about an epitaph that purported to characterise his entire life-experience, aware of the huge differences that are to be found within his waking hours. And yet the same person, if he were a cultural critic, would have considerable confidence in a statement about the people of a certain social group (say, 'the bourgeoisie' – some hundreds of mil-lions of individuals over the last three centuries), about 'Western man', or about the present state of the human soul.

The reasons for this anomalous propensity of the intellectual mind are quite complex. The higher order the general assertion, the more out of focus its referents and the less we are able to test it against what we have experienced ourselves or what we can summon up of our experiences at the time of thinking about its the truth or other-wise. A high-order general statement invites us to put every indi-vidual out of focus so that differences, like the differences between the individual leaves of trees in a fog, are lost. Very high-order gen-eral statements – for examples, those about 'contemporary man' – have the additional advantage of being unimaginable. We can recite them, refer to them, but not reach into them or accommodate them in our minds in any way that would enable us really to test them against what we know and have experienced, as opposed to what we think we know and imagine we have experienced. We cannot, therefore, determine whether they are true or false. However, we may be in-clined to give them the benefit of the doubt for they are very satisfy-ing: they give us a feeling of mental power. The handful of repeatable words seems to confer an ability to rise above our own lives, our own groups and classes and intimates and casual acquaintances, to encompass the vast crowd of mankind in our all-seeing mental gaze. The words are easy to invoke at seemingly appropriate times; and each repetition seems like a kind of confirmation (we tend to believe what we have heard ourselves and others say repeatedly) and strengthens the conviction the words command – rather as well-used neuronal pathways have facilitated activation. Finally, there is the tendency we have to focus on one visible aspect of things and ignore all the rest; to identify a simple distinctive essence and overlook everything else. This is the basis of stereotyping that is so essential to our mental handling of the world and to the prejudices with which we master it. We notice one difference and ignore the complex network of similarities and dissimilarities that form the background against which this difference emerges. The higher-order the diagnosis, then, the more difficult it is to test and the more satisfying it seems. No wonder, then, that global diagnoses enjoy such wide currency and command such respect.

There have been many famous diagnoses of the sickness of contemporary life. To take a few at random:

1. For Marx and his numerous contemporary epigones, man in capitalist society is alienated from himself. That essential property of the human spirit – in virtue of which it meets itself through otherness – is turned into a separation of the human spirit from itself. Man in capitalist society meets himself though the process of production which reduces him to a fragment, a part-spirit, an operative, a functionary. He is divorced from himself through his alienated relation to the means of production and his having to negotiate self-awareness through the mediation of fetishised and non-fetishised commodities.

2. For Eliot and many others contemporary consciousness is marked by a dissociation of sensibility, whereby the thinking and feeling, the intellectual and sensuous, functions of man are divided from one another. Whereas for Donne the smell of a rose and a physical thought are as one, for the modern poet these are quite remote from one another, kept apart in the two halves of his divided soul.[10] The difference between the poets signifies the differences between their respective epochs.

3. Wordsworth stands at the head of a long tradition of Romantic protest against the dissipated, broken condition of modern man, a creature for whom 'getting and spending early and late' has made him forget his true nature, the purpose of life, the beauty of the world:

> For a multitude of causes unknown to former times are now acting with a combined force to blunt the discriminating power of the mind, and unfitting it for all voluntary exertion, to reduce it to a state of almost savage torpor. The most effective of these causes are the great national events which are daily taking place, and the encreasing accumulation of men in cities, where the uniformity of their occupations produces a craving for extraordinary incident which the rapid com-

munication of intelligence hourly gratifies. To this tendency of life and manners the literature and theatrical exhibitions of the country have conformed themselves. The invaluable works of our older writers, I had almost said the works of Shakespear and Milton, are driven into neglect by frantic novels, sickly and stupid German tragedies, and deluges of idle and extravagant stories in verse.

(Preface to *Lyrical Ballads*, 1802)

What would he have said now, in the era of hourly news bulletins, Sky television and the airport paperback? Gurus 200 years his junior tell us.

4. Lawrence's diagnosis is another favourite. According to him, we have lost our natural piety, our contact with and awareness of our animal and instinctive origins, with our primitive ancestry and the dark gods calling in our blood. Though our book-ridden, sex-in-the-head, repressed culture has forgotten these ancestral voices, it still somnambulistically enacts their imperatives in the ritual of mass murder exemplifies in total war and in the perverted and mangled relations between the sexes.

These are but a small (predominantly English-speaking) sample of the shades whose diagnoses haunt contemporary cultural criticism. And they are, of course, only a minute sample of the vast numbers of diagnoses of contemporary man that are available to cultural critics who may select from them, as from a sweet trolley. There are, for example, many varieties of alienation on offer – from the gods, from our selves, from the sense of the sacred, from nature (too much indoors city-dwelling, too much greed, too many books), and even from Being itself. Or, if that is not enough, from Being plus something else: for Heidegger, we are 'too late for the gods and too early for Being', (echoing Matthew Arnold's *Grande Chartreuse*, 'Wandering between two worlds, one dead/ The other powerless to be born').

These diagnoses have many different origins. In some cases, they arise from the visionary dissatisfactions of individuals of genius with their own lives. In some cases from field observations – such as those that compare us with animals and with our primitive selves, to our current disadvantage. In other cases (as in the seminal work of Frazer), from armchair syntheses of the mythical past or (as in Nietzsche and Burckhardt, Toynbee and Vico) of history. These provide the primary texts. But in addition to these, there is an all-pervasive interpretosis that affects all writing, such that there is hardly a novel, a poem, an essay, or even a book review in the Sunday press that does not have a diagnostic swipe at society.[11]

Merquior has classified the humanist attacks on modernity into two major kinds:

One is the 'revolt of the spirit', the protest of the 'disinherited mind' . . . against the utilitarianism and materialism of modern life. It ran, conspicuously from, say, Coleridge to Rilke, from Baudelaire to Eliot. The other is the revolt *against* the spirit – a far more radical, indeed nihilistic branch of humanistic *Kulturkritik*. It goes so far as to reject culture itself root and branch. Germany under the Weimar Republic provided outstanding examples of this, as in Ludwig Klages, Gottfried Benn, or Carl Schmitt, all foes of every positive value. This was also the mood of Heidegger's destruction of metaphysics.[12]

The revolt of the spirit has faded: 'unabashed idealism is out'. And the revolt against the spirit was too closely associated with the Nazi hatred of culture and the Nazophilia of intellectuals such as Benn and Heidegger. The humanistic rejection therefore developed a third position, 'at once less mystical than the romantic revolt and less nasty than the anticulturalism of the nihilist Right' (ibid., p. 409). Merquior calls this Marcionism, after the views of the Gnostic Marcion. For Marcionists, whether ancient or modern,

> the social world, its values and institutions are just a messy rub-
> bish, crying for thorough replacement by good totality . . . their pas-
> sionate revulsion against the world as a whole demands nothing
> less than total rejection.
>
> (ibid., p. 410)

This hostile attitude towards modern civilisation, and the values preva-lent within our social culture at large, is now, Merquior points out, a *fable convenue* in the humanist establishment. And its target is not confined to capitalism, but extends to encompass science, industrial-ism and mass society. Merquior, with his usual felicity for telling epithets, calls it 'hysterical humanism' which, in the last analysis, is at war with progress and civilisation. He predicts that 'as long as our educational revolution goes on swelling the humanist-academe, there will be a ready-made public for professional hysteric-humanists', who will, of course, do very nicely out of this. They will do nicely; but the society from which they draw their nourishment will be badly served by them. Hysterical humanism and other modes of *kulturkritik* are themselves in need of *kritik*.

KRITIK OF KULTURKRITIK: (1) JUMPING TO CONCLUSIONS

It may be helpful at this juncture to be autobiographical – to explain, as the phrase goes, where I am coming from. My profound scepticism of diagnosticians of society comes from my experience as a doctor; or, more precisely as a medical student; more precisely, yet from the experience of the transition from the pre-clinical to clinical years at medical school.

Medicine is a practical art, but until recently the training that led up to the clinical years was based on the ability to digest, remember and regurgitate in one's own words what one had learnt about the relevant basic sciences of chemistry, physics, physiology, anatomy, etc. Gaining admission to medical school and surviving as far as the clinical years, therefore, depended very much on an ability to handle text. A high premium was placed on abstract, verbal skills to be deployed in the presence of a piece of paper. In this respect, the skills that are required of entrants to medical school are not so very different from those asked of entrants to Arts faculties. These skills are, however, quite insufficient for the clinical years of training and in the real world of practice after qualification. Making medical diagnoses and deciding on appropriate treatment demands something rather different: in addition to the application of general principles and the mobilisation of factual knowledge, students must develop a sense of what is a likely diagnosis in the individual case and what is a reasonable treatment for the person who has come to them for help. We call this sense 'clinical judgement'. The discrepancy between the level of the students' theoretical knowledge and their practical clinical judgement shows up in the ludicrous suggestions for diagnosis and treatment they typically put forward. A considerable part of the clinical years is taken up not only with learning how to make clinical observations and draw conclusions from them, but also with acquiring clinical judgement. A necessary part of this process is learning from error. Armed with a little knowledge and a good deal of abstract intelligence, students regularly – indeed typically – make diagnoses, often rare diagnoses, that seem to fit the case perfectly, but are simply wrong. They learn that just because a diagnosis fits the facts they have latched on to, this is far from sufficient to ensure that it is correct. This comes as a terrible and sometimes humiliating shock: it undermines their confidence in the intelligence that brought them into medical school. But, at the same time, it forces them to appreciate the importance of common sense, of developing a feeling for what is probable, of not focusing on just a few aspects of the case, and of being deeply sceptical of their own first and second thoughts – especially where those thoughts are particularly clever.

How different is the experience of the diagnostician of society, whose home discipline will usually be one of the humanities. He or she will be content with a plausible diagnosis cobbled together on the basis of evidence that falls absurdly short of any adequate basis; for example,

'Modern man is sick because of the separation of his thoughts from his feelings' – a statement of enormous scope supported by reference to a few, usually highly subjective, sources. A diagnosis of this kind will usually be intrinsically untestable because it is difficult to know what objective fact or facts would support – or refute – what is essentially a comparative and, even, a quantitative, assertion relating the global present to the global past. Sufficient evidence can therefore be selected or manipulated by a half-way competent essay writer to render the diagnosis proof against decisive contradiction. There is certainly nothing comparable to the biopsy or necropsy that provides the final test of a medical diagnosis. Social diagnosticians thus miss out entirely on the training – the constant punishment-by-error – that leads most medical students to be able to develop clinical judgements that correspond to some kind of reality. Social judgements, being empirical in implication but conceptual in essence, evade both tests of empirical truth and of coherence and plausibility. Their predictive value unlike those of science (which are forever being tested in their practical applications) is zero. Such judgements are neither proved nor refuted; they simply pass into and out of fashion.

The likelihood of global diagnoses of society being true may be judged by the quality of the diagnoses that are applied to smaller-scale social phenomena. Consider the epidemic of middle-order social diagnoses spawned by events such as inner city riots and football hooliganism. The 1981 Toxteth riots in Liverpool will serve as a typical example. The press, radio and television chewed over the events *ad nauseam* and in this respect were ably supported by academics who showed a remarkable willingness to discern long-term trends, patterns and 'inevitable' social forces in this and one or two other similar events. The respective contributions of under-policing and police harassment, of racial tensions and lack of parental control, of poverty and unemployment, of drug barons and politically motivated *agents provocateurs*, in bringing about the riots were debated at length and for every explanation there was at least one authoritative voice. What most of the diagnoses had in common was the assumption that the riots were due to an all-pervading sense of hopelessness and that, unemployment and alientation being permanent features of the landscape of Britain under the Tories, riots would become a permanent feature of mainland Britain. A police state, with a hostile underclass kept under control by a violent police force, would emerge. This prediction has proved untrue: though the conditions that were said to have produced the riots have remained in place, constant and proliferating inner city rioting has not been a feature of the last 15 years in the United Kingdom. No explanation for the lack of rioting, despite the continuing presence of the supposed causative factors, has been sought or offered. But neither has anybody said, 'we got it wrong'. The lecturers who spoke their errors with such authority in 1981 have, in many cases, gone on to be professors, pronouncing on subsequent events with equal authority.

The topic of football hooliganism provides even more startling – and instructive – examples of social diagnostics uninformed by any sense of the appropriate means of establishing causal relationships between events and predisposing conditions. In the late 1980s, this theme became an overwhelming preoccupation of social commentators and diagnosticians. Like the inner city riots, football hooliganism was attributed to the emergence of an alienated underclass; related anthopologically to the need for tribal warfare; and connected with a variety of other things. All of these 'explanations' and the social nostrums that flowed from them were simply beside the point as would have been obvious had any of the commentators been blessed with the slightest numeracy.

Consider these facts. On any given Saturday afternoon, something of the order of a million individuals are attending league football matches. Perhaps 500 000 of these are young men. Over the season there will be about ten million attendances by these young men. Nevertheless, serious violence is rare enough to warrant reporting in national newspapers, though the massive coverage and endless analysis of a small number of episodes may have belied their rarity: repeated reporting of a single event may give the impression of repeated events. This comparatively small number is astonishing in view of the fact that youths are meeting in large numbers, in an explicitly adversarial situation, often after the consumption of quantities of alcohol. After all, at any time, there will be a small minority of violent individuals in society who, in a crowd possessed by partisanship, could well have the capacity to cause continual rioting. It is the comparative *rarity* of football violence, not its occurrence, that requires explanation. This will strike anyone who, after many years of reading about hooliganism at football grounds, actually goes to a match and listens to the polite rounds of applause that greet brilliant moves on the part of the visiting team. There can have been few times in history when large numbers of young individuals gathered together under such circumstances have indulged in so little, or so sporadic, violence. It is arguably not football violence so much as football quietism that should attract comment.

These two favourite topics of social diagnosticians – inner city riots and football hooliganism – illustrate two connected and characteristic flaws in their interpretations of the world. The first is a tendency to focus on some particular event or phenomenon and overlook the million other events or phenomena that make up the society they are

diagnosing. These events are then seen to be larger and more defining than they really are – rather as if an individual was defined by
his ability to curl his tongue axially, so that mankind was divided
into curlers and non-curlers, and an entire theory of personality were
based on this, forgetting that most curlers and non-curlers have many
millions of things in common and that there are many millions of
differences within the population of curlers. The second is a failure
to respect the principles by which causal relationships are established.

This latter is sometimes due to the failure of social diagnosticians
to appreciate that their diagnoses are implicitly causal assertions,
particularly when the diagnoses are cast in the form of assumptions.
'What do you expect lads to do at football matches if, as a result of
unemployment, they feel alienated from the society in which they
live?' Or are disguised as predisposing-circumstance assertions. Or
as 'factors': 'In order to understand football rioting, we must take
into account many factors – unemployment, the breakdown of the
nuclear family, Britain's loss of world power status post-war', etc.
But even where there is some vestigial awareness that the 'explanations' of social phenomena are causal in nature, there is no understanding of the disciplines of numerate observation that need to be
mobilised in order to establish a clear causal relationship. This sense
does not weigh heavily upon social diagnosticians because they know
that their diagnoses will not be subjected to decisive testing. If I say
that A (unemployment) causes B (football hooliganism) then it should
be incumbent upon me to show that:

1. football hooliganism does not take place at times of full employ-
   ment; and
2. it always takes place when there is high unemployment.

At the very least, I should be able to show a numerical correlation
between levels of unemployment and the quantity of football hooliganism (and the latter should be distinguished from the quantity of
reporting about football hooliganism). I should be able to give figures to show that (a) there has been a net upsurge in football hooliganism; (b) that this is not merely due to a transfer of violence from
one setting to another (with no net gain in overall violence); (c) that
this has had a certain kind of temporal relationship to changes in
factors that are being put forward as the supposed cause or causes of
this upsurge; and that (d) there are no other variables whose changes
have a time course that would make them rival candidate causes. All
of these would amount to meeting the necessary, though not the sufficient, conditions of demonstrating a causal relationship between
unemployment and football hooliganism – not sufficient because both

may be markers of other ('underlying') factors that are the place where the causal interaction occurs.

So far as I am aware, no serious attempt has ever been made to quantify football hooliganism and to correlate it in this way with the factors to which it has been attributed, themselves quantified in a valid way. Apart from idleness – it is much more fun to pontificate than to measure – there are several reasons why such a quantitative approach may be unattractive. The main one is that social phenomena are multifactorial in origin, so that it would be unrealistic to hope to demonstrate a quantitative relationship between football hooliganism and a single factor such as unemployment. Many other factors may be operating to predispose youths to riot. Indeed, every historical or social circumstance is, or is the product of, a unique concatenation of factors or events. Another reason for not doing the decent quantitative thing is that there are certain putative factors that do not readily lend themselves to quantification – for example, 'alienation'.

The excuse that social factors are often multiple and rarely quantifiable is valid. But it achieves rather more than those who use it would wish: it actually rules out large-scale assertions of causal relations. And this means ruling out making diagnoses altogether – since aetiological explanations are implicitly causal-statistical or causal-quantitative. In the area of social diagnosis, there is, therefore, little certain knowledge, little genuine expertise, though there are many experts with claims to knowledge and expertise.[13]

The groundlessness of the claims made by diagnosticians of society even in relatively small-scale, highly specific topic areas such as 'football hooliganism' has passed largely unremarked. Even those who challenge social diagnosticians do so only to question the particular aetiological factors the latter have identified, in order to clear a space for their own factors, and not to address the inadequate methodology that undermines the validity of diagnostic claims. In the absence of a sound method, the test of truth becomes the persuasiveness of arguments, not the quantitative facts. Social phenomena become Rorschach ink-blots in which 'experts' and 'culture critics' can read their own preoccupations, and from these mine the necessary theses and other CV fodder.

If the small- and medium-scale diagnoses are almost without foundation, then the larger diagnoses, those in particular that are beloved of the Global Village Explainers, are totally unsupported. This, too, has passed largely unnoticed because most diagnosticians are simply innumerate, having no feeling for the kinds of facts – for the size of the factual base – that would be needed to support their claims. They do not appreciate that the assertion 'this is a greedy age' presupposes a vast quantitative comparison between the human qualities of the present age and other ages. Even if they did, and were aware of the impossible complexity of societies as objects of study, it is not impossible that they would still choose unfounded publishable statements in preference to untenurable silence. So, for social diagnosticians who deal with vague terms and large capacious categories, just as for first year medical students (and for self-diagnosing hypochondriacs and for rambling drunks, religious maniacs and paranoiacs), everything fits their theories.

In the absence of a proper statistical analysis, there is little to separate the assertions of social diagnosticians from the banal prejudices of barroom bores. The hostility to science, and to the quantitative approach, often expressed by social diagnosticians may, however, represent an indirect acknowledgement of the invalidity of quantitative assertions in the absence of quantitative fact or measurement. What is missing is a further recognition that an assertion such as that the present age is 'more materialistic' than its predecessors is essentially an epidemiological-quantitative one about populations; if it is to be taken seriously, supporting evidence is required using agreed measures of materialism and application of those measures to the occupants of the ages being compared. This is clearly impossible, and so the scientific method is rejected, even attacked, as one of the characteristic markers of a materialistic, spiritually impoverished age. Gellner has made this connection between 'the crisis-mongering of humanist-intellectuals' and 'the archaism of their cognitive equipment'. And so has Merquior. As he puts it:

> In the age of cognitive growth and universal literacy, the humanist clerisy is a kind of antique. It no longer holds the monopoly on writing and knows that its expertise, verbal knowledge, is no match for the authority of science. One possible strategy is, therefore, to disparage this unfavourable setting by decrying modernity.
>
> (p. 409)

It seems such a shame to spoil the sport of social diagnosis and the modernity-baiting of the hysterical humanists. Insistence upon a rigorous methodology would seem to make cultural criticism well-nigh impossible. But would this be a bad thing? 'It is better to live on the acorns and grass of knowledge, for the sake of truth,' as Nietzsche said. Or, to put it less grandiloquently, it cannot be a good way to earn one's living by adding to the stock of the world's innumerate prejudices and untruths, howsoever they are dressed up in fancy prose.[14]

KRITIK OF KULTURKRITIK: (2) BITING
THE HAND THAT FEEDS

The exponent of *kulturkritik* is typically a child of an extremely advanced state of culture. It is this that creates the conditions in which he or she is fed, watered, warmed, housed, clothed, employed, heard and admired. And yet it is precisely this advanced state of civilisation that attracts his or her intensest loathing. Only an hysterical humanist could fail to see the contradictions of this – the existential self-refutation built into a hatred not only of capitialism but, indeed, of science, industrialism and mass society. Without science-based mass production, there would not be the spare capacity in the social system to support vast numbers of individuals in jobs, such as that of being a cultural critic, that are remote from the productive processes that support physiological need. Of course, critics of culture are not entirely twentieth-century inventions. But, previously, their survival has depended upon talent rather than tenure or syndication, and their job description, in the case of the earliest examples, has included starvation or lunching in open air canteens in which the only dish on the menu was locusts and wild honey.

To say this is not intended to silence all dissent – put up or shut up – but to question whether it makes sense to dismiss precisely those conditions, in particular the advanced technology of an industrial democracy, that make possible the very things that cultural critics themselves seem to value – a life spent amidst works of art, hours, weeks, decades in a library, etc. Life without the comforts of widely available technology, dependent upon mass production, upon complex, interlocking social organisations and upon science, is almost unthinkable. As I have argued elsewhere, it would be necessary to renounce only a little of our currently taken-for-granted technology in order to lose a lot in terms of material comfort, safety and free time.[15] And this would be true, even if one assumed an implausible ideal case scenario, with technology being renounced voluntarily and universally and withdrawn in a coordinated fashion, by general consensus, with no fear or favour and no special cases, no backsliding and corruption and blackmarkets and blackmail, no *nomenklatura* retaining what they urge others to give up. I shall not go over the arguments again. Let me instead look critically at the idyll of The Simple Life, by returning to Dudley Young and his anti-modernist, hysterical humanist call for 'a return to the primitive', his critique (echoed by hundreds and thousands of others) of consumerism.

The keynote quotation for *The Origins of the Sacred* is provided by Wordsworth. It is the 'getting and spending' sonnet and not only is it quoted at the head of the book, but it is alluded to frequently and is an important element in the ethos pervading the entire book:

> The world is too much with us; late and soon,
> Getting and spending, we lay waste our powers:
> Little we see in Nature that is ours;
> We have given our hearts away, a sordid boon! . . .
>   Great God! I'd rather be
> A Pagan suckled in a creed outworn . . .

> (from *Poems in Two Volumes*, 1807)

Young asks, somewhat rhetorically,

> Would we not still prefer to be pagans 'suckled in a creed outworn'
> than to live, as most of us do, unmoved by beauty? And do we not
> still lay waste our imaginative powers, offering them up as if sacrifi-
> cially in the service of those monster twins called 'getting and spending'?
> (p. xv)

What Young forgets to wonder is whether the pagans had such a
relaxed time in their pre-technological, pre-industrial, perhaps even
pre-agricultural, Eden.

In fact, little is known about the duration of, never mind the quality of, the palaeolithic working day, about how long was spent in
fatiguing and futile chase after elusive animals. Nor do we know how
preoccupied individuals were with the untreated and often infected
injuries they received in the chase. We don't know whether, as they
hunted and gathered, they quarrelled and teased and bullied, and
were teased and bullied in turn. We simply don't know how happy
or miserable they were. Even less do we know how much time they
spent relating to their gods and to the *pneuma* that Young believes
gave them the sense of life more abundant. He seems to assume that
this was the dominant experience: the lawlines that harness the flow
of *pneuma* are 'the grammar of primitive experience'. But the poverty
of the historical record discourages us from thinking critically about
these things; indeed, it is so universally blank that we don't notice
the specific blanks corresponding to subjective experience. Our notion of primitive. hunting expeditions is reduced to a few images of
the crucial and climactic moments. We forget the long, fruitless hours
in the ice and snow, the hungry empty-handed return to a hungry
wife and starving, howling children, the limited delight of laying one's
tired infested body next to that of one's tired, infested wife, or the
even more limited delight of hungrily awaiting the return of one's
earthly master while one struggled to care for those of one's children
who survived, the return of one who could treat you precisely as he
liked. The tedium, the long unremitting labour of technology-free life

on earth, is forgotten. (How much of the history of the world is the history of aching bodies? How much of the history of human con- sciousness is a history of itching?)

The habit of defining primitive experience in terms of magic and ecstasy – catching sight of Proteus rising from the sea, hearing old Triton blow his wreathed horn – as opposed to jealousy, cold, hun- ger, pain, fear, boredom, etc. is such as to make one wonder why, if things were so good in primitive times, people bothered to try to make them better. The truth is that we know little or nothing about the quality of primitive life; and so our forebears remain more or less exotic, and we reduce the lives of exotic people to exotica: ecstatic matings and huntings, magic, closeness to the gods.

We know little or nothing. But we can making an intelligent guess. Such a guess would suggest to us that the urgency and pressure of getting and spending is hardly likely to be less for a hunter-gatherer who lives by the vagaries of hunting, for the farmer who is vulner- able to the weather or the merchant-adventurer who is the plaything of the sea and the trade cycle. An intelligent guess would suggest that, in the absence of the network of contracts that regulate advanced industrial society, custom and charisma and the survival of the strongest would probably favour bullying, cruelty and arbitrary rule by fear in both private and public life, rather than the happy organic community of the academic imagination. But even the violence of remote com- munities tends to be sentimentalised by those whose contact with violence is mediated rather than direct – as if there were something intrinsically more attractive about having an organic fist or a stone axe crashed into one's organic face than being blown up by a bomb. The attractions of the former are illusory; at least the latter would promise a quick death rather than the traditional death-by-putrefac- tion unalleviated by analgesics.

Young's views on primitive life are scarcely unique; that is why they are worthy of detailed examination: their faults are typical of an entire seam of contemporary writing that would criticise the present in terms of the past. Of course, the gilding of the past has been a fault of writers since time immemorial was first memorialised. The primitive world free of the baggage of knowledge, of possessions and of preoccupation with getting and spending may be less comfortable and less spiritually fulfilling and morally admirable than the critics of contemporary civilised life may care to think – if it cares to think. It would certainly be less comfortable for the hysterical humanists themselves – illustrating their tendency to bite the hand that feeds them. The most telling examples of this tendency are to be found in the hostility to (a) individualism and (b) book-led abstraction. These may be dealt with briefly and decisively.

The hysterical humanist vilification of individualism is based upon three assumptions: that individualism is an essentially modern phenomenon; that individualism is an absolute, yes/no phenomenon; and that individualism is a bad, a miserable, thing. The master-myth is that individualism has led to or reflects the fragmentation of society into atoms, so that the breakup of the collective, whose original and ultimate expression is collective religious worship, participation in a self-dissolving Dionysiac frenzy, has now reached its terminus in a colloidal suspension of private lives tapping electively into the collective only via the computer terminal or the satellite dish. With this myth comes disapprobation: modern man is egotistical. Whereas in pre-modern societies relations between people were more important than relations between things, in modern society it is the other way round. This leaves social hungers unappeased: such is the hunger of the denizens of *Gesellschaft* for immersion in *Gemeinschaft*, they will try to recover the sense of a warm, organic society by all sorts of pathological processes, including the blood-and-soil nationalism which received its characteristic expression in and was exploited by Nazism. Each of the three claims about individualism is false.

First, as Merquior argues in his critique of Louis Dumont,[16] Western individualism is not a modern phenomenon at all: it goes a long way back. Most of the elements – the replacement of collective eschatologies as a theme of spiritual preoccupation by that of the fate of individuals, the emphasis on the self and a concern with self-discovery, as reflected in the emergence of the autobiographical genre, the birth of a proto-modern concept of natural rights – present by the twelfth and thirteenth centuries. And, although it was easier to be one's own person when one was not destitute or owned by someone else, this shift was not confined to the aristocracy. After the Black Death, when labour was short, 'many English peasants managed to become non-servile toilers and started acting like rational calculators, evincing strikingly individualistic attitudes towards ownership and family' (Merquior, pp. 310–11). The explicit critique of individualism is also decisively pre-modern – though it does not go back as far as individualism itself. Merquior alludes to de Tocqueville's judgement upon individualism as (to use Merquior's words) 'the uncivic perversion of democracy in an unheroic materialist society' (p. 317).

Secondly, the belief that individualism is a yes/no phenomenon belies its complexity – the many ingredients that it contains. These include: the sense of personal dignity (though this is the most social, or external, aspect of individualism); the feeling of autonomy; the valuing of privacy; and the concern with self-development. As Merquior points out, different elements emerged in different ages – though all of them 'had more than a prefiguration in a more distant past' (p. 319). Each of these ingredients may exist to a greater or lesser degree.

Finally, there is the assumption that individualism is necessarily a bad thing. But why is it a bad thing? Is it because it is associated with selfishness? There are other, more potent, images of selfishness in holistic, hierarchical societies: the notions of the lord, the prince, the prelate do not awaken particularly strong images of self-denial and altruism. And those over whom they lorded, princed and priested did not exactly choose to subordinate their own interests to that of the common good as expressed in the needs of lords, princes and prelates. Is it because individualism is incompatible with collective, cooperative, social action? Nothing could be further from the truth: the collective organisation and coherence of contemporary society exceeds anything that the world has ever known, even though much of it, as in a communication system, is embodied in technology rather than in a vast coincidence of individual intentions. Is it because the members of contemporary individualistic societies suffer an unprecedented isolation and loneliness? We know little or nothing about the inner experience of the individuals of past epochs. Literature can tell us little because its contents are so decisively influenced by what it is customary to speak about: the preoccupations of poetry are filtered through generic conventions. But we may assume that the metaphysical isolation of a man in the sixteenth century contemplating his death, or the physical isolation of a woman in the twelfth century going through the agony of childbirth and its unending consequences, or the social isolation of the hungry, poverty-stricken, exploited masses are not plausibly exceeded by the comfortable inhabitant of the twentieth century engaged in a conversation about a television programme watched by 20 million others at the same time. The deepest solitudes are in the abysses of physical experience and, while these cannot be compared across ages, it seems implausible that they are deeper now than they were in the unrelieved, uncomforted centuries of the past.

Moreover, the examples of anti-individualistic societies are not inspiring. In the previous chapter, we touched upon the caste-based societies whose racialist iniquities were structural and planned rather than episodic and casual. And it is worth remembering that, as has often been observed, traditional, pre-modern societies have other undesirable features. Spite, envy, jealousy flourish where physical proximity is not palliated by privacy, autonomy and an inward-facing concern with self-development – in short by individualism. As Jonathan Elster remarks, many people get their pleasure from making others worse off. 'A depressing fact about many peasant societies is that people who do better than others are often accused of witchcraft and thus pulled down to or below the level of others. Against this background, ruthless selfishness can have a liberating effect.'[17]

I have referred to the hostility to individualism as exemplifying the way that *kulturkritik* tends to bite the hands that feed it. The life of the humanist intellectual is an extreme expression of individualism. It revolves around at least the pretence of independent thought; it involves activities – reading and writing – that are usually performed in solitude or by a consciousness turned away from the collective activities around it; and its characteristic product is a series of publications specifically attached to the name of the individual, that feed into his/her reputation and chances of self-advancement and whose final output is that supreme expression of the care of the self and self-development – the curriculum vitae, the deposit account of the secular soul that grows over time. But even this is not the most striking example of manuphagy. Young, in company with many others, attacks the very substance through which he earns his living and from which he derives much of the interest, the variety and the pleasure in his life: abstract thought. Let us examine his criticism of book-led abstraction.

In the passage cited earlier, Merquior noted one form of the revolt against modernity: a rejection of the spirit. One manifestation of this is a hatred of abstractions, the hegemony of signs, of mediations, of the second order, and a longing for the direct life of unmediated feelings and unpeeled sensations. Few hysterical humanists would go as far as Barthes in asserting that language itself – the material of his lectures, his books, his conversations, his private monologues[18] – is Fascist. But many of them loathe language, seeing it, and abstraction in general, as a prison for the spirit, the imagination, the soul. Dudley Young's hostility to science is of a piece with his hostility to abstract thought and our 'book-ridden' culture. But this is (to put it mildly) a difficult position for a man who has written, and presumably hopes that we shall read, a book (*Origins of the Sacred*) of some 372 pages. It is even more difficult when one considers what a mighty work of scholarship the book is: the bibliography lists some 200 items, most of them book length, many of them massive. And the authoritative tone of the text hints at a deep familiarity with many thousands more books, articles, poems, novels, etc. It is yet more difficult again for a man who quotes with approval Goethe's statement that 'one needs to know about the last 3000 years in order to be acquainted with the things of darkness':[19] how could one acquire such information except through books; indeed, without being, even by the supposedly bookish standards of the present day, book-ridden.

There is a deep and thought-provoking paradox in the fact that the message of *Origins of the Sacred* – itself the product of patient decades spent in the library and an important addition to the stock of books the well-read ought to read – should be to advocate a return to a relatively bookless culture, to what Young characterises as the 'way we knew – and know – the world before we began – and begin – to talk too much, read too many books'. This from a man who has clearly spent much of his life in the capillary mazes of the printed word and has written a massive book that supports its case with bibliographies that none us could hope to shift in a lifetime of leisure! But in this, as in so many other respects, Young is typical of those who hate modern life. One is reminded of D.H. Lawrence writing numerous novels about sex which advocate, among other things, that we should remove sex from the head. One of the main effects of Lawrence's books, of course, has been to spawn thousands of books and millions of conversations about DHL's views on where sex should be. Young identifies as one of his major purposes 'to commend the concreteness of the primitive imagination as an antidote to the bodilessness of modern abstraction' (p. xxxi). His own book, however, is a mass of abstractions – even relatively concrete concepts such as 'Sumerian culture' are remote from the body. And nobody reading his complex, erudite argument or the ingenious, even tortuous, interpretations of Wordsworth, Pound or Stevens that he uses to support his thesis, could have guessed that he was in the business of commending antidotes to 'the bodilessness of modern abstraction'. Little of my life over the last few months has been as rich in bodiless abstractions as the hours I have spent arguing with Young's book.

And just to see how bodiless and abstract Young can get, consider this extract from a dense, three-page analysis of three lines of the *Prelude*:

> At the heart of the mystery is his [Wordsworth's] belief, which brings him nearer to the primitive dancing ground than any other modern poet, that the way to make breathings is to stop talking, and if you do *that* (self-cancellation for the poet), the words (or the music) may come of themselves, unbidden, unforced, and deposit themselves as real presences in the silent spaces you have cleared for them. So certain is Wordsworth of the metamorphic power that he persuades us the breathings have been made by words on the page.
> (p. xxviii)

Anyone wanting to count the number of 'bodiless abstractions' mobilised explicitly or implicitly in this short passage would have to be good with figures. The 'bodiless' count, however, gets worse:

> Enough of mysticism. Let us return to primitive metaphor. As if all this were not sufficient load for three lines to carry, there is yet another theme, and it bears upon our present undertaking. Look again at the run of seven plurals in the first two lines that shrink

into the singularity of 'all that stand single'. Bizarre? For non-Wordsworthians let me offer a sub-textual gloss: the man who wrote this had recently returned from France, almost crazed by desolate perception of what the September Massacres had made of the ardent ecstasies in which the Revolution had begun ...

Young's explanation continues, with deepening complexity and subtlety, for a further half page. The lines in question are as follows:

> Points we have all of us within our souls
> Where all stand single; this I feel, and make
> Breathings for incommunicable powers

Perhaps Young's book – like Lawrence's novels – is simply the wrong sort of medium for his message; perhaps it should have been Aeolian harped in the wind or scratched on bark, or danced – certainly not published by a major house. The messenger himself, the author of other bodiless abstractions, such as a book about Yeats' poetry, with much bookish talk about books about books, undermines the message. If we are to go back to the primitive, would a careful and thoughtful reading of this book, groaning under its weight of erudition (the names of 200 authors in its index, nearly 30 pages of notes), help us in this direction? Does not his massively knowledgeable – and eloquent – case for us to learn to 'recover our innocence' make it more, not less, difficult to do so? Is Young not like someone making peace with the sword?[20]

I repeat: in arguing his case against erudition with such erudition, in writing such a long book against books, Young is typical of many hysterical humanists who argue against culture, civilisation, advanced industrial society, while themselves being its most characteristic flowers – and benefactors. Like many other humanist intellectuals, Young would, in the name of our humanity, deny most of what makes us distinctively human: books and the capacity for abstract thought remote from the body; the great intellectual and imaginative adventure of science and technology (driven at least as often by humanitarian dreams and by a sense of reverence for and the beauty of the non-human world as by the appetite for power and control); and, finally, beliefs that have cognitive as well as affective content. Early in his book, Young argues that 'reacquaintance with our primitive selves may make us *more* human rather than *less*'. This seems a rather unlikely outcome of a bookless, scienceless immersion in the primitive frenzies. Precisely how human that would turn out and how useful a prescription for our troubles this would be in the world as it stands today is highly debatable. In trying to determine whether the sacrificial frenzy would, or would not, make it possible 'to live life more abundantly', we ought to think not only of the delighted dancers, but also of those who, for whatever reason, are excluded from the dance, including the unlucky ones chosen for the blood sacrifice.[21]

HYSTERICAL HUMANISM: BORROWED GRIEVANCES, SYMBOLIC WOUNDS

Merquior has pointed out that humanist intellectuals who diagnose culture, Western civilisation, society, or whatever, have a vested interest in pronouncing the patient to be gravely ill. It is this that justifies their existence, 'as soul doctors to a sick civilisation'. I have elsewhere discussed whether or not the diagnosis makes sense in historical, as opposed to hysterical, terms. Whether, in short, we are justified in concluding that contemporary civilisation is uniquely uncivilised.[22] Of an age that discovered the nuclear bomb and yet managed to confine itself to using it only twice, that witnessed both the horror of Auschwitz and the birth of the Geneva Convention on War, we might at the most say it is 'the worst of times and the best of times'. But this is not enough for hysterical humanists: they need the comfort of knowing that they live in the worst of times. In order to prove their diagnosis beyond doubt, they have to add to the objective horrors – of which, as in any other age, there is a plentiful supply – certain subjective or symbolic horrors that reveal the truly horrible nature even of those features of the present age that seem to redeem it.

The classic text here is Adorno and Horkheimer's *Dialectic of the Enlightenment*, which we discussed in the opening chapter. Although it was written under the impact of Auschwitz, its denunciation of modernity was not based solely upon such unimaginable horrors as concentration camps. It also encompassed ordinary affluence and the 'mass deception' of 'the culture industry'. This discovery of a form of unhappiness worse than toothache, or the savagery of pre-modern surgery, or any of the immemorial woes of mankind such as hunger and thirst and brutal cruelty, has the advantage not only of helping to prove that this is the worst of times, but also deals with the potential guilt of those who seem to be having a fairly cushy time in this, the worst of times. It extends the franchise of suffering. Alienation, emptiness, etc. are available to everyone as the birthright of 'the modern' whose 'predicament' is endlessly described, alluded to, assumed. The Spiritual Void is a kind of common pasture where humanist intellectuals can lead their disciples and there graze with them on a rich grass of self-pity. Those of us who seem to be doing well are really having a terrible time.

When David Morris suggests that

It is as if the numbness associated in the nineteenth century with
hysterical women has become simply a normal condition in a world
where everyone lives on intimate terms with the holocaust, genocide, the greenhouse effect, and prospects of nuclear winter,[23]

the views he is expressing are representative of many hysterical humanists. His further suggestion that hysteria itself may nowadays have
'discovered its ultimate disguise and learned how to simulate the invisible omnipresent malady that passes these days for health' gives
everyone permission to feel sorry for themselves. The assumption that
'everyone lives on intimate terms with the holocaust' is an extraordinary piece of moral cheek, suggesting some kind of equivalence between those who have merely read about the holocaust and those
whose lives were destroyed in or by it. And yet it would not raise an
eyebrow amongst humanist intellectuals where parasitising the suffering of others is commonplace, and abstract and symbolic wounds
are ranked as high as real, fleshly ones.

A parallel ploy is to refuse to recognise gradations of suffering,
injustice or exploitation. This has been most notoriously evident amongst
certain well-off, middle-class radical feminists. So far as they are concerned, their own position is no different from that of women throughout history. As Catherine MacKinnon (a tenured law professor in a
major American university who spends much of her time lecturing
on the international stage) says, the status of all women is essentially
the same and it has not changed throughout history.[24] For her friend
and fellow campaigner, Andrea Dworkin, 'the situation of women is
basically ahistorical'. Male power 'authentically originates in the penis'
which, as 'a symbol of terror', is even more significant than the gun,
the knife, the bomb or the fist. Heterosexual coitus is, from the woman's
point of view, simply rape. When a woman consents, she is merely yielding
to social pressures to collude in her own colonisation, degradation, enslavement, etc. There are no grades: the tenured professor of law is as
much a victim of the sex war and wider female oppression as the female
slave who is raped by her owner and dies in unattended childbirth, or
the Muslim women in Bosnia who were raped as an act of war and
were then rejected by their menfolk as they bore the children that,
through lack of access to abortion, they were obliged to carry.[25]

The capacity to talk up one's own suffering to the level of that of
real victims – so that consensual sex with a partner who is sharing
your life and your mortgage suddenly seems the same as being raped
by a gang of HIV-positive soldiers before you are left to die in the
road; so that to have read about Auschwitz is the same as to have

been a prisoner there – is an essential skill in a humanist intellectual. This has become even more important in an era dominated by Cultural Studies – aptly renamed Grievance Studies – where students and their teachers rediscover themselves as victims. They do this by examining the ways in which society has victimised, silenced, exploited, or at least marginalised, the groups for which they feel they qualify for membership. Being gay, being female, haling from an ethnic minority, having Irish ancestors permit one to take on the mantle of the marginalised and to claim special insights into the viewpoint from which alone the truth about society is visible – the view from below of the downtrodden, who may be badly treated but precisely in virtue of this are existentially privileged to be free of false consciousness. It is a poor scholar who cannot find a category under which to rediscover her/his true identity as a victim.[26] A lamentable consequence of this is that debate about real, unequivocal, crimes is thickly muddied. Real crimes tend anyway to be relatively neglected by Cultural Critics, or simply regarded as logical extensions of the symbolic crimes; they are of lesser interest because they require no hermeneutic skills to uncover them.

To be a hysterical humanist, therefore, carries two advantages: it downgrades modernity and so makes the need for the 'soul doctor' – whose 'work' might seem otherwise unnecessary – more important; and it closes the gap between the comfortable cultural critics and the uncomfortable lives on whose behalf they set themselves up as advocates. Critics needs to borrow grievances since, if the world really is unprecedentedly horrid, it would be morally uncomfortable to be as unprecedentedly comfortable as the historical record would seem to suggest. Generously helping themselves to the riches in the grievance kitty, cultural critics can feel more secure in their feeling that they are not only useful, but also morally superior.[27]

One of the most venerable targets of hysterical humanism has been 'the bourgeois', who is not so much a subspecies of humanity as a child of the power of the mind to generalise and simplify the multitudinous variety of human beings and to demonise the product. The history of this word gives a glimpse into the mental processes of hysterical humanists who are almost defined (despite their objective situation) by the distance they put, or would wish to put, between themselves and the bourgeois condition. Originally, the bourgeoisie comprised an economic and social class that included merchants, entrepreneurs and townsmen who derived their income from commercial and industrial enterprises. With the introduction of mechanical power into industrial production, the distinction between the employers and employees became sharper and the bourgeois tended to be identified with the employers – those who owned the means of production –

and were contrasted with the proletariat who had only their labour to sell. Marxism, of course, reduced the bourgeois to exploiters of the exploited working class. As they became more wealthy, the bourgeoisie emerged as major patrons of the arts. This did them little credit: their tastes did not keep up with those of artists and connoisseurs. Upwardly mobile factory owners and others were consequently mocked for their attempts to appropriate the aesthetic values of the artists and their traditionally aristocratic patrons without possessing the latter's sensibilities. This snobbish spin enabled the bourgeoisie to be despised not only for their concern for material goods, their preoccupation with respectability and their lack of generous concern for the welfare of mankind in general, but also for their philistinism. The bourgeoisie were narrow, selfish and practised in a self-serving sentimentality that enabled them to conceal their rapacity beneath a veneer of philanthropy, morality and even spirituality. Those who did the most famous despising – from Marx to Sartre, from Flaubert to Barthes – were invariably bourgeois themselves and beneficiaries of the very technology-driven revolutions in the means of production that they despised.

The demonisation of the bourgeoisie has ranged from contempt – for example, Barthes' assertion that 'the petit bourgeois is a man unable to imagine the Other' (some feminists would tell us that this is a property of all males) – to out-and-out horror. There is a recent example of the latter response in Bryan Appleyard's discussion of Homais, the pharmacist in *Madame Bovary*. Homais, he tells us, 'is a liberal-minded sceptic, anti-clerical and progressive' and, at the same time, 'the most vivid, unforgettable realization of evil ever to spring from the Western artistic imagination'.[28] 'This colossal monster' is unable to recognise the depth and meaning of Emma's passion and his glib doctrines are simply unequal to her tragedy. His beliefs have, from one perspective 'a terrifying inadequacy' and, from another, an 'equally terrifying adequacy'. All of this is captured in the fact that 'Homais is a bourgeois' – a sober, serious member of society, who can keep things in perspective:

> The precise resonance of the word is important. The bourgeois is not merely middle class, nor is he merely an anti-clerical technocrat. He is not merely materialistic, nor is he merely complacent. He is all of these and yet he is also savage and inhuman in defence of his own complacency,
>
> (p. 104).

'Savage' and 'inhuman': surely the twentieth century and its predecessors have more appropriate referents for these words than the bourgeois's desire to defend his 'complacency'. The complacency and self-assurance of someone, such as Bryan Appleyard, who feels able to see to the bottom of many millions of souls and to summarise what he sees in a few phrases should give pause for thought. Is not over-generalisation and the refusal to accept the Otherness of the Other precisely the deepest spiritual vice ascribed to the despised bourgeois? Be that as it may, the failure to acknowledge that there are grades of evil – to make distinctions as to intensity, gravity and scale – is itself deeply suspect. Just as to compare England to the Soviet Union mocks the memories of the hundreds of thousands who died in prison camps and the tens of millions who were imprisoned, exiled and in a thousand other ways persecuted in the Great Terror. And just as to say, as Barthes said in his inaugural lecture at the Collège de France, that 'language is Fascist' spits in the face of those who drowned in the liquid manure of truly Fascist discourse. But such distinctions are not helpful to the cause of hysterical humanism.

SOME CONCLUDING THOUGHTS ON CULTURAL CRITICISM AND THE PROPENSITY TO GENERALISATION

Perhaps the most striking feature of much cultural criticism, even more remarkable than its self-serving hostility to modern life, is the vast scope of the general statements upon which it depends. Western civil-isation 'since at least Plato', 'European society', male human beings, 'the bourgeoisie', 'the East', and many other comparably large his-torical, geographical and social lumps of the world are effortlessly encompassed by the wide-lensed glance of the exponent of *kulturkritik*. There is something deeply laughable about this propensity for confi-dent over-generalisation. It shows that, despite their hostility to hu-manity in its contemporary forms, hysterical humanists are not as remote from the common run of unthinking humanity as they would sometimes like to think they are. On the contrary, they are human, all too human. This propensity to over-generalisation connects the cultural critic with the barroom bore – with the Alf Garnett figure whose half-truths (or 'Alf-truths) equally effortlessly gather up the world into the small prison woven out of the unchallenged preju-dices he took in with his mother's milk. The pompous guesswork of, say, Barthes' *L'Empire des signes*, in which 'Japan' is reduced to a place 'where artifice reigns, forms are emptied of meaning and all is sur-face', bears a strong formal relationship to Alf Garnett's sweeping animadversions about 'yer average darkie'. In both cases, the asser-tion so exceeds any conceivable empirical evidence available to any consciousness of finite energy and duration, that it is simply ludicrous.

It would be giving either Barthes or Alf Garnett too much credit even to contest them. The only thing to do with such general statements is turn one's back on them; to mock them with some such tease as 'There you go again, summarising India'; or to do what the character in the Thurber story did – collect them, reflecting, as he did, that at least they don't cost any upkeep or clutter the spare room and get thrown out in frenzy of spring-cleaning.

This may, however, be to go too easy on them. After all, making general statements, however ludicrous, is not an entirely empty speech-act, void of illocutionary and perlocutionary force. We do not excuse Alf Garnett on the grounds that the scope of his racially prejudiced statements is so wide as to make them meaningless and consequently harmless. There is no general statement so absurdly over-reaching, so groundless as not to have the power to cause harm. Wagner generalised from his hatred of one particular Jew who had been kind to him (Meyerbeer), but who had committed the unforgivable crime of being successful when Wagner was struggling, to a hatred of all Jews, which ended with suggestions that anticipated The Final Solution.[29] This may have been ludicrous, but it was certainly not funny.

Since the Second World War, it has been more fashionable to be a cultural critic on the Left rather than on the Right. The views of the former are not usually explicitly directed against vulnerable groups. Rather, they are directed against those who are perceived to be culturally, economically or politically dominant: the bourgeoisie, men, white Europeans, and so on. But this switch to a less vulnerable target does not render these views entirely harmless. They add to the stock of hatred in the world and they are counter-educational inasmuch as they conceal the infinite variety of their targets – not all white Europeans are the same – behind a caricature that gratifies the mind's desire for a simple and portable account of things that are neither simple nor portable. The demonising prejudices of so many humanist intellectuals, however complex their elaboration, are an anti-education of the imagination, blocking any nascent impulse to reach into the unimaginable otherness of so many others. There is a deeper and wider educational harm which I have touched upon elsewhere but which is worth drawing out again.[30]

Cultural criticism – cultural studies, cultural theory – occupies an increasingly dominant position in secondary and higher education. Its immense ambition and scope is reflected in its tendency to connect all sorts of disparate social phenomena in general statements that have the form of laws but none of their substance. To see this clearly, let us compare the easy generalising commonly seen in cultural criticism with the painstaking process by which laws are arrived at in science – through examples that may be considered typical of each

sphere. (Whether leaning on single examples to draw such wide conclusions may be considered as an instance of higher-order over-generalisation, I leave to the reader to judge. It seems to be impossible to
make adverse general statements about general statements without
running into something like the Russell paradox!)

For our example from cultural criticism, let us make one final visit
to Dudley Young's *Origins of the Sacred*. There is hardly a page that
does not yield treasurable generalisations about modern life, man versus
woman, the twentieth century, many of which dramatically illustrate
what happens when one is liberated from the discipline of rigorous
argument, itself disciplined by ascertained or ascertainable fact. In a
matter of a few pages, we move from all-inclusive handwaving with
reference to 'the general loss of instinctual and intuitive uncertainty
about our world as the expanding neocortex unlocked our instincts'
being 'one of the important falls from grace into fumbling that our
mythologies lament'; to male abstraction versus female empathy, female-centre, male-margin, stillness and movement, man the ritualist
and woman the pragmatist; and on to the 'schizoid' nature of the
classical music of the nineteenth and twentieth century foreshadowed
in the music of late Beethoven and explained by his deafness. Such
lightning leaps – sublimely forgetful of how little is known about
causality and influence in this context – are so reductive as to make
1066 and All That seem carefully argued and densely documented.
The world, with all its fathomless complexity, its overwhelming variety, is gathered up like a syllabus.

Consider what Young has to say about Beethoven. This is worth
quoting at some length, so that the mad tendentiousness of his argument may be savoured in full:

> when music gives up its 'physiological presupposition', its somatic
> rhythms, the emotions its conjures are no longer channelled through
> and harnessed by the enacting body (which is therefore allowed to
> wander unmonitored into camp and kitsch). Our clinical adjective
> for such a breakage is 'schizoid', and I believe I can hear it fore-
> shadowed in late Beethoven; in the Ninth Symphony clamorously,
> in the pastiche of the *Missa Solemnis* anguishingly, and in the
> discarnating last quartets sublimely. . . .

> My rough sketch of [the Romantic soul of bourgeois Europe] may
> be concluded by briefly considering Beethoven's last piano sonata
> (Opus 111). This piece is a wonder of sanity regained, also a prophecy:
> after some initial argument melody returns gracefully to its home
> in the human body, and the syncopated rhythms point amiably and
> unmistakably forward to ragtime and the jazz age. As he leaves
> the stage, this last of the European grandmasters quietly announces

that he has concluded (consummated and killed) a musical life that had begun in medieval plainsong. As Vico would say, it was time for a *ricorso* to primitive beginnings and Beethoven tells us here that we should look for it not in Wagner's Europe but in the Southern States of America, where the chanting equivalent of monastic plain-song would be the Negro 'holler'. Whereas most nineteenth cen-tury music (post-Beethoven) seems to me only somewhat slandered by Ezra Pound's description of it as 'steam ascending from a mo-rass', the jazz that evolved from holler and folk music has housed as best it could the wandering soul of the twentieth century.

(pp. xxx–xxxi, xxxii–xxxiii)

Should one bother to argue with such assertions? The difficulty would be to know where to begin. To question the clinical dismissal of Beethoven's ecstatically wonderful late works would simply be to invite the charge that one had been brainwashed into appreciating what 'one' has to appreciate. To bridle at the suggestion that Beethoven was the last of the European grandmasters (and that he also knew it) and mention the names of Schubert, Schumann, Brahms, Mahler, Richard Strauss, not to speak of occasional non-Germans such as Berlioz and Chopin, would simply invite the retort that one had not understood the fundamental re-evaluation that is being advanced. To question whether the Opus 111 sonata (a favourite hunting ground for pseuds and diagnosticians since *Dr Faustus*) is so utterly different from the Opus 110 sonata is only to invite the charge of literal or spiritual deafness. This last, in Young's scheme of things, is a grave charge: it accounts for where Beethoven went wrong, why his earlier music gave way to 'a breathtaking mixture of bombast, chaos, and preternatural beauty', why his music became unphysiological, and why the man himself is 'the great modern Lucifer'. (If deafness accounts for so much, it is difficulty to see how, with his deafness unchanged, he should have produced the Opus 111 sonata – 'a miracle of sanity regained' – miracle, indeed.) And to question whether Beethoven in 1827 should have wanted to tell us (through music) that we should go for Negro hollers rather than Wagnerian *leitmotiven* is to risk the charge of lit-eral-mindedness.

Such a fever of retrospection, of Rorschach ink-blot reading, of mad interconnnectedness, is rare even in the world of social diagnosis, though it is customary in the world of those possessed by religiose ideas and the clinically insane. Nevertheless, it illustrates vividly what lies at the bottom of the slippery slope on which cultural criticism stands. The habit of making magisterial and absurd statements, which pre-suppose a range of knowledge and a power of judgement, an author-ity not possible to any human being, becomes inescapable.[31]

Let us now turn to science, where general statements have to be robust if they are not to be immediately refuted – either by the discovery of counter-instances, or by the failure, possibly disastrous, of a practical application. The general statement in question is one that appears in a recent issue of the *Lancet*[32] and it concerns the use of certain drugs, such as ACE inhibitors, following heart attacks. It concludes that ACE inhibitors post-infarction probably reduce mortality by about 10 or 15 per cent at one month and that this reduction is maintained at a year. The authors of the report scrupulously add the 'confidence intervals' which indicate the likely bounds of error of the study. This modest (though important) conclusion was based on a study that involved recruiting a total of 58 000 patients by approximately 1000 contributing doctors in hundreds of centres in a score of countries.

The contrast between a small-scale conclusion patiently arrived at as the result of a huge collective effort and large-scale conclusions seemingly dreamt up by one individual in his armchair of an afternoon is compelling. The contrast is striking because the two statements represent extreme examples of their kinds; but it is a contrast, too, of two cultures. Cultural critics, hysterical or otherwise, are, when all is said and done, humanist intellectuals and while, especially amongst Marxists and post-structuralists, there is much talk of rigour, they share the relaxed approach to evidence of their less hysterical counterparts in history, literary criticism and other human sciences. Innumeracy and naivety are not confined to the beer-swilling herds in the taproom: they affect a sizeable section of the academic community. Why does this matter? Because, I believe, the counter-educational effects of this lack of rigour, which seems itself to be *de rigeur* in the humanities, and the harm that may flow from it, is connected with the low level of debate about social issues evident amongst politicians and journalists. Let me take an example close to my own interests: the reform of the health service.

In 1989, the government announced that it was planning to revolutionise the way health care was funded and delivered. The proposals were largely politically driven: the aim was to download responsibility to the providers for any failures in the service and to move power to control the service upwards. The secondary aim was to move the debate away from arguments about overall levels of funding towards technical issues of the way funds were spent. The government refused to test out the reforms in a pilot study, asserting quite simply that they would work and that piloting would only lead to endless delay. Shortly afterwards, government officials were announcing that the reforms were working. One minister visited a hospital that had achieved self-governing (or 'Trust') status three months before and declared that there had already been improvements in the health of

the population served by the district. This claim was flawed at several levels. First, there were no adequate data (with or without confidence intervals) on the trends in the particular health indicators he
had used. Secondly, there was no way that the changed status of the
local hospital could have influenced cardiovascular health in the period of time in question, as cardiovascular health reflects causes that
act over decades, not months. Thirdly, even if there had been a change
in cardiovascular health in a few weeks and this could have plausibly been related to the reformed management of the hospital, one
could not demonstrate a causal relationship between management
changes and improved cardiovascular health without, for example,
also showing that there were no other changes, no confounding factors, accounting for the changes. All of this is elementary and obvious to anyone who had had any kind of training in science.

The minister in question was not a fool. It is unlikely that, as a
Fellow of All Souls, he was not aware that *post hoc ergo propter hoc* is
a fallacy. (Though whether he saw this as a mere Latin tag to adorn
the speech of the civilised or a principle that should constrain the
production of causal statements is uncertain.) What is certain is that,
if he knowingly deceived the public, his appetite for truth being considerably blunted by his appetite for power, he could count on a public
and a journalistic fraternity that would not be able to see through his
deception. The advent of cultural criticism, with its propensity for
uncontrolled generalisation without the underpinning of the right kind
of evidence, will make the work of ministers bent on deceiving the
public yet easier. Those in higher education will simply have had a
higher counter-education.

The question of how to determine the proper scope for general statements is the most important and one of the least addressed questions
in the human sciences and cultural studies. We need to generalise
even to venture an interpretation of what is going on even in a small
corner of society; and yet most generalisations are absurd. They all
exceed the data available to the individual making them. And whereas
this is, of course, also true of the natural sciences, the human sciences
– sociology, history, economics, political theory, etc. – and social commentary in general have yet to establish a method (or methods) of
reliably generalising from a restricted database. Until they do, there
will be little to distinguish them from taproom gossip, except that
they are better and more persuasively articulated, better referenced,
better articulated and command more respectful attention and better
remuneration.

Plato and Aristotle were the first to appreciate the intimate relationship between generality and the mind: explicit generalisation is
a deep propensity of human consciousness. Even at the level of per-

ception, what we experience is classified as belonging to certain types and this shapes our expectation. From the particular part, we infer the general whole, and from the general whole we draw innumerable general conclusions. Cognition – which links present experience with similar past experience – is re-cognition; and recognition mobilises general beliefs and activates anticipations. We may say that this mode of explicit generalisation – which always exceeds its evidential or experiential base[33] – is one of the distinctive glories of the human mind. We need to generalise in order to recognise the objects about us; we have to exceed the data presented to us in order to make sense of what we see: in order to understand our fellow human beings – even to empathise with them. In our power to generalise lies not only much of the glory but also much of the perniciousness of human beings: while it underpins our ability to imagine into the lives of others and so to deal with them sensitively, it is at the same time the basis of crude stereotyping and prejudice. There are the comparatively innocuous dangers of a reach that exceeds its grasp and the production of hot but meaning-free air. But there are deeper dangers of fomenting prejudices; and even where those prejudices are politically correct – where, for example, the targets are 'the powerful', such as 'men' – they still add to the stock of the world's illwill and confusion. The wild generalisations of cultural critics are not immune from this propensity to cause damage – not only to their targets (who in many cases are well able to look after themselves) – but to the critical sense itself.

Part II
Marginalising Consciousness

Introduction: Rational Consciousness as an Embarrassment

> The researches of psychoanalysis, of linguistics, of anthropology have 'decentred' the subject in relation to the laws of its desire, the forms of its language, the rules of its action, the play of its mythical and imaginative discourse.[1]

Let me recapitulate the story so far. In the Prologue, we examined the essential features of Enlightenment thought and of the critique launched upon it by the major Counter-Enlightenment figures. We concluded that this, largely pre-modern, critique was not decisive. In modern times, however, the critique of the Enlightenment has taken new forms. In many respects, this repeats themes of the earlier critique but visits them at a deeper level: in particular, the modern critique attacks the assumption, thought to be required by the Enlightenment programme, that men and women are deeply reasonable and well-disposed to one another – or could be educated to have these qualities – and that appealing to, and mobilising, that reason and intrinsic goodness is the way to progress. In Part I of this book, we have looked at a major contemporary restatement of the belief that, at the core of every human being, there is not a rational, civic being but an irrational animal, a primitive creature whose deepest needs – addressed and acknowledged in myths – are expressed (or betrayed) in Dionysiac delight in repeating the mutilating sacrifice of the hunt. In Part II, we shall look at trends of thought that have, over the last century or so, rejected not only the assumptions that human beings are essentially good and capable of using reason, but have questioned whether they are genuinely autonomous agents, able to direct their lives, individually or collectively, by deliberate, thoughtful action. I shall examine, that is to say, a handful of thinkers who, whether they fully intended to or not, have been hugely influential in creating a climate of intellectual opinion in which individual human consciousness and the part it plays or might play in bringing about change for the better is marginalised.

Whereas the argument of thinkers such as Dudley Young and those he cites is that humans are too animal to be rational, that of most of the thinkers addressed here is that they are too social, too collectivised to be truly individual, autonomous agents. For the latter thinkers, individual men and women do not act out of individual consciousnesses but out of a collective unconsciousness rooted in history, society, partially transformed instincts, symbolic systems or the cognitive structures of the brain/mind. These thinkers have attacked one of the foundational assumptions of the Enlightenment: that there is such a thing as a genuinely independently thinking individual – the notion (parodied in these thinkers' representations of Cartesian thought) that humans are self-centred agents.

There is, of course, some overlap between the thinkers Dudley Young invokes and the thinkers we shall discuss in this second part of the book. Freud, for example, appears in both roll calls. And Freud is important in other respects. He not only straddles both kinds of anti-Enlightenment thought but he is, in a curious way, a son of the Enlightenment as well – albeit a wildly prodigal son. He took very seriously the Enlightenment notion that man was a piece of nature; but he concluded from this, not the optimistic Enlightenment conclusion that man could therefore be understood and controlled as readily as the nature that had proved transparent to the gaze of physics, but two rather pessimistic conclusions. The first was that, being a piece of nature, man was an animal driven by anti-civic instincts which could be controlled only at a very great price. (That is why he is one of Dudley Young's main witnesses.) The second conclusion was that the cost of overcoming instincts was so great that the suffering had to be hidden from the sufferer: in the repressed depths of humanity lay the savage dialectic between the immemorial instincts of the solitary human animal and the historical constraints experienced by the social being. For Freud, human consciousness is deeply enmired in and compromised by an instinctual unconsciousness, and autonomous agency is consequently severely limited. Our civilisation is hard won and precarious: we are survivors of a fight to the death between our animal natures and the cultural demands that are made upon us and the wounds received in that battle may reopen at any time, dictating behaviour that runs counter to the civic norm and the demands rationalistic society makes upon us. And yet, despite this, Freud believed that he had developed a rational (if irrationalist) account of humanity and a cure for some of the ills of the human psyche based upon the Socratic-Enlightenment notion of the healing power of self-knowledge.

Most of the other thinkers whose influence I examine here – Marx, Durkheim, the post-Saussureans – emphasise the unconscious cultural or culture-making forces that act upon us and so limit our autonomy, preventing us from being the kinds of individuals that the Enlightenment project would seem to require. As Goldmann[2] observed, all the leaders of the Enlightenment regarded the life of a society as a sort of sum, or product, of the thought and action of a large number of individuals, each of whom constitutes a free and independent point of departure.

The leaders of the contemporary Counter-Enlightenment assert, above all, that the individual is anything but 'a free and independent point of departure'.

The marginalisation of individual consciousness was seemingly endorsed by the single most important intellectual development over the last 400 years – the rise of a spectacularly effective physical science. (The fact that this science grew out of the deliberate activities of highly conscious individuals is invariably overlooked.) The success of the physical sciences in providing a unified explanation of an impressive range of phenomena and in enabling physical events to be predicted and controlled, made their methods, presuppositions and procedures the model for the way in which understanding in other fields could be most effectively pursued. The advance of physical science was associated, both as cause and effect, with the retreat of mind from matter. Newton's first law of motion represented the end of a long process by which all traces of animism were removed from the laws of physics.[3] The power of movement was no longer a sign of life, even less of consciousness. The forces that acted on and within matter were not expressions of some immanent soul, an unmoved mover activating objects, nor of conative forces within the objects themselves.

When scientific attention was directed more systematically towards living organisms, mind continued its retreat. This had, of course, been prefigured by the philosophers: for dualists such as Descartes, animals, being without souls, were insentient machines; and monists such as La Mettrie had pronounced even man to be a machine. The demonstration that there was no sharp or even substantive distinction between living and non-living matter – traditionally connected with Wohler's synthesis in 1828 of urea from an inorganic substance, though it seems unlikely that this would have cut much metaphysical ice in the absence of a pre-existing hostility to vitalism – spelled the end of vitalism and of the belief that the approach to living organisms should be fundamentally different from the approach to inanimate matter. Since organisms were essentially physico-chemical systems, biology would, in its maturity, boil down to physics and chemistry.

This gradual elimination of the difference between living and non-living matter was paralleled by a breakdown of belief in a fundamental discontinuity between human and non-human life. This received its most impressive and influential expression in Darwin's assertion that they had a common origin. The blindly groping earthworm and the apparently conscious and self-motivated human being seemingly had more in common than they had differences. Reflex behaviour, tropisms and deliberate action converged, if only in the ultimate purpose they served – that of assisting survival. Mind, as an immaterial influence on the unfolding of events in the material universe, and mentality as a distinctive feature of human beings, retreated further. The scientific picture of the universe came close to the corpuscularian vision of qualitatively neutral matter differentiated only by the size, shape and motion of its atomic or corpuscular constituents, with the addition of force-fields whose nature was uncertain, except in so far as they were definitely not conscious.

The continuing astonishing successes of the methods of the physical sciences in explaining the properties of both living and non-living matter have further encouraged the physicalist outlook. Increasingly, it has been suggested by both philosophers and scientists that the ultimate aim of all systematic investigation is to explain all phenomena in terms of the behaviour of physical systems, as expressed or captured in the laws observed to govern those systems. Consciousness, and events apparently initiated by conscious decisions that seem to have no place in the operation of the laws of nature, have in consequence become increasingly problematic. Even in the absence of a programmatic physicalism it is not surprising, when the methods and objects of study of physical science are regarded as paradigmatic, that scientists and others have tended to sideline consciousness and regard conscious experience as unimportant – as peripheral rather than central to what it is to be human. There is a natural inclination among those who practise a discipline that has universalist pretensions to deny the existence of things that cannot be readily studied using the methods of that discipline or, even less, accounted for within the framework of its established principles and laws. Methodological positions and problems dictate ontological prejudices.

Just as the physical sciences progressed by focusing on the objective, measurable aspects of the physical world and overlooking its subjective experiential face (to the point of denigrating characteristics such as colours, tastes and smells as mere 'secondary qualities'), so it seemed that the way forward for the human sciences – the direction they would have to take if they were to become truly 'scientific' – must be to ignore and, indeed, deny subjective experience. One of the defining characteristics of scientism is the tendency to import into human studies (sociology, psychology etc.) the habit, established in the physical sciences, of ignoring subjective experience and to regard that methodological decision as being ontologically binding as well.[4] Physical science progressed by privileging objective observation over subjective feeling. The human sciences, if they are to progress, must do the same: it is a simple mistake to imagine that, in the case of the human sciences, subjective feeling is the heart of the matter, the essential objective of study. Subjectivity must not only be set aside in order that truly scientific observations can be made; it must be denied to exist.

If this attitude seems unimaginable, it is necessary only to recall the long period when psychology, whose subject one would have thought would have been, inescapably, consciousness and subjective experience, passed through the absurdities of behaviourism, in which methodological puritanism – 'We must confine ourselves to what can be objectively observed and measured' – merged with an ontology that denied the reality of the usual contents of consciousness.[5]

The connection between the bid for scientific status and the denial of the centrality of subjective experience, of individual human consciousness – either as an important reality in itself, or as a testimony as to what is really going on in the individual's world, or even in the subject herself – is set out very clearly in some of the foundational documents of the human sciences – political economy, sociology, linguistics and psychology. (There is a profound irony in this, inasmuch as many of the same social sciences have been, to say the least, sceptical of the claims of science and sometimes bitterly hostile to its positivist outlook and its physicalist and naturalist assumptions.) Those human sciences which have contributed most decisively to the marginalising of consciousness have typically indicated their elevation to the status of a real science with a declaration of the autonomy of the object of study (language, society, etc.), its objective reality, and of the special procedures that would be appropriate to investigating it. The object under study was pronounced to be a 'thing' – as real as the things investigated in the physical sciences. The thing-like nature of the object placed it beyond the reach of common sense, intuition and introspection – beyond subjectivity, beyond the unmediated consciousness of the subject. The true character of language, the psyche, political behaviour or society could not be revealed merely by thinking about it. Although each of these phenomena has a subjective, individualistic dimension, this is held to be of less importance than the objective, systematic aspect.

The 'scientification' of the human sciences thus not only placed them beyond the reach of the layman's armchair; it also dismissed the validity of reflection upon experience as a guide even to what is going on in an individual. Where the findings of objective science were counter-intuitive, this was simply evidence of the unreliability of the individual (individual language speaker, individual psyche, individual citizen) as a source of information about what is happening in herself.

The reasons and intentions that individuals imagine lie behind their actions must be treated with suspicion or even contempt – or as further objects for (objective) study. The only sure guide to truth is the objective, logico-empirical, hypothetico-deductive method developed in the physical sciences. Common sense and introspection no more give us access to ourselves and the society of which we are a part than they tell us about the relation of the earth to the sun, about gravitational fields or about the nature of matter.

The rejection of introspection and of the testimony of individuals is evident throughout the 'reformed' humanities (following their attempted transformation into human 'sciences' and their practitioners' strenuous claims to be scientists). It is expressed with particular clarity by Durkheim:

> I consider extremely fruitful this idea that social life should be explained, not by the notions of those who participate in it, but by more profound causes which are unperceived by consciousness, and I think also that these causes are to be sought mainly in the manner according to which the associated individuals are grouped. Only in this way, it seems, can history become a science, and sociology itself exist.[6]

And it is spelled out also in his *The Rules of Sociological Method*:

> We assert not that social facts are material things but that they are things by the same right as material things, although they differ from them in type. . . . A thing differs from an idea in the same way as that which we know from without differs from that which we know from within . . . their characteristic properties, like the unknown causes on which they depend, cannot be discovered by even the most careful introspection . . . One might even say in this sense that every object of science is a thing. . . . In the case of 'facts' properly so called, these are, at the moment when we undertake to study them scientifically, necessarily unknown things of which we are ignorant; and any 'representations' which we have been able to make of them in the course of our life, having been made without method and criticism are devoid of scientific value. . . . Consciousness is even more helpless in knowing [social facts] than in knowing its own life.[7]

It is the mark of a science that it does not consult its object in order to find out the truth about it. After all, Newton's laws of motion were not discovered by interviewing lumps of matter. Only those who recognised this would be in a position to fulfil Comte's dream (which he himself saw as a completion of the Jacobin cult of reason) of a truly scientific sociology – of the science of man and society that the *philosophes* had believed would be possible once the implications of man's status as a piece of nature were fully understood.

The theories associated with this marginalisation of consciousness may have different motives, but they have in common a belief, adopted for both methodological and other reasons, in the hegemony within human life and society of the unconscious over consciousness, or of unconscious processes over conscious ones. What follows in this second part of *Enemies of Hope* does not pretend to be anything like a systematic account of this tendency towards marginalising consciousness. Instead, I alight on a few seminal thinkers who have, over the last 150 years or so contributed to the now widespread belief that individual consciousness is unimportant in human affairs – or at least far less important than has hitherto been believed – and have undermined or been interpreted to undermine the notion that humans typically act in a deliberate, non-automatic way in accordance with a formulated or at least formulable reason; putatively scientific theorists who marginalise consciousness by denying its influence, its autonomy or its sovereignty in ordinary affairs.

In my examination of these thinkers, I have not attempted to encompass the richness of their massive *œuvres*, focusing rather narrowly upon their relation to my fundamental concern. It follows from this that my treatment reflects less their intrinsic subtleties than the way their thoughts have been received by others and in this particular regard: how the notions, derived from them, have been influential in marginalising consciousness. Although some of them have been long dead, the present influence of these thinkers is still immense and, for this reason, they remain key figures in the contemporary Counter-Enlightenment. My specific focus on the marginalisation of consciousness has also emphasised the similarities between these thinkers rather than their very real differences – of preoccupation and worldview. And of quality. Only in the present context is it unimportant that Freud and Marx are usually wrong and/or unhelpful, whereas Durkheim and Helmholtz are often right and even more often helpful.

One final preliminary point. Although the marginalisers of consciousness have been dominant, there have been important, though far less widespread, trends in the opposite direction. Neo-Kantian idealism, especially in its phenomenological manifestations, has gone to the other extreme of making consciousness not merely central but all-encompassing. For Sartre, against whom the structuralists were in most direct rebellion, the Cartesian *cogito*, in the modernised form of the *pour-soi*, was not merely autonomous but responsible for everything, including

its own world which it synthesised through its value-investing activity. (The conflict between the 'philosophers of consciousness', who
enclose the world within the individual consciousness, and the
marginalising 'philosophers of the concept', is discussed in the Appendix to chapter 10.) It is perhaps hardly necessary to point out that
the purpose of the present critique is not to reinstate the transparent,
self-possessed, controlling Cartesian *cogito*. My aim is the more modest one of reasserting the centrality of individual consciousness, of
undeceived deliberateness, in the daily life of human beings. We are
not absolutely transparent to ourselves but we are not utterly opaque
either; we are not totally self-present in all our actions but nor are we
absent from them; we are not complete masters of our fates, shaping
our lives according to our utterly unique and original wishes, but
neither are we the empty playthings of historical, political, social,
semiological or instinctual forces.

5

Marx and the Historical Unconscious

My dialectic method is not only different from the Hegelian, but is its direct opposite. To Hegel, the life-process of the human brain, i.e., the process of thinking, which under the name of 'the Idea', he even transforms into an independent subject, is the demiurgos of the real world, and the real world is only the external phenomenal form of 'the Idea'. With me, on the contrary, the ideal is nothing less than the material world reflected by the human mind, and translated into forms of thought[1]

Marx took from Hegel the supremacy of general historical forces over the deliberate actions of individuals in determining the course of events. History for Hegel is the unfolding of the Universal Mind or Spirit in its progress towards absolute self-knowledge and its final return from alienation in the identity of Knowing and Being. Marx accepted the notion of the inevitable laws of the unfolding of history but deposed the Mind or Spirit from a sovereign position in the historical order of things. He turned Hegel on his head and discovered 'the rational kernel within the mystical shell' of Hegelian thought (*Capital*, p. 20). For him the motor of history was not the intrinsic dynamic of Universal Spirit dialectically quarrelling with itself, but the complex and enormous consequences of the fact that man, uniquely among the animals, produced the means of his own subsistence. History was the dialectic arising out of the material forces of production and the relations of production. Historical idealism was consequently replaced by an historical materialism in which the economic structure was the fundamental basis of human society. What Marx retained from Hegel was a view of history and the affairs of human life in which individuals had very limited ability to influence the course of events – although in Marx's view, the shots were called not by the Universal Mind but by the productive process. The individual's relation to the latter was a crucial determinant of his or her self-consciousness: it intervened decisively between consciousness and self-consciousness. The unconscious (and often irrational) forces of history outweighed the forces

229

of reason in the very processes of reasoning itself. Consciousness – in particular reflective consciousness concerned with our own and others' own nature and rights, our sense of who and what we and 'they' are – is doomed to be false consciousness. How?

The relations of production are such as to make men into alienated beings. The labour of the worker, which produces surplus-value he cannot himself enjoy, devalues him and estranges him from himself. Economic alienation is associated with political and, more generally, ideological alienation which misrepresents the world. Consciousness, above a certainly elementary level, is an emanation from the material conditions of existence, in particular the relations of production. Our sense of who and what we are, of how things are with others and of how things should be is determined by the economic infrastructure in which we are immersed. Consciousness (or at least that of the non-proletarians before the advent of communism) is thus both passive and, since inevitably infected by ideological alienation, false, self-deceived, opaque. At the root of social consciousness – and, therefore, of individual self-consciousness, since according to Marx man is not only a social animal but an animal that can individualise himself only within society – are the relations of production and the class consciousness arising out of this.

The classic statement of this belief is in the preface to *A Contribution to the Critique of Political Economy*:

> In the social production which men carry on they enter into definite relations that are indispensable and independent of their will; these relations of production correspond to a definite stage of development of their material powers of production. The sum total of these relations of production constitutes the economic structure of society – the real foundation on which rise legal and political superstructures and to which correspond definite forms of social consciousness. The mode of production in material life determines the general character of social, political and spiritual processes of life. It is not the consciousness of men that determines their existence, but their social existence that determines their consciousness. . . . With the change of the economic foundation the entire immense superstructure is more or less rapidly transformed. . . . Just as our opinion of an individual is not based on what he thinks of himself, so we cannot judge of such a period of transformation by its own consciousness; on the contrary this consciousness must rather be explained from the contradictions of material life, from the existing conflict between the social forces of production and the relations of production.[2]

Few thinkers believe that the transformations in the social order are brought about by individual initiative. 'The great man theory of history' – which Tolstoy set out to refute in *War and Peace* – has few adherents. But Marx's disbelief in the capacity of individuals to bring about social change went much deeper than that. No change could be effected without a prior change in 'objective conditions' and *that* was not the result of anyone's activities:

> No social order ever disappears before all the productive forces, for which there is room in it, have been developed; and the new higher relations of production never appear before the material conditions of their existence have matured in the womb of the old society. (p. 31)

Individual consciousness – and individual agency – reflects social forces of which it is largely unaware and by which it is unconsciously manipulated. Aron encapsulates Marx's position as follows:

> men enter into definite relations that are independent of their will . . . so that we can follow the movement of history by analysing the structure of societies, the forces of production, and the relations of production, and not by basing our interpretation on men's ways of thinking about themselves.[3]

Or, to quote the *Manifesto*:

> in every historical epoch, the prevailing mode of economic production and exchange, and the social organisation necessarily following from it, form the basis upon which can be built up, and from which alone can be explained, the political and intellectual history of that epoch.
>   Man's ideas, views and conceptions, in one word, man's consciousness, change with every change in the conditions of his material existence, in his social relations and his social life. . . . What else does the history of ideas prove, than that intellectual production changes its character as material production is changed. The ruling ideas of each age have ever been the ideas of its ruling class.[4]

In Marxist thought (and the relationship between Marx and Marxism, between Marx's writings and those of the Marxists is a vexed one – but we are talking about influences here) man is therefore opaque to himself, separated from consciousness of what he is, not only because of the effect of the division of labour under capitalism, which alienates him from his true vocation of becoming universal (he no longer recognises himself in his activities or his productions), but also because his position in the productive process (except perhaps in the case of the proletarian) prevents him from thinking correctly about things: at every level of conscious thought, he is largely in the grip of ideology or false consciousness. False consciousness is linked to class consciousness. The latter will be fashioned not by personal, biographical experience of the individual, but by the objective conditions of his life. These conditions will be mainly concealed from him; or rather, they will be implicit because he cannot see the world except in terms of the situation of his class. As Sartre would say, the bourgeois sees a world defined by the rights he possesses in it; and the proletarian, correspondingly, sees a world defined by the rights he lacks or others have over him. Neither perceives the world except through the medium of class position and, being unable to see the alternative worlds, cannot see the constraints upon the formation of the world-picture, just as the eye cannot see itself as a determinant of the visual field. That is why the conflictual nature of society and of history ('The history of all hitherto existing society is the history of class struggles', Marx, ibid., p. 40) and the reality of political power ('Political power, properly so called, is merely the organised power of one class for oppressing another' (ibid.)) is obscured from those who are caught up in them.

Marxist thinking had a decisive impact on the way in which individual consciousness and self-consciousness were regarded. Not only our sense of justice and fair play and our ideas about history and society, but our very sense of who and what we are, were seen to be influenced, indeed determined, by things to which our consciousness did not usually have access. By introducing the notion that in many crucial areas our apparently fully conscious and rational thoughts are the products of a political or historical unconscious operating through us, Marxism opened the way for others to deny the influence of rational, deliberate processes of thought over a much wider area. Once it is accepted that ideas (about the world, about society, about ourselves) are not powerful because they are true, rather that they seem true because they emanate from the powerful – 'The ruling ideas of each age have ever been the ideas of its ruling class' – the way is open for the undermining of reason in argument and, more profoundly, for the decentring of the self. If the ideas we entertain derive their authority not from the reasons we appeal to in developing and advocating them, but from their role in maintaining the position of those who are benefiting from the way the productive process is organised, then our apparent self-knowledge is typically self-deception. The careful consideration of the facts and reasons that seems to lie behind our beliefs are so much surface show, concealing, from ourselves as much as from others, the forces that are in fact at work in us when our views, our deepest beliefs, our entire world-picture, are being developed.

As Berlin has pointed out, Marxism shared with Romantic, pessi-mistic conservatives such as Carlyle, Dostoyevsky, Baudelaire and Nietzsche the view that 'rationalism in any form was a fallacy de-rived from a false analysis of the character of human beings, because the springs of human action lay in regions unthought of by the sober thinkers whose views enjoyed prestige among the serious public.'[5]

Marxists have always had trouble reconciling the radical relativisation of truth that seems to be an inescapable consequence of Marx's ideas with the absolute truth of Marx's own claims – about the direction of history and the determinants of ideas – and, even more pressingly, the absolute truth of the viewpoint of the emergent working class.[6] The latter, it seemed, were spared ideological delusion in virtue of their situation: the truth about society spoke to them through the howling misery in which it forced them to live. This did not, how-ever, account for the special insights of the, largely bourgeois, intel-lectuals to whom Marxism owed its origin, its propagation and its development. The fact that Marx was not himself drowned in the political and ideological alienations that should have followed from his class position remains inexplicable in Marxist terms. If all knowl-edge is class-conditioned, how can knowledge acquired from the point of view of the working class lay claim to universal validity? How could Lenin maintain that 'Marx's theory is all-powerful because it is true rather than that it seems true because it is powerful' in the light of Marx's own views of the relationship between knowledge and power?[7]

In fact, Marx did not entirely rule out the decidedly un-Marxist idea that there were absolute, or at least trans-historic, truths that could not be reduced to the status of battle cries in the class struggle. As Raymond Aron has pointed out,

> Marx believed that there were domains in which the thinker can arrive at truths that are valid for all and that there are domains in which the intellectual and aesthetic products of societies have value and importance for men of other societies . . . and he conceded the possible universal truth of certain sciences and, on the other hand, the possible universal value of works of art.
>
> (Aron, p. 179)

These contradictions, which have haunted Marxism almost from the beginning, have not prevented Marxists from elaborating Marx's atti-tude of suspicion towards rational discussion to the point where, as in Althusser, it becomes an almost global paranoia. The lamentable results are are all too clear in the twentieth century. Camus has traced this trajectory:

> That the demands of honesty and intelligence were put to egoistic ends by the hypocrisy of a mediocre and grasping society was a misfortune that Marx, the incomparable eye-opener, denounced with a vehemence quite unknown before him. This indignant denunciation brought other excesses in its train which require quite another denunciation.[8]

According to Althusser, whose attitudes and methods typify recent influential Marxist thought (influential in universities, it must be conceded, rather than in the real world of politics), human consciousness is manipulated by the collective in order to serve a particular aim: the reproduction of the relations of production, specifically the capitalist relations of exploitation.[9]

One does not have to be a Marxist, even less a structuralist Marxist, to accept that what is experienced as 'reality' is not reality-in-itself but a version of it related only indirectly (and rather problematically) to what is really 'out there', to 'how things really are'. We do not confront reality solely as isolated individuals but also as members of collectives. The encounter between the individual and what is 'out there' is to a great extent mediated by the other individuals who constitute those various collectives. For what is 'out there' is not simply matter-in-itself but what is *acknowledged* to be out there; pieces of matter are 'there' not simply in virtue of their existence but because they matter to someone. Reality is what counts as real to someone who is in part representing or adopting the viewpoint of a group or somehow under its influence. Reality, in other words, is what 'they' – or 'I' in so far as I am one of them – acknowledge. This is certainly and obviously true at a higher level – as when we are talking of political or social reality. But even apparently direct or unmediated sense perception is socialised; and so *a fortiori* is the judgement, based upon perception and conception, that something or other is 'real'. There will be competing versions of, but no direct, unmediated or unbiased access to, reality. In practice, however, there is considerable consensus. Much appears to be 'indisputable' and 'obvious' and is consequently 'taken for granted'.

So much is common ground. Where Marxists, and in particular neo-Marxists such as Althusser, part company from others is in their confidence about the basis for the consensus as to 'what is there'. This consensus is – according to them (and other philosophers and social theorists for whom power and politics are the ultimate social realities) – the achievement of ideology. Ideology in this extended sense is that

in virtue of which one account of, or one part of, reality becomes
reality *tout court*; or by which an historically derived reality presents
itself as eternal and extra-human, as given rather than made, as objectively sensed rather than intersubjectively constructed by a particular interpretive community. Ideology 'privileges' one version of
reality over all others and suppresses any suspicion that things might
be ordered or perceived differently. (Controlling the order of things
and controlling the way that order is perceived are separate tasks;
but it will be obvious from the discussion so far that they are not
easily separable, especially as the latter is essential to the former.)
Ideology is the medium or mediator which guarantees that I identify
myself with the dominant group so that my version of reality coincides with 'their' version of reality: it ensures that I – the subject –
construe reality as 'they' construe it, or that we all intuit the world
'out there' as 'one' intuits the world out there. Reality is defined by
those who have control over the means through which realities are
given public acknowledgement – by, that is to say, the ruling class.
Itself privileged, the ruling class assists in maintaining the privileged
position of those who endorse it. The seeming manipulators are, of
course, generally as unconscious and unwilled as those who are
manipulated.

If the idea that 'reality' is defined in this way seems implausible,
this is precisely (it would be argued) because it is of the essence of
ideology to efface itself and to conceal all evidence of the struggle
whose outcome it influenced so decisively. Reality will carry few marks
of the processes by which it was derived or produced. It will appear
to be simply, incontestably, irrefutably 'there': the historical and transient
will seem natural and permanent. Indeed, it is the central task of
ideology to naturalise social phenomena and to confer upon them the
objectivity of the material world, to make that which has been constructed by human beings seem to confront them as naturally
given.

Ideology, according to Althusser, 'interpellates individuals as subjects', and in so doing constitutes them as subjects. We are individuals in so far as we have separate bodies, but this is not in itself sufficient
to make us into subjects. A subject is someone who 'works by himself' and contributes – without asking, or being able to ask, radical
questions – to the process by which the state ensures the reproduction of the conditions of production:

> In fact, the State and its Apparatuses only have meaning from the
> point of view of the class struggle, as an apparatus of the class
> struggle ensuring class oppression and guaranteeing the conditions
> of exploitation and its reproduction.[10]

This is inescapable because 'man is an ideological animal by nature' (Althusser, p. 58), and our status as subjects 'is an ideological effect, the elementary ideological effect'. Our acquiring the status of subjects is linked to our becoming political subjects: the price of 'recognition' as a subject is to be subjected, to submit to a higher authority, 'to be stripped of all freedom except that of freely accepting his submission' (Althusser, p. 45). Thus subjected, the individual will 'work all by himself'.

Few would dispute the general thesis, developed most convincingly by Mead in the 1920s and 1930s, that the sense of who and what one is comes from without rather than from within; that, even if the sense that one is a self cannot be so derived but is presupposed in every moment of consciousness,[11] the detailed content of that sense of self is decisively determined from without. The formation of a self-image, even self-recognition, consists to some extent of locating ourselves on a grid of attributes, concepts, comparisons, judgements, etc., that belong to the collective consciousness. Self-recognition and being recognised or acknowledged by society at large are intimately, even dialectically, related: self-awareness is a kind of other awareness from within. As Mead would put it, the self is in part constructed out of internalised glimpses of the Generalised Other. Some of our most personal, private and intimate decisions have a public or external origin; their intelligibility is a general intelligibility: falling in love is no more personal than getting married; and no one would have devoted himself to literature or to healing the sick if (to adapt La Rochefoucauld's aphorism) 'he had not read about it first'. The very process of 'searching for one's true self' is an activity whose origin, rules and destination are largely social. Many thinkers, too, would concur with the Durkheimian perspective – to be discussed presently – that locates most, though not all, apparently 'internal' concepts outside of the individual psyche in 'collective representations'.

We may grant all of this without having to agree that self-presence, as the condition for differentiated self-awareness, is entirely social in origin; or, more to the point here, that all thought and action, in so far as it is intelligible, is subordinated rather particularly to reproduction of the processes of production and consequently to perpetuating a situation in which the few exploit and oppress the many. Nor is one obliged to subscribe to the rather improbable view that metaphysical subjectivity must be replaced by social and political abjectivity (sic), that the heart of consciousness is occupied by a political unconscious.

This, however, is Althusser's position, which is less an individual aberration from, than an expression of, an essential tendency of Marxist thought. This passage shows the connexion with mainstream Marxist philosophy of history:

the human subject, the economic, political or philosophical ego is not the 'centre' of history – and even . . . that history has no 'centre' but possesses a structure which has no necessary 'centre' except ideological misrecognition.

(ibid., p. 56)

Althusser does, however, address a real problem. And one that seems particularly urgent if one assumes an individualistic standpoint, however faint or implicit its Cartesianism. Once religious explanations – a pre-established harmony of perception, individual minds cohering in the mind of God, and so on – are abandoned, then the social consensus seems to be a case of an inexplicable coincidence, a chance dovetailing of literally millions of different viewpoints. We are obliged, therefore, to postulate that there are 'social forces' ordering the developing consciousness so that it may participate in, understand and operate within, the intelligible order that has been agreed upon collectively. It does not, however, follow that these forces can be expressed entirely in narrow political or politico-economic terms or summarised so easily as Althusser seems to imagine. It is unlikely that the fundamental process of establishing and maintaining an intersubjective reality can be encompassed by the class struggle. After all, a common world, a shared or public reality, must be logically prior to the dimmest intuitions of class interest, not to speak of the ideas of oppression, justice and relative poverty in that world which lie at the base of Marxist thought.

The contradictions in Althusser's position are in part those of Marx writ large. Once it is assumed that consciousness is inescapably riddled with self-deception and miscognition, and reason is the plaything of forces it is unaware of, then there is no basis for assuming that one conviction, however well argued, is closer to the truth than any other: we have no grounds for believing Althusser any more than those whose views he opposes. This is a particular problem for Althusser because he sees ideology as both coterminous with all that is intelligible – it is almost 'the form of intellectual intuition' (analogous with the Kantian forms of sensible intuition) – and at the same time necessarily 'miscognition'. His ideas, as much as those of their opponents, must therefore be the products of history; they, too, must be passive reflections of changes in the productive process. There is no 'outside-of-ideology' from which expressions of class interest can be perceived at all, never mind be seen for what they 'really' are.

It is one of the stranger, but most telling, weaknesses – but, as will be evident from the above discussion, inescapable consequences – of Marxist historicism that it assumes that the changes in history are not idea-led. It overlooks – or takes as given – the technological advances that lie behind the changes in 'the material forces of production'

and the consequential changes in the relations of production. The transformations of the economic infrastructure are crucially influenced by
ideas. What is technology – the major source of revolutions in the
process of production – if not the application of carefully thought-out
ideas to the process of production? Technology is the supreme expression of 'Man, the explicit, self-conscious, deliberate animal' (see
The Explicit Animal, chapter 6, note 4). And technological advances
are rooted in science and the visions, insights, dreams and skills of
individual scientists and technicians. Through science and science-based
technology, man has transformed the productive process and so driven,
rather than merely being driven by, the historical process. Moreover,
the truth of technological ideas does not seem to be class-conditioned
or historically dependent; on the contrary, these ideas condition history, at least in part because they influence the creation of those very
classes whose antagonism Marx identified with the essence of history.

The unexplained privileged status of scientific ideas in Marxist thought
– according to Althusser the sciences were distinguished from ideology 'by virtue of being rationally constructed truths that served no
interest at all except that of science itself'[12] – is deeply significant.
Not only does it suggest that there are, after all, other influences than
the (unconscious) material dialectic driving history and that individual
consciousnesses, as well as the ideologically misrepresented material
forces of production, may have had in the evolution of society. It
also, implicitly, points up the fundamental contradiction in Marxist
thought. For no one could accuse Marxism of serving no interest except science itself. Marxism, as revealed in the *Theses on Feuerbach*,
had its roots in a desire to change the order of things ('Philosophers
have hitherto only described the world; the point, however, is to change
it'); and it wanted to change them in a certain direction – one that
would abolish the exploitation of the working class by those who owned
the means of production. Marxists were not themselves unaware of
the need to give their own views a special status; and the rhetoric of
science – the universal talk of 'scientific socialism', etc. – was freely
deployed with respect to their own theories in the forlorn hope of
raising them above the ideology enmired condition of 'bourgeois liberalism' and other rival politico-economic theories.

In the context of Marxist theory itself, this claim of Marxism to
scientific status looks like magic thinking: 'if I call Marxist thought
"scientific socialism" often enough, then it will by this means become
scientific'. And the status of truly scientific ideas – in the real world
and as reflected in Marxist thought – casts a sharp light on the most
deeply unresolved conflict within Marxism: the role of ideas held by
conscious individuals (as opposed to economic forces acting in or on
them) in bringing about the very revolution that is the *raison d'être* of

Marxism. There is a supreme irony in the way Marxists simultaneously assert the unimportance of ideas as motors of change and assert the importance of Marx's ideas. The assumption that the laws of
history and the implicit ideas inscribed existentially in the immiserated
proletariat are sufficient for the enabling cataclysm that will bring the
classless society to birth is contradicted by the huge intellectual, political, military (and largely unsuccessful) efforts made by bourgeois
intellectuals to make the revolution happen. One would have thought
that the inexorable processes of history would not have to be preached
in order to unfold – even less by those whose class position has conditioned them to misread the entire world around them.[13] It is quite
an achievement to have failed to bring about the inevitable.

6

Durkheim and the Social Unconscious

In his perceptive and illuminating introduction to the English translation of Saussure's *Course in General Linguistics*,[1] Jonathan Culler draws
attention to the 'happy coincidence' that placed Saussure's year of
birth (1857) between that of Freud and that of Durkheim. He identifies fundamental insights common to the founders of, respectively,
modern linguistics, psychology and sociology. All three rejected genetic explanations based upon personal history – or the sum of personal histories – in favour of analyses of systems: language as a system
in Saussure, the psyche as an interpersonal system in Freud, and the
system of collective norms and beliefs in Durkheim. In contrast with
positivist thought, in which society was merely the sum of the feelings and activities of individuals, society for these three thinkers is a
reality in itself; moreover, it is in a sense prior to the psyche. Social
institutions are the very conditions of individual experience. Social
sciences should consequently study those interpersonal systems of norms
which, internalised by individuals as the culture within which they
live, create the possibility of a wide variety of meaningful activities.

This way of thinking has four related elements. The first is the shift
from what linguists call an 'item-centred' to a 'system-centred' approach. This is most clearly evident in the case of Saussure, where it
was central to his thought. The second is the rejection of conventional, historical-causal thinking: of the Comtean belief that really to
understand an object it is sufficient to know its history in the sense of
its individual vicissitudes; of the hegemony of the diachronic over
the synchronic. At the very simplest level, according to the new view,
an act does not refer to, and draw its meaning from, antecedent acts;
rather, its meaning is rooted in a profound, hidden, timeless structure. In some cases, it might appear that the rejection of individual
history and of causal determinism is incomplete – a rejection only of
immediate history as an adequate explanation of actions, with more
remote events being seen as influential antecedents. Freud, for example, according to Sartre, 'refuses to interpret the action by the antecedent moment – i.e. to conceive of a horizontal psychic determinism'.[2]

240

He does, however, aim at 'constituting a vertical determinism' which refers to the subject's distant past and 'the history of the subject that will decide whether this or that drive will be fixed on this or that object' (Sartre, *Being and Nothingness* p. 458). Even so, personal history is not the crucial determinant. The search for the latter, he points out (in agreement with Freud) refers us to a wider (cultural) system. The remote antecedents of behaviour make sense only within that system. The Oedipus complex, he claims, could not arise among, say, people indigenous to the Coral Islands in the Pacific.[3] The third feature is a profound undermining of the individual's sense of his or her own uniqueness – of the foundational assertion of Romantic individualism, set out in the first chapter of Rousseau's *Confessions*:

> But I am made unlike anyone I have ever met; I will even venture to say that I am like no-one in the whole world. I may be no better, but at least I am different.[4]

Reference to the system which is expressed through us weakens the individual's claims to uniqueness: each of us looks more like an instance, a copy of a general type that the system generates. Uniqueness, such as it is, is merely that of the combination of non-unique elements. An individual who was unique through and through, as Rousseau – and every romantic and adolescent since him – imagines he is might be not only socially unintelligible but, perhaps (since the social self is mediated through the system) unintelligible to himself.

The fourth feature, and the most important one for our present purposes, is a consequence of the first three: a severe reduction of the scope of 'personal' sources of meaning based upon individual experience, and a decisive 'Copernican' (to use Freud's analogy) displacement of the individual from the centre of his or her world.[5] Meaning comes from without, or at least is guaranteed by an external system, and is independent of the history of the individual. Individual consciousnesses do not discover their own meanings in the world: meanings are in the keeping of, indeed are the product of, objective systems sustained by the collective. These collective or system-based meanings, moreover, do not necessarily become available to the individual as a result of his or her own history – by a process of, say, association of (private) sensory experiences. And they are most definitely not formed by any such private experience; as Durkheim believed, it is society, not private experience, that forms human minds and is the ultimate origin of our behaviour.

In consequence, the meanings of an individual's actions may be incompletely available to him or her. The role of individual consciousness in the activity of the social being is radically diminished: the individual is born of society and not society of the individual. The whole has priority over the parts and the individual is social through and through. Like Marx, but at a much deeper level, Durkheim believes that 'Man is human only because he is socialised'.[6] Of course, 'the general characteristics of human nature participate in the work of elaboration from which social life results'. But

> they are not the cause of it; nor do they give it its special form; they only make it possible. Collective representations, emotions, and tendencies are caused not by certain states of the consciousnesses of individuals but by the conditions in which the social group in its totality is placed. Such actions can, of course, materialise only if the individual natures are not resistant to them; but these individual natures are merely the indeterminate material that the social factor molds and transforms.[7]

Durkheim's vision of the domination of social over individual sources, determinants and loci of meaning is tested and developed in various key areas, notably in the study of suicide, in the interpretation of religious belief and in the analysis of the origin of concepts and categories.

At first sight, no act would seem to be more individualistic, less socially determined, than taking one's own life, as a result of which one literally drops out of society. The decision to kill oneself would seem to be the ultimate assertion of an unsurpassable solitude. For this reason suicide looks like a supreme test case for Durkheim's claims about the social determination of actions.

His examination of suicide begins with the epidemological facts. If the suicides committed in a given society during a given period are taken as a whole, the total 'is not simply a sum of independent units, a collective total, but is itself a new fact *sui generis*, with its own unity, individuality, and consequently its own nature'.[8] This social fact is independent of individual decisions. The evidence Durkheim adduces for this is the remarkable stability of the annual suicide rates in different societies. From this he infers that even the apparently least socially determined suicides – those that he calls 'egoistic' and which 'spring from an excessive individualism' – are essentially socially determined, the rates varying inversely with 'the degree of integration' of domestic, religious and political society. 'The incidents of private life which seem the direct inspiration of suicide and are considered its determining causes are in reality only its incidental causes. The individual yields to the slightest shock of circumstance because the state of society has made him a ready prey to suicide.' Suicides occur because of 'suicidogenic factors' in society:

The conscious deliberations that precede suicide are purely formal, with no object but confirmation of a resolve previously formed for reasons unknown to consciousness.[9]

Suicide, then – this seemingly most individualistic and anti-social of acts – seems to be socially determined. What of religious belief? This again would seem to be an impressive test case for the claim that we are social in essence and that our selves are social products. For religion, at least according to the Protestant tradition, seems above all a matter of a deep personal vision, of inner conviction. Although many of its expressions are public and collective, these count for little if the impulse behind the individual's participation in them is not derived from places of the self that are most remote from society. Moreover the ultimate object of religious worship – God – transcends society, indeed transcends life itself, localising society itself as a mere temporary house for the soul, and the very life of the individual as only a transitional state.

According to Durkheim, this view of religion is quite without scientific foundation; indeed, it is part and parcel of the very misconceptions necessary to make religion work, to enable it to serve its social function: that of maintaining the cohesiveness of society.[10] Participation in religious rituals is not rooted in personal vision and private experience; nor do religious convictions stem from personal decisions. At the heart of religion is the sense of the sacred; and the emergence of this category, over against the much larger category of the profane, originates in an ill-defined feeling, experienced by the members of a collective that there is something superior to their individuality. This superior reality is the force of society which precedes, transcends and will outlast them. Religious sentiments are society's self-worship expressing itself through individuals: the worshipped God is collective social reality transfigured.

For Durkheim religious interests are merely the symbolic form of social and moral interests; and this is as true of the contemporary religions of salvation as of the totemic religions with which Durkheim began his investigations. The sanctity of totems is due to the collective sentiment of which they are the object. Far from pointing beyond society – either into some transcendental sphere beyond temporal reality or into some society-proof fastness of the self where the individual confronts his situation in a mode that transcends social self-definition – religious sentiments simply reiterate those of the collectivity. Religion, instead of breaking the closed circle of society, reiterates it in a kind of tautology. 'Religions are the primitive way in which societies become conscious of themselves and their history. They are to the social order what sensation is to the individual.'[11] Through the expression of religious sentiments, the individual worships society without appreciating what it is that he worships. He is aware only that it is superior to his individuality but he does not know that it is society itself. The Kierkegaardian existentialist Protestant who believes himself to be furthest from society when he is alone before his Maker is utterly self-deceived.[12]

Durkheim's explanation of suicide and his interpretation of religion are a direct challenge to the belief that consciousness knows what it is about and that self-consciousness is a reliable guide to the nature of consciousness.[13] So little are we the *fons et origo* of our behaviour that we are unable even to see that we are not: we overlook not only its social basis, falsely intuiting that it originates from within ourselves, but we also are unaware of the extent to which a dim intuition of the superior force of society lies at the heart of what is supposed to be deepest within us. Until Durkheim revealed their true character, individuals were doubly mistaken as to the nature of their religious beliefs, imagining them to originate from within a trans-social self when, in fact, they came from society; and believing them to have as their object a society-transcending God, when in fact their object was society itself.

An even more intimate assault on the individual autonomy of consciousness comes from Durkheim's consideration of the origin of the concepts and categories that mediate the interactions between consciousness and the world and under which it classifies what it encounters. Nothing could seem more 'intra-psychic' and centred on the autonomous, individual self than concepts and yet, according to Durkheim, they derive from the collective.

The empiricist account of concept acquisition is essentially an individualistic one: concepts are acquired through sense experience and are rooted in this; and they are the means by which personal experience is ordered. The organising principle, according to the empiricist psychology dominant when Durkheim was writing, was that of association: our individual 'impressions' are organised by the principles of resemblance, opposition, etc. that psychologists, inspired by David Hume, invoked to explain how the haphazardness of experience could be combed to the order of a shared world. According to Durkheim, this account must be false because it seems implausible that the enormously complex network of concepts and categories through which we order our individual and shared worlds could be derived from the disorderly and haphazard data supplied by our senses, regulated only by association which is, as Gellner expresses it, 'boundless, unconstrained and undisciplined'. As Coleridge pointed out (when he was describing his own recovery from associationist theories of the mind), associationism does not enable us to distinguish between ordinary, organised experience and delirium. There is a huge gulf between personal sensations and their echoes in memory on the one hand and shared concepts on the other. Durkheim's charge against the empiricists, Gellner says,

> was that they could not explain the pervasive compulsive constraints, the remarkable congruence of ideas within any one society; and, indeed, that they barely noticed it.[14]

'Associations' it would appear, 'are born free but are everywhere in chains' (Gellner, *Reason and Culture*, p. 34). Concepts, therefore, do not stem from individual consciousnesses and subsequently achieve collective status as a result of the concordance of individual consciousnesses with one another (itself the consequence of the common physical reality to which they are exposed); on the contrary, concepts originate in the collective life. Concepts are collective representations, impersonal in origin, and imposed on individuals from without.

This, according to Durkheim, is particularly obvious in the case of large concepts, such as space and time. Although we encounter a spatialised and temporalised reality, the succession of our experiences would not give rise to the abstract and impersonal framework within which we live and whose manifestations include the map, the calendar and the chronologies of history. Likewise, the idea of causality comes from society; it is not our sensory experience but the experience of the collective life that gives rise to the idea of force:

> the space which I know by my senses, of which I am the centre and where everywhere is located in relation to me, cannot be space in general; the latter composes the totality of dimensions, and these are co-ordinated by impersonal guide lines which are common to everybody. In the same way, the concrete span of time which I feel passing within me could not give me the idea of time in general: the first expresses only the rhythm of my individual life; the second must correspond to the rhythm of a life which is not that of any individual in particular, but which everyone shares. In the same way, finally, the regularities which I am able to perceive in the manner in which my sensations succeed one another may well have a value for me; they explain how it comes about that when I am given the first of two phenomena whose concurrence I have observed, I tend to expect the other. But this personal state of expectation should not be confused with the conception of a universal order of succession which imposes itself upon all minds and all events.[15]

Society, finally, has provided the very space in which logical thought works.

In part, Durkheim's position on concepts and categories stems from his difficult idea that all general knowledge is religious in origin:

> No known thing exists that is not classified in a clan and under a totem, and hence there is nothing that does not in some degree possess something of a religious character.

(p. 255)

These primitive classifications are, in turn, modelled upon social organisation; they have taken the framework of society for their own framework. Thus, 'the fundamental ideas of the mind, the essential categories of thought' are not at all what the mind thinks they are but 'the product of social factors'. The categories, being of religious origin, must share the nature common to all religious facts; they are social things, the product of collective thought. 'No doubt', Durkheim adds, 'when an individual utilises the concepts he receives from the community, he individualises them and marks them with his personal imprint; but there is nothing impersonal that is not open to this type of individualisation' (ibid., p. 268).

Thus, even what is individualised is 'pervaded by social elements'. The soul itself is pervaded in this way; indeed, it is a social construct:

> Our inner life has something like a double centre of gravity. On the one hand there is our individuality and, more particularly, our body in which it is based; on the other there is everything in us that expresses something other than ourselves.

(ibid.)

It is this latter which corresponds to the soul: we are ensouled in so far as we are socially possessed; individuality is relegated to our body with its sensations.

The cumulative impact of Durkheim's ideas upon the notion of a controlling, self-possessed consciousness at the centre of the individual's life is devastating. It goes far beyond exorcising the Cartesian ghost in the machine. In the collaboration between the individual and society that determines the individual's understanding of the world, the individual is a minor partner. 'The individual is born of society and not society of individuals.' Indeed, the sense of individuality is a late entry on the human scene. Those societies in which individuals resemble one another and are lost in the whole, in which they are entirely external to themselves, have historical, and, indeed, genetic, precedence over societies whose members have acquired both awareness of their individuality and the capacity to express it. Society cannot be reduced to the sum of its elements: the whole has priority

over the parts, the structure over the individual, the social type over
particular phenomena. Individual consciousness is permeated through
and through by collective consciousness. The latter – most simply
expressed as the body of beliefs and sentiments common to the average members of a society – has a life of its own: it evolves according
to its own laws rather than being the expression or effect of individual consciousnesses.

Parallel to this externalisation of the origin, the field and the content of consciousness, is Durkheim's methodological distrust – to which
we have already referred – of the account consciousness gives of itself. A truly scientific sociology dismisses the validity of introspection:

> We assert not that social facts are material things but that they are
> things by the same right as material things, although they differ
> from them in type.... A thing differs from an idea in the same
> way as that which we know from without differs from that which
> we know from within ... their characteristic properties, like the
> unknown causes on which they depend, cannot be discovered by
> even the most careful introspection ... One might even say in this
> sense that every object of science is a thing ... In the case of 'facts'
> properly so called, these are, at the moment when we undertake to
> study them scientifically, necessarily unknown things of which we
> are ignorant; and any 'representations' which we have been able to
> make of them in the course of our life, having been made without
> method and criticism are devoid of scientific value ... Conscious-
> ness is even more helpless in knowing [social facts] than in know-
> ing its own life.
>
> (Durkheim, *The Rules of Sociological Method*, pp. xliii-xliv)

For Durkheim, 'understanding, being the highest and therefore the
most superficial part of consciousness, may be rather easily modified
by external influences like education, without affecting the deepest
layers of psychic life'. That part of consciousness which is most self-
conscious, in other words, is most likely to be out of touch with the
true intra-psychic realities and anyway carries no weight. To find out
what really is going on in someone, you should not rely on introspection, direct insight or intuition – least of all on their own introspective reports, insights, intuitions, etc. Such intra-psychic phenomena
should be treated as social phenomena external to individuals and
consequently regarded not as scientific data but as objects for scientific investigation:

> The facts of individual psychology ... although they are by defini-
> tion purely mental, yet the consciousness we have of them reveals

to us neither their nature nor their genesis. It permits us to know them up to a certain point, just as our sensations give us a certain familiarity with heat or light, sound or electricity; it gives us confused, fleeting, subjective impressions of them but no clear and scientific notions or explanatory concepts. It is precisely for this reason that there has been founded . . . an objective psychology whose fundamental purpose is to study mental facts from the outside, that is, as things.

Such a procedure is all the more necessary with social facts, for consciousness is even more helpless in knowing them than in knowing its own life.[16]

The true meaning of such phenomena does not correspond to what we can intuit directly; this meaning can be discovered only by objective, scientific exploration. Individualist and psychological explanations are inappropriate: every social fact is the result of another social fact and never of a fact of individual psychology.

Nor are introspections a true guide to the origin of our ideas. Individual consciousness is opaque at the centre and even the formulation of ideas and the development of 'deep' beliefs and 'personal' convictions are the result of extra-psychic influences. The true functions of our apparently voluntary behaviour – revealed only in objective exploration – may be quite different from the purposes that it seems to have for us.

Durkheim's theories have cast a long shadow over twentieth-century thinkers reflecting on the relation between the individual and society and the availability to individuals of the meaning of their actions, their ideas and their beliefs. Most importantly, Durkheim, like Freud, accustomed thinkers to expecting a dissociation between the meanings attributed by individuals to their actions and the true meaning of those actions; to the idea that we individuals are not authorities on ourselves, even in those areas where, hitherto, we had thought that we were. Of course, it had long been accepted that our intuitions were no guide as to the nature of the processes that are going on inside our physical bodies as when, for example, we are ill. Now the validity of our intuitions about the meaning of what is happening within us is questioned in the area of everyday activity. The actual function of a certain form of behaviour – visible only to the objective gaze of science – is quite different from the purpose apparent to the behaver. But it is possible to take this process further by dissociating actions even from those functions that are visible only to the scientific gaze of the sociologist and locating their meaning and origin in a hidden social system, understood as a structural unity. Durkheim paved the way to a structuralist sociology, which takes social science further along the path leading from a recognition of the sociality of individuals (society being the result of individual interaction which then reshapes individuals) to one of the reduction of individuals to functions of society.[17]

The structuralist elaboration of the Durkheimian vision takes an extreme form in the work of Lévi-Strauss, for whom society is a closed system of signs that signify only within the context of the system. Armed with the Saussurean system-centred conception of language and with the Saussurean postulation of a more general science of signs based upon language understood as the model for all meaningful activity, Lévi-Strauss interpreted kinship structures, totemic systems, myths and the ordinary rituals of everyday Parisian life in non-naturalistic and non-utilitarian terms – in terms, that is to say, which suggested that the true meaning of these things was not accessible to the consciousnesses of those who were caught up in them.

This is made plain in a famous passage from the Overture to *Le Cru et le Cuit*, where he writes of his analysis of myth:

> We are not claiming to show how men think the myths, but rather how the myths think themselves out in men and without men's knowledge.[18]

Every item – every element of every myth and the myths themselves – makes sense only in terms of the 'matrix of meaning' to which it belongs, the structure within which it has value. The final abandonment of an item-centred analysis of cultural data means a rejection of a functional – if not a functionalist[19] – analysis of human behaviour: the true occasion of our actions is unrelated to their apparent purposes. Or they are not to be analysed in those terms. The analysis of human behaviour shifts from an investigation of its utilitarian functions to investigating its signifying function: the place of this behaviour in a hidden symbolic system. And the interpretation of myth is no longer in terms of content; rather the interpreter seeks to uncover the hidden form, the meta-mythic system, of which the individual myth is only a small part.

Social systems are analysed in a similar fashion. For example, kinship networks merely embody abstract relations of consanguinity: women in prescriptive marriage systems are compared to verbal 'messages' sent from one clan to another; and incest taboos are not designed to prevent the mayhem that would result if males routinely paid sexual attention to their mothers and sisters, but are subordinate to the needs of maintaining the linguistic order that preserves the meaning of these messages. Totemic species are 'goods to think with', tools for assisting the mind's classifying play, helping it in its fundamental project of segmenting the continuum of experienced reality.

This play is very complex indeed: for instance, the resemblance presupposed by totemic differences is not between individuals and the
totemic signs but between systems of differences.

This is not the place to embark on a critique of Lévi-Strauss's methods;[20] he is mentioned only to indicate the possible consequences of
Durkheim's vision of the unequal relationship between individual and
collective consciousness. Structuralist anthropology lies at the end of
the road that Durkheim opened up. J.G. Merquior, however, has pointed
out that Lévi-Strauss 'stood Durkheim on his head': whereas for
Durkheim mind mirrored society, for Lévi-Strauss society mirrored
mind.[21] Our present concern, however, is concerned less with their
differences than with what Lévi-Strauss and Durkheim have in common: a profound conviction that mind does not know itself; that individual consciousness has an opaque heart, namely the collective
unconscious. Durkheim helped create the climate of opinion – crucial
for the development and acceptance of Lévi-Strauss's views – in which
it could be taken for granted that individuals were ignorant of the
origins of their own behaviour. In both, the individual consciousness
is decisively marginalized. What arguments can one mount against
Durkheim's notions?

First of all, it should be pointed out that it is possible to dissociate
the objective meanings of the actions – their functions – from their
subjective meanings only when they are remote from physiological
need, and from the sensations associated with such need. The reality
which is crucial to the individual is, indeed, the social milieu, a system of rules and norms internalised by the individual, rather than
the physical environment when the activities in question are those
such as religious ceremonials, marriages or football matches, whose
meaning is obviously primarily social, rather than rooted in bodily
need, and whose reference is obviously collective. But these are not
the whole story; and such purely social behaviours should be contrasted with the more fundamental activities that are rooted directly
in experience and relate to trans-historical, culturally invariant physiological need, such as the pursuit of food to satisfy hunger. Now it
might be argued that man has never in his history – i.e. since he
evolved from his animal ancestors – been a physiological animal. Even
the most apparently direct satisfaction of physiological need in the
most primitive society is mediated through rituals which give them a
thick overlay of meaning.[22] None of this, however, should obscure
the absolutely fundamental relationship between an appetite and its
satisfaction, which cannot be superseded. Hunger and thirst are not,
ultimately, mistaken as to their own nature. Consciousness, therefore,
at bottom, is able to know itself in this crucial area; why should it be
so self-ignorant elsewhere?

It might be argued that physiological need and its satisfaction is not sufficient to create or sustain a coherent self or a sense of our destiny and purpose – the feeling, in short, of who we are. However, it is possible to distinguish direct, personal, self-rooted experiences and meanings from the collective ones that Durkheim argues are available only to the objective sociological sciences even in the case of activities that are remote from the immediacy of physiological need, and are therefore ripe to be sociologised. Take a game of football. Within the meshes of the rules of the game and the social conventions that give it meaning, there is a rich, and finally basic, continuous substratum of experience that constitutes my personal involvement: my physical experience of the game (weather, fatigue, the fear of the tackle, the pain of injuries, excited apprehension of goal-scoring possibilities, etc.); what I wish to achieve through the game; the place it has in my biography (where it fits into my day, my life, my framework of hopes and fears). In other words, there is a continuum of consciousness that fills all the interstices of the rule matrix. In so far as the game is socialised to the point where its meaning is withheld from me, this occurs only at the level of the rules, not at the level of the experience of the realisation of the rules. Granted, the social norms are what gives football its meaning; but they do not create the experiences, nor do they determine the precise meanings that come with the experiences. The Durkheimian vision of a self-ignorant and marginalised consciousness is most persuasive (and even then not fully convincing) when applied to our consciousness of activities whose meanings manifestly originate outside the individual, activities whose significance depends on a context of which he or she can know only a small part, and when it describes such activities at a level that makes the rules and the social norms seem paramount.

And what of the individual's internalisation, interpretation and application of the rules? While it is valid to say, along with Sartre, that 'the meaning of the act is not the simple effect of an immediately prior psychic state and does not result from linear determinism', it is also true, as Sartre goes on to say, that the meaning of the act 'is integrated as a secondary structure in global structures and finally in the totality which I am'.[23] This integration is a supremely and inviolably individual act. It is clear that the individual's identity, as a framework within which life is experienced, has sources that are independent of collective experience.

The collective genesis and individual internalisation of concepts seems even less likely to be accounted for by Durkheim's theories. These are wonderfully and devastatingly mocked by Gellner, who is sympathetic (as I am) to Durkheim's question, but sceptical of his answer. 'Conceptually and verbally, we are astonishingly well-disciplined

and well-behaved. Both our capacity to communicate, and the very
maintenance of social order, depend on it' (Gellner, p. 34). There must,
therefore, be some mechanism for bringing us conceptually to order,
or for inculcating concepts in the form of 'comunally shared inner
compulsions'. This mechanism is the ritual dance:

> In the crazed frenzy of the collective dance around the totem, each
> individual psyche is reduced to a trembling suggestible jelly: the
> ritual then imprints the required shared ideas, the collective rep-
> resentations, on this malleable proto-social human matter. It there-
> fore makes it concept-bound, constrained and socially clubbable.
> The morning after the rite, the savage wakes up with a bad hang-
> over and a deeply internalized concept.
>
> (ibid., p. 37)

It is evident that Durkheim's account accounts for nothing. In particular, it cannot account for the genesis of vast numbers of concepts,
including those innumerable homely ones that it is impossible to imagine
anyone dancing into being: not 'King', 'God', 'Man', but 'red', 'toilet
paper', 'boredom', etc. And, moreover, it seems to presuppose the
very thing it would wish to explain. For the ability to dance an abstract concept into being and to propagate it through dancing assumes
a rather high level of communication, so that the dancers know what
is being danced. As for the grammatical relations between items – for
no one can deny that language has a systematic aspect – it is difficult
to see how much of this could be danced. The language of Durkheimian
society would be at best a concept-heap. More fundamentally, it does
not address the relationship between universal, social, collective concepts and individual consciousness; as we shall discuss in the Appendix to chapter 10. Any account of such a relationship must take
note of the facts that concepts do not really exist outside of consciousness
(nor social concepts outside of socialised consciousness) and that a
truly human consciousness does not exist totally outside of concepts.

Although, therefore, we accept the validity of Durkheim's challenge
to the empiricist account of a Crusoe-like consciousness, beginning as
a *tabula rasa* and constructing its own world out the material served
up to its senses, and would question the 'personal' account of the
genesis of the socialised human mind, it is difficult to accept his answer,
that tries in Gellner's words to reunite 'cognition, ritual and social
order' (p. 214). His answer to the puzzle of how it is that the collective world is not a colloidal suspension of cognitive city-states each
with a single citizen, gives birth to a multitude of questions.[24]

And this is not the end of Durkheim's troubles, or those of any
functionalism in which the true purpose of an activity is concealed

from the actors. As David Ramsay Steel has pointed out,[25] most cultural practices are so complex, it would be impossible to see how they could be transmitted unconsciously, never mind modified and developed. This is particularly obvious in the case of technologies – 'It would be futile to try to explain the evolution of bow-and-arrow technology without reference to the *intentions* of hunters' (ibid., p. 129) – but it must also be the case in relation to institutions that have some purpose, even when, as in religious institutions according to Durkheim, that purpose is misunderstood. 'The fact that humans do things on purpose limits the scope for explaining the determination of culture purely by unconscious transmission processes' (p. 129). Moreover, 'in order to imitate something effectively, you have to possess some understanding of it' (p. 130). Without this understanding, it would not be able to separate the relevant from the irrelevant, as illustrated by the cargo cult 'scientists' of Papua New Guinea – or the reflexologists and other pseudo-scientific practitioners of alternative medicine.

Much of the thinking of the structuralist heirs to Durkheim betrays a dim awareness of this problem. They seem to get round it by assuming that the system, in which the unconscious consciousness of the individuals absorbed into the collective is deposited, can not only look after itself, but that it has a kind of agency or even consciousness of its own. The passage from Lévi-Strauss that we quoted earlier betrays this:

> We are not claiming to show how men think the myths, but rather how the myths think themselves out in men and without men's knowledge.

I had always thought that this counter-intuitive idea of myths thinking themselves out in men who are themselves unable to think was meant to be a shocking or teasing paradox. It seems possible, however, that it is an attempt to find thought somewhere, given that it has been banished from the one place where it should normally be expected to be found – the minds of individual men. Likewise, the claim that 'language speaks us' – the central thesis of the post-Saussurean thinkers – may not just be a paradox designed to upset the slow-witted, but actually be an attempt to rediscover agency. This would not be all that surprising, in view of the commonness of the Fallacy of Misplaced Explicitness and the related Fallacy of Misplaced Consciousness that so often entraps thinkers who try to drive consciousness away from the usual places.[26]

We cannot, finally, allow to pass without challenge Durkheim's assumption that society can be accepted as a given that exists prior to individuals and does not itself require explanation. Ought we to agree that society simply explains everything else – suicide, religious sentiments, concepts and categories, etc. – whilst it can be assumed explanatorily to look after itself? That, for example, institutions simply emerge independently of the experience of the individuals over whom they are supposed to hold such sway and that they endure without effort on the part of their constituents? It seems not unreasonable to assume that institutions – and their propensity to evolve or remain unchanged – reflect the experiences, and needs, of individuals. At the very least, they must be based on past experiences and meanings. It could be argued that past experiences are not active experiences; they are experiences that belong to others – or to the collective other. They are a 'dead weight'. But the meanings based upon those experiences are present meanings; they are not opaque to those who experience them and they cannot be in the keeping of anyone other than the individual. The deposit from the past cannot of itself keep institutions alive in the manner requisite for Durkheim's purposes without the active collaboration of individual consciousnesses. At the very least, the individual and the collective are equal partners in the process of keeping collectives – rituals, concepts, etc. – alive and meaningful. One could turn the tables on Durkheim, as Raymond Aron does, when he says that Durkheim 'mistakes the social milieu for a *sui generis* reality, materially and objectively defined, when it is in fact merely an intellectual representation' (Aron, p. 98). Society, in other words, is a construct of the intellect, and not vice versa. Social facts, according to Aron, are not things out there, but abstractions in the consciousness of individuals. One need not, however, go this far to show that Durkheim's displacement of individual consciousness to the margins, with the centre-stage being occupied by 'society', has little foundation; or, at any rate, little explanatory force.

Durkheim's genius has uncovered questions that cannot be ignored. He challenges us to demonstrate that individual identity, as the framework within which life is experienced, has sources that are independent of the collective experience, of society. To respond to this challenge by arguing that experience is primarily private and that no two individuals have the same set of experiences may be inadequate. For it might be argued with equal force that my experiences make sense to me only within the framework of their social intelligibility. It is 'they' who tell me what my experiences mean, or at least what they should feed into my (relatively) stable sense of self, so that I am presented to myself through the mediation of general (collective) interpretations. This counter-argument, if accepted, would be important in relation to the question of to what extent I am simply a copy of the types that society generates – a mere instance of a class. This is a question that bears heavily on our *amour propre* and the sense, indispensable to our existential well-being, that we are irreplaceable. Even the tightest model of the social mediation of intelligibility, however, would allow considerable room for uniqueness: the categories remain baggy and each of us could belong to a unique constellation of different categories.

The model of language is a useful one here: all English speakers speak English and, in order to be intelligible – to themselves as well as to others – they have to use elements common to all and conform to rules to which all subscribe. Even so, there are huge variations in the way individuals speak English. So much so, that they seem genuinely to express (relatively unique) selves through language. And when it comes to the question of the *extent* to which the self is socialised and the extent to which it is deposited in a collective consciousness that is somehow unconscious of itself, there is yet more freedom and private self-presence. For language, by definition, is social in essence; it is shared intelligibility. Whereas experience has at its core an incommunicable content of sensation that may be interpreted – and so given over to the collective in accordance with rules that even the collective does not know – or may remain uninterpreted – or may undergo a succession of nascent interpretations. In short, at the core of experience is an irreducible remainder that cannot be dissolved into interpretations, into concepts generated in the collective; in such things lie self-presence and the distance between the conscious self and the unconscious of society.

7

Freud and the Instinctual Unconscious

Amongst the marginalisers of consciousness, amongst those thinkers who have advocated that consciousness should be, in Nietzsche's words, 'more modest', Freud is the most ambitious and influential. The reach of his (conscious, superficially rational) thoughts about the limitations of thinking, rational consciousness is enormous, encompassing sanity and madness, sexual and non-sexual love, war and peace, art and politics, life and death. Indeed, his example is so obvious, his ideas so well known, his *œuvre* so vast and his commentators so numerous, that it is tempting to bypass him altogether. He is, however, unignorable, and his case for the marginalisation of the conscious mind must be answered, or at least assigned its correct place.

I shall not attempt here a comprehensive critique of Freud. Over the last decade and a half, the long overdue appraisal of his contribution to our understanding of the psychobiology of the human mind, of the place of reason and passion in human affairs, and of the aetiology and treatment of mental illnesses has at last been undertaken. The verdict, alas, is uniformly negative: Freud as a scientist, metapsychologist and diagnostician of society emerges as a quack, and his influence and example seem to have been largely pernicious. Grunbaum[1] has been foremost among those who have demonstrated the groundlessness of Freud's claim to be a scientist: his method of acquiring and interpreting data have not the slightest resemblance to the scientific methods that have proved elsewhere so effective in permitting the discovery of the powerful truths that have transformed human life. Masson[2] and Esterson[3] have shown how unscrupulously Freud manipulated the little evidence upon which his theories were founded. The picture of Freud that emerges from the studies of Thornton[4] and Webster[5] at the time when he first formulated and propagated his theories amongst an inner circle of disciples shows that the main influences upon him were an old-fashioned quasi-scientific *naturwissenschaft* (particularly evident in his foundational *Project for a Scientific Psychology*), the lunatic numerological notions and mystical fantasies of Wilhelm Fliess, and his own cocaine addiction. Their portrait of a ruthlessly ambitious man desperate for fame, a brutally insensitive and unscrupulous clinician, quite unrepentant over his terrible diagnostic errors, a master manipulator of friends and colleagues engaged in an endless quest for self-promotion, is convincing, chilling and unforgettable. It is complemented by Wilcock's[6] detailed account of the rhetorical techniques Freud used, not only within his work, but also in shaping his reputation to support the belief that his totally unscientific ides were the products of a master-scientist. They show that Freud was the greatest of all Image Consultants. This, it hardly needs saying, is somewhat at odds with the Shilling Life and the hagiographies (notably that of the obsequious Ernest Jones) in which he is portrayed as the Great Healer, as a pure scientist consumed only by a desire to know the truth, as a prophet who saw deeper and further into the soul of human beings and society, and who, with no regard for his personal advancement, spoke fearlessly of what he saw.

The cumulative effect of the studies referred to above and many others – for example, those of Gellner[7] and Sulloway[8] – has been to show that there is hardly any aspect of the man or the *œuvre* untainted with the scandal of scientific incompetence and moral failure. Moreover, people are starting to count the cost of the 'depth psychology' that he pioneered and marketed. The long-standing observation that psychoanalysis is merely expensive – of money and time – and inefficacious has given way to a recognition that in many cases it has been dangerous and destructive. Psychoanalysts have often imitated their master in attributing to psychological causes serious illnesses that have all too organic origins, with often fatal consequences. Even where they are not criminally negligent, their crazy ideas often confuse and undermine desperately vulnerable individuals. Few psychoanalysts are as nakedly psychopathic as Lacan,[9] Freud's most prominent French disciple, but many seem happy to manipulate the affections of their clients to ensure continuing lucrative commitment to their quack remedies. The scale of the damage has recently become manifest in the United States, where it has been estimated that since 1988 a million families have been affected by therapist-inspired charges of sexual molestation, supposedly uncovered by the awakening of repressed memories. Freud's once unique ability to suggest to his patients the very 'facts' that he needed to support and fulfil his theory-fantasies, reinforced by his aura of wisdom, is now disseminated among hundreds of thousands of disciples who believe, like him, in the central importance of certain kinds of repressed memories and their own privileged access to them. They enjoy power greater even than his to do harm, by persuading individuals that they may have been forced to participate in or submit to disgusting and appalling events ranging from simple – and sickening – molestations to extraordinary Satanic rituals.[10] During the twentieth century, as Webster (p. 525) has pointed out, 'many women have suffered immensely as a result of orthodox psychoanalysts construing real episodes of abuse as Oedipal fantasies'. The skills Freud deployed to persuade people that their problems were due to their once wanting to have sexual relations with their parents are now used in persuading people that they have been subjected to sexual abuse of which they have no recollection. Beyond the damage to individual patients – now reaching epidemic proportions – there has been the more widespread intellectual damage: there is hardly a discredited idea in social work, child psychology, pedagogy and therapy that has not had some input from Freud's ideas.

It is, therefore, not only cost-conscious health insurers who will support the theoretical critics of Freud in hoping that the days of the Viennese witch-doctor (as Nabokov called him) are numbered. Anyone who has a respect for justice and the rules of evidence will be appalled at the witch-hunt his theories have licensed. The tide against Freud, Freudianism and psychoanalysis should, therefore, probably be irreversible. Why, then, bother to address, and criticise, his fundamental notions? Because these notions seem likely to outlive Freud's specific nostrums and his specific claims about the origin and genesis of the developed human mind. Auden, famously, said of Freud that

> To us he is no more a person
> Now but a whole climate of opinion
> Under whom we conduct our differing lives.

It is difficult to refute a climate of opinion, especially if, as in the case of Freud, the climate is constructed out of fog. Freud's theories, notoriously, have the added inbuilt survival kit in the form of the idea that disagreement with them is 'resistance' and this resistance is itself a sign of pathology in the resister and, moreover, pathology that can be understood only in Freudian terms: hostility to the theories – particularly if it goes so far as reasoned refutation – is yet more evidence in favour of the theories. Freud, like Marx, can rise above the bricks of many a ruined Berlin Wall. Indeed, his indestructibility is reminiscent of Rasputin: he was shot and still kept running; he was thrown into the icy waters and came up time after time; he was given an arsenic cake and asked for a second helping. Part of this indestructibility lies in Freud's penetration of quasi-disciplines whose practitioners are unable to judge him as a clinician and are entirely innocent of the methods of science and the means by which reliable scientific knowledge is acquired: the new, eclectic humanities gathered under the general banner of Theory.[11] He remains, therefore, a powerful marginaliser of consciousness and worth addressing.

The concept of the Unconscious is central to Freud's account of the mind. It was elaborated in connection with the theory of repression. An idea, or feeling, or experience that is repressed disappears from consciousness but remains active in determining behaviour, attitudes, feelings and other thoughts. In consequence, actions and responses that seem to result from rational intentions are, in fact, the expression of unconscious forces and feelings, or of forces and feelings rooted in unconscious forces and feelings. The rationale of certain behaviours is mere rationalisation of irrationality. The presence and power of the Unconscious becomes explicit when it is manifested in slips of the tongue, absentmindedness, jokes and all the other small change of what Freud, with conscious provocation, called the 'psychopathology of everyday life', in dreams and – when repressive mechanisms break down or the tension between unacceptable instinctual wishes and acceptable overt behaviour becomes insupportable – in major psychopathology. In short, it becomes apparent and accessible to knowledge when it is symptomatised in behaviour that could not be explained without reference to ideas, thoughts and feelings of which the individual had no awareness – in dreams, in neurosis and in the psychopathology of everyday life.

Freud, as Gellner (p. 207) points out, 'did not discover the Unconscious. What he did was to endow it with a language, a ritual, a church.' He went further than that: he reified it, gave it the status of an active force, an ensemble of forces and things, engaged in perpetual hidden civil wars re-enacting the Titanic struggles that attended the transformation of the instinctual infant-animal to the rule-governed child-citizen. In Freud's hands, the Unconscious became the dominant but hidden influence of the remote and archetypal past in the ordinary present.

Most importantly from our point of view, Freud endeavoured, more systematically than anyone before him, to demonstrate that not everything that is mental is conscious. In its extreme version – and Freud, like the Unconscious, was able to tolerate self-contradiction and put forward many inconsistent versions of his theories – the Freudian worldview holds that the conscious mind is a small island lost in a sea of unconscious mental processes:

> The more we seek to win our way to a metapsychologically real view of mental life, the more we must learn to emancipate ourselves from the importance of the symptom of 'being conscious'.[12]

Ideas might be present in the mind, in the sense of being available for retrieval, or active in determining our feelings and behaviour, whilst being only 'latent in consciousness'. Such ideas do not merely provide an underground substrate ensuring the continuity of consciousness, but are also active or dynamic – most obviously in the case of psychiatrically ill patients (as, for example, the hysterical patient whose behaviour is 'ruled' by unconscious ideas) but in reality in all of us. For psychopathology merely presents the economy of the normal psyche in italics. The material in the Unconscious is unconscious because it is unacceptable to consciousness. It has not merely failed to reach consciousness as many things do – such as the neural mechanisms which make consciousness itself possible. It has been banished to the Unconscious, having once been conscious. The conscious individual cannot admit unconscious ideas because she cannot allow herself to entertain what the social conventions she has internalised will not admit. The limitations of consciousness are even more severely defined by Freud, inasmuch as the process of repressing such ideas is itself unconscious: consciousness does not deliberately put things out of consciousness.

The domination of the Unconscious over the conscious places severe limitations on the role of fully formed intentions in the governance of behaviour and of reason in the shaping and initiation of actions. Apparent or conscious or avowed reasons are mere rationalisations of actions prompted by forces other than rational intentions. The politician who declares war is not safeguarding the interests of his or her country, as he or she states and believes, but indirectly enacting a repressed wish deriving from unresolved childhood conflicts. The artist who paints a landscape is not celebrating and preserving a piece of the world, as she thinks, or experimenting with colour; she is servicing her libido in several complex but related ways. The woman who trains to be a doctor is not, as she imagines, following a wish to help others by relieving their physical suffering, but is giving way to the deep-rooted sadistic appetites which her medical licence to witness and inflict suffering will enable her to satisfy. And so on.

In short, we live our apparently rational lives in the octopus grip of forces whose nature is such that they cannot be acknowledged by our 'higher' – indeed 'ordinary' – selves. Irrespective of whether those forces are expressed or, more commonly, repressed, they direct our actions. The ebb and flow of our feelings – even, or especially, those which seem to have a rational basis – is commonly an indirect expression of illicit desires, often sexual feelings reaching back to infancy, that have no other means of expression. Introspection gives us only a very limited access to the real motives and purposes of our behaviour and even the most sincere and honest account of our actions must be treated with the utmost suspicion. The subject's reports on, and beliefs about, his or her actions are always open to correction by objective analysis.

Even if one takes at face value the evidence upon which Freud has based his views, and accepts the interpretations he places upon it – and there are, as we have already noted, excellent reasons for doing neither of these things – it will be clear that, at the most, he has actually demonstrated only that conscious behaviour and thoughts may have their roots, or be shaped by, unconscious processes. Freudian theory undermines psychological theories that overestimate the transparency, autonomy and controlling power of consciousness. The autocratic self-transparent *cogito* of Cartesian individualism is made irretrievably opaque by the unconscious ideas that manipulate it. Does this, however, diminish consciousness – that is to say, the central place of deliberate, conscious, action in human affairs and daily experience? To reassert the latter is to go beyond the trite point that, even if seemingly conscious behaviour is manipulated by things that it may be unaware of or wish to deny, it is no less real or complete for being so. Freud's observations and arguments may justify a certain attitude of suspicion towards consciousness, especially when it claims to be the sole or central determinant of actions, and to have privileged and complete access to the causes of those actions – when, in other words, it is suggested that conscious intention is the sole motor of an action and its entire occasion. The question still remains as to the extent to which such suspicion is justified in diminishing the central position of consciousness – of deliberate, self-aware and undeceived thought and action – in human life.

The fact is that Freudian theory has been taken to imply – even to prove (since to believers, if Freud said it, or is thought to have said it, it must be true) – that rational consciousness driving deliberate action has a drastically reduced role in directing human affairs. It is difficult to know how much this is Freud's own position, but it is certainly that of many of his followers and, in view of the statement quoted above, that 'we must learn to emancipate ourselves from the importance of the symptom of "being conscious"', it may indeed be Freud's own. One absurd consequence of extending Freud's views to encompass the whole of conscious life is well illustrated by Freud's own claim that

> in the life-history of the individual, everything was originally unconscious, and it is only under the continual influence of the external world that some of the mind's contents become pre-conscious and so, should the occasion arise, conscious.
>
> (quoted in Wollheim, p. 173)

The passage could be made acceptable by being given a banal interpretation. Indeed, *everything* was originally unconscious in the life history of the individual, if we think of that life history as beginning with the embryo. But it would seem the Freud was trying to do more than reiterate the obvious point that the conceptus from which we developed is rather less on the ball than we are. The assertion that 'it is only under the continual influence of the external world that some of the mind's contents become pre-conscious and so, should the occasion arise, conscious' seems to imply that consciousness is a rather late, hesitant and intermittent entry into the order of things. This seems at the very least to incorporate a rather curious ontological inversion: if it is the influence of the external world that prompts the mind's contents to become conscious, by what criterion is 'the external world' 'external'? Is it merely 'external' in the sense of being outside the organism? This seems insufficient to sustain the kind of separation and relatedness implied by the terms 'mind' and 'external world' – except in a child's spatial ontology in which 'mind' is 'in here' (in the body/organism) and the 'external world' is 'out there'.

The more fundamental objection is that it is inadmissible to assume the existence of 'mind' and 'external world', and of the distinction between them, without presupposing the viewpoint provided by (conscious) consciousness. Consciousness cannot be a late entry into a scenario in which mind and external world are already presupposed. Freud's explanation that 'an idea can only attract consciousness to itself if it has a high sensory content' does not help. It illustrates only the general principle that, once you are in a hole, you should stop digging. It seems hardly necessary to point out that, in the absence of consciousness, it is difficult to give a meaning to 'sensory content'. 'Unsensed sensory content', perhaps? Sensory content cannot be that which provokes the mind's contents to move from unconscious to a pre-conscious and conscious mode: it is their conscious (and pre-conscious) content. The point becomes clear if we actually think of something specific, such as the sensation of red: the suggestion that the sensory experience of red moves the mind's content 'red' from the unconscious to the pre-conscious and the conscious mode is manifestly daft.

The Freudian notion that unconsciousness is continuous whilst the conscious is a matter of scattered occasions, that the Unconscious is an ocean from which consciousness emerges like a minute atoll, is thus without foundation. Consciousness may not reign supreme but it reigns continuous – even in the case of 'set-piece' expressions of the Unconscious, as for example manifestly neurotic behaviour: the patient who suffers from obessessive, irrational worries has to deploy a good deal of highly organised, consciously driven behaviour in order to enact them. Moreover, the present, conscious experience of the feelings and motivations, howsoever they are interwoven with self-deception, is the reason the neurosis matters to the patient and so desperately needs treatment. The fact that clinical, disabling neurosis is not universal, that we normally distinguish between dream and ordinary wakefulness and that our tongue does not usually slip, confirms the focal nature of the expression of unconscious forces and shows that, in ordinary waking life, the atoll is the Unconscious and the ocean consciousness, and not the other way round. There is also a fundamental contradiction between Freud's conception of the individual as an irrational dreamer and his vision of the same individual unremittingly mobilising reasons (rationalisations) to excuse/explain his/her somnambulant behaviour.

Let me illustrate this point with an example from my own experience in the short period when I trained in psychiatry. The patient was a woman in her forties who suffered from a recurrent anxiety that she might have breast cancer. She kept finding lumps in her breast and when she found such a lump she felt compelled to go to her GP at once – however inconvenient this was – so that he could examine her and reassure her that the lump she had felt was part of the normal anatomy. This behaviour was attributed by the consultant psychiatrist to a deeper anxiety arising from the death of her father, by suicide, when she was 12 years old. The visit to the fatherly GP who took her complaints seriously was a way of reassuring herself of her worth – a permanent requirement since her father's decision to desert her by electing to die had left her feeling worthless. As an explanation, it seemed – and still seems – reasonable to me. It assumed that the breast examination had a symbolic significance and that the need for it did not have the same basis as the patient attributed to it: the true object of her anxiety was her own sense of worthlessness and her fear of desertion and not the possibility of breast cancer. It is worth noting, however, that such an explanation fitted with common sense and did not require the mobilisation of a baggage of psychomythology about the successful or unsuccessful negotiation of Oedipal fantasies by her infant self. More important still, the patient's irrational, unconscious behaviour, far from being a sea engulfing rational behaviour aware of its own true purposes, was itself just a small island in a sea of rationality. In order to get to the doctor's surgery and be seen by him, she had to make an appointment, make arrangements for child care, negotiate her way with an adequately fuelled car to the surgery and await her turn like any well-behaved citizen. On the way back, she said ruefully, she would usually take advantage of the fact that the GP's surgery was near the cheapest shops and she would buy in food for the week. The central irrational act that had brought her to psychiatric attention was made possible only through the medium of extremely complex rational activity, remote from the wild associative behaviour of the dreamer.

The customary rationality of behaviour – including those components of behaviour necessary to bring about the successful expression of neuroses – is precisely what one might expect if one begins from the very biological standpoint that Freudian psychoanalysis believes it is rooted in. Biological survival demands that action should be dominated by the Reality Principle and be a response to an extra-psychic reality, that is, at some level, rational. The pursuit of basic needs requires non-neurotic perception of what is actually out there. Even those appetites that are unconnected with physiological survival – so that they can proliferate into infinitely complex *desires* (as in the case of the sexual appetite, which can elaborate into boundless love or ingenious perversity) – demand a similar responsiveness to external reality. The lover in frenzied pursuit of his mistress requires more than the unconscious instinctual forces at the root of his socialised desire to secure the practical arrangements necessary for the longed-for meeting. A 'physiological tempest' (to use Sartre's phrase) filtered through a neurosis would be insufficient to ensure that he caught the right bus to his assignation. Likewise, the successful prosecution of even the most brutal and irrational war requires a high level of conscious, organised, rational behaviour. The framework of our activities may often be irrational; but, even in such cases, once the agenda has been set by need and desire, the stretches of our lives lived out within them must necessarily be guided by reason and conscious deliberation. No matter how irrational our needs are, how crazily we are driven in our frenzy to satisfy them, we have to behave with a very complex rationality even to look as if we are trying to satisfy our needs, never mind to succeed in doing so.[13]

This is not to claim that the means by which we carry out even ordinary (non-neurotic) activity and the motors underlying it are transparent to introspection. There is only a limited sense in which we can be said to choose our desires; and at a deeper level, that of physiology and of the physics in which physiology is in turn rooted, we do not run our own bodies. Most of the bodily systems that make rational as well as irrational behaviour possible are not under the direct control of our conscious deliberations; they are not even accessible to our introspection. (We shall return to this when we examine Helmholtz's ideas.) Ordinary perception is founded upon (to use Helmholtz's perhaps unfortunate term) 'unconscious inferences' and is influenced by cerebral mechanisms that can be revealed only indirectly by careful experimentation. But to show that consciousness is rooted in that which it is not is not to show that consciousness is a fiction, nor is it, even, to diminish it.

There is, of course, a fascinating problem about the relationship between the unconscious background of our behaviour and our actions as conscious, responsible agents. If we set aside the concept of the all-powerful Unconscious calling the shots (and isn't it odd how the Unconscious, despite its refusal to be constrained by the Law of Contradiction seems to enjoy a coherence and purpose and unity that consciousness lacks) in the present from the darkness of our monotonously similar pasts (pasts that seem to verge on being a collective or archetypal past, at once ontogenetically and phylogenetically determined), we might be able to think about this in a fruitful way.

There are ways of thinking about the relationship between what is and what is not conscious in our behaviour, between the conscious and the non-conscious determinants of what we do as conscious agents. There is no doubt that on any given occasion, there is a focal centre of awareness – the bit of our sensory field that is commanding our attention, the specific aspects of specific events and situations that capture our interest – and a vast penumbra or periphery of things that have been, or might be capturing, or are just failing to capture our full attention. We are not conscious at any given time of everything we might be conscious of – the entire field of possible sensation, the entirety of our retrievable past. And yet we are – and have to remain – open to anything that might be, or become, relevant or interesting or important. Of course, even our attentionally focused self is not totally self-aware: we lose ourselves in activities – in reading books, in conversation, in dreams and fantasies – and are surprised at times by ourselves when we 'come to'. We carry out actions with different degrees of deliberation. And so on. But none of this corresponds to the Freudian vision of everyday man and women being manipulated by the Unconscious, by those members of the 'psychic government-in-exile' (to use Gellner's phrase) who have been expelled from memory by a process of repression covering up memories of infantile sexuality and our Oedipal longings for our parents.

Once one sets aside the mystic machinery of the Freudian Unconscious, it becomes possible to think about those things that are conscious and those things that are not yet or no longer conscious in everyday experience. We can even think rationally and helpfully about unconscious thoughts, as Chapman and Chapman-Santana have in their recent paper.[14] One of the points that these authors make is that there is no way of demonstrating the connection between a present thought, of which I have just become conscious, and an unconscious thought – 'or between the same thing in a current, perceived state and a previous, unperceived state' – if you cannot examine it by sensory methods in one (the unperceived) state. This is true of simple translations; for example, my 'recognising' when I learn of Uncle Fred's dishonesty that I had previously had unconscious thoughts along these lines, suspicions that I had repressed because it was painful to entertain them. It is even more evidently so in the case of Freudian translations

– for example, when my dream about a wolf attacking a rabbit is interpreted as referring to my feeling of powerlessness as a child under the thumb of a tyrannical father, and this in turn as being a translation of perceived castration threats emanating from my rival for the lustful affection of the wife-mother. Once we set aside the Freudian interpretation which lies remote from ordinary intuitions and observations, we can recognise what is central to ordinary consciousness: the ability to make almost anything explicit. I shall return to this presently.

Ultimately, it is not clear how much Freud believed in the irrationalism implicit in his account of the human mind. But there is little doubt about his irrationalist influence. Let me take a few examples. The first two illustrate the vogue for a quasi-Freudian irrationalist interpretation of political and social history; and the third the equally widespread fashion for asserting that the self, after Freud, is de-centred.

We have already encountered the first example: Adorno and Horkheimer's critique of the Enlightenment on the grounds that it mistook the nature of human beings. Writing under the twin influence of Freudian theory (especially as expressed in *Totem and Taboo* and *Civilisation and its Discontents*) and the recent horrors of European war, they argued that anti-Semitism proved that 'the dark impulse' is closer to mankind than reason. 'The rational island' (their telling phrase) is readily overwhelmed by a generalised urge to destruction. Anti-Semitism, for example, expresses something permanent and general in the human psyche that is only accidentally directed against Jews:

> The psychoanalytic theory of morbid projection views it as consisting of the transference of socially taboo impulses from the subject to the object. Under the pressure of the super-ego, the ego projects the aggressive wishes which originate from the id (and are so intense as to be dangerous even to the id), as evil intentions onto the outside world, and manages to work them out as abreactions on the outside world; either in fantasy by identification with the supposed evil, or in reality by supposed self-defence. The forbidden action which is converted into aggression is generally homosexual in nature. Through fear of castration, obedience to the father is taken to the extreme of an anticipation of castration in conscious emotional approximation to the nature of a small girl, and actual hatred of the father is suppressed. In paranoia, this hatred leads to a castration wish as a generalised urge to destruction. The sick individual progresses to archaic non-differentiation of love and domination.[15]

Anti-Semitism is, according to the authors, a reflection of this and, more generally, a demonstration of the fact that 'paranoia is the dark side of cognition' (Adorno and Horkheimer, p. 195).

This diagnosis is not only baseless – inasmuch as Freud's theories are without empirical foundation – it is also destructive – inasmuch as it gives no grounds for hope of improvement. If Auschwitz is an inevitable consequence of a combination of historical forces (which no one can regulate) and the unchanging features of human nature, the fundamental dynamics of the human mind, then we can do nothing to prevent future Auschwitzes. The post-Freudian vision of mankind – assumed by so many humanist intellectuals reflecting on the twentieth century – is that seeming reason is irrational; objectivity is an illusion; there is no beneficence, only vested interests rooted in unreflecting instincts; and that humans are essentially moved by unconscious forces. In short, they are not rational agents, being neither truly rational nor sufficiently in command of themselves to count as agents. They are either animals or metaphysically fallen beings who cannot be trusted in a secular society.

This vision is deeply pessimistic and, since it is likely to bring about the bleak scenarios for human futures that it predicts, should at least be firmly based on incontrovertible evidence. No such evidence in support of the psychoanalytic understanding of the human mind is forthcoming. Allan Esterson's revelation (in *The Seductive Mirage*; see note 3), that the only direct evidence Freud ever adduced for the Oedipal complex (according to his own estimate, the cornerstone of psychoanalysis and his own greatest contribution to knowledge) was a single doubtful memory of a journey in which he may have seen his mother naked and become sexually aroused is instructive.

The bleak vision of mankind being manipulated by deep and simple forces overlooks something very important about collective human behaviour – which is that it is mediated by extremely complex and inescapably conscious and deliberate coordinated action. This omission is evident in the discussions of the human propensity to violence. Let me illustrate this by returning briefly to Dudley Young's *Origins of the Sacred*. His prescription for solving contemporary spiritual confusions is, it will be recalled, a quasi-Freudian one: we need to remember the deep past:

> modern man no longer knows who he is, and this is largely because he has forgotten where he has come from; and just as a neurotic individual can be strengthened by retrieving some repressed or disavowed aspect of his past, so can we all, both individually and collectively, be strengthened by remembering our sense of the sacred and the animal powers from which that sense arises.
>
> (p. xvi)

These animal powers were 'concerned with managing sex and violence which we currently mismanage unlike the primitives whose religious approach was more effective'. To deny our past and the animal powers within us, is to remain in the grip of the past and of the animal powers, just as, according to orthodox psychoanalytical theory, to deny our instincts is to be more helplessly enclosed in them:

> The mutilating cruelty that opens our eyes to the monstrous epiphanies of the sacrificial divinity is properly cancelled, preserved and redemptively transformed in the elaboration of a *harness* for our sacrificial instincts, not in some rationalist attempt to close the door on them; and moreover, that the harness is to be sought *within* those instincts, and not elsewhere.
>
> (p. xx)

The question then arises whether some kind of active acknowledgement of our sacrificial instincts would help us to manage sex and violence better. There is the possibility that it may make things worse.[16] It is, however, more likely to be irrelevant; for very little large-scale aggression as expressed in modern warfare is concerned with the kind of violence – frenzied tearing of the flesh – that affords the ecstasies of the sacrifice or the hunt. And this is the important point.

War is obscenely violent, yes. This much is self-evident. What is not appreciated is how little of the experience of war in developed countries is the experience of inflicting violence, even as immediate and intimate as in a playground fight or street riot, never mind the ritual of the hunt. In modern warfare, most of those involved will not be directly concerned with damaging the flesh and property of others. Communications staff, supply staff, transport staff, planning staff, etc. far outnumber the actual combat troops. As for the combat troops, little of their time is spent in actual engagement with the enemy: from the moment the call-up papers arrive to the moment of discharge from the army, from hospital or from life, war, for them, consists largely of hanging around in various states of apprehension and boredom, waiting for something to happen. In the short periods of actual combat, the troops (even less the airmen) will have little idea of what they are doing, why they are doing it and the consequences of what they have done. Those who cause direct damage that is visible to themselves for reasons that are apparent to them are a (peculiarly) privileged minority and the period during which they are doing it is a (peculiarly) privileged few moments. Very few of those engaged in war get involved in it through bloodlust, through pursuit of something comparable to the mutilating ecstasy of sacrifice. Perhaps they may be nudged a little in that direction by training, waiting, injury, death of colleagues, exposure to propaganda. Even so, My Lai and the 'Turkey shoot' that came at the end of the Gulf War, are exceptional moments of hideousness. It is surprising how little it is appreciated that, such is the power built into the system of society, we don't have to be violent in order to create violent havoc.

In short, the violence that is expressed in modern war rarely takes the form of the satisfaction of impulses of those caught up in it. The proportion of ecstatic infliction of damage to fear, boredom and frustration, of free self-expression to circumambient limitation, is minute. Nothing could be less orgiastic than modern warfare. In a sense, smashing one's organic fist in one's conspecific's organic face is a more intense and direct expression of violence than dropping bombs that destroy the lives of many thousands in a city. Talk about violence without further analysis – as if the violence of a war were the same as the violence of a street fight – is inadequate: in war, the feeling, the expression of violence must be separated from its consequences. There is as little that is Dionysiac about a bombing raid briefing as there is about the debriefing. Contemporary violent wars are remote from violent passions. Modern wars, with the exception of those that take place in relatively primitive societies (for example Somalia and Rwanda), do not grow out of blood-feuds but out of complex and abstract political considerations. It is at this level that modern mega-violence (which involves many millions of people and not a few hunters seeking the ultimate recreational fix in the twitches of prey in its death throes and the taste of freshly spilt blood on their lips) must be addressed. Contemporary wars are more plausibly and more helpfully seen as the result of instabilities built into huge complex systems amplified by technological power than by the collective expression of ancient unexpressed feelings denied either expression or even recognition.

The Freudian analysis is thus a positive hindrance to understanding the forces that move modern societies. Our violence is enacted through the collective actions of explicit, rational consciousnesses rather than the merged passions, the upwelling frenzies, of the collective unconcious. A diagnosis that addresses human conflict as if it were a displaced expression of frustrated appetites will be worse than useless: it will lead to the kind of 'wise' (actually complacent) acceptance of the necessity of war and persecution and sacrifice expressed in Auden's lines:[17]

Without a sacrifice (it must be human, it must be innocent)
No secular wall will safely stand.

As for the de-centring of the self, many contemporary thinkers have taken their cue from Freud's claim to have brought about a Copernican revolution in human psychology, displacing the conscious, rational

self from the centre of daily life, as Copernicus displaced the earth
from the centre of the universe. Althusser has expressed most clearly
what Freud's thought meant for him and the first generation of 'de-
centerers'.[18] 'Freud', he reminds us,

> did not simply discover the existence of the unconscious; he re-
> jected the notion of the psyche as a structured *unity centered* on
> consciousness; instead he conceived it as an 'apparatus' composed
> of 'different systems' irreducible to a single principle . . .
>
> The apparatus is not a *centered unity* but a complex of instances
> constituted by the play of unconscious repression. The splitting of
> the subject, the decentering of the psychic apparatus in relation to
> consciousness and to the ego, is accompanied by a revolutionary
> theory of the ego; the ego no longer considered the seat of con-
> sciousness, becomes itself to a great extent unconscious.[19]

Freud, Althusser claims, 'rejects the primacy of consciousness not only
in knowledge but in *consciousness* itself', in so far as he 'rejects the
primacy of consciousness in psychology in order to think the 'psychic apparatus' as a whole in which the ego or 'consciousness' is only
an instance, a part or an effect' (p. 18). And he links Freud with Marx,
arguing that they both called into question 'a certain "natural", "spontaneous" idea of "man" as "subject". The *unity of which is assured or
crowned by consciousness*' (p. 23):

> The ideology of 'man' as a subject whose unity is assured and
> crowned by consciousness is not just any fragmentary ideology; it
> is quite simply *the philosophical form of the bourgeois ideology* that has
> dominated history for five centuries.[20]

In the absence of good evidence – as opposed to unsubstantiated claims
– from Freud's writings that this is how things really are with individuals, their consciousnesses, their egos and their daily life, we have
to appeal to intuitions and ask whether this is how things seem to
be; whether, in short, we seem to be the kind of somnambulating
plaything of forces and apparatuses that Althusser claims us to be.
The answer is that we do not and that a small, but telling piece of
evidence supports our intuitions here: the painstaking and (relatively
lucid) arguments that Althusser himself mobilises in favour of his
post-Freudian position. Does it seem likely that the kind of entity
that Althusser regards us as being would write the kind of things
that we see written in the passages quoted above? Is an ego that is
no longer 'the seat of consciousness' and itself 'to a great extent unconscious' likely to write that 'the ego no longer considered the seat

of consciousness, becomes itself to a great extent unconscious'? And when, as so many thinkers like to do, someone writes: 'It thinks, therefore I am not', as a way of summarising the implications of Freud's work, who is writing this – 'it' or some 'I'?[21]

This might prompt us to reflect more generally upon the status of psychoanalytic theory itself. What more striking sign of the power of explicit (ego) consciousness could there be? Freud's mighty *œuvre* arguing the reality of the Unconscious and describing his ever-more complex and far-ranging notions of how it interacts with and dominates consciousness – in short, bringing it into consciousness at the highest level – constitute an impressive refutation of his own position that the conscious must be subordinate to the Unconscious. After all, it is consciousness and not the Unconscious that has discovered the supposed Unconscious, its origin and its interactions with the conscious. What Freud – in common with other theorists who would displace consciousness by an Unconscious that they have, by the most ingenious and complex mobilisation of consciousness, uncovered and characterised to their own satisfaction – always overlooks is the part played by consciousness in uncovering the Unconscious and in drawing the kind of conclusions Freud teased out in the many volumes of his *Collected Works*.

Many actions are, of course, performed mechanically, unthinkingly, unconsciously. We cannot be conscious of everything all the time; nor can we be conscious of the entire background to our conscious acts, of the things that have made them seem imperative or at least appropriate – the social pressures, the learned patterns of expectation, etc. The things that we do consciously ultimately depend upon physiological mechanisms we are not conscious of and certainly cannot intend. Even where we do things deliberately, we do not, as we do them, 'foreground' the frame of reference, the presuppositions, that give them meaning and purpose or make it possible for us to enact them. It would be absurd to deny this unconscious context of, background to and basis for, conscious, deliberate behaviour. But it does not add up to something called the Unconscious which, being composed of repressed memories, regulates our behaviour in a way that lies beyond the reach of the consciousnesses of all but those who have trained in psychoanalysis or read the necessary Penguin Books. Freudian theory is not only groundless but its notion of the Unconscious does not even begin to address the fundamental puzzle of human psychology: how it is that we are able to forge deliberate, voluntary rational action out of the mechanisms in which we are caught up. This is the true mystery of man, the Explicit Animal.[22]

8

The Linguistic Unconscious: Saussure and the Post-Saussureans

Language is huge, extraordinary, ungraspable; mysterious in its instances, stupendous in its range. It is easy to see why thinkers over the millenia have so often defined humanity in term of it, describing Man as the Speaking Animal. As I have argued elsewhere,[1] this is mistaken: language is the most striking thing about us, yes, but it is in turn symptomatic of something else that is also expressed non-linguistically; namely, the propensity to make things explicit. Although it is almost absurd to put it thus, we may see language, at the very lowest estimate, as a kind of tool, an infinitely pliable instrument for ensuring, among other things, maximum cooperation between members of a species engaged in common purposes. Language enables individuals to draw on a powerful communal consciousness to serve their needs better than their own unassisted minds. The extension of oral communication by means of writing and other systems of enduring signs enables information to be stored, not only outside the moment of its production, but also outside the human body.[2] This has permitted humankind to develop at an extraordinary rate, quite independently of changes in genetic structure, bodily composition or the physiological parameters within which the body operates. As George Steiner put it 'Man has talked himself free of organic constraint.'[3] Language has driven a widening wedge between human history and animal evolution, between culture and nature.

The fact that written language outlives its moments of utterance allows it to be used as a means of asserting group solidarity among those who share a common tongue against those who do not. Moreover, written language is the supreme means of raising self-awareness. It is as if consciousness, in depositing itself in objects and events – written signs and spoken sounds – that are outside of itself, is able to encounter itself more clearly and so become more self-conscious. Self-knowledge and self-awareness, the individual and collective sense of self at the highest level are mediated through language. This was

272

why, in the previous paragraph, I hesitated to speak of language as a
mere 'tool': it is not merely used, and constituted by, subjects: it is
also, in a sense, constitutive of them.[4]

The protean manifestations and influences of language tempt one
to think of it as coterminous not only with human society but with
human consciousness above the level of the most primitive, formless
sensations. 'My languages are my world', to modify Wittgenstein's
dictum. The intimate relation between language and consciousness
and self-consciousness works both ways. Not only does language seem
to be the bearer of most of human consciousness, but linguistic acts,
under the usual conditions, seem to be uniquely the product of consciousness, of deliberation and voluntary choice. Discourse is signally
infused with consciousness and self-consciousness and is, apparently,
deliberate and non-automatic to a unique degree. It seems, therefore,
to be a conspicuous sign of a self-present individual, a final redoubt
against the assaults from thinkers such as Marx, Durkheim and Freud
on the centrality of the conscious, rational agent.

This view of the special status of linguistic acts has, however, been
challenged by certain post-Saussurean thinkers, notably Derrida. Post-
Saussurean thought – which encompasses not only such post-structuralists as Derrida but also structuralists such as Lévi-Strauss,
Benveniste and Todorov, and writers such as Foucault, Barthes and
Lacan who seem to fit into both categories – minimises the role of
intention and the status of the individual human being as a free, deliberate agent. More comprehensively than Marx, Durkheim or even
Freud, post-Saussurean thinkers question whether the individual can
ever know what she is doing. Some, such as Lacan – who claimed to
read Freud correctly by structuralising the Unconscious, making it
into a language – even have a theory of how the individual acquires
the illusion that she does know what she is doing, and the deeper
illusion that she is a 'she' at the centre of her knowings and doings,
to which the various knowings and doings of her life can be attached.[5]

The Saussurean revolution in linguistics, like the Durkheimian revolution in sociology, was marked by a shift from an item-centred to a
system-centred approach. The post-Saussurean revolution in thought
about man and society was characterised by a similar shift, based on
the belief that language was the paradigm of society as a whole, the
latter being simply a larger system of signs that included language.
The analysis of society and of the individuals within it thus became,
in Edmund Leach's words, a kind of 'cultural linguistics'.

Post-Saussurean thinkers drew their inspiration from Saussure's
writings on theoretical linguistics. To what extent do Saussure's writings license these large assumptions about human nature? I have
previously argued that there is nothing in Saussure to justify post-

Saussurean thought.[6] I concluded that a handful of careful distinctions would have caused roomfuls of post-Saussurean steam to condense into a few drops of water. It is necessary to reiterate some of the main points of the argument against the post-Saussureans here.

At the heart of Saussure's vision of a system-based, as opposed to an item-based, analysis of language is a recognition of the arbitrariness of the linguistic sign. Individual linguistic signs have no natural connection with that which they signify; 'dog' does not look or sound like a dog, and so on. Saussure was not the first to draw attention to this feature of language, nor did he say that he was; he claimed only to be the first to appreciate 'the numberless consequences' that flow from it. Since linguistic signs do not derive their signifying power from being naturally associated with the things they are used to signify, they must derive it from somewhere else.

According to Saussure, the source of the signifying power of signs is the *system* to which they belong. A sign has two aspects – the signifier, corresponding roughly to the phonetic aspect, and the signified, corresponding roughly to the semantic aspect – and it consequently belongs to two systems, the system of sounds and the system of meanings. Neither the signifier nor the signified enjoys an independent existence outside of the system of language. The linguistic system is *a set of differences* and its component signifiers and signifieds, being purely differential, are essentially negative. They can be grasped only through the network of other units. This crucial point warrants further elaboration.

Consider first the signifier. According to Saussure, this is a set of contrasting features realised in sound opposed to other contrasting features realised in other sounds: it is a bundle of phonic *differences*. The realisation that actual words (spoken or written tokens) served as the physical realisation of abstract, contrasting sound-features opened the way to numerous advances in our understanding of phonological relationships between languages and of the way the brain extracts verbal tokens from the acoustic material served up to its ear.

The signified, too, is not a naturally occurring entity but a 'concept' (a term Saussure uses in a special sense), and one whose boundaries are determined only within the linguistic system by its opposition to other concepts. It is not, however, intra-psychic – a mental entity such as a mental image: like the signifier, it is not a thing but a *value*; and it has value only within the system where it coexists with other opposing or different values. The denial of the pre-linguistic reality of the signified is the most revolutionary aspect of Saussure's theory. The signified is not a 'thing' 'out there'; nor is it a pre-linguistic psychological entity. The signified is purely relational:

The conceptual side of value is made up solely of relations and differences with respect to the other terms of language . . . differences carry signification . . . a segment of language can never in the final analysis be based on anything except its non-coincidence with the rest. *Arbitrary* and *differential* are two correlative qualities.[7]

So verbal meaning is specified not in virtue of an external relation between a sound and an object, but of an internal relation between oppositions at the phonetic level and oppositions at the semantic level.

So far so good. Few people would dispute that language is more of a system than a word heap and that its component signs are arbitrary in Saussure's sense. And it is perfectly obvious that the semantic catchment area of individual terms rarely corresponds either to patches of space-time or to 'natural kinds' (or 'types of patches of space-time'). Most people would be prepared to accept that linguistic value, as Saussure meant it, is negative or differential and that it is the differences between linguistic units rather than their positive contents, that carries verbal meanings.

Most of the errors in, and much of the excitement of, post-Saussurean thought derive, paradoxically, from thinkers overlooking Saussure's fundamental doctrines: for example, the fact that the signifier does not strictly correspond to a sound and the signified does not strictly correspond to the meaning. The signifier–signified relationship should not be confused with that between a sign – an uttered sound or written mark – and an external, extra-linguistic object. Therefore, the intra-linguistic and arbitrary nature of the former cannot be taken to imply the intra-linguistic and arbitrary nature of the latter. To think that the one implies the other is also to confuse the language system with the use of that system by an individual on a particular occasion in generating a particular speech-act.

This seems obvious; nevertheless, as a result of this confusion, several un-Saussurean conclusions have been drawn from Saussure's work. They underpin a radical assault on the central role of consciousness (and consciously formed intentions) in everyday life and, more particularly, on those actions – speech-acts – that seem supremely volitional and governed by intention. Saussure, it is correctly argued, has shown that behind every speech-act – indeed, every discourse act – lies a language system of which speakers are largely unconscious. But then it is incorrectly argued that, since the signifier and the signified are internal to the linguistic system, discourse is closed off from the extra-linguistic world. From this it follows that discourse is only a series of moves within a system of which speakers and writers are unaware and that such meaning and reference as discourse has is to this hidden system. The meaning of individual utterances (and other linguistic and non-linguistic signs) is not merely achieved through the implicit or hidden system, but indeed *is* that system. Speakers don't know what they are talking about. Derrida, for example, has argued that behind every specific text – written or spoken, a novel or a chance remark in a conversation – is 'the general text'. This text is not a collection of words but a condition of the possibility of discourse; since this condition cannot be specified, all specific texts remain 'undecidable'.[8]

One of the best-known applications of what is taken to be post-Saussurean, but is actually non-Saussurean, thought is in the writings of Lévi-Strauss, on whose ideas we have already touched. Saussurean phonology licensed his structuralist analysis of myths. We have seen how, according to Lévi-Strauss, the meaning of myths lay not in their individual, manifest content, but in the group or system to which they belonged. This group had an enormous scope, stretching across millenia and straddling continents. Myths were recounted by individuals not, as those individuals may have thought, in order to convey their content, but in order to realise part of a form whose totality, hidden from everyone except the structuralist mythologiser, was the structure of the human mind itself.

Less spectacular, but of more direct importance for the present discussion, is the post-Saussurean dissolution of the speaking (or discoursing) subject. The scope of the claims advanced under this heading varies from writer to writer, but we may identify three types of claim:

1. The denial (associated in particular with Barthes) of the originality and unitary nature of the author.
2. The denial (associated with Derrida) of the presence of the speaker (or his/her intentions) in the speech-act.
3. The assertion (associated with Benveniste) that the self, the self-present I, the centred ego, is the product of language and (according to Lacan and Derrida) is therefore illusory.

I should like to deal with each of these in turn.

THE ABSENT AUTHOR

According to Barthes,[9] readers of texts, especially officially approved texts that belong to the literary canon, believe that behind the text there is a rather special person called an 'Author'. The Author has several characteristics: he/she is the true origin of the text (he/she is an original creator); he/she has a unity represented in his/her distinctive presence throughout his/her text (or, indeed, his/her *œuvre*);

and he/she is the authority on the definitive meaning of his/her text. In short, he/she shares most of the properties of a deity and is in truth an Author-God. Barthes denies the author these extraordinary properties and, on the basis of this argues that there is no such thing as an author of a text. The text is merely 'a multi-dimensional space where a multitude of styles blend and clash'. And the author is merely a site through which language passes: he/she does not use language to write his/her text; rather, language uses him/her to write the text.

The dissolved Barthesian Author is a straw person, who may be alive and well in the French educational system, but has little life elsewhere. His/her death would go unremarked in most parts of the world. The ordinary everyday author remains alive and well. Roland Barthes may not have created his texts *ab initio* (he cannot take exclusive credit for all the dubious ideas in them), but he is infinitely closer to being the author of them than I who quote from and comment on them. No one would wish to claim that he has the last word on their meaning or that his unique personality is uniformly present throughout them. Most literature (even a short, much revised poem) is put together out of scattered moments of inspiration that have no more claim to be uniquely expressive of their author than do the products of other, less literary moments.

THE ABSENT SPEAKER

Barthes' attack on the Author – once regarded as revolutionary bordering on scandalous because of his semi-deliberate confusion between the Author-God and the author-in-ordinary – and on the presence of the author in the text is not too upsetting to ordinary thought. What writer would wish to claim that she is fully present in passages she wrote a decade, or even a minute, ago? One of the benefits of writing is that it enables us to clarify and extend our ideas and distance ourselves from them. (This would not be possible for Barthes' all-pervasive Author-God.) The text is a product in which the craftsman is only metaphorically present. At best, past moments of consciousness are embodied in this product.

However, one of the arguments used by Barthes to bump off the Author – namely, that we do not use language to speak/write, rather that language speaks/writes through us – has become the basis of a more radical attack on the presence and self-presence of the speaker/ writer:

once the conscious subject is deprived of its role as a source of meaning – once meaning is explained in terms of conventional systems which may escape the grasp of the conscious subject – the self can no longer be identified with consciousness. It is 'dissolved' as its functions are taken up by a variety of interpersonal systems that operate through it.[10]

This implication is at work in Derrida's critique of the idea of expression and self-expression in discourse. Derrida begins with a radicalisation of Saussure. He draws out what he considers to be the logical consequences of Saussure's vision of language – consequences from which he believes Saussure shied away. Language, Derrida points out, is difference-riddled: the meaning of a signified is owed entirely to its difference from, its non-coincidence with, the other signifiers. Signification therefore depends on the domination of absence over presence – the absent system dominates over the present signifier or signified. Derrida then takes three steps which enable him to get to his position that speakers are not expressed in (are indeed absent from) their speech:

1. He extends the domination of difference (absence, negativity) from the signifier and the signified taken singly (where they are indubitably the playthings of absence) to the sign-as-a-whole (a step specifically warned against by Saussure);[11] and thence to the completed speech-act. This is nonsense, of course – the speech-act does not belong to the systems of signifieds and signifiers. It uses the systems, but is not part of them. The confusion between the system, and its use, between *langue* and *parole* (especially unforgivable in writers who claim to be familiar with Saussure, as it was one of the latter's great achievements to distinguish these things), has complex consequences that one can trace through the whole absence–presence argument.

In fact, presence in a speech-act is due to the actual presence of the speaker. This is not merely a question of bodily presence (otherwise one would have to consider an individual vocalising in a coma as being self-present) but also of being-consciously-here. The latter is also the necessary basis of the closure of general meaning to specific reference to particulars. The implicit *deixis* of the physical situation of the speaker, and of 'the story so far', provides the coordinates of a closed universe of actual discourses, which is quite unlike the boundless universe of Derrida's disembodied text floating freely in a limitless sea of other texts. The *system*, of course, lacks deixis, knows no referents and has only the general possibilities of meaning at its disposal. It is quite without presence, achieved meaning and reference.[12]

In order to draw the conclusion that discourse and discoursers are empty in the way that Derrida argues that they are, it is necessary, as has already been said, to conflate the system with the actual discourse acts that utilise it. In order to do this, it is necessary to conflate the components of the system with the use of the system on particular occasions; to conflate the meaning and reference of signs with the means by which meaning and reference are achieved; to conflate the system (which is hidden from users) and the use of that system in particular speech-acts; to conflate signifier and signified with the-sign-as-a-whole, the type with the token, the sign with its meaning and/or reference on a particular occasion of use, etc. All of this Derrida and his post-structuralist followers gladly and repeatedly do.

Consider, for example, this argument from a prominent British literary theorist, Terence Hawkes.[13] All systems, he says, including language, 'act to maintain and underwrite the intrinsic laws which bring them about, and to "seal off" the system from reference to other systems.' Consequently, language

> does not construct its formations of words by reference to the patterns of 'reality' but on the basis of its own internal and self-sufficient rules. The word 'dog' exists, and functions within the structure of the English language, without reference to any four-legged barking creature's real existence.
>
> (ibid.)

If this were the case, we would expect the strings 'The dog is barking', 'The dog is quacking' and 'The dog is reading *Of Grammatology* with pleasure and profit' to occur with approximately equal frequency.[14] As it is, their respective frequencies seem to reflect the frequencies with which the extra-linguistic situations to which they refer occur in real life.

2. Derrida introduces a new term – *différance*, which compounds both difference and deferral: the meaning of an utterance is never reached, so that deferral is indefinite. There is thus a double absence in discourse – the difference upon which it is based and the endlessly deferred meaning that it never reaches.

On closer examination, however, the meaning that is eternally deferred by Derrida is a special sort of meaning – an absolute closure that would bring the search for further meaning to an end. Ordinary meaning, by contrast, does seem to be available within the lifetime of the speaker – indeed, within the sentence-time of the speaker – and so is not a serious source of absence. And even if Derrida were referring to the ordinary delay that occurs in the fulfilment of the meaning of an utterance (having to wait to the end of a sentence or paragraph to arrive at some more or less definite meaning), this would not make speech-acts any different from other acts such as walking. When I walk to the pub, I undertake numerous acts which, though they contribute to the goal, the overall meaning, of my behaviour, are not themselves endowed with the full meaning. Crossing the road to avoid someone who will delay my arrival at the pub has a relationship to the overall meaning of my behaviour no less remote than that of a participle to the final meaning of the sentence.[15]

3. Derrida considers the fact that the meaning of an utterance is determined not solely by the words of which it is composed and the order in which they are placed, but also by their social context. The utterance 'I do' has a very particular meaning in the course of a marriage ceremony, which it does not have outside of this. The felicity conditions that will ensure that the utterance 'I do' is valid as a marriage vow are extraordinarily complex and largely hidden from the utterer. Generalising from this, Derrida argues that the context that determines, or gives determinacy to, the meaning of an utterance is boundless and, consequently, is unavailable to the speaker. More specifically, since I cannot know the infinitude of conditions that would ensure that my utterance had one meaning rather than another, I am not in the control of the meaning of my utterance. How, then, can I say that I am expressed in my utterance, when the determinants of its meaning lie beyond my ken?

One can answer this question by considering the fact that I don't know whether my legs are going to work or, even less, how my legs work. In the present state of ignorance of physiology, nobody, least of all myself, could state the conditions necessary for walking. Specifically to address Derrida's point, we need only to remember the distinction between the context that makes an utterance felicitous and the meaning meant by the person uttering it; between the conditions of meaning and the meaning that is meant. The fact that my utterance turns out to be invalid (the vicar had been unfrocked, the woman was already married, an edict has just been issued banning all marriages, etc.) does not in any way attenuate my intention to get married, sever my speech-act from that intention, or diminish the extent to which I am expressed in, or present in, the act. Indeed, these things, precisely by being so obviously unknown to me, can be clearly separated from my intention. Just as the referential opacity of most terms (for example, water which is also H_2O), does not prevent me from referring to water.

THE ABSENT SELF

We may conclude that Derrida's grounds for denying the presence of the speaker in the speech-act – so that absence is the hallmark of discourse, and writing, in which the originator is explicitly absent, becomes the paradigm of discourse – and for denying that speech-acts realise intentions are to say the least dubious. Nevertheless, his ideas have been extremely influential and represent the most important contemporary attack upon the centrality of the individual's consciousness in ordinary life. For if even speech-acts are not importantly conscious and deliberate, what is? The displacement of the speaker's expressive intentions from the centre of his/her speech-acts becomes the basis for a wider displacement of the individual and his/her intentions from his/her acts. Intention is pushed to the margins of behaviour. Because we cannot formulate the context that gives our actions their meanings, we cannot be said truly to intend what we do.

A further assault on consciousness and the centrality of the conscious self deriving from Saussurean linguistics is the assertion (associated with Derrida, Lacan and the linguist Benveniste) that the self, the self-present I, the centred ego, is the product of language. According to Derrida, the illusion of self-presence, and so of unmediated presence, is based on the accident of hearing oneself speak. The closed circuit whereby speakers hear themselves without mediation creates a (false, illusory) sense of being present to oneself. For Lacan, the unconscious is structured like a language and the sense of self is inscribed in 'an endless chain of signifiers' to which there corresponds no signified. The *symbolic* self is the successor to the *imaginary* self derived from catching sight of oneself in a mirror. The infant enters the realm of the imaginary by identifying (incorrectly, according to Lacan) with its mirror image. This imaginary self subsequently enters the symbolic realm, at least in part as a strategy to resolve the Oedipus complex. These views (which I have discussed extensively elsewhere)[16] entrain the errors common to all post-Saussurean notions about the relationship between the speaker and the linguistic system plus faults of their own. The most explicit attempt to found the self, the sense of I, in language is that of the influential French linguist Emile Benveniste.[17]

For Benveniste, the act that constitutes the 'I' is a linguistic act. Catherine Belsey succinctly summarises his position:

> it is language which provides the possibility of subjectivity because it is language which enables the speaker to posit himself or herself as 'I', as the subject of a sentence. It is through language that people constitute themselves as subjects.[18]

Since individuals do not invent the language that they use (and are anyway used by, rather than users of, it) they are constituted rather than constituting. The 'I' is consequently caught up in the play of difference that is the essence of language, a play of which nobody is master.

In the essay in which he most explicitly advances his views about the relation between the self and language,[19] Benveniste begins by

pointing out that discourse consists essentially of an *exchange*. In contrast
with Derrida et al., he recognises that the exchange, which lies between
individuals, cannot be due to the properties of language itself – of
the system rather than of individual speech-acts. But he draws a rather
extraordinary, and distinctly post-Saussurean, conclusion from this:

> 'subjectivity', whether it is placed in phenomenology or in psychol-
> ogy . . . is only the emergence in the being of the fundamental property
> of language. 'Ego' is he who *says* 'ego'. That is where we see the
> foundation of 'subjectivity', which is determined by the linguistic
> status of 'person'.
>
> (ibid., p. 224)

Subjectivity, however,

> is not defined by the feeling which everyone experiences of being
> himself (this feeling, to the degree that it can be taken note of, is
> only a reflection).
>
> (ibid., p. 227)

A reflection, that is, of a subjectivity primarily established in and through
language. Linguistically constituted subjectivity, however, accounts for
the self as 'the psychic unity that transcends the totality of the actual
experiences it assembles and that makes the permanence of consciousness' (ibid., p. 224). This is what holds together all the various experiences attributed to the self and makes them experiences had by that
self. The 'establishment of subjectivity in language creates the category of person both in language and also . . . outside of it as well'
(ibid., p. 224). Thus, according to Benveniste, being a subject, along
with some aspects of feeling that one is a (unified) 'self', result from
one's entry into language as a speaker. The individual's engagement
in dialogue constitutes him as a person. Conscious personhood is an
inner reflection of the oppositional status the speaker has when he
engages as an 'I' in dialogue:

> Conciousness of self is only possible if it is experienced by con-
> trast. I use *I* only when I am speaking to someone who will be a
> *you* in an address. It is this condition of dialogue that is constitu-
> tive of a *person*, for it implies that reciprocally *I* becomes *you* in the
> address of the one who in his turn designates himself as *I*.
>
> (ibid., pp. 224–5)

The physically or bodily differentiated individual becomes *I* by positing himself as 'I' opposed to 'you'.

For Benveniste, then, the actual feeling of subjectivity, of being *this* self rather than another, of being *this enduring thing*, is a secondary aspect of subjectivity, a mere reflection of actual subjectivity, which has an origin (in language) unappreciated by the subject. The fact that the subject should be mistaken as to his/her true nature suggests another severe limitation placed on consciousness and self-consciousness – one yet more drastic than those suggested by Marx, Durkheim and Freud.

Even if one allows the starting assumption – that the self is experienced not positively but only differentially, that it exists (like a linguistic unit) by contrast with the not-self – it does not follow that it is experienced, or exists, only in opposition to the *verbal* 'you'. For the not-self includes non-verbal objects (material things outside its body) as well as entities apprehended verbally. Benveniste's position would not hold up even if he were to retreat to the claim that his 'I' were only the grammatical 'I' understood as the subject of a sentence. For the purely oppositional and commutative relationship that Benveniste refers to is not confined to *you*: 'I' may also be opposed to *it, he, we, that (over there)*, etc.

It seems that Benveniste does at times intend the 'subject' to be construed in the narrower sense of the grammatical subject, in which case, his achievement is even less impressive. For it is, of course, easy to decentre the self linguistically if one either begins with the assumption that the I is a linguistic entity; or if one takes it for granted that the self is to be identified with the grammatical subject. The assertion that it is through language that people constitute themselves as subjects becomes a mere tautology if the term 'subject' is employed only to indicate the grammatical subject of a sentence. You couldn't make yourself the subject of a sentence if there were no sentences to be the subject of.

Benveniste draws some surprising conclusions from the status of the first-person singular personal pronoun:

> What does *I* refer to? To something very peculiar which is exclusively linguistic: *I* refers to the act of individual discourse in which it is pronounced. . . . The reality to which it refers is the reality of the discourse. It is in the instance of discourse in which *I* designates the speaker that the speaker proclaims himself as 'subject'. And so it is literally true that the basis of subjectivity is in the exercise of language.
>
> (ibid., pp. 226)

It is, of course, untrue that 'I' refers to a particular linguistic act; on the contrary, it mobilises the <u>deictic</u> coordinates in order to arrive at ‖ ? its true referent, which is the in<u>div</u>idual generating the act – the speaker.

The conclusion that language itself is the referent of the pronoun is based upon a mistake we have already noted as being widespread and crucial for establishing many of the startling ideas associated with post-Saussurean theory: that of confusing the *referent of discourse* with the *means by which reference* is achieved. If the means of access to the referent of I is confused with the actual referent, then it only requires the observation that the means of access is linguistic (the speech-act itself) in order to arrive at the conclusion (quoted above) that the referent of 'I' is linguistic and that 'the basis of subjectivity is in the exercise of language'.

That there is an elementary error at the root of Benveniste's claim that subjectivity has a linguistic origin will come as no surprise to those who feel that, whatever construction one puts on the 'subject' or the self, it seems implausible that the totally fortuitous and unexplained existence of handy pronouns used to refer to oneself should play such a central part in the creation of the self. The unconvinced will be puzzled as to how language – and especially dialogue – could arise in the absence of pre-existing subjects. Or how a speaker could appropriate to himself an entire language by designating himself as 'I' if he were not already a self.

Benveniste himself seems undecided on this crucial point; for at one place in the essay he says: 'Language is possible only because each speaker sets himself up as a *subject* by referring to himself as *I* in his discourse' (ibid., p. 225). From this, it would appear that the subject, far from being 'a tropological construct' (to use a favourite post-Saussurean term) – the mere product of language – is itself the source of language; for language seems to depend upon the subject 'referring to himself'. The speaker, it seems, is not only prior to the subject but also to language, and is, indeed, the very condition of its possibility. Language is predicated upon the prior existence of speakers: a most curious state of affairs that raises innumerable questions. Amongst them is the question of how the speaker could refer to himself or herself – or want to do so – unless he or she existed in the first place: self-reference must presuppose some kind of pre-existing self to refer to.

The Benveniste circuit – language creates the self that creates language; self-reference generates the pronouns that in turn create the self that is referred to by pronouns – illustrates the problems that arise when one wishes to dissolve the self entirely into a social system or institution such as language. For those systems and institutions would seem to require the interaction of subjects to bring them about and keep them in play; and such an interaction between subjects must imply that they pre-exist the system, however much they are bound up with or shaped by it.

The primacy of self-presence and of consciousness – that I am this thing, that I am here – is not, therefore, impugned. For, as already noted in the discussion of post-Saussurean thinkers, successful linguistic acts depend, ultimately, upon determinations of meaning and reference that cannot be performed through the system alone. In the end, deictic coordinates have to be brought into play. These are implicit in the individual's bodily location and historical situation; but they are explicit only in his or her sense that 'I am this thing' – and its consequences: 'I am here and now', etc. The centred self tethers language to the actual; it is essential for reference to be secured. As we have already noted, the system itself has no referential powers.

It would be difficult to exaggerate the significance of the tendency of system-based discourse theorists – those who give the system priority over the speaking subject, so that the latter is subordinated to the former ('Language speaks us', 'We are not the authors of our sentences, we are constituted through the General Text') – to overlook deixis. It is this symptomatic omission that confers a temporary plausibility upon even the most outrageous claims about the relationship of the self to language. When Dragan Milovanovic asserts that 'Language speaks the subject, providing it with meaning at the cost of being',[20] he is only following out the consequences of this fatal omission. A moment's thought would reveal that, in order to speak meaningfully, we have to mean (transitively) what we say. As Grice[21] has pointed out, meaning is active. The meaning of what I say is not merely the meaning my uttered signs happen to have, as clouds may mean rain or spots mean measles. It is the meaning *I intend*; and in trying to determine what I mean, you will be interpreting what I intended to mean. And this interpretation will be heavily dependent upon what you know of me as an individual and certain specifics, such as our shared immediate context, my tone of voice, etc. This presupposes my presence in what I say. The meaning of utterances emerging from an absence possessed by the discursive system would be literally undecidable. Meanings, in short, are produced as well as consumed; and without the notion of the individual producer and without reference to the deixis implicit in his/her situation, to the existential context of the production, the meaning of an utterance would be simply undecidable. Communication cannot take place without both meaning-producer and meaning-consumer mobilising the deictic coordinates in the speaker's situation.

To say this is to envisage the spoken word as primary and the written word as secondary, and so to fly in the face of the entire post-Saussurean framework of assumptions. Notoriously (and counter-intuitively), Derrida asserted that Writing, in which the apparent origin of the discursive chains was absent, had primacy over Speech, in which the apparent origin of the discursive chains was apparently present. For Derrida, Speech was merely an instance of the former or, rather, of 'Writing-in-General' and the absence of an originating source of the discursive chain, evident in writing, was actually a characteristic of all discourse.

I have set out Derrida's arguments, and my own view of the fallacies upon which they are based, elsewhere[22] and will not repeat them here. Suffice it to say that even the most fervent post-Saussureans feel uneasy about them. For example, Milovanovic seems to wobble uneasily between the speaking subject as the patient and the agent of discourse:

> Much contemporary semiotic analysis is divided between a self-referential and a referential view. The former arguing for a system in which the reference of the sign is exclusively the system of signs itself and the latter for some external referent to which the sign points. The constitutive elements of the sign – signifier and signified – are also implicated in the subject taking up a position as an *I* in an utterance. Benveniste argued that what appears as an *I* in a statement is but a present of an absence, the absence being the producer of the discursive chain itself.
>
> (ibid., p. 9)

Milovanovic, who allies himself with the self-referential party (indeed, he entertains 'visions of an empowered democracy in which a new concept of subjectivity reigns', based upon Lacan and Benveniste's ideas), is clearly uncertain as to whether or not the discursive chain – a particular utterance – is produced, or simply happens, or is generated by the system, understood as a general absence instantiated in, or alighting upon, a particular individual. The latter is a difficult and even self-contradicatory notion, being that of a not-here (or even a not-anywhere, or a not-anywhere-in-particular) taking root in a here and still retaining its status as a not-here.

The whole question of presence and absence in speech cannot be answered in general, unless one believes that anyone who participates in any rule-governed activity is somehow absent from the acts that result – that they have no choice in what occurs and are not represented personally in their acts. The untenability of this position becomes evident when one compares, for example, my comparative presence in my words when I make a statement calculated to impress, convince or persuade someone else, with my comparative absence from: the words I shout out in a dream; the sounds I babbled associatively as a child and may babble meaninglessly if I am demented in my old age; the words I have taped on a dictaphone and someone plays many years after my death; Homer's words (which I have not yet read). Post-Saussurean thought does not allow for these distinctions: we are all ego-less babblers, soluble fish in the sea of language.

Contrary to the claims of the post-Saussureans, the systematic nature of language does not deny us freedom of expression through it or insist that we are first dissolved in it before we can have freedom of expression, so that our freedom is freedom only to express the selves that are constituted through language. As a speaker, I am always outside of language: my existential situation is that of someone whose being is not fully consumed by the meanings that are embodied in the utterance I am engaged in. At the very least, my bodily weight and position, its numerous unexpressed sensations, the physical world around me with its endless source of novelty, and its inexhaustible particularity, distance me from the meanings that I offer for consumption and make me, as producer, the source of meaning and distance from it. There is no question of vanishing from actuality into the world of possibility and the system of language, of giving myself up to general meaning at the cost of my particular being. The rules are enabling constraints, not straitjackets, that permit communication. They are necessary, just as the friction that makes walking into hard work is necessary in order that walking should be possible. We do things with words, but we can do them because we obey the rules that make communication possible.[23]

This is illustrated if we consider the most commonplace, off-the-shelf utterances, the kinds of things that would seem likely to show us as speakers at our most limited, constrained, system-bound and 'constituted in language'. Saying 'Hello' would seem to fit this description: it is almost a linguistic reflex and, as a phatic word, free of informational content and triggered by certain stereotyped situations; it would seem, if any utterance is, to be something that chooses the speaker rather than vice versa, an example, if there are any examples, of language speaking us rather we speaking language. And yet we have the option of exerting an extremely complex and deliberate freedom through saying, or refraining from saying, 'Hello'. We have a choice of degrees of warmth. We may assume a variety of accents to impress or amuse. We may speak it quietly so as not to be heard saying it and betray a relationship we would rather conceal ('strangers when we meet') or so as not to wake the children or out of respect for the occasion in which the word is uttered. We have a choice, when we are approaching someone, of the distance at which we say the word and we exercise our judgement – based upon the respective relationships between us, actual or to be asserted or underlined – as to who we think should say it first. We may say it in a certain way to mock the pretensions of others; or to remind us of the occasion many years ago when we said it in a certain tone of voice and so to assert our solidarity, through this signal, of the duration of our relationship. In summary, this simple greeting may express, or be used to express, an infinitely elaborated and highly personal consciousness of another person.

From this, we may draw two non-post-Saussurean conclusions. First, language is used by us to express selves that are not anticipated by, or nodes in, the language system. And, secondly, we are explicit animals whose ability to develop our explicit consciousness of things, include what we feel ourselves to be at a particular time, or over time, may be elaborated indefinitely. We are free agents when we use language: the rules of language do not specify what we shall say in a given situation – except within very broad constraints – and certainly not the way in which we say it. This is at least in part because language does not tell us what the situation is – what, for example, my relationship is with this woman coming towards me down the road that will, along with many other things (such as how I am feeling at present, my recent and remote history), determine if, when, and how I shall say Hello to her. Using language – even the most stereotyped elements of it – we can unravel more and more complex modes of explicitness.

Saussure himself emphasised that the act of speech (*parole*) is an individual act of intelligence and will[24] in which the speaker's freedom of choice is only loosely constrained by the possibilities available in the linguistic system (*langue*). The choice is still the individual's, and the choosing is still conscious, or part of an act that is conscious. Far from decentring the self, *parole* requires a centred self in order that the speech-act shall be spoken and enacted. The post-Saussurean claim that it is language which speaks (presumably *langue* that *paroles*!), which we have seen cannot be sustained even for stereotyped utterances such as 'Hello', sounds even less plausible (and, indeed, downright odd) when one thinks of ordinary discourse: whether one says or wishes to say 'Pass me the salt please' versus 'Pass me the pepper please' versus 'Get your elbows off the table' is obviously not determined by the system, which could neither provide a reference for 'me' nor legislate over its use on a particular occasion. *Parole* – actual talk – is always rooted in particular occasions, and those occasions are not intra-linguistic – as if everything that was said was a series of stereotyped responses between priest and congregation. The rules of language do not specify what we say, even less how we should say it, precisely because so much of what we say is prompted by events whose occurrence is not regulated by the rules of discourse.[25]

In placing the arbitrariness of the linguistic sign at the centre of his understanding of language, Saussure's instincts were surely correct. And his intuition that the consequences of his 'first principle' of language are 'numberless' absolutely right. It is therefore a shame that so many of his followers have misunderstood the arbitrariness of the linguistic sign and have focused on its non-consequences. For there is even more to be understood about arbitrariness than is indicated in Saussure's writings. It is connected not only with the nature of human language, but also with the nature of human consciousness. The arbitrariness of the linguistic sign sets discourse off from the associative net that would embed language in the causal nexus of the physical world. Arbitrariness is a pervasive manifestation of the status of linguistic signs as *explicit signs* and of their role in making explicitness (consciousness) explicit. Arbitrariness creates the essential distance across which reference is possible and on the basis of which expression, rather than reaction, representation rather than duplication, can be enacted. If linguistic signs were not arbitrary, they could not carry an increasing burden of explicitness and many things, amongst them the formulation of theories of language, would not be possible. Thus, as I have argued in more detail elsewhere, a true understanding of Saussure's thought would see that it emphasises, rather than dissolves, the relationship between discourse and individual consciousness, between speech-acts and individual intentions.[26]

9

Unconscious Consciousness: From Behaviourism to Cognitive Psychology

I observed at the beginning of this discussion of the 'marginalisers of consciousness' that consciousness seems resistant to scientific treatment and is consequently an embarrassment to any discipline that is in the process of establishing itself as a science. So long as the procedures and, indeed, the results of the physical sciences are seen to be the paradigm of science itself, then a science of consciousness is going to start at some disadvantage. For, at the very least, ignoring the subjective viewpoint, overriding the testimony of the objects of study, seem to be the *sine qua non* of any systematic enquiry that pretends to the status of science. A genuinely scientific political economy will dismiss the reasons individuals give for their beliefs and ideas and focus instead on the objective conditions of production – which will explain everything from the most abstract ideas those individuals have about society as a whole to the actual behaviour they exhibit. Thus Marx. A truly scientific sociology will look not to the reported experiences of individuals when trying to understand either individual actions or mass social phenomena but to objectively observable social forces. Even apparently deeply personal sentiments and actions – such as suicide, religion, the formation of concepts, etc. – are to be explained by reference to these things and not in terms of intra-psychic events. Thus Durkheim. The motives behind my actions are not those available to introspection – these are mere rationalisations – but those revealed to an analytical psychopathology guided by a science of the mind rooted in the already established sciences of zoology, physiology and anthropology. Thus Freud. Finally, a scientific linguistics will be based not on the intuitions of individual speakers about their actual speech-acts, but on the objective study of the linguistic system largely hidden from those speakers. Thus Saussure – or, at least, post-Saussurean theorists.

The extent to which the emergent human sciences aped the actual procedures of the physical sciences, and the degree to which they regarded the actual laws of physics or principles analogous to them (e.g. Spencer's talk of Social Statics and Social Dynamics) as the asymptote of their field of enquiry, varied from discipline to discipline and from practitioner to practitioner.[1] What did not vary was the emphasis on objective analysis at the expense of the deliverances of intuition and introspection and placing the search for general laws above investigation of individual cases – the nomothetic over the idiographic. These tendencies marginalised the subjective view and the individual consciousness – the desired outcome for disciplines aspiring to be sciences on the model of the physical sciences.

The rejection, or at least extrusion, of consciousness was not, however, an entirely easy option for the one discipline whose original object of study was consciousness itself – namely, psychology. You can't really do psychology – the science of the mind – without trying somehow to come to terms with consciousness. The history of post-introspectionist psychology has been characterised by two ways of dealing with the awkward problem of consciousness: the problem, that is, of becoming scientific on the model of physics without losing sight of the distinctive subject matter. The *behaviourist* approach has been to deny the importance – or even the reality – of consciousness altogether (and relegate mind to the events in an uninteresting black box). The tactic adopted by *cognitive psychology*, on the other hand, has been to admit that there is such a thing as mind while limiting the extent to which mind is conscious, by redefining mind as 'central processing', the vast bulk of which has little or nothing to do with consciousness – in other words, by making mind as unmindlike as possible.

Cognitive psychology now dominates contemporary psychology. Things were not always so. For the greater part of the history of experimental, as opposed to speculative, psychology, scientific psychology in the English-speaking world was almost synonymous with behaviourism. In those years, consciousness was *persona non grata*. Behaviourists rejected consciousness primarily on methodological grounds: it was difficult stuff to observe and even more difficult to measure – especially in animals where first-hand reports were unavailable. This unmeasurability was fatal from the viewpoint of an emergent science that felt it could earn its place in the sun, along with physics and chemistry, only if like them it, too, could make reliable, repeatable, objective measurements. Measurement of consciousness was especially unreliable if, as in the case of introspective psychology, the measurer and the measured were the same thing. At the very least, the process of self-inspection and self-measurement altered the findings; at the worst, the expected results determined the findings that were obtained. Introspection was clearly no kind of

business for a grown-up science to be involved in. Consciousness –
which was directly accessible only through introspection – was, therefore, scientifically unrespectable.

Watson, for example, explicitly rejected the technique of introspection because investigators could never agree about their findings;
behaviour, however, was public. There could be no disagreement about
how long it took a rat to get to the end of a maze or how much time
individuals spent in face-to-face conversation.[2] Moreover, there was
the feeling that behaviour was the thing that really mattered – in
evolutionary/survival terms and in everyday life – and the relation
of introspectable mental phenomena to behaviour was uncertain. From
the point of view of useful, practical, adaptive behaviour, much consciousness seemed gratuitous. Introspectable consciousness had little
connection with visible 'cash value' activity. In the case of animals,
in whom it could be inferred only indirectly from behaviour, consciousness seemed an unnecessary hypothesis, a superfluous middle
term between observable stimuli and observable responses. Scientifically dubious as objects of study and measurement in their own right,
unnecessary as mediators between scientifically respectable inputs and
outputs that could be measured, intentions, ideas, emotions and
propositional attitudes were best ignored by an upwardly mobile discipline determined to take its place on the high table along with the
physical sciences. From a tactical decision to ignore consciousness, it
was but a step to deny its existence. Methodological decisions gradually slid into ontological positions: the unstudiable became the dubious; and the dubious became the non-existent.

Crude, undisguised, behaviourism as a serious movement in psychology is now dead.[3] Aseptic descriptions of the input-output relations of organisms under controlled conditions produced results of
increasing banality that seemed to be more and more remote from
human life or, indeed, from the life of the observed organisms. Lashley
pointed out in his seminal paper[4] that no collocation of stimulus–
response relations could account for the ordering of the components
of (ordinary) complex behaviour, in particular for the serial order
necessary to accomplish the simplest of tasks. The organism would
seem to have know what it was doing in order to do it. He gave
some examples to illustrate this. Hearing the sentence 'Rapidly righting with his uninjured hand, he managed to prevent the canoe from
sinking', the listener would interpret 'righting' correctly only when
he reached the word 'canoe': in the light of the latter word, he would
re-read – or re-hear – 'righting' as 'righting' and not as 'writing'. Talking
with a hot potato or a pipe in my mouth, I make totally different
movements and slightly different sounds from those I make when I
am speaking unencumbered. The differences – which are differences

in absolute movements and actual sounds but not in the formal relations between the sounds and movements – can be handled by the producer and by the recipient only by assuming that the behaviour is goal- and meaning-led and has an abstract form that can be realised in numerous different ways.

This near-recantation by Lashley, a leading behaviourist, was impressive. The decisive event in the decline of behaviourism, however, was the renewed attention paid to linguistic behaviour and the consequent appreciation of the fact that no schedule of reinforcements or other unconscious shaping could account for the acquisition of language observed in human beings. Chomsky's savage attack on Skinner's *Verbal Behavior*[5] sounded the death knell of behaviourism. The wider implications of this were understood: what makes bodily movements count as behaviour is not their physical characteristics but their *goal*, some idea of which had to be present to, or represented in, the behaver. This insight – a revelation to a science committed to denying the obvious – was embodied in the landmark text published in 1960 by Miller and his colleagues.[6] Even if psychology were still too embarrassed to call itself the science of the mind, and still called itself the science of behaviour, it now recognised that it could not escape reference to mind-like or mind-affiliated things, such as goals, intentions and meanings.

What, however, of cognitive psychology, seen as the now-triumphant rival of behaviourism? Cognitive psychology coped with the apparently conflicting demands of being scientific and of addressing itself to consciousness, not by denying the reality of consciousness but by redefining consciousness – more precisely, by giving a rather narrow role to consciousness as normally understood whilst, at the same time, emphasising the importance of 'inner' or 'central' processes, interpreted in a non-threatening way. Consciousness was not, in other words, abolished; it was, however, still marginalised. This marginalisation was – and is – all the more remarkable for being conducted in the heart of mental territory: in relation to the planning of actions and, most remarkably of all, in relation to conscious perception.

For cognitive psychology, mind undubitably exists, but very little mental activity enters consciousness: mind is the site of 'mental processing' which is overwhelmingly unconscious. This vision of consciousness deeply interpenetrated with the unconscious goes further than Marx, Durkheim and Saussure, though perhaps not certain versions of Freud, would have countenanced. For it is not a question merely of a false social consciousness; or of a false construction placed by the individual upon the relationship between himself, his society and his larger visions and decisions; or of the possession of an adult present by an infant past. It is a matter of the invasion of the seemingly most

self-present moment of consciousness, of the here-and-now of perception, by the unconscious; of a darkness at the heart of light.

The emergence of cognitive psychology was not simply the outcome of internal development of psychology wrestling with the inadequacies of behaviourism. 'The cognitive turn' was the result of many
other important influences from outside of psychology narrowly interpreted. Amongst these, one must include: the anti-psychologistic
approaches to logic, concepts and meaning originating with Frege's
quarrel with Husserl (see the Appendix to chapter 10), which was
transmitted via the fathers of symbolic logic (Russell, Quine, Church)
to influential computer theorists (and part-time philosophers of mind)
such as Turing and von Neumann; the hostility of analytical philosophers towards mental concepts transcendentally understood; the detailed teasing out of the neural theory of perception; and major advances
in computational practice, particularly in its pursuit of the will-o'-
the-wisp of Artificial Intelligence. Nevertheless, the most important
influences came from within psychology itself. Amongst these were
ideas advanced in the decades before psychology grew into a full-
blown autonomous discipline, separate from 'mental philosophy' and
physiology and became preoccupied with its status as a science.

Helmholtz's brilliant writings on perception[7] – published from the
1850s onwards – were the first to emphasise how ordinary perception is based on more than present or actual sensation. For example,
the world revealed to visual perception is one of three-dimensional
objects set out in three-dimensional space; the images of that world
on the retina, however, are two-dimensional. How do we get from
the two-dimensional world of visual sensation to the three-dimensional one of visual perception? According to Helmholtz, this takes
place by means of *unconscious inferences*:

> The psychic activities that lead us to infer that in front of us at a
> certain place there is a certain object of a certain character, are gen-
> erally not conscious activities but unconscious ones. In their result
> they are equivalent to a *conclusion*, to the extent that the observed
> action on our senses enables us to form an idea as to the possible
> cause of this action; although, as a matter of fact, it is invariably
> simply the nervous stimulations that are perceived directly, that is,
> the actions but never the external objects themselves . . .
>
> These unconscious conclusions derived from sensation are equiva-
> lent in their consequences to so-called *conclusions from analogy*.
>
> (Warren and Warren, p. 174)

Since in the overwhelming majority of cases retinal activity is due to
three-dimensional external objects, *all* retinal activity is attributed to

such external objects, at least in the first instance. These inferences are examples of what modern cognitive psychologists would call 'top-down processing': a higher concept operating on raw, and often insufficient, data to produce a conscious perception that is only partly justified by those data. In this case, the concept is that of a three-dimensional object, and the data the two-dimensional images on the retina. Helmholtz believed this concept to be innate. For his opponents, the concept that 'completed' the data, and in accordance with which the inference was performed, was not innate but derived from past experience. Previous encounters with three-dimensional objects – successive sightings from different angles, tactile experiences in association with visual ones – generated the concepts that enabled the data, and so the objects, to be rounded off. Past experience justifies the attribution of three dimensions to the object that is causing the two-dimensional image.

Most contemporary psychologists of perception would hold that both Helmholtz and his opponents had a share of the truth: the nervous system does appear to have 'feature detectors' 'hard-wired' into it (though these perhaps hardly correspond to the Kantian built-in intuitions – 'the forms of sensible intuition' – invoked by Helmholtz against his adversaries); on the other hand, ordinary perception does require a good deal of experience and training. Moreover, the wiring of the nervous system is greatly modified by experience and, in the absence of appropriate experience, some feature detectors fail to develop. Nevertheless, Helmholtz's fundamental idea that our perceptions are based upon unconscious inferences still seems to hold sway. Indeed, such inferences – under the broader heading of 'computations' – retain their place as the central concepts of perceptual psychology. As Helmholtz pointed out, they seem to explain many of the most striking features of perception, in particular perceptual illusions.

The latter, it is usually stated, are the result of the application of the unconscious inferences to data that normally warrant them produced by objects that do not:

> The general rule determining the ideas of vision that are formed whenever an impression is made on the eye ... is that such objects are always imagined as being present in the field of vision as would have to be there in order to produce the same impression on the nervous mechanism, the eyes being used under ordinary normal conditions.
>
> (Warren and Warren, p. 172)

Visual illusions result, for example, when a two-dimensional image is produced by a two-dimensional stimulus and the unconscious inference leads to the assumption that it has been produced by the three-dimensional object that would normally produce such an image. Other illusions – visual, tactile, auditory, etc. – are likewise the result of similar unthinking application of the general principles, based on previous experience and training, linking sensory impressions with their usual sources. It is precisely because the inferences are unconscious that they cannot be overcome by refutation. The demonstration that the object is not as it is perceived to be does not reform the perception, does not cure the illusion. To be shown that Muller-Lyer lines are in fact of equal length does not prevent one from continuing to see them as being of unequal length.

Helmholtz's views have been enormously influential in psychology; and his fundamental notions justify his status as the founding father of modern cognitive psychology, a century or so *avant la lettre*. One may trace his influence through figures like Edwin Boring[8] to recent major experimentalists and theoreticians such as R.L. Gregory and David Marr. Indeed, it might almost be thought that with the advent of 'cognitive science', Helmholtz has come into his own. Yet this may not be the case. His views are considerably more careful than those of many of the present generation of perceptual psychologists.

We may draw out two features of Helmholtz's views on perception:

1. Ordinary perception is based on *inferences* – from (insufficient) sensory impressions to useful, practical information about the world.
2. These inferences are *unconscious* – so much so that they are hidden (except when they are rendered visible by visual illusions) and insuperable.

These views, and the observations upon which they are based, still retain their validity. The question is whether they license, as they are claimed to do, some of the ideas that are current in modern cognitive psychology. For example:

1. The assumption that mind is largely unconscious; that only a small proportion of mental activity is conscious; that unconsciousness reaches to the very heart of consciousness.
2. The belief that a genuine inferential logic can operate in the absence of consciousness. If there can be unconscious inferences, then inference is not importantly conscious: the logic of the brain is remote from awareness. Indeed, we can think of brain logic in the way that we think of logic in the 'logic circuits' of a computer.

These ideas taken together lead to the kinds of claim exemplified by Philip Johnson-Laird's statement that 'vision is rather like the problem of finding the value of X in the equation $5 = X + Y'$.[9] For the cognitive scientist, perception itself is a series of unconscious inferences (and calculations), rather like those that appear to take place in the arithmetic-logic unit of a computer. At the heart of much cognitive psychology is the implicit or explicit assumption that *consciousness is unconscious processing*; or (if that seems blatantly absurd) that *to understand consciousness it is sufficient to describe the unconscious processes upon which it is supposed to depend.* This belief is expressed in fashionable computer models of the mind in which mechanistic and algorithmic accounts of unconscious processes drift towards being accounts of consciousness *per se*.

The most brilliant of Helmholtz's heirs was David Marr, whose ideas and investigations into the nature of vision were published in 1980 after his premature death.[10] For Marr, vision 'is the *process* of discovering from images what is present in the world and where it is' (Marr, p. 3); the underlying 'task' of 'visual processes' 'is to reliably derive properties of the world from images of it'. The visual process is a series of transformations in the brain of the neural codings of the initial image formed on the retina. All of this is unconscious and, because of this, renders the perceiver vulnerable to visual illusions – to misperceptions based upon unconscious visual hypotheses.

The fundamental error of Marr's approach – the characteristic error of the standard approach of cognitive psychologists – is that he confuses seeing with 'machine-vision'. As Hacker (op. cit.) points out, Marr's analysis will be very helpful for developing ever better imaging devices, but will tell us nothing about vision. The crucial point, as Hacker emphasises in another article,[11] is that what is on the retina is not an image at all – if it were, then we should already be seeing, and nothing more would be required. It is an image only to someone who is looking at the retina from outside. If, however, we are careless enough to think of perception as being the processing of unperceived images or data, then perception seems to be essentially unconscious and consciousness a peripheral, even superfluous, feature of perception.[12]

The transition from Helmholtz's position to that of contemporary cognitive psychology involves two illegitimate steps:

1. The assumption that, because perception involves inferences, it consists of solely making inferences; or (as it is more usually put) that perception consists of problem-solving. Richard Gregory (who regarded his book *Eye and Brain* as 'simply an extended commentary on Helmholtz's passage about unconscious inferences')[13] asserts that the main theme of his work is 'that perception is a kind of problem-solving'. Yes, of course, there is a problem-solving element in perception. Perception is the result of a decision based on inferences in turn rooted in previous experience. But perception is not itself the decision-making

process. Actual perception goes beyond the process of problem-solving: it is a state that we, the perceivers, are in. In being encompassed, as we are, by a perceived world, we are not surrounded by 'finding the value of X in the equation $5 = X + Y$'. The elements of the perceived world are neither the components of an equation, nor the process of solving it. Perception begins when the equations have been solved; it is beyond the problems. It is not a matter of finding the solution but of *inhabiting* it, of being deployed in it.[14] Events may occur in my visual pathways which could be plausibly described as calculations of spatial frequencies, in order to determine whether what is in front of me is likely to be an object or a shadow behind an object; but these are not the actual visual experience of the object.

And even this objection is not a strong enough rejection of the jargon of 'problem-solving'. For the process of solution-finding involved in, say, disambiguating an ambiguous figure, relates only to the transition from sensations to perception. In other words, it assumes a basis of consciousness in the form of sensations. These latter are necessary to give perception something to choose between. In short, we can localise the 'problem-solving' aspect of perception to a narrow liminal zone between sensation and achieved perception. Problem-solving at best only gets us from sensory impressions to achieved perceptions. It does not displace (or supersede) sensations or constitute the achieved perceptions.

2. The assumption that, since there are unconscious inferences in perception, we can talk easily of 'logic' in the absence of consciousness. Helmholtz does not share that ease, and his reference to 'unconscious inferences' caused him (and his contemporaries) considerable disquiet. A little further on from the passage quoted earlier, he qualifies the idea of unconscious inference:

> But what seems to differentiate [the unconscious conclusions] from a conclusion, in the ordinary sense of that word, is that a conclusion is an act of conscious thought. An astronomer, for example, comes to real conscious conclusions of this sort, when he computes the positions of the stars in space, their distances, etc., from the perspective images he has had of them at various times and as they are seen from different parts of the orbit of the earth. His conclusions are based on a conscious knowledge of the laws of optics. In the ordinary acts of vision this knowledge of optics is lacking. Still it may be permissable to speak of the psychic acts of ordinary perception as *unconscious conclusions*, thereby making a distinction of some sort between them and the so-called conscious conclusions. And while it is true that there has been, and probably always will be, a measure of doubt as to the similarity of the psychic activity in

the two cases, there can be no doubt as to the similarity between the results of such unconscious conclusions and those of conscious conclusions.

What a world of difference between this cautious statement and ~~and~~ the careless contemporary talk of 'logic circuits', 'brain logics' and 'inference engines'.[15]

Helmholtz's ideas and observations do not license the cognitive scientist's idea that the unconscious dominates consciousness even in perception, in the very heartland of waking consciousness. They do, of course, underline the extent to which perception is the result of *processes*, many of them unconscious. They emphasise that the basis of perception is itself largely unperceived; that the processes that give consciousness its final form are themselves unconscious; that ordinary awareness is founded upon mechanisms which are not available to intuition or introspection, but can be revealed only by careful, often ingenious, experimentation. But by drawing attention to those processes, Helmholtz has also, by implicit contrast, underlined the extraordinary nature of consciousness – a phenomenon that may require the support of, but is not interrupted by, unconscious processes.

The sharp distinction between the unconscious inferences embedded in perception and the conscious inferences we make in everyday life, made even sharper by the fact that perceptual inferences, such as those that lead to visual illusions, cannot be reversed by knowledge of their invalidity, demonstrates the great distances travelled by consciousness from its roots in unconscious processes. For conscious inferences can be invalidated by knowledge and argument. Helmholtz's own observations and arguments are a striking demonstration of the fact that consciousness is able to look into its own origins; that it is possible to make a conscious study of the unconscious inferences behind perceptions, of the extent to which perceptions are hypotheses and our gaze on the world is theory-laden; that consciousness may be rooted in the unconscious but not drowned in it. That, in short, there is no limit to what can be made explicit.

Enough has been said, I hope, to make plain that a would-be scientific psychology embarrassed by consciousness cannot get rid of that embarrassment by treating consciousness as if it were a matter of (unconscious) inferences or (to use the favourite contemporary term) 'mental processing'. Helmholtz (perhaps because he brought to his work in psychology impeccable credentials earned in the physical sciences) felt no such embarrassment. He saw that a science of perception did not require a denigration of consciousness; there was a place for both the physicalist and the intellectualist approach to perception:

The basic problem which that age placed at the starting point of all scientific knowledge was that of the theory of perception: 'What is truth in our intuition and our thinking? In what sense do our concepts correspond to reality?' Philosophy and natural science attack this problem from opposite sides; it is a common problem to both. Philosophy, which studies the intellectual side, tries to eliminate from our knowledge and our conceptions everything derived from the influence of the physical world in order to present in pure form that which belongs to the specific activity of the mind. Natural science, on the other hand, tries to distinguish definition, designation, forms of conception, and hypothesis in order to retain in pure form that which belongs to the world of reality, the laws of which it seeks.

The fallacious consequences of a cognitive psychology that does not recognise this are two-fold: losing consciousness from the places where it should be; and finding it in places where it shouldn't be. The latter – the Fallacy of Misplaced Consciousness (or explicitness)[16] – has attracted less notice than the Fallacy of Unconscious Consciousness. The two go hand in hand: any attempt to machinise consciousness is invariably associated with a (half-conscious) tendency to confer consciousness upon a putative machinery (of conciousness). Johnson-Laird (p. 114) tells us that 'the machinery of [visual] identification is unconscious in the Helmholtzian sense' while also informing us that that same machinery seems to be quite purposefully busy in the way that we might expect of a highly conscious entity:

> Your visual system constructs a description of the perceived object and compares it with some sort of mental catalogue of the three-dimensional shape of objects. It can recognise them from particular viewpoints and then make automatic extrapolations about the rest of their shapes.

> (ibid.)

According to Gregory, Penrose's Impossible Triangle illusion occurs because 'the perceptual system assumes that the two ends of what appear to be sides of a triangle are joined'.

This account of what 'the system' 'assumes' is a perfect illustration of the homunculus fallacy, as Hacker (who quotes this passage) points out. The homunculus is available to mop up the consciousness that the machine approach lets spill. The close relationship between machine (conscious-denying) and homuncular (consciousness-retrieving) thought is exemplified in this passage from Daniel Dennett's recent book 'explaining' consciousness. Here one of the most prominent contemporary philosophers of mind influenced by cognitive science gives his Thumbnail Sketch of the Mind:

> There is no single, definitive 'stream of consciousness', because there is no central Headquarters, no Cartesian Theater where 'it all comes together' for the perusal of a Central Meaner. Instead of such a stream (however wide), there are multiple channels in which special-ist circuits try, in parallel pandemoniums, to do their various things, creating Multiple Drafts as they go. Most of these fragmentary drafts of 'narrative' play short-lived roles in the modulation of current activity but some get promoted to further functional roles, in swift succession, by the activity of a virtual machine [a temporary struc-ture made not of wires but of rules] in the brain. The seriality of this machine . . . is not a 'hard-wired' design feature, but rather the upshot of a succession of coalitions of these specialists.[17]

The vision is homuncular, despite the routine, passing jibe at Descartes, but the little man has split into lots of littler men, specialist circuits who 'try . . . to do their various things' and, if they are lucky, or if they pool their resources in a coalition, get promoted to the slaves of a virtual machine. The oscillation between consciousness-talk and mechanism-talk – or between machinomorphism and anthropomor-phism – is dizzying; so much so that one might almost suspect Dennett of trying to demonstrate the non-existence of the Central Meaner by his own example.

Interestingly, Dennett brings together many of the themes that have haunted this discussion of the 'marginalisers' of consciousness. For it is clear (I think!) that Dennett is very much of the 'It thinks, therefore I am not' school. He clearly does not believe in a unitary self, and his dissolution of the self is yet more radical than Freud's. For Freud, the ego was not master in his own house; for Dennett, there are a thou-sand not-quite-egos uninhabiting a shifting estate of virtual houses. It will be recalled that, as Althusser put it, Freud 'rejected the notion of the psyche as a structured *unity centred* on consciousness' and that instead 'he conceived it as an "apparatus" composed of "different systems" irreducible to a single principle'. In Dennett's writings – and those of many other cognitive psychologists – Freud's three or four centres have given place to flocks of passing clouds. None of this can explain how it is that we have any sense of self: our feeling of instan-taneous unity ('I am here, now') that encompasses the innumerable perceptions, memories and feelings that we might identify within ourselves. Nor does it explain our feeling of continuity over time, so that not only do we remember things that happened to ourselves and distinguish them from things that happened to others, but we also formulate and execute long-term plans and feel responsible for things that took place in the past – always separating the things we did from the things that merely happened to or around us – and feel bound by promises, contracts, roles, etc. The succession of fugitive virtual machines envisaged by Dennett is remote from the rational agents we usually take ourselves to be and also from the continuing responsibility-bearing individuals we count on others being.

Dennett's vision of the mind has important points of overlap with the biological and ethological vision advanced by Dudley Young and many others, and criticised in Part I of this book. The virtual machine is the upshot of the succession of coalitions of specialist circuits:

> The basic specialists are part of our animal heritage. They were not developed to perform peculiarly human actions, such as reading and writing, but ducking and predator-avoiding, face-recognising, grasping, throwing, berry-picking and other essential tasks.
>
> (Dennett, 254)

Dennett does, however, allow that they have been modified by culture: 'Thousands of memes [units of cultural transmission], mostly borne by language, but also by wordless "images" and other data structures, take up residence in an individual brain, shaping its tendencies and thereby turning it into a mind' (ibid.). One would have thought that a brain would have to be pretty mindful in order to gobble up memes – data structures – in the requisite way.

It is abundantly clear that the science of the mind is in trouble trying to reconcile being a science in a rather traditional sense with accommodating the mind, the subjectivity that the most successful sciences have typically set aside. Dennett – whose own starting point is 'the objective, materialistic, third person world of the physical sciences' and who believes that 'this is the orthodox choice today in the English speaking world'[18] – wobbles between machinomorphic accounts of conscious human behaviour and anthropomorphic accounts of neuronal circuitry, illustrating how the cognitive turn has only deepened, not eliminated, the conceptual problems of psychology. At least this is *seen* as a problem – if only one to be brushed under the carpet – by many cognitive psychologists, rather than as a discovery or a solution, as triumphant proof that we are not the rational beings, the coherent selves, the responsible agents, we thought that we were.

10

Recovering the Conscious Agent

The theories I have discussed have, directly or indirectly, had an enormous influence on contemporary thought about what it is to be human. They have in common a tendency to discredit the notion of the self-possessed individual choosing at least some aspects of his or her life, and to downplay the role of conscious decision-making, deliberation, and indeed consciousness itself, in everyday life and behaviour. A specific consequence of this is a sceptical attitude towards the role of reason in personal and public life – in behaviour *tout court*. My avowed reasons for actions or beliefs or principles are taken to be rooted in self-deception or self-misrecognition: they are mere rationalisations of behaviour whose true origin lies in some place hidden from me. According to Marxist thinking, the hidden hand that shapes my behaviour – my political beliefs, my sense of morality, my choice of ends in life – is a largely unconscious, historically determined class consciousness, created by the objective conditions of production. For Durkheim and his sociological successors, the hidden hand is the society to which I belong and whose larger outline and deepest tendency is concealed from me. For the Freudians, the ultimate source of my behaviour is to be found in the asocial instincts refracted through the ancient repressive structures of civilisation – an origin I could not fully acknowledge, even if it were revealed to me. Much of my most reasonable, civil, practically useful activity is in fact the enactment of buried damage sustained in the battle during my infancy between the warring imperatives of natural instinct and cultural constraint. For the post-Saussureans, it is the system of signs that gives my behaviour its meaning; and this is so deeply buried that only the smallest part of it has been revealed even to semiologists. And so I am simply unconscious of the true significance of what I do, say or think.

The cumulative impact of these theories – which have assigned reason, deliberation, the conscious, retrievable intention and finally consciousness itself, an increasingly marginal status – has been to suggest that human beings are to a greater or lesser degree automata acting in accordance with laws of which they are unaware. While not all these

303

theories impinge with equal directness on the question of the reality
of consciousness *per se*, they all contribute to an intellectual climate
in which ordinary consciousness is granted at best a relatively minor
part in the drama of ordinary private and public life.

Although it is obviously counter-intuitive to deny the centrality of
consciousness and deliberate choice in everyday life – to deny, for
example, that there is any fundamental difference between falling down
a cliff and going shopping or between formulating a treatment plan
for a patient with epilepsy and having an epileptic fit oneself – it is
not easy to rebut the overall tendency of these revised accounts of
what it is to be a conscious human being. After all, not even the most
passionate defender of consciousness can ignore the ever-present background of *mechanism*, against which even voluntary action based on
full consciousness takes place. Much of what we do and most of what
happens in our bodies, during the course of carrying out ordinary
actions, is not accessible to introspection; below a not very deep level
we do not have control over our activity. The goal towards which I
am moving, or in accordance with which my actions are taking place,
is only small part of what is going on in me when, for example, I
rush down stairs to answer the doorbell. It is an even smaller part of
what is going on when, missing my step, I make a series of lightning
adjustments to prevent a nasty fall. We have no way as yet of describing, or even conceptualising, the dialectic between mechanism
and conscious choice in actions that are ordinarily seen as free. Moreover, it is very difficult to underline the autonomy of consciousness
without seeming to espouse a naive, unreformed Cartesian vision of
a transparent, auto-regulating self. The best one can do, perhaps, is
to reassert the obvious – the centrality of consciousness and deliberate choice in everyday activity – and then show why the theories that
question this are mistaken and how the general project of finding
reasons for marginalising consciousness is ultimately self-contradictory.

As regards the latter, it is necessary only to consider the status of
the theories themselves: they manifestly exceed the very limitations
they place on consciousness. Freud's own works – his act of writing
them, others' acts of reading them with something approaching comprehension and assent or disbelief – justify the observation that humans in their waking life are so far from being in the grip of dreams
that some of them, at least, are able to formulate theories of dreams
and of their relationship to, and significance for, waking life. The many
pages of *The Interpretation of Dreams* are evidence of the domination
of consciousness of a high order over the unconscious. And what more
potent counter-argument to Marx's claim that the social existence of
men determines their consciousness and that their ideas, views and
conceptions passively reflect the material conditions of their existence,

than the pages of this petit bourgeois's *Capital* or the spectacular transformation of his own views in the 1840s and 1850s? (The very fact
that *Capital* has been so influential is itself a tribute to the independent power of the superstructure. Marxist theory is not only unable
to explain why the Marxist revolution began in Russia – this is a
point that has often been made – but is also unable to explain why
there is such a thing as Marxism and why Marx was such a major
force in history rather than being simply a wild footnote to intellectual history, like Bakhunin.)

The problem, in short, with any theory that wishes to curtail the
importance of consciousness is that such a theory itself embodies higher
levels of consciousness than the theory would seem to allow. For Freud,
Marx, Durkheim and some other (though by no means all) apostles
of the unconscious, this must mean, if one believes their theories to
be true, that those very theories entrain the greatest measure of self-
deception about their own nature and origin. To put this another way:
unless the very existence of these theories is accepted as bearing witness to the reality of higher, disinterested reaches of consciousness –
and of our ability to get at truths that are not invalidated by their
ultimate roots in unconscious processes – they seriously undermine
their own credibility. There is an additional irony in the fact that those
theories which have most radically questioned the originality and
innovativeness of the individual, which have, indeed, undermined the
whole idea of the *individual* as a *source*, have as their own source
immensely fertile, innovative individuals. To the unprejudiced mind,
Freud, Marx, Durkheim, Derrida et al. are clearly defined individuals
not anonymous voices arising out of history, society, the Unconscious
or the language system.

It may seem odd to choose something as narrow and specific as the
theories themselves to counter their claims about the nature of consciousness and to use the theories to reinstate consciousness in the
middle of ordinary life. The choice is not, however, mere polemic
mischief. For what the theories exemplify, despite themselves, is the
general tendency of human beings to become ever more self-conscious,
and the seemingly unlimited capacity human consciousness has for
becoming ever more explicit. None of the 'unconsciousnesses' discovered – or allegedly discovered – by the theorists can, despite what is
claimed, either deny or account for the fundamental and inescapable
fact that explicitness, consciousness, deliberation, lie at the heart of
human life. The actual process of discovery of the various brands of
the unconscious – political, social, psychological, linguistic, perceptual – itself underlines this capacity of human consciousness for unfolding explicitness. It is, therefore, no minor or insignificant observation
that consciousness can even bring its own unconscious to consciousness!

Moreover, as I pointed out in the discussion of Freud, when we consider the characteristic examples of the operation of the unconscious in the naive subject – the invocation of reasons that turn out, on more sophisticated inspection, to be mere rationalisations – we still do not diminish the presence of explicitness, the role of consciousness. It is difficult to think of anything less machine-like or more remote from automaticity than the ability to produce a flattering but untrue reason for some action one has performed. Or the ability to show that that reason is flattering but untrue.

The thinkers I have discussed are united in their rejection of the world-picture that Lucien Goldmann identified as underpinning the Enlightenment hope of progress:

> all the leaders of the Enlightenment regarded the life of a society as a sort of sum, or product, of the thought and action of a large number of individuals, each of whom constitutes a free and independent point of departure.
>
> (Goldmann, op. cit., p. 32)

Our contemporary anti-Enlightenment figures see society, or one of its proxies such as history, socially mediated instincts or language, as being anterior to, and transcending, the individuals that make it up. Consequently, they deny those individuals any space outside of society from which they could engage it as 'free and independent points of departure'. The vision, promoted by these Apostles of the Unconscious, of the individual as helpless, blind, self-ignorant has undermined the notion of agency and relativised the Enlightenment's most trusted weapon against ignorance: the reason deployed by men of honour, justice and common sense. For Marx, the individual is a cork swept along in the sea of history, her consciousness and self-consciousness being shaped by the laws of social development of which she is unaware, except through the distorting lens of ideology. *Homo politico-economicus* has no personal or private space outside of, or transcending, the collective dream. As Camus expressed it:

> To put economic determinations at the root of all human action is to sum up man in terms of his social relations. There is no such thing as a solitary man, that is the indisputable discovery of the nineteenth century. An arbitrary deduction then leads to the statement that man only feels solitary for social reasons.
>
> (Camus, op. cit., p. 167)

Durkheim, too, had uncovered society – unobserved social forces – at the heart of individuality. Freud had found the self to be rooted in a collective past, instinctual and mythical, and the ordinary intercourse between individuals to be deeply influenced by the damage sustained during forgotten prehistoric battles between nature and culture. The post-Saussureans (many of whom had also incorporated the pessimistic messages of Marx, Durkheim and Freud into their thinking) saw selves vanish completely into nodes in sign systems or soluble fish in a boundless sea of discourse.

In placing these thinkers under the single rubric of 'marginalisers of consciousness', I have overlooked important differences between them. And in focusing on their contributions to contemporary *Zeitgeist*, I have most certainly been guilty of simplifications and, in some cases serious injustices. I have already made clear that Saussure cannot be blamed for the influence his ideas have exerted through the post-Saussureans. And it would be most unfair to lay the misconceptions that vitiate much thought in cognitive psychology at the door of Helmholtz. It is unlikely that Durkheim would have accepted the elaboration of his functionalism into the wild speculations of Lévi-Strauss. On the other hand, Freud surely is to blame for some of the ideas – not to speak of the practices – of the post-Freudians; and attempts to distance him even from the 'abuses' of psychoanalysis by the 'recovered memory' therapists in contemporary America require a good deal of active forgetting of Freud's own clinical methods and the grotesque irresponsibility of his means of obtaining and interpreting 'facts' supportive of his theories. Even Freud, however, is capable of being misrepresented and his message distorted; for example, Lacan's 'structuralising' of psychoanalysis produced results that The Master would have resisted. The transformation of his view that 'the ego is not master in his own house' to the dissipation of the ego altogether into an illusion that is kept alive through the constant pursuit of itself through the generalised absence of language is a travesty of Freud's travestying account of human nature. Likewise, although Marx did not advocate the evils done in his name in, say, the Soviet Union, his theory rationalised the abuses of human rights and the crushing of free speech that Lenin made a matter of policy. Marx inaugurated 'the hermeneutics of suspicion',[1] which made generosity to those who disagreed with one's own views merely a sign of lack of revolutionary will. The connection between the totalitarian practice and the Marxist theory is set out very clearly by Isaiah Berlin:

What Lenin demanded was unlimited power for a small body of professional revolutionaries, trained exclusively for one purpose and ceaselessly engaged in its pursuit by every means in their power. This was necessary because democratic methods, and the attempts to persuade and preach used by earlier reformers and rebels, were ineffective; and this in turn was due to the fact that they rested on a false psychology – namely the assumption that men acted as they did because of conscious beliefs that could be changed by argument. For if Marx had done anything, he had surely shown that such beliefs and ideals were mere 'reflections' of the condition of the socially and economically determined classes of men, to some one of which every individual must belong.[2]

Another important difference between the thinkers I have discussed is the different extent to which they undermined the notion of the autonomous, rational self as agent. (What subsequent writers made of them is, as I have indicated, another matter entirely.) Marx and Durkheim focus less on everyday life than on political actions and collective beliefs. It is arguable that neither saw ordinary human beings as going about in a kind of daze when pursuing their daily business. Men would be capable of pursuing their ends rationally. However, those ends might be seriously misconceived. For Marx, the bourgeois is simply unable to see the naked rapacity behind the respectable business of commerce and industrial production and the way the law and morality are so ordered as to ensure the reproduction of the conditions of production and the perpetuation of the ruling elites. And for Durkheim, the pious man freely expressing his deep and private religious sentiments, his intimations of transcendental otherness, is unaware that he is simply affirming the greatness of society. Once such major misconceptions have been conceded, of course, then the way is open to questioning the rationality of all behaviour; and, although Marx and Durkheim did not do this, they did make it easier for others to take this step. Marx's and Durkheim's views, that is to say, were ripe for transformation into a more radical critique of the notion of individual agent. As we have seen, 'structuralised' Marxism deeply undermined the notion of autonomous action on the basis of transparent reasons; for Althusser, Marx had shown that the unified self crowned with consciousness was a bourgeois illusion. In Lévi-Strauss's hands, the insights of Durkheim could be married to those of structural linguistics to create a structuralised social anthropology which saw all behaviour as being subservient to the fulfilment of unconscious structures reflecting the fundamental structures of the human mind.

Freud's undermining of the rational self was more thorough than that of Marx or Durkheim (prior to radicalisation by their followers); after all, in his view irrationality, indeed psychopathology, pervaded everday life; and it was that much easier for his followers, inspired by his hints of the ubiquity of the Unconscious, to see behaviour as largely symbolically driven: whatever I do, I am not really doing what I think I am doing; whatever I feel, my feelings are not what I think they are. Smoking a cigarette, arriving late, on time or early for an appointment, getting angry with someone who has stolen something from you, becoming a doctor, playing tennis, eating a meal, writing a book: these are all motivated by drives that I cannot be aware of and would deny if they were pointed out to me. (The passion and sincerity of my denial is, of course, a measure of the truth of the ascription.) The apparent function and rationale of an action has nothing to do with its true reason.

There is, as we have discussed, a serious problem with this irrationalist view of the springs of human behaviour: in order to bring about the set-piece irrational actions, numerous subsidiary rational actions are necessary. The means to the irrational ends have to be composed of rational actions; the latter are rationalised in only one sense. In order to act out my irrational anxiety about my health, I have rationally to fill up my car with petrol or, even more rationally, make sure that the garage has repaired the engine fault to meet the standards required by the Ministry of Transport. The arias of Freudian irrationality and magic thinking are, in short, embedded in a recitative of rational activity driven by non-magic thinking; the acting out of neurotic symptoms requires the support of a densely woven canopy of non-neurotic behaviour and calculation. Freud's vision of the nature of the human animal cannot accommodate this difficult notion of a plain of rationality sustaining a few peaks of irrationality. For once we accept the ubiquity of hidden, unacknowledged forces and motives in behaviour, then it is difficult to see how even dysfunctional behaviour can get off the ground. Why in short, there should be any kind of limit to the operation of irrational forces in our everyday life; why we should be able to wake up out of the nexus of potent symbols that is supposed to have us in its grip? Irrationality should infect everything that we do, since the forces that drive it seem to have such powers of penetration. Why, since everything we think of, say or do is at such a small remove from the dark forces that are supposed to motivate us, does the tongue not always slip, do angers not always possess us in a seemingly inexplicable manner, and so on?

Of course, Freud the metapsychologist and metaphysician of the Unconscious was generally much more radical than Freud the clinician; the latter, therefore, could not base himself in the former. The healer of suffering souls should have had nothing to do with the prophet of the Unconscious. Nevertheless, the influence of the prophet has been the more potent and widespread. The radicalism inherent in his thought has been picked up by many writers and the claims of his more lunatic followers – people like Groddek and Lacan – that they have drawn out of his writings no more than what was implicit or inherent in them are justified.

As for the post-Saussureans, they have emphasised rather than denied the extremity of their attack on the rational, conscious individual capable of undeceived self-awareness and purposeful agency in private and public life; indeed, they have celebrated 'the death of the subject' and the myth of actions informed by the intentions of agents.

In some respects, the project (and the assumptions) of cognitive pyschology – with the computerisation of mind, or its reduction to unconscious mechanisms, culminating in the elimination of difficult elements such as qualia – should have seemed a yet more radical undermining of the rational self and marginalising of consciousness. But cognitive scientists are themselves instinctive believers in rationality and do not have the courage of their convictions. They recover consciousness in the places where it should not be[3] and they feel that the mind is obliged, for biological reasons, to be rational: rationality is built into the very structure of the brain to ensure the survival of the human organism.

It is post-structuralism – or, more widely, post-modernism – that has brought to a climax the anti-rationalist, anti-individualist and, indeed, anti-humanist strains in the Counter-Enlightenment thought of the present century. 'Post-modernism', as Edward Said asserted, 'stresses the disappearance of grand narratives of emancipation and enlightenment.'[4] Although there has been a good deal of discussion as to whether Marx and Freud are simply more radical *aufklärer* underneath – after all, Marx believed in a universal human nature, a 'species being', and Freud shared the Enlightenment vision that man was a part of nature and understandable through science, and both Marx and Freud hoped to shed the light of human reason even on irrational human behaviour[5] – there is no room for doubt about the attitude of those post-modernists who cite these marginalisers as the grounds for their anti-individualist, anti-liberal, anti-humanist discourses – and their 'hermeneutics of suspicion'.

The collective impact of the marginalisers, as gathered up in postmodern thought, is to induce a sense of helplessness in anyone who would wish to help bring about improvement in the world. I have already noted this in the Introduction, with specific reference to Foucault. It will be recalled that he rejected the very idea of progress in the treatment of criminals and regarded as baseless the belief that Enlightenment ideas had led to a humanisation of the law and, specifically, of punishment. So far as he was concerned, the pre-Enlightenment live dismemberment of Robert-François Damiens was not morally different from the regulated life of the prisoners recommended in 1838 by prison reformer Faucher. The difference between the two modes of punishment – the one unspeakably barbaric, the other relatively humane – is, for Foucault, merely one of style, a change in the discursive formations gathered about the notion of punishment. Post-Saussureans are equally pessimistic about the future of politics. Language can never express purely human intentions, wishes, feeling, aspirations because the system dominates over the speaker and it produces effects which are not signs of anything other than the operation of its own rules. It follows from this that, according to Paul de Man, 'the political destiny of man', which is 'structured like and derived from a linguistic model' does not lie in his own hands:

> 'society and government' are neither natural nor ethical nor theological, 'since language is not conceived as a transcendental principle but as the possibility of contingent error'. Hence, de Man concluded, political activity is 'a burden for man rather than an opportunity'.
>
> (Lehmann, p. 219)

Even those who turn to post-modern thought as an instrument of liberation are not very certain about its power to foster desirable change. Take Joel Handler's 1992 Presidential Address to the annual meeting of the Law and Society Association – a bastion of Critical Legal Studies and of the anti-establishment semiotic approach to the law. He begins with some cheerful noises:

> And what does postmodernism have to do with society and the law? The major theme that I emphasise is subversion, the commitment to undermine dominant discourse. The subversion theme – variously described as deconstruction, radical indeterminacy, anti-essentialism, or antifoundationalism – whether in art, architecture, literature or philosophy – seeks to demonstrate the inherent instability of seemingly hegemonic structures, that power is diffused throughout society, and that there are multiple possibilities for resistance by oppressed people. The postmodern conception of subversion is a key part of contemporary theories of protest from below and the new social movements. ... Deconstruction ... seeks to destabilise dominant or privileged interpretations.[6]

The Kingdom of the Instantaneous is at hand! However, the cheerful noises are soon drowned by cries of despair, which recognise the consequences of accepting the post-modern world-picture. In the very same lecture, Handler quotes Rosenau who points out that since 'Postmodernism questions causality, determinism, egalitarianism, humanism, liberal democracy, necessity, objectivity, rationality, and truth. ... [It] makes any belief in the idea of progress or faith in the future seem questionable' (Handler, p. 726).

Questionable indeed, especially since those who espouse the belief
that we are the helpless playthings of discursive systems, of a society
composed of semiotic formations, of an unconscious that is structured
like a language, are deeply suspicious of those systems, those formations, that language. All discourse – even liberal discourse – is, Foucault
would have us believe, about power and, in particular the retention
of power by the powerful. There is no difference, it seems, between a
drunken bully imposing his views on a helpless and weak opponent
and John Stuart Mill arguing for tolerance in *On Liberty*. It is all simply a matter of the exercise of power – though it remains unclear as
to where the power is. (At times, the drunken bully, John Stuart Mill
and the helpless opponent seem equally powerless before the all-powerful, autonomous activity of the discursive formation.) Semiotic formations are committed, according to Barthes, to the misrepresentation
of History as Nature and so support the process by which ideology
conceals itself, with the inevitable consequence that, again, the dominant and the powerful may pass unchallenged. The very language
we use in ordinary life, we learn from Barthes and the thousand writers
imitating his easy omniscience, wears jackboots: it is Fascist because,
'even in the subject's deepest privacy, speech enters the service of
power' (Barthes, Inaugural Lecture, p. 460).[7] Our selves are socialised
without residue; our sociality is deposited in the systems of signs
that is language; language, which speaks us, is either opaque (or its
signs are totally indeterminate as to meaning) or malign ('Fascist');
and so we are helpless to choose, to change ourselves. We are caught
up in a net of all-powerful forces and unreliable signs. Thus does
pessimism becomes totalised and, with the assumption that the totalitarian state is already in place within us, does the return of totalitarianism come a little nearer. Or would, if anyone took this kind of stuff
seriously.

There is considerable evidence that nobody, least of all those for
whom they have brought tenure, jet travel and adulation, truly believes in the ideas that have so much currency amongst humanist
intellectuals. When de Man asserted that it was never possible to
determine the meaning of a text because all meaning was indeterminate, he did not therefore stop writing, or give up trying to mean one
thing rather than another. Nor did his official position of total doubt
about truth, meaning and reference prevent him (as Lentricchia has
pointed out) from having sufficient command of 'the rhetoric of authority' to say what literature – and indeed the world-text – has been,
is and must always be:

> Even while, in *Blindness and Insight*, he was telling us that there
> was no truth, or if there was, that it could never be known, he

spoke transcendentally of 'the foreknowledge we possess of the true nature of literature'.[8]

And while Barthes was confident that language (including that of Roland Barthes, presumably) was 'Fascist', he saw the role of intellectuals (wordsmiths like himself) as 'taking action against *powers*' (p. 459) – a difficult task, one would think, if what he said about language were true – like trying to dry up a stream by adding water to it or cleaning out the Augean stables with a few buckets of liquid manure.

Foucault, likewise, could not see any way out of the nexus that bound together power and discourse. Nevertheless, he too saw his works as unmasking the mystifications of power, even if they abjured 'the grand narratives of emancipation and enlightenment'. And he certainly seemed to believe that his own writings were less oppressive than the dossiers produced on suspects by the Gaullist police. Foucault's deeply pessimistic and anarchistic views – his profound suspicion of order and authority of all sorts – did not prevent him from scheming to be elevated to a prestigious Chair; nor did it make him very tolerant of any breakdown of order that might adversely impinge on his ability to pursue his scholarly pursuits. His biographer (Macey, p. 413) notes that this savage critic of bureaucratic, rational society became increasingly frustrated 'with increasing delays in the book delivery service' at the Bibliothèque Nationale and that this resulted in bitter personal quarrels with the library's director. (Luckily he found a cosy billet in a very well-appointed private library.)[9]

In many cases, post-modern ideas are anyway literally unthinkable, if only because of their limitless scope (cf. 'Language is Fascist') and, related to this, their failure to respect or even to retain the kinds of distinctions that not only critical intelligence but also common sense would demand. Yes, there is a sense in which speech is in the service of power – or quite a lot of the time. But there is the world of difference between the performative, persuasive, manipulative use of speech to soothe a frightened child or to raise help for someone who is drowning, and its use by a bullying gang directing humiliating taunts at a victim, by an interrogator in a police state or by Joseph Goebels furthering the cause of state-sponsored anti-Semitism.

Nor are these ideas greatly helped by being, as has repeatedly been pointed out, self-contradictory.[10] Perhaps this is why they do not, in everyday life, seem to influence the behaviour even of those who profess them. Perhaps it is a bit much to expect that the moral outrage you might feel when someone infringes your person, property or rights – taking your parking slot, overcharging you in a restaurant, clapping you in gaol without due cause, or threatening the life of your child – should be dissipated by thinking that you and he are simply corks floating helplessly on the stream of history or soluble fish in a sea of discourse. And it is understandable that, despite earning their living and world-wide renown for denying the difference between magic and scientific medicine (merely rival discourses propagated by different power groups), post-Saussureans prefer to have a ruptured appendix abscess treated by conventional surgery (which has advanced enormously over the last few decades) – supported by antibiotics and the recent generation of anaesthetics and intravenous fluids – and so (usually) survive, than by a witch doctor who, barring a miracle (or a lucky misdiagnosis) would supervise the patient's avoidable metaphysical translation to the sunless realm.[11]

Trilling's observation is again apposite: 'it is characteristic of the intellectual life of our culture that it fosters a form of assent which does not involve actual credence.' The mitigating plea that post-Saussurean and, more widely, post-modern, ideas have little effect, that they are totally insignificant in the real world outside the Academy, that they are harmless fun, should not, however, be upheld, in order to let the post-Saussureans and other posties off the hook. They may have long-term adverse consequences in muddying the waters, in their anti-educational effects on students who are force-fed on them and in the resources they consume elsewhere. The overriding concern must be that this 'form of assent' and these glamorously pessimistic ideas undermine others' efforts to see things differently; to see, for example, that an even relatively autonomous thinking agent might contribute actively to a process of change – the necessary condition for individuals deliberately trying to bring about progress – or, if this sounds too grand and too global, to try to improve things. The communal trances in which individual consciousnesses are supposedly dissolved allow no room for individual initiative or even for change that anyone actually wills. The marginalisers of consciousness deny the possibility that human beings might be able to reform human institutions in order to ensure the more efficent production of goods and a more equitable distribution of power in society. They foreclose the future and say that, if the latter is different from the past, it will not be the result of individual's efforts or of deliberate collective effort. Where there is unequivocal evidence of progress, the totalising pessimism of the marginalisers of consciousness will oblige them to deny this evidence: it was not simply a spirit of perversity that moved Foucault to maintain that penal reform was merely a change of style aimed at inserting social control deeper into the criminal's soul and that the reformed prisons of Faucher were therefore no advance on the public dismemberment of criminals. The anti-individualistic and anti-humanist works upon which his fame had been built required that he should take this position – as well as other anti-progressive positions, such as preferring the theocracy of Iran to the democracy of France.

The system-trance theorists are not able to explain change; even less are they able to explain beneficial change; and even less still are they able to explain beneficial change brought about by the will of individuals. The subject is simply the effect of a nexus of signs or power relations. And this makes science and science-based technology a particularly hard nut for the marginalisers to crack. For science has brought about the most dramatic changes in the nature and conditions of human life (including the conditions and forces of production); it has been the result of willed effort (technology is the most effective application of will and intelligence in the pursuit of generalised goals); and many of the advances can be attributed to the genius of named individuals. Although the ant-heap of anonymous toilers has supported and developed the great advances contributed by those individuals, the latter have formatted the disk on which the rest of us have written. Newton, Ampère, Faraday, Clerk Maxwell, Boltzmann and Einstein cannot simply be read as nodes in discursive systems, as soluble fish in the seas of language, even though they were steeped in the languages and notions of their time.

Let us take a particular example. The observation of the relationship between electricity and magnetism, which has transformed the world in the century and a half since Faraday made his crucial experiments and surprised himself with the results, was not simply the sign system acting out its eternal tunes. The first demonstration of electromagnetic induction by a changing magnetic field was a specific event that took place on a particular day – 29 August 1831. Following up its practical and theoretical implications was an heroic ten-year effort, which brought Faraday close to breakdown.

Nor is it valid to claim to recover the system within the individual by sociologising away science, seeing its results as simply the emergence of a dominant herd rhetoric about 'nature', itself understood as a term mobilised by an interpretive community to support their claim to having captured objective truths. Scientific knowledge cannot be merely a matter of fantasies induced by social or economic or depth psychological forces because its technological applications *work*. A computer that reliably executes its routine miracles of data-handling proves the differences between group fantasy and natural reality. Technological success is the proof of the non-historicist basis, the extra-human truth, of science; evidence of the difference between the methods of the scientist and of the public relations man. Contemporary system-trance thought denies that margin of freedom which makes the willed future, so evident in the internal progress of science, possible – the freedom in virtue of which 'L'homme passe infiniment l'homme'.[12]

Many of those thinkers who, with varying degrees of sincerity, deny that an autonomous agent may make an independent contribution to society, do so because they are prone to a rather simple fault: a propensity for gross exaggeration. More specifically, they share a tendency to unpack the whole truth about man and society from a few grains of truth about both. Without those grains of truth, their implausibilia would not have generated so much excitement, would never have been so widely accepted. Let us look at some of these grains of truth. It is true, for example, that the most seemingly altruistic of us may have political views, which, although they are sincerely held, and seem to promote the general welfare, are to our personal own advantage and to the advantage of those with whom we are most likely to identify. In short, we are sometimes likely to present our interests to ourselves as if they coincided with the general interest. This does not, however, mean that the origins of our political views are buried away from us in an historical unconscious sealed into our souls. And it is also true that we – our thoughts, our feelings, our attitudes, our convictions, our 'deepest' sense of self – are more socially determined than our adolescent selves, intensely aware of the distances that separate us from others and acutely conscious of our differences from them, would readily acknowledge: things that seem most personally chosen may well merely reflect the implicit social framework within which our presuppositions operate. But this does not mean that we are entirely dissolved in society. It is, further, true that we may sometimes do things for reasons that are objectively different from those that we ourselves give for our actions. But it does not follow from this that we are ordinarily irrational, or that the true reasons for our actions are usually hidden from us. Finally, it is true that there is a good deal that is automatic in our speech, and individual linguistic responses in certain circumstances may show a high level of predictability; but it does not follow from this that we are simply sites where language speaks itself through us. As I have already discussed, even the use of a predicable phatic word – a linguistic near-reflex – such as saying 'Hello' is rarely if ever 'the system speaking itself' through the mouth of the speaker. The decision to say 'Hello' is very often calculated, as is the tone in which it is said, not to speak of the endless elaborations of 'joky' consciousness that may be expressed through it. We may summarise the process by which thinkers marginalise consciousness, the individual agent, the rational self as follows: grains of truth are expanded into the whole truth. And when a small part of the truth is jacked up to the whole truth an untruth results.

What, then, is the truth about the self, about individuality, about rational agency, and about the relationship between the subject and society? The first – and most important – truth is that there is not going to be a *single* truth about these things. To forget this is to replicate precisely the error of those thinkers we have been criticising who have, as we have said, strayed into untruth by elevating part of the truth into the whole truth. There are, however, important truths (in the plural) about the self, society and rational agency that need to be reaffirmed in order to correct the unbalanced and deeply pessimistic account of humanity and society that contemporary Counter-Enlightenment thinkers seem to delight in.

Any credible attempt to restore the notion of the individual, the subject, the self as in some sense the centre of a world in which she/he acts as a rational, self-present agent must take account of some obvious limitations on freedom, self-presence and unity. It is worthwhile setting these out, if only to avoid the charge that one is naively unaware of them:

1. *Human lives are enacted within certain physical and biological limits.* There are the laws of physics under whose sway we fall, as do other pieces of matter. Most of the time these law are implicit – as the framework within which our lives are enacted – but they may come to the fore under certain circumstances – as when we engage in physically strenuous tasks or fall down stairs or are involved in a car crash. On top of these physical limits, there are the biological constraints placed upon our performance by our species characteristics. These latter are represented not merely in, for example, our inability to fly without the assistance of an aeroplane or run at 100 m.p.h. They are also reflected in the biological determinants of our needs and of the means that our body provides for satisfying them. These needs give rise to reflexes, tropisms, instincts, appetites, etc., which to some extent regulate our behaviour. The biological constraints are most dramatically illustrated in the almost total absence of control we have over the developmental processes by which we unfold from the potentiality of the gamete to the actuality of the adult human being: we are, for the most part, the passive site of this process rather than en-actors of it. Moreover, we carry within us, as important influences on our behaviour and experiences, the particular vicissitudes that we have as individuals experienced during this journey. Intra-uterine growth retardation, an early injury, malnutrition will irreversibly influence our feelings and be major biological limits added on to the general physical and species-specific contraints upon our range of self-choice that we share with our conspecifics.

This natural 'facticity' will, in the very broadest sense, determine the agenda our lives will be obliged to address. Nevertheless, this is a very loose determination. The entirety of cultural evolution, the distances between humans and nature (set out in my *Explicit Animal*)[13] measure the extent to which (in Steiner's words quoted earlier), 'man has talked himself free of organic constraint'. This is widely appreciated and there are relatively few sincere biological determinists for whom the self is simply dissolved into a sea of physiological events or waves on such a sea. Yes, we are animals inasmuch as we indulge in feeding behaviour; but in human animals such behaviour undergoes extraordinary transformations – for example, into concern not to embarrass one's host (especially if one perceives him to be less well off than oneself) by ordering an expensive dish when one is dining out at his expense. And, yes, like other animals, humans learn. But the 'bump-into and explore' learning of animals is a long way from the kind of way we learn. To appreciate this, it is only necessary to think of a mother planning to gain some points in the babysitting circle this year so that she will be able to attend an evening class next year without having to worry about childcare arrangements.

2. *There are (more or less local) cultural constraints on one's experiences, feelings and behaviour.* Our desires, our sense of what we ought to do, want to do and can do, are multiply determined by the macro- and micro-environments within which we were born, have developed, are enjoying (or not enjoying) our lives, and will decay. What is expected of me will influence what I do; what 'they' think will influence what I think; what 'they' want will influence what I regard as desirable and so desire. The scope of 'they' will vary – ranging from the history of my culture (itself influenced by the history of preceding and surrounding cultures); through the implicit and explicit pressures applied by the groups within which I have landed or with whom I most closely identify; to the individual examples set by, and the approval sought from, the specific 'significant others' who have been co-actors on the stage of my particular life. A cultural constraint may arise from sources as disparate as the transient disapproval of a friend or the Judaeo-Christian tradition of which I am a late product.

No one, then, in his or her right mind will deny the potency of biological and cultural constraints on the self, the individual. Our self-possession is limited by our deepest selves being to some extent in the keeping of forces, fields, influences of which we are incompletely aware. I do not invent myself; to an extent, the self I live out is implanted in me by things I am not fully aware of. For this reason, the notion of the individual as an utterly transparent Cartesian self, and as an absolute point of origin in the world, is unsustainable. Once the Cartesian self had lost its theological support (the truth of rational thoughts was underwritten by a non-deceiving God and so could rise above any formative influences such as custom and tradition and gain access to the truth by the pure operation of reason) and once it had lost the backing of epistemological idealism (the Cartesian self

could always master its own world once it shed the other partner in
the duo – the material world), there remained only the transcendental *cogito* of the phenomenologists, which ultimately required the absurdly exaggerated claims of existentialism to keep it viable.

For Sartre,[14] the self – in the form of the pure current account *pour-
soi* – is not only utterly self-possessed and transparent, but enjoys
absolute freedom. The only constraint on its freedom is its obligation
to avoid the bad faith of denying its freedom. Otherwise it is in charge,
inserting into a world of not-quite-fact the very values by which it
lives and which make up the agenda of its existence, making of its
circumstances (such as they are) what its freely chosen values wish
to make of them. The pure current account *pour-soi* is not even limited by its own history; it is not cluttered with the baggage of permanent characteristics (these latter are the alibis of the scoundrel wishing
in bad faith to escape from the responsibility of the *pour-soi*, into the
inert condition of the *en-soi*); and in so far as it is affected by (biological
or cultural) history – its own or that of the collective to which it belongs only electively – it is marked by it but not determined by it.

This is manifestly false – as Sartre himself recognised eventually
(though his recognition was so deeply buried in his *Critique of Dialectical Reason* that he managed to execute his U-turn without anyone,
least of all himself, feeling the centrifugal force). We do not choose
that we should exist at all; we do not choose our date of birth or
(usually) our date of death; we do not choose the kind of families we
are born into, the talents or disabilities that we have – to mention
only the coarsest of the meshes in the net that constrains us. And
although it may be argued that we freely choose to evaluate painful
stimuli negatively and pleasurable ones positively, we do not choose
that it is easier to enjoy an orgasm or even a cup of coffee than it is
to enjoy having one's foot sawn off without an anaesthetic; or that it
is easier to exercise free choice in the usual sense of the term when
we are awake than when we are seriously obnubilated after a head
injury – still *pour-soi*, of course, but somewhat confused.

Sartre's twentieth-century updating of an ahistorical, absolute-free-
dom Cartesianism is a sufficiently compelling bad example to discourage anyone from swinging, when trying to defend the autonomy
of the self against post-modernists and others, from system-trance half-
truths to neo-Cartesian half-truths. In fact, the post-modern marginalisers
of (rational, free) consciousness and the existentialist globalisers of
(rational, free) consciousness between them indicate the limits within
which any viable theory of the self and of the role of the rational
agent in daily life must be found.[15] Interestingly, these opposing visions
are vulnerable to the same criticism: they fail to recognise that self-
presence, self-possession, freedom, rationality, consciousness have

degrees. Neither the dissipation of self into an unchosen and hidden system on the one hand nor, on the other, its elevation to an all-powerful legislator over a world created out of its synthetic activity, can accommodate the ordinary facts of life; that, for example, we seem to be more free, more self-possessed and more rational on some occasions than on others – when we are awake than when we are asleep, when we are sober than when we are blind drunk, or when we are walking down the stairs as opposed to falling down them. The Sartre of *Being and Nothingness* could not accept that freedom is contingent and limited by history,[16] and that agency is dependent upon and operates through mechanism, that it is both limited and made possible (as well as being given its content) by 'the given'. The system-trance theorists, conversely, cannot accept that there is freedom and self-possession outside the regulating power of the system, and that individuality contains something real that cannot be dissolved, without remainder, into the system.

Let us examine this freedom and self-possession, and do so through a specific example, one we have used already – that of greeting another person. This is, as already suggested, a particularly useful example, if only because saying 'Hello' seems a classic instance of near-mechanical system-driven behaviour into which nothing personal is inserted. If any speaker is a mere node in the system, a rule-driven non-agent, it is surely the Hello-sayer. If what is going on *here* doesn't yield to post-Saussurean system-trance analysis, such a mode of analysis won't apply *anywhere*. Well, let us see.

Imagine, as not infrequently happens, I catch sight of the potential recipient of my 'Hello' some way off. This gives me time to make a series of decisions. The first decision is whether or not I should respond at all; the second, whether I should respond with something middle-of-the-road such as 'Hello' or something more informal (such as 'Hi!') or more formal (such as 'Good morning'). My decision whether, and how, to respond will be most crucially dependent upon whether or not I recognise the potential recipient. 'Recognise' here has a variety of meanings: there are degrees or levels of recognition. I need to recognise the approaching person as being a human being, not some other physical or biological entity. One could imagine this recognition process being automated and the greeting-response, if appropriate, being secured automatically – rather as a robin will automatically peck at red patch against a brown background. However, being human is not in itself sufficient to earn a greeting from me. The oncomer will need to belong to certain categories: the next level of recognition, in other words, is classification, allocation to a familiar (or unfamiliar) category of human being, a category towards which I am likely to deport myself in a certain way. This will depend upon things that are specific to myself, though some of the simpler considerations may be cast in general terms. I may be more likely to greet a male of my own age than a child (I don't want, after all, to be accused of being a child-molester) and a middle-aged man rather than a young woman (again, I don't want to be accused of invading the female's personal space, and worse). In other words, my classification of a stranger as an appropriate recipient of my greeting depends upon an enormous number of considerations that could not be captured in even the most complex set of general rules. For their application is dependent upon all sorts of things that are specific to me and/or specific to that moment. What sort of place we and the oncomer are in, what (inner or outer) business I am engaged in, what I feel like today (how grumpy or good-willed, well- or ill-disposed to humanity or particular sub-categories of humanity), etc. will all influence whether and how I greet. The vast majority of these factors is highly personal: the joy I am feeling at having secured an MRC grant, my awareness from reading about it in the newspaper that this ('lonely', as I now appreciate it) place is where a woman was stabbed a few years before, my tooth-ache, my attitude (as this person draws nearer and I see how smart a suit he is wearing) to something I choose to classify as 'businessmen' at present, etc. The number of considerations, and the way their application is crucially determined by so many things that are specific to me, clearly demonstrate how no finite set of rules could capture my likelihood of greeting the individual, predict the way in which I would greet them, or regulate my greeting behaviour in general. Only if I belonged to an exotic tribe and were being observed by an anthropologist ignorant of what was going on inside me, could these things seem predictable or tightly rule-governed. To a truly informed observer (for example, myself), it is evident that the greeting-rules, and the rules for the application or waiving of those rules, are infinitely complex.[17]

The complexity that attends the greeting of classifiable strangers is, however, only the beginning of the story. Many people I greet fall under unique categories – the single member-class of their unique selves. I recognise exactly who it is that is coming towards me in the proper-name sense: this is not merely 'middle-aged businessman', but Fred whom I met on the train a few years ago, after we hadn't seen each since we had been schoolboys together. Such specific – token-rather than type- – recognition mobilises a vast additional army of considerations arising out of our actual (not general) shared experiences, but whose potency will depend upon how I am feeling at this moment – cheerful-communicative or grumpy-withdrawn. Again, none of these influences on my decision will belong to 'the system' – of language, of society – or to history in general. Our shared (or unshared)

history will be utterly individual: it will belong to our actual episodic, autobiographical memories, whose elements may have certain,
general, culturally-specifiable characteristics, but whose overall character and whose combinations will be unique. And it is that which is
unique that will be decisive. It will be the interaction between our
unique histories (as perceived by us now) and between those histories and our present states that determine whether or not we say
'Hello' and, more specifically, the way we say 'Hello'. Even this is
not the whole story. We may be aware of, and resist, the pressures to
say 'Hello', or subvert them. The tendency of human beings to parody
anything stereotyped in their own behaviour, especially linguistic
behaviour, and so to distance themselves from it, or to reclaim the
behaviour for the actual moment they are sharing with the other person, has not been sufficiently remarked upon. It is a symptom of something very deep and distinctive about us – the tendency to make things
explicit, as I have argued in *The Explicit Animal* – but also an uneasiness about anything that looks mechanical. And so, the encounter with
the friend – particularly if that friend is close enough for courtesy,
attention, communicative intent, etc. to be taken for granted – may
prompt a jokingly assumed voice for the greeting – say a 'posh' voice
or a rural dialect – perhaps with reference to a television programme
we both of us admit to enjoying.[18] This, in turn, will refer back to
shared experience, to a common world or a common set of assumptions about and attitudes to the world, or all three.

The potential complexity, consciously engaged in and profoundly
individual – not necessarily in the sense of being original, but in the
sense of being rooted in individual knowledge of the relations between the greeter and the greeted, in actual individual experience –
can unfold indefinitely. None of this complexity could have been generated by the system; indeed, it represents distances (often consciously
underlined) from the system.

Thus even the mechanical 'Hello' is rooted in considerations that
are far from mechanical; and its use is influenced by the utterly personal history upon which the recognition and mode of acknowledgement of the other are dependent. It might be objected that while, yes,
the determinants of whether or not I do something as seemingly
mechanical and steroetyped as saying 'Hello' are autobiographical and
specific to me, this is not the end of the story. The autobiographical
element simply provides a substrate upon which the general rules
can operate. For example, it is a personal, not a system-based, constraint that I shall not say 'Hello' to you again on the very specific
grounds that we have already met twice this morning. However, the
constraint is still a general cultural constraint: in taking notice of the
fact that we have already said 'Hello', I am applying general principles that I did not choose, and to this extent I am caught up in the

system, yielding to its constraints. More generally, it could be argued that it is only in virtue of its obeying the general rules that our communicative behaviour is intelligible and actually communicates. If I said 'Hello' at intervals throughout a conversation, my interlocuter would not know how to take it; my greeting would be unintelligible.

This objection is not decisive; indeed, it clarifies and strengthens the point I am making. For in order to apply the rules (or to break them to some specific communicative purpose) I have to be able to relate them to the absolutely specific situation I am in. No rule could determine or automate the application of the general rules to particular situations. Such applications require full consciousness of the particular situation and an understanding of the point or purport of the rule – in short, a sense of what is going on, where I am and what I am trying to do. That sounds very much like a fully developed conscious agent.

Someone might argue that the contents of the 'personal' history that influences the decision to say 'Hello' and the manner in which it is said are themselves impersonal. This, too, is not the case. First, although the elements that make up an encounter with another may be descriptively general ('We first met at a station, when we were both heading for the London train') they are perceptually unique. The elements of the stylised encounter (e.g. Boy meets Girl) have a unique position in the series of experiences that are the lives of the individuals encountering one another and are themselves experientially unique. Secondly, the perceptions are based upon sensations that lie beneath and escape description. More specifically, sensations do not belong to *any* system. This is not merely a restatement of the epistemological point about the incommunicability of the actual content of experience. No, the content of experience goes beyond the epiphenomenal fact of its incommunicability: its very reality, its being, lies in that which is incommunicable. That in virtue of which I am present here, present to this bit of the world and thus present to myself, lies beyond and beneath the mere forms of discourse – language or other discursive systems.[19] This is the core or basis of the self that is not open to dissipation in the system; it is the basis of the personal histories through which our relations with others are mediated, through which we engage with history, society, culture, etc.

That our personal histories – a unique succession of experiences that are not fully communicable – should lie at the core of our social being seems so obvious that it may seem surprising that others have overlooked it. This is because when we reflect on the self, or try to classify ourselves, we factor out what we can neither express nor generalise. In this way, the meetings between us, the history of the meetings between us, my decision to say 'Hello' on this occasion, and the manner in which I say it, are made to seem like positions in a network of general possibilities. But, of course, a meeting between you and me is composed of more than the factors that made this meeting probable or definite. When Goldmann writes:

> We have long since learned from the social history of ideas that *every* mode of human thought and feeling is determined by mental structures which are closely related to the objective life of the particular society in which they develop
>
> (Goldmann, p. 15)

we have to read him carefully: he says *mode* of thought and feeling, not the thought and feelings themselves. This is what the post-Saussurean dissipation of the self overlooks: the fact that my thoughts and feelings (and goals and actions) are not types but actual specific thoughts and feelings (and goals and actions), and these are not determined by 'the objective life of society'. The latter cannot specify the specific; and everything that exists is utterly specific. It is in the utterly specific that our uniqueness and actuality lies; and in this, too, we find the ontological weight of the non-substitutable self.

No system can handle – determine, absorb, dictate – actual individuals: the system of *langue*, for example, cannot legislate over the occurrence of actual *paroles*. And just as *langue* does not determine when and how I actually say 'Hello' (though it may indicate where it is more appropriate for people in my culture to say 'Hello'), so the Unconscious, or the laws of history, do not determine who or what I should like or want to do. There are two related reasons for this: they are not tightly enough drawn to be able to specify what is to take place to the precise degree necessary to secure passage from the possible to the actual; and it is necessary for me to experience them in my own life, in my own body, to be affected by them. Of course, behaviour is rule-governed and culturally constrained; but a knowledge and understanding of the constraints and the ability to apply the rules to particular situations (and, since all situations are particular and have their own quirks, this means to apply rules *tout court*) requires a profound and complex understanding of an individual situation and a tacit but sure sense of the point of the rule. The rule-governed nature of behaviour not only permits individual creativity and the unique input from the rational, conscious, self-present individual – as a kind of 'personal spin between the meshes' – it actually demands such an input. The rules for applying, breaking, playing with the rules cannot, therefore, be mechanically applied or enacted through a helpless node in the system.[20]

To reassert the reality of the self is in essence to reassert the centrality of individual experience and the individual organisation of those personal experiences through which the general influences, common to a group, a nation, an era, are ultimately transmitted. Although we do not choose the cultural and historical context in which we live out our lives and find our own meanings, any more than we choose the rules of the language through which we speak, we are no less present in the lives we enact and the meanings we find in those lives. For at the heart of the *quidditas* – the coarse general framework within which we think of ourselves as living – there lies an unclassified *haecceitas* which is the only true reality. This is always present and it will always fail fully to coincide with, and hence will subvert, the general structures. The incompletely classifiable and communicable first-person sensations that make up our classifiable and communicable experiences (and classification gets easier as we ascend the scale of generality, to the level at which we touch the laws that post-Saussureans see as swallowing up the self) continually subvert the generalities of history, culture and the rest – just as Winston Smith's aching varicose ulcer was evidence of his non-solubility in Big Brother's closed universe of general discourse. Actual experience is constant proof of our distance from the collective. The uniformity of a company of soldiers standing to attention on parade – the supreme example of the reduction of the individual to a social atom externally determined in all respects – is deceptive. Even during the moments of the parade, the individual soldiers are possessed by different sensations and by different preoccupations arising from the utterly different personal histories to which they are inescapably attached. They are commutable, equivalent signs only with respect to the particular occasion – the parade to which they each contribute one unit of soldierliness – but beyond that they are profoundly different.

If we exaggerate this inner distance and think of the individual as being composed of incorrigible sensations, we run the risk of espousing, as Sartre did, a neo-Cartesian vision of a transparent self, absolutely self-possessed and self-present. If, on the other hand, we overlook this distance, we run the opposite risk of seeing the self dissolve into objective systems. The first error leads to the fantasy of absolute freedom and delusions of omnipotence; the second error leads to a fantasy of total passivity and impotence (omnimpotence, perhaps), a fantasy in which the individual is seen to be so much part of the system that it is difficult or impossible to partition responsibility for actions between individuals and systems, in which we deny the ontological weight of the individual as an agentive centre, as a place where a difference is made.[21]

I am not self-created – after all, I did not make, nor could I run, the body which is my *sine qua non*, and I did not choose many of the conditions in which I act out my life. Nor am I absolutely transparent, totally revealed to myself; not at least at the level of talk about the self – the level at which I am acknowledged by others and at which I seek acknowledgement – for the process by which I am disclosed to myself at this level is mediated by language and society which structures (though, importantly it does not form the actual content of) even my deepest desire. Nevertheless, ultimately, the content of the self, including even its larger-scale aspirations, is held at a level at which I am both present to myself and distant from, or at least distinct from, the anonymous, external world of *systems*. The actual – actual experiences of the world – is rooted in something that lies beyond the system, beyond the mechanism: the pure system, and the mechanism abstracted from its realisation in a particular event or act, are only schemata of possibilities. The actual is what is suffered, or chosen – in short, experienced. This is the content that fills out the form and it is inseparable from the individual considered as a unique and irreplaceable being with a unique biography. Every actual experience, every real choice, is rooted in that individual and goes beyond the system.[22]

For all these reasons, I believe it is possible to make sense of the notions of self and individuality that do not, on the one hand, elide the differences between mechanisms and agency[23] or, on the other, destroy the notion of responsibility and of the individual as an entity making an independent contribution to the world in which she finds herself. It is necessary only to recall how an individual is both particular and generalisable. At the root of this notion is the tautology of the embodied individual which may be expressed as 'I am this thing'. This notion is suspended between the particularity of presence and generality of absence: I am present and coinciding with myself inasmuch as I am embedded in my experiences and have assimilated their general face; and am absent from myself inasmuch as I am classifiable in accordance with objective criteria that are invisible to me. It would be naive to think of 'the life of a society as a sort of sum, or product, of the thought and action of a large number of individuals', as Goldmann said the leaders of the Enlightenment did – if only because the mass of individuals create, sustain, respond to, are part of, something that the individual cannot know in its entirety. Ignorance, as Hayek repeatedly emphasised, is the most significant fact about human nature in relation to society as a whole and legislators and reformers should take account of this when they are preparing to impose their Utopian dreams on their fellow citizens. There is, however, good reason for seeing those individuals as at least in some sense being 'a free and independent point of departure'. This would suffice to rescue the Enlightenment project – or at least the hope of progress – from the contemporary Counter-Enlightenment's gloating and self-satisfied (not to say self-contradictory) metanarrative of its irretrievable demise.

Appendix
Philosophies of Consciousness and Philosophies of the Concept, or: Is There Any Point in Studying the Headache I Have Now?

It shews a fundamental misunderstanding, if I am inclined to study the headache I have now in order to get clear about the philosophical problem of sensation.[1]

Much has been made of the differences between so-called 'Anglo-Saxon' philosophy, practised mainly in the United Kingdom, Scandinavia and the United States of America, and so-called 'Continental' philosophy, practised predominantly in mainland Europe. They are not, however, so exclusively rooted in their putative provenances as has been widely assumed: 'Anglo-Saxon' philosophy owes as much to Frege, Wittgenstein, to the Vienna Circle and to logicians such as Popper and Tarski, as it does to a supposedly British analytical or empirical tradition transmitting the messages of Locke, Berkeley and Hume through Mill and Russell; and much contemporary Continental philosophy may be seen as an indirect response, via Kant and the idealist philosophical tradition he inaugurated, to the challenges of British Empiricists, in particular Hume. And there are other close connections: for example, some contemporary French philosophers have been greatly influenced by their own, admittedly eccentric, reading of 'Anglo-Saxons' such as Peirce and Austin.

There has also been an over-emphasis on the contrast between 'daring' (or 'wild' or 'meretricious' or 'irresponsible') Continental philosophy – exemplified in phenomenology, existentialism, structuralism and post-structuralism – and 'rigorous' (or 'narrow', 'pusillanimous' or 'trivial') Anglo-Saxon philosophy – exemplified in logical positivism, the philosophy of language, ordinary language philosophy and possible world semantics. But there are deep similarities – at least of preoccupation

– and deeper connections; and these are more illuminating than contrasts in style. The deepest, and in the view to be developed here, most important, link is to be found in the tension, apparent in both traditions over the last century or so, between philosophies of the concept and philosophies of *consciousness*.

A philosophy of the concept is one which approaches fundamental questions about the nature of human experience and human beings themselves through a consideration of the language in which the philosophical questions are posed and, more generally, the sign systems through which humans express, identify, recognise, make sense of, structure and understand themselves and their world(s). A philosophy of consciousness, on the other hand, begins with the individual mind, in particular its conscious contents – the lived moment, sensations, perceptions, etc. – which are thought to be pre-linguistic, indeed, pre-conceptual, deriving directly or ultimately from physical encounter with the world, and ordering them according to principles that are either *a priori* (for example, 'the forms of sensible intuition') or, more often, inherent in the extra-linguistic world.

The relationship between Anglo-Saxon and Continental philosophy in the twentieth century is best captured in the criss-crossing rhetorical figure of the *chiasmus*. Although this is a crude generalisation and ignores important details, it is at least more accurate than the notion of two separate and parallel traditions connected only by the contempt they feel for each other's style and preoccupations. In France, for example, the philosophy of consciousness – whose supreme exponents were Bergson and Sartre – was displaced by the philosophy of the concept expounded first by the structuralists and then by the post-structuralists. This occurred in the 1960s, at a time when many Anglophone philosophers were moving in the opposite direction: language-based philosophies of the concept inspired by Frege via Wittgenstein[2] were about to lose unchallenged domination, and there was a return to metaphysics and epistemology as traditionally construed, and to the philosophy of mind – understood at least in part as an attempt to understand consciousness in itself, independent of the language of both everyday and philosophical discourse.[3]

The multiple tensions between philosophies of the concept and philosophies of consciousness have been evident in much of twentieth-century Western philosophical writing; but there have been certain moments when the tensions became strikingly explicit. I should like to examine two such moments. The second I have already alluded to – the overthrow of existentialist phenomenology by structuralism. The first is the argument between Frege and Husserl about the nature of numbers.

FREGE, HUSSERL AND THE ATTACK ON PSYCHOLOGISM

The birth of phenomenology (the dominant influence in twentieth-century philosophies of consciousness), marked by Brentano's 1874 *Psychology from an Empirical Standpoint,* and the birth of symbolic logic (with structuralism, one of the two major elements in the twentieth century philosophies of the concept), signalled by Frege's 1878 *Begriffschrift,* were almost contemporary. Brentano's influence was most importantly mediated through Husserl, and Frege's devastating review of Husserl's first book *Philosophie der Arithmetik* I[4] was a crucial encounter between the philosophy of the concept and the philosophy of consciousness. What Frege took savage exception to was Husserl's attempt to import a Brentanian philosophy of consciousness into his analysis of the seemingly most objective of all concepts – the numbers.

The plot is a bit more complex than this: *Philosophie der Arithmetik* I was not only Brentanian, but also a continuation of J.S. Mill's psychologist programme, set out in his *System of Logic,* for reducing even mathematical axioms and logical propositions to psychological entities – intuitions which we arrive at by introspection upon our experience. Psychologism, 'the theory that psychology is the foundation of philosophy and introspection the primary method of philosophical enquiry',[5] is a powerful assertion of the primacy of consciousness, of individual conscious experience, over concepts: the latter are derived from the former. To extend psychologism to such 'pure', 'abstract' concepts as numbers was brave indeed and Husserl was to regret his bravery.

Frege argued that the attempt to provide a 'psychological' foundation for arithmetic was doomed from the start. The grounds of Frege's attack, as summarised by David Bell, are that

> such a doctrine will be quite unable to provide any cogent explanation of (a) large numbers, (b) of the objectivity of arithmetic, (c) of the universal applicability of number, (d) of the necessity of arithmetical truths, (e) of the nature of arithmetical proof, or (f) of the relation between arithmetic and logic.[6]

Frege's hostility to Husserl's apparent attempt to reduce the meaning of numbers to events in individual consciousnesses was not merely the result of a clash between two complementary approaches to philosophy. It expressed a bitter disagreement about what was foundational in philosophy, about ontological hierarchies and the order of things, and about the proper way to advance our understanding of the world.

Husserl's psychologism was seriously defective even as an account of how we acquire an understanding of numbers – as opposed to the nature of number itself. We cannot rely on mental acts (for example, 'collective associations') to help us to determine the extensions of concepts. How, for example, would cumulative and synthesised experiences enable us to arrive at the correct extensions (i.e. the agreed extension that enables us to use them properly) of the terms 'earth', 'economy', '1000, 1001, 2000', etc.?

For Frege, the failure of the psychologist programme demonstrated that logic was prior to epistemology (and for the analytical philosophers that followed him, that ontology was merely an extension of logic: ontologies were rooted in the logical grammars of language). For Husserl, the emphasis on logic placed too much weight on calculus and symbolic operations and too little on experienced, or understood, meaning. Behind Husserl's early 'genetic' approach to numbers and Frege's preference for approaching them via the logic of classes, using concepts such as 'extension', 'equinumerosity' and 'one–one correspondence', there were, therefore, deeper divisions. Frege regarded mathematical concepts as transcendent realities, Platonic objects that went beyond human experience; as such they were our one hope of escape out of the imperfections of the human sensorium and heart. Husserl's attempted psychologisation threatened to drag them back down into the 'foul rag and bone shop of the heart'.

Michael Dummett's *Frege: Philosophy of Language* makes clear just how strongly Frege felt the need to drive out the ghosts of psychologism from philosophical thought. His background as a mathematician partly accounted for his emphasis on symbolic systems over and above individual consciousness and for the revolution he inspired that turned philosophy away from epistemology to language and logic. However, he went beyond his mathematical base in his explicit commitment to expelling psychologism not only from mathematics but also from logic and, most influentially, from the philosophy of language. Again and again, he argued (as did Wittgenstein after him) that the sense of a word does not consist of a psychological entity such as a mental image. He rejected 'the empiricist conception of sense as consisting in the propensity a word may have to call up in the mind of speaker or hearer and associated mental image' (Dummett, p. 158). The reason for this was straightforward: there would be a huge gap between any postulated mental image and the actual use of any word whose meaning the image was supposed to embody, a use defined and constrained by conventions which could not be imaged. An image could not show how it was to be applied – could not show what it was an image *of*, could not exhibit its extension, the scope of the 'type' which it was to stand for. And if, as Dummett points out, it does not show that, 'then it does not contain within itself the sense of the word' (p. 158). An agreed use for words was essential if language was to be the vehicle for expressing objective truths – homely ones such as 'The dog is next door', or less homely ones such as 'The inertial and gravitational mass of an object are identical'. In order to ensure such an agreed use, the meaning of a linguistic expression must be independent of the psychological state of, or a psychological content enjoyed by, any individual. To identify the meaning of a word with a mental image is to place it at the mercy of accidents of recollection and personal association and so undermine the word as a medium of reliable communication.

There was another important consequence of the rejection of the psychologist account of all symbolic systems such as language. The imagist theory of meaning was, as Quine expressed it, an attempt to construct an atomistic account of the senses of words: 'to each word is correlated an idea, and to a complex expression, including a sentence, a complex idea compounded out of the ideas correlated with the constituent words' (Dummett, p. 597).[7] Frege denied that individual words could have senses in isolation. On the contrary, the *sentence*, not the individual word, is the primary vehicle of meaning; for 'we can give an account of the senses of words only in terms of their relation to sentences of which they form a part'. No image could portray the role of a word in a sentence. Words – and other expressions – have meaning only in the context of a sentence expressing a completed thought that is then able to carry a truth-value. The independence of the truth of sentences from the psychology of the person who utters them, as well as the need for agreed senses between speakers, should discourage any attempts to reduce the meanings of concepts to psychological entities. Concepts belong to the system – of numbers, of symbols, of the language – which confers upon them their extensions. And so Husserl's psychologism fails not only to show how numbers, words, symbols in general actually have the meanings they do; it fails even to explain the way we grasp meanings. Its failure as an attempt to psychologise logic – and to give priority to genetic epistemology – is complemented by its failure even to account for the acquisition of concepts.[8]

Frege's attack on the psychologistic notion that symbols (numbers, words, etc.) carry with them a discrete package of psychological material, such as an image, and that this counts as their sense or meaning, mediated most importantly through the advocacy of the later Wittgenstein, had a huge influence on the course of Anglo-Saxon philosophy. Driving consciousness out of concepts had the effect of discrediting mental entities more widely. There was also a drastic reordering of philosophical priorities. Symbolic systems became the focus of attention. Logic – and, in particular, Frege-derived symbolic logic – moved to centre-stage, both as a tool for analysing philosophical problems and as an object of philosophical investigation in its own right. Epistemological approaches to ontology were displaced by a philosophy of language, of meaning understood in non-psychological terms. Philosophy itself became more explicitly conscious of the terms in which its traditional problems were couched.

This, in turn, had two consequences: attention was shifted from the problems – or from postulated puzzling entities with which they were associated such as 'mind', 'time' – to the terms in which the problems were posed and their meanings; and these meanings were no longer seen to be entities referred to by the words (most words were not names and none had senses in isolation anyway), and even less to be some psychological content. Instead, they were *the uses to which the words were put* in language. Analytical philosophy saw its function as being to determine the relationships between concepts ('the logical geography' of concepts, as Ryle described it). This tended to be descriptive rather than prescriptive or revisionary: philosophers, far from being able to reform the logical geography of concepts, appealed to ordinary usage to validate their own analyses; this way they could avoid the pseudo-problems that had vexed their predecessors.

Thus the displacement of traditional metaphysical questions by an investigation of the language in which they were framed and the suggestion that they were rooted in the misuse of language ('Philosophical problems arise when language goes on holiday' – Wittgenstein) led, via an emphasis on the philosophy of language, from the original analytical programme (exemplified in the work of Bertrand Russell) of pursuing traditional philosophy in mathematically transparent terms, cleansed of confusing connotations, and based in the relevant sciences, to Ordinary Language Philosophy. The 'linguistic turn' in philosophy, which may be seen as the long-term consequence of Frege's consciousness-free concept-philosophy, had, therefore, two aspects: a shift of interest from the conventional ontological and epistemological preoccupations to a concern with the way symbolic systems carried or achieved meanings ('the philosophy of language' in the widest sense); and a change of approach to the traditional philosophical problems, focusing on the terms used and abused in generating and addressing them ('linguistic philosophy').

One of the major achievements of the 'linguistic turn' was Gilbert Ryle's *Concept of Mind*. This was precisely what it said it was: an attempt to treat mind as if it were a concept. The role of the philosopher of mind was to determine the 'logical geography' of the various concepts – such as sensations, feelings, memories, moods, etc. – that had caused so much puzzlement and anguish over the centuries to those who had struggled with the traditional problems of mental philosophy. Although Wittgenstein was disparaging about Ryle, the latter was in tune with Wittgenstein's assertion that understanding the nature of sensation was a grammatical not a causal matter. Wittgenstein's often cited claim that it showed a fundamental misunderstanding if I study the headache I have now in order to get clear about the problem of sensation was, consciously or not, a defence of a conceptual against a phenomenological, consciousness-based approach to philosophical problems, extended even to the nature of consciousness itself.

The very antipodes of psychologism – of the view that the meaning of concepts is reducible to the experiences of the individuals who use them – is reflected in Wittgenstein's characteristic assertions about grammar and meaning in *Philosophical Investigations*:

371. *Essence* is expressed by grammar.

373. Grammar tells what kind of object anything is.

II xi (p. 203e) Our problem [understanding what is happening when we move between aspects of an ambiguous figure] is not a causal but a conceptual one.

II xi (p. 218e) Meaning is not a process which accompanies a word. For no process could have the consequences of meaning.

The journey from Husserl's interpretation of even abstract geometrical concepts in terms of the phenomena of individual consciousness to Ryle's approach to the phenomena of the individual conscious mind through the logical geography of concepts represented a triumph of Frege's philosophy of the concept. No wonder the reviewer of Michael Dummett's mighty book on Frege's philosophy of language saw Frege as having accomplished the greatest revolution in philosophy since Descartes – and as, indeed, having overthrown the Cartesian emphasis on epistemology, not merely the 'official' doctrine, 'the ghost in the machine' image of mind, that had been Ryle's major target.[9]

The Concept of Mind was a high-point of Anglo-Saxon philosophy of the concept. The latter, however – especially in its Ordinary Language Philosophy guise – was starting to earn a reputation for trivialising philosophical problems. Focusing on the meaning of terms seemed to be a second-order activity; and Ordinary Language Philosophy, which examined the meaning of terms as used in ordinary discourse, seemed a further remove yet from a direct attack on the problems that, for all their being dismissed as pseudo-problems arising out of the misuse of language, were still real to some. Ernest Gellner's scathing and witty attack on Oxford Philosophy,[10] inspired by Popper's rejection of scholastic nominalism and supported by an approving preface from an octogenarian Russell, was only the most polished of many hostile responses.

The call for a shift to a philosophy of consciousness was at first inseparable from a call to return to metaphysical problems and was associated with a revival of a willingness to address 'real philosophical issues'. This confusion arose because, in Wittgenstein-dominated Anglo-Saxon philosophy, it was difficult to shake off, perhaps because difficult to separate, the two ideas that (a) concept-based philosophy was not merely a better approach to solving some problems, but also (b) reflected a mature recognition that many of them were not real philosophical problems at all, but were due to a mistaken understanding of ordinary terms: philosophical questions could not be asked without transgressing the rules of language or moving outside the frame of reference within which language could operate meaningfully. The linguistic turn has wobbled between being a new approach to philosophical problems and a denial of their seriousness; between suggesting how they might be solved and indicating how they might be dissolved.

After a few false starts, the philosophy of language was gradually displaced by the philosophy of mind as the central preoccupation of Anglo-Saxon philosophers. This was more than a change of theme: it was not just a question of no longer believing that language was the great theme of philosophy but also reflected a growing feeling that the linguistic approach was unlikely to yield much in the way of new insights.

THE FALL OF THE *COGITO*: STRUCTURALISM OVERTHROWS EXISTENTIALISM

The Continental story has had a different ordering of revolutions and counter-revolutions. The 'old-fashioned' philosophy of consciousness lasted much longer. Phenomenology – through the influence of Husserl and Heidegger, and a transformed Hegel, mediated by Sartre, dominated French philosophical discourse to the mid-1960s. It was then that, through the emergence of structuralism – in Marxist political theory (Althusser) in anthropology (Lévi-Strauss), in linguistics and semiology (Barthes), in psychology and psychiatry (Lacan) and in history (Foucault, despite his protests to the contrary) – that the philosophy of consciousness was routed by the philosophy of the concept. As in the Anglo-Saxon revolution, this was no mere methodological shift. With the change in the method and content of philosophy came a profoundly different sense of the nature of its objects.

Structuralism came, as Merquior points out,[11] 'as an onslaught on the frame of mind associated with existentialism'. It was consciously opposed to what both Descartes and Sartre (for all their differences) were thought to have in common: the idea of a centred self, a focal point of awareness at the heart of the universe. Whereas French existentialism was explicitly a form of humanism, structuralism was anti-individualistic and anti-humanistic. It was also in some respects anti-historicist. In so far as the fixed, static structures permitted change, these changes took the form of abrupt replacement of one type of structure, of one closed system, by another – *ruptures epistémologiques*, as Foucault would term them – or a shift from one crystalline array to another.[12] Even where historical evolution was allowed, history was not the sum of the results, intended and unintended, of events brought about by agents, but the expression of forces whose nature was unknown to those agents. For the classical (1940s, French) existentialists, each of us was 'responsible for eveything in this world'. For the structuralists who rebelled against them, nobody was really responsible for anything: human institutions had purposes that were concealed from those who worked in, with, for or against them. According to Althusser, for example, they were 'ideological state apparatuses', designed to ensure the reproduction of the conditions of production. Ideology was the all-pervasive principle of misrecognition built into ordinary human consciousness, the consciousness of the subjected subject. Actions had meanings that could be understood only in relation to the systems that made them possible and gave them their value, systems so complex that no actor could have an adequate awareness of them. Structural anthropology, it was claimed, had shown us[13] that a particular action (the telling of a myth, the giving of a gift, the permitting of a marriage) was an enactment of a fundamental but unconscious desire of the universal mind to make the world its own thing by finding or realising or affirming certain structures in it. Even that most self-conscious and seemingly self-possessed of all actions – deliberate speech – was simply the enactment of part of a system that in different ways determined the content of thought. We can no more intend what we say than we can intend the whole system in which our sayings make sense and the schema from within which they derive their appropriateness. Language speaks (or, in the case of an author, writes) us, and not vice versa. What we do, say, believe, act, discover, perceive, delight in, get cross about, vote for, etc., is part of and derives its meaning from something – a system of signs, an episteme, a universal unconscious mental structure – that lies beyond our consciousness. Consciousness is to be understood in terms of systems of signs, of concepts, that are outwith consciousness; it is not to be understood in terms of (self-present) consciousness.

The existentialist emphasis on individual consciousness was felt to be self-indulgent as well as methodologically unsound. Lévi-Strauss

dismissed the phenomenological subject as 'an *enfant gâté* of reflection, getting in the way of a serious search for mental structures which
are by definition beyond the purview of consciousness'.[14] Merleau-
Ponty's *cogito* was no better than Sartre's *pour-soi*:

> For structures cannot be 'lived' in any state of awareness – they are
> just experienced, 'undergone', without ever becoming objects for
> the conscious mind.
>
> (Merquior, p. 4)

Getting rid of consciousness was essential for a rigorous science, for

> rigorous analysis entailed an abandonment of the focus on inten-
> tional, even conscious, action, for the sake of identifying hidden
> springs of human conduct.
>
> (ibid., p. 5)

The struggle, as Merquior expressed it, was 'between the philosophy
of cogito and a more "combinatorial" mode' – a battle between the
rival heritages of Descartes and Leibniz – and it provides an interesting parallel with the Husserl–Frege battle over psychologism.

Structuralism was displaced by post-structuralism for a variety of
reasons and a greater variety of causes that go beyond reason in the
strict sense. The frozen, totalising image of structuralist thought was
unattractive to those who believed in history both as a force for change
and as a way of relativising (and so undermining) absolutes of morality, rationalised hierarchies that were merely self-perpetuating power
structures, etc. The merest hint that the structuralist contrast between
structure and event, between sign and system, between signifier and
signified, might run into difficulties was enough, along with the appetite for the new, to make structuralism *passé* overnight.

Post-structuralist thought retained the decentred self and a preference for concepts over consciousness. It, too, regarded the Cartesian–
phenomenological–existentialist sovereignty of consciousness with
contempt. The polycentric post-modern universe does not count the
conscious individual as one of its centres – even less as its definitive
centre. Even the idea of self-presence is an illusion based upon the
accident of hearing oneself speak. In reality, everything runs off in
all directions along chains of signifiers that do not reach a signified.[15]
Admittedly, for some philosophers (or philosophoids), such as
Baudrillard, Lyotard, etc., the failure of the flow of signifiers to reach
a signified is merely a hazard of modern or post-modern life. Rather
than being a permanent condition of humanity, it is the post-modern
condition (or the condition of a mind conditioned to think of itself as

post-modern) in which, due to proliferation of representational systems and, in particular, the all-pervasiveness of the media, there is no direct encounter with any (signified) reality, least of all ourselves.

HUSSERL AND WITTGENSTEIN

Husserl was deeply affected by Frege's critique of his *Philosophie der Arithmetik* I. He rejected psychologism: *Philosophie der Arithmetik* II was abandoned after four or five years of futile struggle to complete the programme set out in the first volume; and he switched direction, a change that gave birth to his first major work, the *Logical Investigations*. As he himself wrote in the Foreword to the first edition of this book:

> I became more and more disquieted by doubts of principle, as to how to reconcile the objectivity of mathematics, and of all science in general, with a psychological foundation for logic ... and I felt myself more and more pushed towards general critical reflections on the essence of logic, and on the relationship, in particular, between the subjectivity of knowing and the objectivity of the content known.
>
> (translated and quoted in Bell, op. cit., p. 83)

The failure to give a convincing psychologistic account of numbers and his inability even to relate the genetic psychology which purports to explain how individuals acquire numerical concepts to the intrinsic properties and the validity of the number system, came to stand for a wider failure in his thinking:

> I was tormented by incomprehensible new worlds: the world of pure logic, and the world of act-consciousness ... *I did not know how to unite them*, yet they had to have some relationship to one another, and form an inner unity.
>
> (quoted Bell, ibid., p. 83)

This inability led him to abandon a philosophy based upon an intentionalist theory of consciousness in which intentionality was 'an exclusively intra-mental relation holding between mental acts and their "subjective" contents' (ibid.). And it defined his task over the next 30 years of his life – that of reconciling the 'psychological' and the 'logical' aspects of knowledge; or, as Bell describes it, of reconciling the subjective and objective aspects of rational knowledge, by the provision of a theory of mind, on the one hand, and a theory of truth on the other, and by showing how each is related to and compatible with the other'. His philosophical methods and theories were successive responses to the challenge of the following questions (as posed here in his middle-period statement, *Philosophy as a Rigorous Science*):

How can experience as consciousness give or contact an object? How can experiences be mutually legitimated or corrected by each other, and not merely replace each other, or confirm each other subjectively? How can the play of consciousness whose logic is empirical make objectively valid statements, valid for all things that exist in and of themselves? ... How is natural science to be comprehensible ... to the extent that it pretends at every step to posit and know a nature that is in itself – in itself, in contrast to the subjective flow of consciousness?.

(quoted and translated by Bell, p. 84)

The questions were to evoke a series of enormously complex answers – at first set within a naturalistic framework (*Logical Investigations*), then invoking a transcendental, solipsist, idealist framework (*Ideas*, etc.), and, finally, accepting the ordinary, intersubjective *Lebenswelt* assumed by Everyman as an unsurmountable given. After 40 000 pages of writing, he came to a conclusion, three years before his death, that 'Philosophy as science, as serious, rigorous, indeed apodictically rigorous science – *the dream is over*' (quoted Bell, p. 232). In my ending is despair. And yet no one had set out more clearly the challenge of reconciling the philosophies of consciousness and the philosophies of the concept – from the standpoint of consciousness.

Like Husserl, Wittgenstein spent much of his philosophical career arguing against his early self. Unlike Husserl, he was predominantly a philosopher of the concept, as one might expect from a disciple of Frege. Innis has expressed this difference as follows: although both Husserl and Wittgenstein were concerned with 'sorting out' the phenomena, for Wittgenstein this sorting was a grammatical sorting, while for Husserl it was a detailed inventory of the forms of consciousness.[16] Wittgenstein, as we have seen, gave priority to concepts in the additional sense of seeing philosophical problems as problems internal to sign systems, and philosophy itself as an activity that consists of ironing out, eliminating, dissolving, recovering from, these problems, by removing misunderstandings of the nature and scope of signs systems. Philosophy is a 'critique of language' and its main outcome will be the discovery that most of the problems of philosophy are meaningless and the deepest problems are not problems at all.

Wittgenstein's quarrel with himself was directed mainly at the atomistic and name-centred philosophy of the sign that he advanced in the *Tractatus* and against the implicit dualism of that book. The *Tractatus* did not allow consciousness, the self, the subject a place in the outer world; but it did allow it a marginal existence as a 'limit':

> 5.632 The subject does not belong to the world: rather, it is the limit of the world.

> 5.641 What brings the self into philosophy is the fact that 'the world is my world'.

> The philosophical self is not the human being, not the human body, or the human soul, with which psychology deals, but rather the metaphysical subject, the limit of the world – not a part of it.[17]

For the later Wittgenstein, the *Tractatus* was wrong to allow even this marginal metaphysical subject-as-limit-to-the-world. More importantly, it was wrong because it had failed fully to incorporate Frege's message about signs: that they have no sense in isolation, only in the context of a sentence. The Wittgenstein of the *Philosophical Investigations* had fully assimilated the Fregean insight that signs made sense only as part of a complete sentence; and their sense was the use to which they were put. These uses were different in kind: not all words were used as names, for example. Consider the different uses of 'the', 'boy', 'and', 'walked' and 'home' in the sentence 'The boy and the dog walked home'. And the use to which 'and' was put was clearly dependent upon the other words with which it was working to convey the sense; this was no less true, though perhaps less obviously so, of the other words such as 'boy'. A name cannot name except in the context of an assertion.

The later Wittgenstein went further than Frege, however, in seeking the sources of the sense of words beyond the completed sentence and other well-formed formulae of the sign system. Language was part of life, and sentences made sense only in the context of a life: as Wittgenstein famously said, 'an expression has meaning only in the stream of life'; it was not sufficient for an expression to be in the stream of expressions, or the stream of sentences.

There seem, therefore, to be two opposing views emerging from the writing of the later Wittgenstein: that the concept had primacy over consciousness – 'Essence is grammar'; and that consciousness, or life, had primacy over the concept – languages being part of life-worlds. The opposition could, however, be read as a convergence of two views, of concept philosophies and consciousness philosophies: yes, essence is grammar; but grammar cannot be understood in isolation from life, separate from the experience of speakers. Innis (op. cit.,

p. xii) has neatly encapsulated this convergence within Wittgenstein and between Wittgenstein and Husserl by speaking of the former's notion of the 'world' (analogous to Husserl's 'life-world') as 'a complex matrix of referential totalities' and Brand complements this by describing Wittgenstein as 'a phenomenologist of the life-world'. Innis also emphasises the fundamental shared interest of the two thinkers in *expressions*:

> what both men emphasized . . . was that one of the central keys to the structure of the mind, and of the self, was a reflection upon *expressions*. Expressions became the mediating term of access not just to the objective realm, to the various domains of cognitive, practical and affective life, but also to the subjective realm, the domain of the source, of the self as limit, to, in Husserl's term, world-constituting subjectivity.
>
> (Innis, p. ix)

Moreover, as Innis points out, Husserl's philosophy of language, of the concept, reached into the heart of his philosophy of consciousness:

> The operations of subjectivity were already thematized by Husserl in *Logical Investigations* along the model of expressions. One went from the expression which manifested its own objective logical grammar, to the lived *act* which generated it, which was its origin. Thus, in this sense, language – the realm of the said – became the transcendental clue to subjectivity.
>
> (ibid.)

And Wittgenstein, too, was aware that grammar was not a terminus of inquiry. For him, questions of grammar ultimately opened, as we have said already, on to the 'form of life'. But, in addition, there were very specific questions about signs, that cried out for 'consciousness' as an answer:

> Every sign *by itself* seems dead. *What* gives it life? – In use it is *alive*. Is life breathed into it there? – Or is the *use* its life?
>
> (*Philosophical Investigations*, paragraph 432)

This intimate relationship between signs and consciousness – consciousness as the life of a sign, as well as the sign as 'shaping' consciousness, mediating the encounter of consciousness with itself – was one of Husserl's most enduring preoccupations.[18] He was acutely aware of the implication of the fact that any event or object may or may not be a sign, depending upon whether or not it was animated by an interpreter using it as a sign. Wittgenstein, perhaps, should have dwelt more upon this: that language *has to be made into language* by conscious beings; that the material basis of discourse requires something else to make it live as the bearer of intended meanings.[19] Perhaps he shied away from making too much of the role of consciousness because of his aversion to making consciousness itself a theme: it was too fertile of what he considered to be pseudo-problems.

INTEGRATING THE PHILOSOPHIES OF CONSCIOUSNESS AND THE CONCEPT

Neither philosophies of the concept nor philosophies of consciousness are sufficient in themselves: each leads to a cul-de-sac, even though there may be much incidental illumination. Let us begin by rehearsing the different ways in which philosophies that focus on individual consciousness but ignore concepts fail; then address the inadequacies in philosophies that focus on concepts but ignore individual consciousness; and finally look forward to a philosophy that takes account of both consciousness and concepts.

First, then, philosophies of consciousness. These typically approach consciousness either atomistically (building up mind and the world it knows out of elements such as sensations, perceptions and 'ideas' – as in the English empirical tradition)[20] or holistically, when consciousness is seen to be given as a whole, even if a stratified whole (as in the Continental phenomenological tradition). Each approach encounters problems which the other seems to solve: atomistic consciousness does not seem to offer any means whereby experience adds up to a complete person acting meaningfully in a synthetically unified world; while holistic consciousness (for example, the *pour-soi* of Sartre's existential phenomenology) does not seem to have a means by which it engages particulars, in order that the world it reveals should have specific, individual features. It is problems they have in common that we are concerned about here, however. The most important of these is that they seem to have difficulty explaining how individual consciousnesses gain access to, or have any sense of, a public, shared, objective world. If we think of that shared world as, at the very least, intersubjective, even if it is not considered to be objective, then knowledge of it has to be mediated by concepts: we share in the same world not because we share sensations (we cannot), but because we share concepts. We need to have some kind of regulatory mechanism to ensure that we form the same concepts out of our experience. It seems as if consciousness needs to have concepts, or a mechanism for forming them, built into it from the start. It is difficult to see how this can

be ensured if we begin with the individual consciousness, irrespective of whether that consciousness is given as a Continental whole
(for example, the *pour-soi*) or built up piecemeal (along British Empiricist lines). The accidents and vagaries of experiences and of our
ways of integrating and synthesising them seem unlikely to be the
basis for forming common concepts without there being some kind of
prior regulating influence – a pre-formed conceptual mediation. As
Coleridge pointed out, against the empiricist account of the mind,
association of ideas (as a result of conjunctions or similarities of experiences), would not easily be demarcated from delirium. And modern genetic psychology offers nothing to explain how the (relatively)
tidy order of the common conceptual schema through which we participate in a common world could be unpacked from the untidy pell-
mell of experience. There is no problem, of course, so long as we are
blessed with innate general ideas – implanted by God (as in Cartesian
linguistics) or simply built into the structure of the mind (as in
Chomskeian neo-Cartesian linguistics); but the notion of innate ideas
presupposes an essentialism that few would subscribe to. For a start,
it goes counter to the recognition that most, perhaps all, ideas evolve
historically and are influenced by accidents of circumstance: that the
emergence of ideas in the collective consciousness seems to *require*
the accidents of experience. An alternative essentialism, a doctrine of
natural kinds – in virtue of which we all form the same concepts because we are all exposed to the same natural kinds which our concepts passively mirror – likewise does not address the temporal and
geographical relativity of concepts, nor the randomness of our experiences, nor the fact that many of our concepts are abstract or, at any
rate, cannot be related to recurring experiences occasioned by recurrent exposure to those objects.

It seems, then, that we cannot begin with consciousness – as a kind
of general possibility of experience – and proceed to build up concepts from this. We are obliged, therefore, to appeal to ready-made
concepts. It is natural under such circumstance to look to the stabilising effect of language as a guarantee that we shall all be inducted
into the same concepts, and so into a common, intersubjective, collective, world. Unfortunately, this is a ploy that has to be taken radically if it is to deliver what we require; in which case, consciousness
becomes entirely displaced or pre-empted by language, as a result of
which we fall into the realm of the philosophies of concepts. Otherwise, it turns out to be less of a solution to, than a displacement of,
the problem. For if the stabilising effect of language is not rooted in
the priority of concepts over consciousness, then we are again faced
with the difficulty of accounting for the fact that we all subscribe to
the same associations between linguistic terms and (unmediated) ex-

perience. How do we acquire the same idea of 'today', of 'green' of 'economic trends' as other people? The exposure to both the linguistic terms and the corresponding experiences seems too haphazard to be a reliable way of ensuring that we all subscribe to the same world, or to worlds organised in a sufficiently similar way for us to be able to participate in shared lives and undertake collective ventures. The Chomskeian near-mystical, nativist idea of a Language Acquisition Device built into the brain, enabling us to extract the rules of language from rather degraded material served up to us, if it solves anything solves only the genetic syntactic problem by which we acquire grammar, not the genetic semantic problem of concept acquisition. Any further claim for the Language Acquisition Device would require innatist assumptions about the brain/mind and an essentialist view of concepts that, we have already noted, would be at odds with the observed cultural and historical relativity of many of them.

So a philosophy that begins with consciousness – understood as the moment of experience or a succession of moments of individual experience – has great difficulty with concepts and, *a fortiori*, with the notion of a common, shared world accessed, suffered and enjoyed by individual consciousnesses as their scene and setting. Some philosophies of consciousness simply ignore this problem and even, it seems, the shared world: the most spectacular case is Sartrean existentialism whose *pour-soi* (specifically targeted for abuse by the structuralist conceptual philosophies) shapes, by uncovering, the world it encounters. It does this, moreover, whole worlds at a time; worlds are revealed to it as totalities. Although, for Sartre, consciousness does relate to particulars – the *pour-soi* is the *néant* that posits the existence of particular objects, of instances of the *en-soi*, as those things which it is not – it reveals them as parts of a world to which they, in an indissolubly spatial and instrumental sense, belong.

What, then, of the alternative – the philosophies of the concept? They have an immediate appeal because consciousness does, after all, seem to be conceptualised all the way through: concepts – rather than some elusive asemic, concept-free awareness or 'moments of experience' – seem to lie at the root of consciousness. To some, most notably the post-Saussureans, concepts (or the system from which concepts derive their signifying values) are the essence of consciousness, its ultimate reality, so much so that consciousness understood as 'personal', 'individualised' experiences, is a secondary rather than a primary notion. The individual experience, the lived moment, concedes primacy to the concepts which shapes it; conceptual form has priority over phenomenal content; the particular consciousness, the self, the individual viewpoint, runs off into general concepts, formations, schemata, horizons, which belong not to the individual, but to the systems of which she is part, though she has an incomplete, or no, inkling of it.

At one level, a de-individualised philosophy of the concept does seems to capture some truths about us and about the world of which we are a part and to which we are addressed. There is much evidence to suggest that once we rise above the level of sensation to perception, we theorise through the mediation of concepts.[21] Concept-free perception is seen only in pathological conditions such as agnosia, where sensations fail to have meaning. When we see not a mere pattern of light and dark but a 'cat', we are (at least according to one interpretation) the beneficiaries of an interaction between concepts and sensations. In short, our experience, at all levels, is conceptualised and, since the impersonal system of language is deeply involved in this process, we must concede that 'personal' consciousness is socialised, in short, 'impersonalised'. Much of our thinking – and, deeper than this, our very self-presence – is mediated through language. Our thoughts about the world we are in, right down to the fine detail of very local preoccupations, are shaped by the set of possibilities that have been linguistically made available to us. There is hardly an aspect of ourselves that has not come from the Great Elsewhere of language. What we are is who we are telling ourselves we are. And the language through which we disclose ourselves to ourselves is not a customised idiolect (or only superficially so, and on special occasions) but made of ready-to-wears, awaiting anyone's use. As we talk to ourselves, in the most private moments of self-communion, we enter generality.

In the light of these observations, the philosophies that give priority to concepts over moments of consciousness seem not merely reasonable but self-evident. Even the structuralist vision – according to which the most intense self-presence is the presence-absence of any-one-to-an-anyone occupying certain available positions in the matrix of the system – seems almost plausible. The sense of self is a sense of the other; or the general Other (to use Mead's phrase) instantiated. Concepts are not merely added on to experience, as a second-order accretion that gets thicker as we get older and move from infancy to university: they are the very stuff of lived human reality.

Just as the philosophies of consciousness run the danger of isolating the individual from the collective and the natural world – from shared, 'objective' reality – philosophies of concepts run the risk of doing away with the individual altogether. I come upon myself through a system of signs in which I am absent: 'it thinks, therefore I am not'. Philosophies of the concept fail to take adequate notice of things that lie at the heart of the experience of being a human: my continuous sense of being *this thing now*, of being rooted in the particular (world, place, body), of having some responsibility for the thing that I am, of my constant presupposition of a viewpoint corresponding to the implicit assumption that I am THIS.

Admittedly, the ego may elude itself when it tries to find something that is uniquely its own, a thought that is cast in its own idiolect, or (to address the problem from the direction that Hume approached it) when it seeks out a self that goes beyond the succession of its individual experiences:

> For my part, when I enter most intimately into what I call *myself*, I always stumble on some particular perception or other, of heat or cold, light or shade, love or hatred, pain or pleasure. I can never catch *myself* at any time without a perception.[22]

But it is equally true that the condition of being (pre-conceptually) this thing is inescapable, particularly when it is a matter of suffering pain, of feeling guilt or personal responsibility, or of fearing some future, such as death. If the dream-like intangible self of the empiricist philosopher is inadequate to the reality of inescapable and inescapably personal destiny – the solid being-here of this thing – how much more inadequate is the absent entity of the system-self.

The philosophies of the concept are seen to overlook, and cannot handle, the real presence of moment-to-moment consciousness, just as the philosophies of consciousness find it hard to deal with generality, with the massive, invisible, conceptual, partly systematic, context of all experience, especially (but not exclusively) when the latter is mediated through discourse.

As I have presented it, the fundamental conflict in mainstream Western, twentieth-century, professional philosophy is not between Anglo-Saxon and Continental approaches, exemplifed in, respectively, the linguistic and analytical traditions on the one hand and several varieties of phenomenology and post-Saussurean thought on the other, but between philosophies that give primacy to the immediate data of consciousness and those that give this primacy to concepts and the sign systems to which they belong. This battle has been fought with equal vigour both sides of the English Channel and both English-speaking sides of the Atlantic Ocean. Benveniste is closer to Wittgenstein on the nature of the self, the ego, the 'I', than to Sartre; Barthes' denial of the reality of the author ('the ghost in the text') is closer to Ryle exorcising 'the ghost in the machine' than to Husserl; the structuralists are closer to the English and American Fregeans in their approach to matters metaphysical and in their attitude to metaphysics than they are to Heidegger.

This is not quite as trivial a point as may seem at first sight. For it suggests that the divisions in philosophy are not as incurable as some have thought; that they are not a question of national or cultural styles; and that individual philosophers may not be as inescapably imprisoned in epistemes as is traditionally assumed. More excitingly, it suggests an approach to philosophical truth that may transcend the familiar party-lines. It raises the possibility that philosophies of consciousness and philosophies of the concept may be complementary aspects of a higher – or at least more satisfying and more interesting – philosophy of the real. Let us look at this possibility through a brief glimpse at language and speech-acts. (The discussion that follows is heavily dependent on arguments and theories – of universals, of reference – that I have earlier presented in *Not Saussure*.)

Structuralism is a philosophy of the concept *par excellence*. The moment of consciousness is subordinated to the system to which the signs utilised by, and which utilise, the conscious being belong. The essential notion is that consciousness, in so far as it is meaningful – and it is inescapably caught up in meaning – is mediated by signs, and those signs derive their value from differences, which belong to the system. Structuralists give priority to the structure over the event, to the form over content, and so elide, or ignore, or deny the content, of the individual consciousness or agent. This applies, as we have already noted, even in the case of what is seemingly the most deliberate and conscious set of acts: speech-acts. The post-structuralists take this further; so far as Derrida is concerned, it is the absent system, all those other elements that the linguistic unit is not, that give it its status as a linguistic unit, its value. Its presence is in fact fashioned out of an absence. In the case of a sentence, or a stretch of discourse, it is the boundless context that provides the discourse with its meaning. Its presence, too, is fashioned out of an absence. The value, the concept, reigns supreme; the present individual does not really have presence; he/she is deposited in concepts, in signs, and so, through them, in the absent system. He/she exists in so far as he/she is spoken by language or, more generally, discursive systems.

There are two major faults in this analysis of language and of the model of the relationship between the individual as agent and his/her acts. First, it overlooks the fact that the sign, an uttered or written token, has a material presence and as such is manufactured by, and originates from an individual, from a material body located at a particular point in time. While *langue* is a locationless, boundless system, occupying no point in space or time and unrelated to an individual understood as a literal 'point of view', *parole*, composed of actual speech-acts that utilise the system of *langue*, are precisely located. Whilst it is true to say that nobody speaks the *langue*, everybody speaks the *paroles* that use the *langue*. Even if it is true, as post-Saussureans claim, that *langue* speaks us (rather than vice versa), it is certainly not true to say that we are spoken by *paroles*. Actual speech-acts occur at particular points in time and are generated by individual agents. When it comes to real speech-acts (and they are real in the sense that *langue*, a purely notional system, is not), the conscious agent is to the fore. The concept does not dominate over, or elide, consciousness.

Secondly, it fails to notice that the meaning of a speech-act depends upon what is meant by it – or what it means to someone; and this requires both the producer and the recipient of the meaning to be conscious. The transmission and reception of meaning both correspond to moments of consciousness. And quite complex moments, too. They include, for the recipient (seemingly the minor partner) the following: receiving the sensations corresponding to the sounds; recognising the sounds as examples of certain word-forms; and bringing the word-forms together with one another and with an awareness of their source and their context in order to generate a semantic interpretation. As Grice pointed out, receiving the meaning of an utterance depends upon decoding what one feels the other means; at the very least, this requires treating the other as if she were an agent producing meaning. You do not mean anything unless I see what you mean; more precisely what I take it that you mean to mean. Ordinary linguistic exchanges depend upon an understanding generated between two interacting consciousnesses reading each other.

The ultimate specification of the meaning of an utterance will depend on an actual, extra-linguistic context; and, where meaning is meant to mediate, and to be mediated by, reference to a particular, to a singular (as it so often is), that context will be literally the spatio-temporal setting of the individual. Structuralist, and conceptualist, accounts of discourse forget the implicit or explicit deixis that is crucial to so much ordinary discourse. Pronouns such as 'I' unite the existential with the grammatical: the grammar of 'I' is not all grammar. And the same applies to a vast number of words that have the property of being deictic shifters. These have definite reference only when a universe of discourse is specified and its specification will be mediated through the co-presence of two conscious individuals in the same (token-) place.

The meaning and reference of *paroles* thus require consciousness, irrespective of what *langue* or the system needs. A philosophy of the concept is therefore inadequate even to that human activity which is seemingly made most purely of concepts, in which humans are apparently most completely given over to concepts – namely, linguistic acts or, more generally, consciously signifying behaviour.

However, consciousness cannot itself being adequately characterised aconceptually. We have already noted how conceptual generality permeates consciousness at every level above brute sensation. Our formed experiences refer beyond themselves to a network of experiences which in turn make sense through being conceptualised. The close relationship between language and ordinary perception is dramatically illustrated by cases of brain-injured individuals in whom language centres are disconnected from higher sensory processing, when there is a failure of recognition of objects, which are also denuded of the aura of familiarity. Just as there can be no fully formed communicable concepts without consciousness, so there can be no fully developed consciousness without concepts. This is, of course, particularly obvious at the 'higher' levels of consciousness, such as those where the sense of self, the idea of others, notions about the world, political, religious and other beliefs are formed and found. Our most personal and deepest convictions have something deeply impersonal (even off-the-shelf) about them. They may not belong to a system, but they have been derived from a system (even if that system is ragged, incomplete, notional); or at least they have origins and relations that are incompletely revealed to us. There is a systematic, structural, formal darkness at the heart of the content of consciousness.

This is, not surprisingly, especially evident in relation to our philosophical opinions. The words we employ to think about philosophical matters cannot be used in isolation: they, too, belong to a nexus of language that is steeped in unchosen connotations. There are rules governing their correct and incorrect use. A word has meaning only as part of a well-formed sentence, as Frege pointed out; and 'a proposition has meaning only in the stream of life', as Wittgenstein said. That is why seemingly isolated introspection and 'studying this headache I have now' may not reveal answers to epistemological problems or the relationship of the mind to the body; for the discourse one mobilises to study, i.e. to think about or articulately introspect upon, the headache will carry residues of meaning that are not made fully explicit; and the rules for proceeding from introspecting about headaches to drawing philosophical conclusions are not as transparent as we would like – as the difficult and unresolved discussions of *cogito ergo sum* have revealed. The ideal of a transparent philosophical discourse in which the thinker thinks first only to herself loses most of its appeal or credibility once it is appreciated that the closer that discourse approximates transparency, the less meaning it can carry, and the more completely it is condemned to generate only tautologies. This was the intuition that lay behind Wittgenstein's assertion that mathematical and logical propositions were tautologies.

The conflict between the philosophies of the concept and the philosophies of consciousness is not, it will have become apparent, merely a methodological issue. Nor is it even a question of the division of labour between philosophers on the one hand and, on the other, linguists, cultural historians, anthropologists and psychologists, whose primary concerns may be thought to be with 'the concept', understood as signs and systems of signs. It is about the nature of human reality, of what it is to be here, about the kind of creatures we are. In particular, it is about the extent to which we are available to, visible to, transparent to, ourselves; and about the degree to which we may consider our actions (or the actions attributed to us by others including ourselves) as emanating from us as deliberate agents; about whether or how much we are our own people. At one extreme, we have a philosopher of consciousness such as Sartre, for whom the *pour-soi* is utterly transparent and chooses not only its own actions but posits the world into which those actions are inserted and from which they derive their meaning and significance. At the other, we find post-Saussurean philosophers for whom the ('de-centred') self is merely a site where sign systems, discursive practices, historical and social forces (largely hidden from the self) operate and have local instantiation.

Each of these two visions emphasises one of two inseparable and contrasting aspects of human consciousness: its particularity and its generality. Every consciousness is a unique, irreplaceable centre, referring back to and (through the mediation of its body and the material world in which that body is located) caring for, itself: it can be seen as a point of view located in a particular point of space-time, sampling and being sampled by, and unfolding as, a particular stretch of space-time, a singular world-line. At the same time, each consciousness is pervaded throughout by the general. What we perceive and how we perceive it, our most public and our most intimate thoughts, are utterly specific to us and at the same time instantiations of something common to many. Out of the generality of consciousness arise classified objects, intersubjectivity, communication and the common theatre of action called the shared world. And yet that consciousness is referred back to a unique individual – one that has a unique and monopolous relationship to a given body and to these (token-) thoughts. Consciousness uniquely gives itself over to a public, common reality of signs and objects and projects and collective actions. This dichotomy – between uniqueness and universality – is replicated at many levels; the recognition of the opposition within the particular between its existential singularity and its status as an instance of a type, and the philosophical problem of universals, both reflect the paradox of the intimate, personal, generality of human experience.

In the last analysis, then, the conflict between philosophies that focus on phenomena of consciousness, such as sensations and 'personal' memories and thoughts, and those that take the concepts and sign systems of the collective as central, is absurd. The two types of philosophy simply address separately the two inseparable faces of experience. Concepts cannot exist outside of consciousness, nor social concepts outside of socialised consciousness; and (human) consciousness cannot exist totally outside of concepts. Experiences are both particular and general: about, for and to a singular someone and a singular something and yet cast in universal terms. We 'come to' our unique selves through generality: we are at the same time instances of generalities and also irreplaceable singulars. My perceptions – in which concepts are at least inchoate, though they are resorbed back into the particular object (to see is to theorise, as Goethe said) – and my thoughts, which belong to a series, an ambience, a context which is uniquely mine and are unique inasmuch as they contextualise *me*, are both general and individual, and owe their existence to association with me and their meaning to their place in the conceptual scheme. At every level at which consciousness makes sense, it points up a generality that exceeds the moment of consciousness and, indeed, the conscious individual. The view from that uniqueness, looking outwards from within ourselves, corresponds to the Cartesian or Sartrean philosophy of consciousness; while the view that looks from without, the objective view, sees consciousness as participating, even in its most intimate and subjective aspects, in things that go beyond it. This is an intuition dimly glimpsed in 'the holism of the mental', which acknowledges that mental contents cannot be specified or even have reference to an external object without reference to a grid of concepts.

The philosophy of the future will take into account both aspects of human experience and will be neither a philosophy of consciousness, in the sense that we have known hitherto in, say, phenomenology, nor a philosophy of the concept, as exemplified by structuralist thought or linguistic philosophy, but something that unites both approaches and reconciles both visions. Such a philosophy will be animated by the tension between the general and the particular and (not quite the same thing) between the collective and the individual, in human experience. This is a tension that takes its extreme forms in, on the one hand, solipsist subjectivity and, on the other, the asymptote of objectivity that Thomas Nagel dubbed 'The View from Nowhere'.[23] The philosophy of the future will be founded upon a recognition of the great mystery of the self-revealing generality of the human mind – its ability to see classes as well as individuals and to map (and so encounter) itself upon an network of types, and ultimately to see itself as a mere instance, though it is all-in-all to itself. Such a philosophy will mediate between the founding Cartesian tradition captured in 'I think therefore I am' and the modern pessimism of the de-centred self captured in 'It thinks, therefore I am not'.[24]

NOTE

I am aware that, as a history of the dialectic between philosophies of consciousness and philosophies of the concept, the foregoing is far from complete. Like many other aspects of intellectual history, this quarrel has many 'cunning passages'.

For example, Russell's opposition to psychologism – which he first learned from F.H. Bradley rather than Frege – was not unwavering, as instanced by his notorious attempt to relate, or even reduce, the meaning of logical operators such as 'or' to the sense of hesitation. (Of course, logical operators operate irrespective of the content of consciousness – as evinced by a calculator working with truth tables.)

Nor were even the early phenomenologists as prone to naive psychologism as is generally supposed. The Polish logician Twardowski, taught by Brentano, emphasised the distinction between what is meant or intended by a mental act – its objective noema or noematic 'intentional object' – from its corresponding noematic *meaning* or subjective 'content', the correlated characteristic or structure by which it 'intends' its 'object' or 'objective' – i.e. *means that* such-and-such (is so). This differentiates, that is to say, the process from the product of meaning, its content from its object. Twardowski's position may be seen as standing at the crossroads between psychologism and anti-psychologism: the former emphasised content and tried to reduce logic and meaning to mental processes; and the latter emphasised the object and tried to eliminate the processes and mental contents.

Finally, the more recent arguments around the question of whether meanings are or are not in the head may be regarded as a re-run of the battle between psychologism and anti-psychologism. Those who argue for the 'holism of meaning', and consequently argue that there are many extra-mental determinants of what we think we means, are advocating a philosophy of the concept.

In summary, I have only touched upon a small part of what has been perhaps one of the most potent, if often unacknowledged, sources of dissent in the philosophy of the last hundred years or so.

AN AFTERTHOUGHT ON THINKING ABOUT CONSCIOUSNESS

Before one talks about consciousness, one should be astonished at it – to the point of being embarrassed to talk about it. Against a background of continuous superficial sense, it is difficult to experience true astonishment and thus to be persuaded of the absurd insufficiency of the available accounts of consciousness. Truly to contemplate one's own consciousness is to stand on the threshold of ecstasy and terror; or at least of unease and delight. The unease and terror makes even a starkly materialistic vision of consciousness paradoxically comforting: it provides a refuge from the utterly unexplained – even at the cost (so eloquently described in Thomas Nagel's *The View from Nowhere*) of embedding subjectivity itself in a great ocean of objective reality. The appropriate state in which to contemplate consciousness – that in virtue of which matter matters – is vertigo; spinning down a maelstrom of amazement, one should be speechless. Such speech as one is capable of should be mere crystallisation of broken sentences out of a supersaturated solution of amazement.

The problem of consciousness, however, has the habit of becoming far too respectable for what is an essentially metaphysical question – for one of the fundamental questions. The fundamental questions, George Wald said, are the questions a child asks and, receiving no answers, stops asking. Supreme amongst them are these three: Why is there something rather than nothing? Why is there life? Why is there consciousness? These are questions that can be recited without really being asked. They cannot usually be really asked at will; more often, if they occur at all, they descend upon one. They tend to do so for the first time in early adolescence – as so memorably described by Richard Hughes in *A High Wind in Jamaica* – as a state of philosophical grace, and then to recur with diminishing frequency and intensity thereafter.

In secular circles, the questions about the origin of life and the origin of consciousness are more respectable than the one about the origin of existence. Not even the new cosmogony that is reaching towards the very beginning dares to address the question as to why there should be the initial singularity out of which everything – matter and its laws – unfolded. 'Why is there something rather than nothing?' seems to invite an answer that invokes the ultimate purposes of a Mega-Someone who somehow managed atemporally and spacelessly to pre-exist existence itself. The intentions of the inscrutable Creator have been of little interest of late to critical minds all too aware that what we think we know of them has come to us from other men. Hence, perhaps, the unpopularity of the question 'Why is there something rather than nothing?'

In contrast, the question of the origin of life promises to yield to a scientific approach. The fundamental assumption is that living forms can be derived from non-living matter through the operation of laws that apply to the latter; life is a subdivision of matter and a particular expression of its properties. This assumption has encountered some unforeseen difficulties. The laws that determine the properties of matter such that it could, eventually, sustain life seem intrinsically improbable. For example, the element carbon, the basis of all living matter, seems to have a nearly zero probability of emerging from the conditions that prevailed in the early, decisive seconds of the universe. The various forms of the Anthropic Principle[25] have not mitigated the sense of the improbability surrounding the emergence of the elements, such as carbon, essential for life. And even if we overlook this problem, and take carbon for granted, bioscientists have travelled a long way into confusion and despair from the heady days in the 1950s when it was thought that amino acids synthesised by electric storms passing through ammonia and carbon dioxide could eventually give rise even to the self-assembling miracles of organisation we call primitive organisms. Despite these substantial difficulties, it is still assumed that, once the basic laws of matter have been been determined, it will be only a question of time before the peculiar parishes of matter called life are fully understood and their emergence seen as inevitable. So the explanation of life itself is almost respectable.

The question of the origin of consciousness is, alas, even more so. It is now often assumed without question to be a problem for biologists (and hence physicists and chemists); to be, like life, an emergent property of matter evolving according to physical laws. The emergence of consciousness out of matter according to physical laws is simply helped along by the competitive interaction between forms of living matter.[26]

My own view is that the origin of human consciousness, far from being a respectable theme for biological or physical investigation, should be regarded as being questionable a topic for scientific investigation as the emergence of existence itself. More generally, the sequence should not be thought to portray a succession of events of lessening mystery. Once the uselessness, the unoccasionedness of consciousness is grasped, we shall see that it is totally mysterious.

NOTHING → EXISTENCE → LIFE → CONSCIOUS LIFE

or

BIG BANG → MATERIAL WORLD → LIVING MATTER →
CONSCIOUS LIFE

There are many ways of extinguishing astonishment at our own consciousness, our conscious existence. Bodily suffering, fear and concern for our own existence, the routines of daily labour and daily intercourse, the pursuit of ambition and appetite, are, of course, the most universal and continuous and are usually sufficiently potent to wipe out any deep excitement about ourselves. When they fail and we are in danger of realising *that we are* and *that we are explicitly*, off-the-shelf answers may be provided by religion, which neutralises questions even in advance of our asking them, bringing with it new fears, new concerns, new duties and new appetites. If religion fails us, then there are secular answers which shoot down with lead pellets of quasi-explanation the rising lark song of amazement and joy. Foremost amongst these is contemporary materialist thought, of which evolutionary theory is a crucial element, that places a horizon around our meditations on our own nature.[27] Evolution can seem like an explanation of consciousness – its origin and its nature – only when consciousness is reduced to a localised mystery – such as, for example, a property of the brain.

To recognise that evolution – the deepest and widest attempt to describe and explain our place in the universe – does not capture us, is a gain. To suffer loss of explanation is also, as I have said, deeply disturbing. To reflect upon our consciousness, to turn our thoughts upon that within ourselves that makes it possible to think, is to court a peculiarly terrifying vertigo. However, the sense of the mysteriousness of consciousness is exhilarating and liberates us from the dead feeling of the obvious that often pervades the daily life of which we are conscious. Or should do, anyway.

Epilogue

The Conscious Agent and the Hope of Progress

In 'The Apotheosis of the Romantic Will', Isaiah Berlin, while recognising the horrors let loose upon the world by 'the extravagances of romantic irrationalism', concedes that

> this at least may be set to its credit: that it has permanently shaken the faith in universal, objective truths in matters of conduct, in the possibility of a perfect and harmonious society, wholly free from conflict or injustice or oppression – a goal for which no sacrifice can be too great if men are ever to create Condorcet's reign of truth, happiness and virtue, bound 'by an indissoluble chain' – an ideal for which more human beings have, in our time, sacrificed themselves and others than, perhaps, for any other cause in human history.[1]

It is difficult to imagine a more savage indictment of the *philosophes* and their dreams of a better world brought about by the efforts of men and women of good will, motivated by a hatred of suffering and injustice, and guided by the light of a reason in which common sense joins hands with science. There are more charitable assessments of the legacy of the Enlightenment thinkers:

> The intellectual power, honesty, lucidity, courage, and disinterested love of the truth of the most gifted thinkers of the eighteenth century remain to this day without parallel. Their age is one of the best and most hopeful episodes in the life of mankind.[2]

What makes these two passages a particularly dramatic illustration of the ambivalence of contemporary attitudes towards the Enlightenment is that they come from a single pen – that of one of the most profound historians of ideas of our time. We cited these passages in the Introduction and, in a sense, they set the agenda for the subsequent discussion.

This ambivalence has extended beyond the Enlightenment programme for social reform and the amelioration of the human condition, to a radical questioning of the role of reason, and conscious human agency, in human affairs. Whilst much contemporary Counter-Enlightenment thought has simply recapitulated some of the arguments of the early opponents of the Enlightenment, twentieth-century attacks on the dreams of the *philosophes* have often been much more fundamental. The currently dominant version of anti-Enlightenment irrationalism asserts that all discourse undermines itself through a systematic indeterminacy of meaning or inescapable self-contradiction; and it embraces a deeply pessimistic vision of the absolute omnimpotence (sic) of individuals.

And this is all the more telling for originating within a secular framework: the contemporary Counter-Enlightenment cannot be dismissed as a mere rearguard action by those whose vested interests lay with the hierarchies underpinned by religion. When neo-Marxists and post-Freudians mock the would-be rationalism of human beings, they are not doing so on behalf of the superior wisdom of a Supreme Being, but on the basis of an analysis of what they see to be permanent in the human condition. And those ethologists and anthropologists who diagnose the ills of modernity as being due to the frustration of our instincts or our unmet need to assert collective identity through ritual, are doing so from within a scientific, or quasi-scientific, not a religious, framework. For them, Man is not a fallen being – uniquely sinful but also uniquely redeemable – but a particular part of the animal kingdom and, as such, subject to the laws that govern primate behaviour.[3] The radical anti-humanism of post-Saussurean and other contemporary Counter-Enlightenment thought has developed in the heart of secular humanism.

This may explain in part why the contemporary Counter-Enlightenment enjoys such wide currency. Far from being 'against the current' (to use the title of one of Berlin's books dealing largely with Counter-Enlightenment thinkers), hostility to the Enlightenment, and mockery of the hope of progress, is the current. I have already cited Fukuyama's rueful observation that

> pessimism has become something of a fashion, a kind of intellectual pose to demonstrate one's moral seriousness. The terrible experiences of this century have taught us that one never pays the price for being unduly gloomy, whereas naive optimists have been the object of ridicule.[4]

And I have noted the widespread assumption, emanating from the Global Village Explainers, that 'the grand narratives of emancipation and enlightenment' have disappeared. Even allowing for a good deal of hypocrisy – irrationalists expect and demand the everyday world to be regulated by contractual reason as strongly as rationalists do – and a lack of insight that prevents many post-modern thinkers from appreciating the extent to which the climate of assumptions in which they conduct their lives has been created by the *philosophes*, there can be no doubting the dominance of Counter-Enlightenment thought amongst humanist intellectuals. *Kulturpessimismus* and 'hysterical humanism' are the order of the day.

What the two passages from Berlin challenge us to do is to try to gain a clear view of the Enlightenment and its dream, most poignantly expressed by Condorcet, of 'a reign of truth, happiness and virtue', and of a gradual, if asymptotic, progression towards the perfection of life on earth: to try to see the object whole – as Arnold would say – and attempt to separate what is benign in the Enlightenment project from what is malign or at least dangerous. Since the *philosophes* wrote, we have learned that Utopian dreams – even, or especially, rationalistic ones – may license massacres; and that the net result of the actions of men and women of goodwill, inspired by revulsion at human suffering and anger at the injustices people suffer at the hands of nature and of humankind, may well be greater suffering and greater injustice – especially if those actions are orchestrated by ruthless and self-righteous opportunists justifying their means by ends which they themselves define.

Condorcet himself knew this only too well from personal experience; after all, his *Sketch for a Historical Picture of the Progress of the Human Mind*, in which he famously argued for the indefinite moral, intellectual and physical perfectibility of man, was completed while he was in flight from his erstwhile colleagues in the Convention. The mean-spirited like to focus on the bitter irony of the philosopher of hope, a believer in the essential goodness of mankind, being hunted to his death by his fellow-believers.[5] Those with more generous dispositions can only be moved at the idea of a man rising above his own suffering and composing, as his last will and testament, a document sounding the themes of equality, tolerance and liberty and proclaiming the potential capacity of man in the future infinitely to surpass what he has been in the past.

The challenge set by Berlin's 'Yes–But' to the Enlightenment is to separate and develop what is fundamentally decent in the Enlightenment ethos from the profound dangers of its Utopian dreams; to retain and foster its vision of justice and its hope of progress on the basis of the cooperative action of rational agents, informed by a passionate sense of fair play and justice, without thereby helping to create a monstrously oppressive society in which tyranny wears the mask of virtue and despotism sponsors persecution for the good of mankind.

The present book has tried to meet this challenge by critically examining two major elements in contemporary Counter-Enlightenment thought. The first is a form of pessimistic irrationalism which argues from biology, ethology and anthropology that mankind is still enmired in its collective past – in the past of our hominid forebears and, beyond this, in the yet more remote past of the primates from which we have evolved. This pessimism has many strands: that rational thought is no advance upon (and indeed may be a backward step from) primitive, pre-logical thought; that science no better than magic – either because science is simply the dominant magic of the modern era or because it focuses on means rather than ends, facts rather than values; that modern life, modern society and modern thought do not answer to our true needs so that material prosperity has brought with it spiritual emptiness; that knowledge has displaced something richer and more powerful than it – namely wisdom; and that, because of all of these things, and our lack of insight into them and into ourselves (we have forgotten our true – primitive, animal – selves and our true needs), we are in great danger of falling into a collective fatal tailspin. The catastrophes of the twentieth century will be the merest foretaste of the catastrophes awaiting us in the twenty-first century if we do not abandon the rationalist delusion that we are reasonable beings and that reason can continue to be our guide in public and private life. The second element is also pessimistic and irrationalistic and derives its inspiration from the belief that conscious human beings are subject to forces, of which they are largely unconscious, emanating from the systems or structures within which they find meaning and enact their lives. These beliefs reached their climax in the writings of the post-Saussurean thinkers, who have all but denied the reality of the rational human agent.

It has, as I have said, been the purpose of this book to criticise this irrationalist and pessimistic critique in order to make it possible to separate what is of enduring value in the Enlightenment project for the improvement of the lot of man from what is dangerous. A full-blown defence of the defensible in Enlightenment thought – its belief in the evidence of the senses, in the appeal to objective facts; in reason; in the analytical approach; in the autonomous self; and in the rejection of arguments from authority – is beyond the scope of this book and beyond the scope of the author. I should, however, like to end with some reflections on what I believe to be the Enlightenment's most important contribution to human thought – the hope of a better future for humanity brought about by the efforts of human beings, based upon the application of reason to human affairs, driven by a sense of the right of individuals to just treatment, and guided by an ever-more powerful understanding and control of the natural world derived from advances in science and technology. I shall address some of the concerns that still remain when the task that has been the central concern of this book – that of exposing the emptiness of contemporary, mainstream Counter-Enlightenment thought – has been accomplished.

My ambitions in this final chapter are modest. I shall not even attempt some of the most pressing questions that need to be addressed by others more qualified and more intelligent than myself. Among them I would include:

1. *Determining that level of government or judicial regulation necessary to maximise the extent to which scientific advance and material enrichment benefits mankind.* I shall not try to address questions such as: How much should the state interfere in the lives of individuals? What form should this interference take? What should be the relative proportions of direct (by edicts and regulations) and indirect (through, for example, education) interference? These questions now have an added urgency for liberals since the minimal statist thinkers seem to have had all the running in the United Kingdom and the United States since the 1980s. As Jeffrey Friedman notes:

> The minimal statist target is no longer a sitting duck, socialism, but the amorphous, immensely variable and ambiguously democratic, interventionist, redistributory state. Behind the terminological question of what to call minimal statism lies the dawning realization that its relevance is problematic, now that the fall of Communism has stripped away the comforting idea that opposing the latter meant embracing the former.[6]

2. *Determining the point at which the application of scientific knowledge and scientific methods to human affairs passes over from science to scientism.* The botched attempts to implement 'scientific' socialism – Marxist, Saint-Simonian, Comtean, Maoist – have almost completely discredited the very notion of a scientific approach to social reform. It has become synonymous with the arrogance and insensitivity of 'experts' whose reach exceeds their grasp and whose inevitable failures have wreaked such havoc as will take centuries to repair. The question boils down to the size of the pieces that are likely to lead to successful 'piecemeal' social engineering. As was noted in the Introduction, the advent of dynamical system theory, which emphasises the non-linearity of the relationship between input and output in complex systems and thus the partial unpredictability of the outcome of interventions, has provided a further weapon for those who would attack 'the fatal conceit' of holistic social engineers. It is important, however, that revulsion against the 'fatal conceit' of untrammelled interventionism should not license 'the fatalistic conceit' of those who would deny the value of all interventions, and urge us to trust to custom and tradition to generate the spontaneous order that ignorance of true social forces prevents us from shaping ourselves. Defining the boundary between arrogant and excessive intervention and what is effectively

paralysis or passivity before the evils of the world is a task that will have to be attempted again and again.

3. *Determining the appropriate boundary between the realm of public control and accountability and that of private choice.* In the twentieth century, totalitarianism has shown how the state can expropriate, or attempt to expropriate, most of the territory formerly assigned to the soul, the conscience and the self. The collapse first of Fascism and then of communism has not quite reversed these trends: the resurgence of fundamentalism and the restoration of theocracies are obvious examples of a continuing trend in some parts of the world towards politicising every aspect of private life. Even some seemingly progressive movements contain totalitarianising tendencies. After all, it was feminists, not Muslim fundamentalists, who emphasised that 'the personal is political', by which they meant not only that personal life was cradled in a larger political framework that determined its meanings, but that this should be recognised and exploited: the way to reform the personal sphere was through making explicit the politics regulating the assumptions implicit in that sphere. The question raised by John Stuart Mill's *On Liberty* – 'how to make the fitting adjustment between individual independence and social control' – remains unanswered.

These are crucial questions if only because one of the sharpest ironies of the history over the last two centuries has been that the Utopian dream of the 'withering away of the state' and 'the replacement of the government of people by the administration of things' has often resulted in ever more oppressive and obtrusive control of private individuals by the state or by those who presume to speak for the state. As Camus said, 'All modern revolutions have ended in a reinforcement of the power of the state'.[7]

WORRIES ABOUT UNIVERSALISM

The most persistent criticism of Enlightenment thought is that it is (arrogantly) universalistic. The charge is that, beneath the differences between individual thinkers, there was a central dogma to which they all subscribed:

the reality of natural law . . . of eternal principles by following which alone men could become wise, happy, virtuous and free. One set of universal and unalterable principles governed the world for theists, deists and atheists, for optimists and pessimists, puritans, primitivists and believers in progress and the richest fruits of science and culture; these laws governed inanimate and animate nature,

facts and events, means and ends, private life and public, all societies, epochs and civilisations; it was solely by departing from
them that men fell into crime, vice, misery.[8]

Although the Enlightenment thinkers were prepared to accept that,
since men lived in different circumstances, the same ends might be
achieved by different means, their cultural relativity stopped short at
relativising those ends themselves:

> they . . . retained a common core of conviction that the ultimate ends
> of all men at all times were, in effect, identical: all men sought the
> satisfaction of basic physical and biological needs, such as food,
> shelter, security and also peace, happiness, justice, the harmonious
> development of their natural faculties, truth, and, somewhat more
> vaguely, virtue, moral perfection and what the Romans called
> *humanitas*.
>
> (Berlin, p. 3)

It is this – and the related assumption that the *philosophes* and their
heirs know what is best for all mankind – that is credited with generating the deepest and most enduring resentment. And Berlin, in
common with many less temperate contemporary critics of the Enlightenment, seems to think that this resentment is justified: the assumption of universality inevitably leads to oppression; at the worst
the often violent imposition of modern European values – or eighteeth-
century Parisian values – upon the entire world.

This criticism of the Enlightenment has been, I believe, a serious
source of confusion; and that confusion is evident in the list of 'ultimate ends' that Berlin gives. They are actually a mixture of (a) basic
or universal *means* (to life) such as food and shelter, (b) arguably
universal ends, such as happiness, (c) possibly culturally-specific means,
such as justice, and (d) manifestly culture-specific ends, such as moral
perfection. We need to keep these separate if we are going to evaluate the Enlightenment justly and separate what is enduring and beneficial in its programmes and dreams from what is either of historical
interest only or an insolent and dangerous imposition of Eurocentric
or Parisocentric values upon the world. Keeping these things separate, while at the same time noting the interrelationships between them,
is essential if our argument about the Enlightenment – and, indeed,
the hope of progress – is not going become enmired in irresolvable
disputes about cultural relativism and ideological and axiological
colonialism.

Let us begin with groundfloor values, with means–ends pairings
that seem to be universals. Can we take it for granted that physiological

survival is, other things being equal, a basic desire? It would be seem, at first sight, that we can. Of course, there will be circumstances in which death will be preferable to life – where, for example, an individual is suffering so much pain that life is not worthwhile, or where there has been such loss of face or status that death is preferred to dishonour. We shall assume, however, that there are no cultures or societies where unbearable pain or insupportable dishonour are the norm. If there were such societies (present-day Rwanda may be approaching this condition) we would be inclined to regard them as gravely dysfunctional rather than as a serious challenge to the cultural universal of the desire to live.

We may think of the will to life, then, as a core value, invariant across cultures. If we accept this, then we may derive other cultural invariants; for example, the value of secure, adequate supplies of safe food and water. Ensuring the provision of safe food and water for all creates technological and logistic challenges that, historically, have been most successfully addressed by rational and scientific means rather than, say, by means of magic or prayer. Consider, for instance, the reliable and universal provision of clean water supplies. Historical experience would suggest that this essential requirement is met most effectively not, say, by 'wise men' preaching irrationalist doctrines of 'pollution' but by an approach based upon contemporary scientific understanding of microbiology, the physics and chemistry of pipes, etc., and also upon a clear understanding of the importance of individual accountability amongst those involved in ensuring the supply of water. Not all of this may be explicit – when I select the sort of pipes I am going to buy on behalf of the people on the basis of disinterested advice rather than 'backhanders', or on the basis of sound technical advice rather than the visions of the elders, I do not need to know much about the relevant metallurgy but this knowledge lies in the background of the consumer item in question – but it does need to be implicit. We can thus move very quickly from a small handful of incontestable value-universals to a vast nexus of value-imbued and ethics-imbued technical consequences which, themselves, acquire the status of secondary or consequential universals.

Sharing a universal wish that one should not die prematurely of thirst or diarrhoea already commits one to subscribing to a complex, subtle and boundless nexus of rational, technological and contractual thought and behaviour. Although some of those universals will seem to be purely technological, behind or implicit in them, or their application, will be other things that are not purely technological. It is no use taking on board the principles behind water purification, based upon a knowledge of bacteriology and toxicology, if these principles are not implemented by individuals who feel personally responsible for their own part in a very complex interlocking network of activities involving large numbers of people and agencies. Science-based provision of safe water supplies will not be possible in a world where the individuals employed by the water companies have a fatalistic attitude to life, do not value their work, do not have any conscience with regard to it, or are appointed and promoted on the basis of nepotism and/or bribes rather than on the basis of competence or genuine commitment. Buying into technology means buying into the collective and personal values necessary to make technology work.

We seem now to be well advanced in defending quite a lot of universal values, and thus allaying the fear that even a greatly reduced universalism deriving from, or implicit in, a reassertion of the Enlightenment hope of progress, would involve an oppressive value-colonialism. Before proceeding any further, however, we should go back and question our starting assumption: that there *is* universal consensus at the level of basic biological need originating from agreement about the value of life. Is life, or a life not dominated by pain and privation, an absolute value? Is the desire for such a life a cultural invariant? We noted the fact that there are, in certain cultures, circumstances in which death may be preferred to life, and dealt with this by suggesting that any society in which a preference for death or continuous pain was the norm should not be considered merely as having different values but rather as being pathological. De Maistre, however, has suggested that such cultures may not be all that unusual; on the contrary, they may themselves reveal the deep truth about mankind. As Isaiah Berlin expressed de Maistre's view:

> men's desire to immolate themselves is as fundamental as their desire for self-preservation or happiness.[9]

> The rational man seeks to maximise his pleasures, minimise his pain. But society is not an instrument for this at all. It rests on something much more elemental, on perpetual self-sacrifice, on the human tendency to immolate oneself to the family or the city, or the church or the state, with no thought of pleasure or profit, or the craving to offer oneself upon the altar of social solidarity, to suffer and die in order to preserve the continuity of hallowed forms of life.
>
> (ibid., p. 123)

There are several possible responses to this. The first is to repeat the point already made in the Prologue, that Maistre, by making such general statements about human nature, is committing precisely the essentialist, universalist error for which he so ferociously criticised the Enlightenment thinkers. His pessimistic view of mankind as irrationally committed to evil and self-immolation is no less universalist than the *philosophes'* vision of mankind as essentially rational and beneficent. The second is to note that, although humans do seem sometimes so to order their affairs as to increase rather than decrease the sum total of happiness, the suffering and death consequent upon human actions, even war, is frequently not anticipated even by the chief agents. The English and German Establishments who sponsored the First World War on a wave of patriotic excitement and moral indignation expected a rapid victory, not a protracted blood-and-mud bath in which they lost their own lives or those of their children. (Most of those who participated did not, of course, freely choose to do so.) Immolation, in short, is rarely a primary aim or value, but more frequently an undesired consequence of pursuing other aims or protecting other values. And finally, if it were a primary value, the question arises whether it should be simply accepted as such, as part of the rich pattern of life, a thread in the multicultural variousness of the world, or whether, on behalf of other universal values, it should be opposed. The answer must surely be that it should be opposed.[10]

The reason for this is worth examining in more detail because it opens on to the larger question of the relations between different values and different value-systems – of how, from a non-universalistic standpoint, from a point of view that does not claim transcendent authority, we can adjudicate between fundamentally different, but conflicting, values or value-systems. It may well indeed be the case that for some men, immolation is preferable to life and that the short path of slaughter and glory leading to the grave while crusading on behalf of some higher principle of honour – the glory of God, the glory of the nation, the glory of Glory – may be better than the long littleness of ordinary, comfortable existence. One can even imagine a society in which *all* adults subscribed to this view: Jonestown was a microcosmic example of such a society. Is not this *weltanschauung* entitled to its share of respect? There are two very specific reasons for answering this question firmly in the negative.

The first is the minor practical reason that the self-immolating sword wielded by such societies has to be forged by other societies – or other individuals within that society – who do not subscribe, or have in the past not subscribed to, this value-system. The humane methods of self-destruction, the life leading up to the climactic act of mass suicide, is parasitic upon the work of others who work within a rationalistic, contractual framework. Consider the terrible events at Jonestown when the Reverend Jim Jones persuaded all the several hundred members of his sect to kill themselves and their children. The deaths were brought about by poisoning with cyanide. Cyanide doesn't happen: it has to be made and bought. Expert chemists, transport systems, etc., were therefore necessary to bring about the realisation in Jonesville of this ultimate expression of irrationalism. This tends to be overlooked, just as, in a less dramatically horrible context, principled and articulated rejections of progressive, organised society presuppose a background of progressive, organised societies to be able to achieve the expression they need. Dependency upon a context of reason is true of all irrationalist movements and philosophies, and usually to a much greater degree than is true of that small number of them that, like the Jones cult, are committed to a climactic apocalyptic end.[11]

The second, more important, reason is that the sword that is wielded will not usually be directed exclusively inwards. The Jonesville massacre was untypical in this regard. More commonly, apocalyptic romantics do not confine themselves to their own society or their own circle; they do not bring the temple down exclusively upon the heads of the faithful. The Jungerian Storm of Steel was not an internal affair of the Club of Junkers. The active nihilism of the dispossessed and militaristic right-wing aristrocrats, obsessed by the lost honour of their Fatherland, who brought the Apotheosis of the Romantic Will to its realisation in Europe in the 1930s and 1940s of this century, had grievous consequences that were borne by others than themselves, by others who did not share either their preoccupations or their value systems. The freely chosen values of these apocalyptic romantics inspired the most appalling oppression, often unto death, of many millions of others whose freely own chosen values were thus trampled over.

There are two elements in this second argument. First, the values achieving predominance in a particular society may not only not be of material benefit to many members of that society, whatever moralistic or altruistic dress they wear (this is the Marxist point), and secondly, they may be explicitly not shared by all its members. Even if we take a small, closed community like the People's Temple commune in Jonestown, while it is not impossible that the mother gassing her six-month-old child during the collective immolation was doing it as a free agent, it is equally possible that she might have not been acting as a free agent. What is certain is that the six-month-old child was not acting as a free agent.

In larger communities, this disparity of values and wishes is much more evident. So a progressive, liberal commitment to multi-culturalism may well involve not only respecting the different values of a society other one's own, but incidentally result in colluding with the tyranny of the majority (or, not infrequently, a powerful minority) in that society. My wish not to judge what goes on in your country X by the values predominating in my country Y may mean that I refuse to pass judgement on the way the predominant values in X oppress some individuals in that country. The well-founded fear of being arrogant over values may give rise to a culpable quietism when others' values in other cultures are being trampled on. The judgement as to the point at which laudable tolerance gives way to culpable quietism depends upon two things: one's definition of a society as being other one's own and therefore outside the scope of one's moral or axiological jurisdiction (was Jonestown part of America or not? Who is my neighbour?); and, related to this, the sense that one has of the scope of one's responsibility. 'It's none of my business, ducky' may or may not be an admirable response to learning of actions that, although they contravene one's own value-system, are consistent with the values and assumptions of the society in which they take place.

The definition of the society or community to which one belongs is rather more difficult nowadays in a world that has – though to a lesser degree than some of our cultural prophets would claim – moved towards being a global, electronic village. Cultures intermesh through communication and trade. The sale of arms, the conduct of any trade, the provision of Overseas Aid – all of these directly or indirectly support or undermine the value-systems of the countries on both sides of the transactions. In profiting by the sale of arms to an evil regime, one becomes in some degree evil oneself – by sponsoring or supporting their evil. That is obvious. What is less obvious is that, in selling technology to a country that does not have the rationalistic value structure to enable that technology to run effectively (cf. the discussion about unpolluted water supplies above), one is being dishonest, even if the sale is not lubricated by backhanders funded from overseas aid. The non-fraudulent sale of technology presupposes some commonality of values – the values that will create the context in which the technology will work effectively.

In short, one cannot in the modern world avoid some universalism, if one is at all honest. It is implicit in one's dealings with other cultures, even if it not made explicit as a condition of those dealings. This is not always fully appreciated by those who preach a politically correct value relativism. It may be intellectually pure and morally self-satisfying to jet round the world pointing out that there cannot be universal individual rights because neither the idea of the individual nor that of rights is universal. But it may also be deeply hypocritical because those who are able to do this enjoy freedom of travel, wealth beyond the dreams of most individuals on the planet, and many other things that come from the idea of contractual rights. They would be outraged and inconsolable if they were deprived of them – gaoled, debagged, impoverished at the behest of a theocrat who disapproved on religious grounds of jet travel. And they might be rather anxious if they learned that the quality of maintenance of the engines in the jets were also subject to culturally relative variation and were

not subject to a universal agreement on contractual and personal responsibility and the superiority of rational science over magic.

We need some way, therefore, of working out how to separate benign from malign universalism: distinguishing between on the one
hand, extending to all human beings the status one automatically grants
one's own fellow citizens and, on the other, an arrogant imposition
upon other cultures of one's own culture-specific beliefs; separating
rational respect and concern from missionary fervour, moral superiority and patronising arrogance. For we can no longer be neutral as
to what goes on in other cultures, because we are so deeply implicated
in them as to make it difficult to define the boundaries of our own
culture and hence the boundaries of our responsibilities. This is not
merely a matter of knowledge – modern methods of newsgathering
and dis-semination prevent us from pleading ignorance about what
is going on in the wider world – but also a matter of action and
interaction. Whereas we may choose to regard tyranny in remote places
as being, unlike tyranny at home, none of our business, we are no
longer entitled to do so: it has everything to do with our business in
the narrow and the wider sense. Our choices directly or indirectly
support or undermine the value-systems in other cultures. And this
also has everything to do with our more obvious responsibility for
tyranny nearer to hand. Cheap imported goods manufactured under
conditions of slave labour in a distant Hell sooner or later contribute
to the unemployment that will add to the underclass in our own country
and make that Hell a little less distant. We can no longer retreat to
cultivating our own garden because the garden belongs to everyone:
our spades always dig into global soil. (This globalisation is not an
accident, or the product of specific technologies such as those related
to communication and travel. It is an inevitable consequence of the
mobilisation of undubitably universal scientific knowledge in the service
of human needs and wishes.)

Finding a criterion for demarcating benign from malign universalism
is, therefore, a matter of some urgency in a world in which diverse
cultures are becoming more and more intimately intermeshed – particularly as liberalism has lost self-confidence, in response to the
criticism that liberal thought simply cannot take on board the illiber-
alism that, by its own principles of value-relativism and tolerance, it
should tolerate. Many erstwhile liberals purport to believe that a
Declaration of the Rights of Man is a Eurocentric (and hypocritical)
imposition on cultures in which individual rights are not recognised
as absolute indefeasible values.[12] And yet they feel unhappy about an
attitude of complete moral and cultural *laissez-faire*, or a mere 'live
and let die' *modus vivendi* that does not even aspire to shared moral
and political beliefs.

How, then, shall we arrive at the desired criterion? I believe that we can do this by making a very small number of assumptions. Indeed, it may be that we need to assume only two things:

1. That an individual has a right to choose whether he/she lives or dies – except in those circumstances where he/she has committed a crime for which the penalty in his/her culture is death. This does not pre-empt any discussion as to whether there should be a death penalty at all (I personally am deeply opposed to judicial execution); or for what crimes the penalty is appropriate.

This assumption is not vulnerable to the charge that it presupposes that the individual has the right to enjoy rights and values that override those of the culture or society or value-community in which he/she is born and so smuggles in moral or value universals that may be used to judge a culture from the outside. It does not require this further presupposition; for it is difficult to imagine any culture in which all individuals, or even the majority of individuals, would fall foul of the death penalty. Nevertheless this will prove, as we shall see, sufficient to establish an important distance between the individual and the culture of birth without invoking 'individualistic' 'rights' and incurring the charge of imposing an implicit Eurocentric, 'Occidental' liberal individualism upon cultures to which this is alien. My criterion is less vulnerable to this charge than, for example, the Canadian Charter of Rights and Freedoms (quoted in Stephen Sedley's lecture 'Rights, Wrongs and Outcomes', *London Review of Books*, 11 May 1995, pp. 13–15) which asserts that 'Everyone has the right to life . . . and the right not to be deprived thereof in accordance with the principles of fundamental justice'. 'The principles of fundamental justice' are, implicitly, cultural invariants that in fact cannot be assumed as a given.

2. That, given a multicultural world and multicultural societies within that world, in which people will have values that are deeply and irreconcilably at odds, we should, if we have to adjudicate between them, favour those values that still leave individuals who do not share them most free to express or act in accordance with their own different values, above those values that do not leave that freedom intact. This is more likely to minimise culture wars – in which no one's values will be respected – but I think, as I shall show presently, it goes deeper than mere pragmatism or political expediency.

With these two assumptions, I believe that we could make considerable progress towards deriving very general criteria permitting a distinction to be made between benign and malign universalism. I also believe that these criteria are sufficient for a reformed Enlightenment programme to proceed without being a Eurocentric imposition. What grounds do I have for believing this?

Once it is accepted as basic that an individual has a right to choose to live, except where he/she has earned judicial execution, then it follows that society has a duty of care towards its citizens; for it has an obligation to ensure as far as possible that its members do not incur avoidable death – at least in the years before they can make a free choice for or against life – in, for example, the years of infancy. (This right, of course, brings with it reciprocal duties: I have a duty to contribute, directly or indirectly, as far as I can, to the process of ensuring that other members of my society do not die an unchosen, non-judicial death.) It might be thought that little could be derived from this criterion for separating benign from malign universalism: motherhood and no arsenic pie and that's about it. It does not tell us, for example, how far the duty of care extends. But it has a very important aspect. It cuts beneath specific, culturally-relative values and mores to something more fundamental: to the individual's existence or non-existence. And it is this that permits one to find a place outside of particular cultures from which some judgements can be made.

Using this criterion, we can accept the cultural relativist's position that, once we are in a culture, we are part of it – it is in a sense constitutive of us – and it is insulting for others to try to impose upon us (or even to save us from) values that are not the values of that culture. And yet, despite being good relativists, we can still retain a universal principle which is derived from the fact that existence is a *necessary* condition for us to enjoy (or not enjoy) the *accident* of being part of one culture rather than another. There is, in other words, the pure fact of one's existence that has logical, ontological, existential priority over one's status as a member of a particular culture. One's belonging to a particular culture is an accidental inflection of one's existing at all. The question of the life and death of its members thus cuts beneath other value questions. I am an Englishman and it would be patronising for someone from Chad to try to rescue me from the nexus of belief-systems that comes with being an Englishman – except inasmuch as my existence itself is threatened by those beliefs – when, in other words, I am in danger of becoming nothing. Being nothing is a greater, a more fundamental danger than being a Chadised Englishman. The right *to be* – or to choose to be, to continue choosing my own existence – is a groundfloor right that lies deeper than, in the sense of being a necessary condition of, my right to be in the English manner that I choose or have had chosen for me by the fact of having been born in an English culture. The tautologous right to be this particular entity goes deeper than the semi-tautologous right to be the kind of entity my culture commits me to being. The fact that you exist (or do not exist) is not culture-relative; it is the one indubitably extra-cultural datum; for it is a pre-condition of being able to participate in a particular culture and accept or reject its values. A person from Chad would thus be in a strong position to disapprove of the English set of values if it resulted in the death of most English people in infancy, before they had the opportunity to choose for or against life and for or against life English-style. The contingent fact that I exist is a necessary condition of the contingent fact that I exist as an Englishman with an English set of values. The fact that my existing at all is a necessary condition of my existing as a certain type of being reveals that there is something deeper than those things that are culturally relativisable.

Several things follow from my right to be able to choose whether I live or die. The most important is that society should be so governed and, as far as possible, so ordered as to minimise, for example, infant mortality (for we are all infants at some time or other). This at the very least will require attention to the provision of basic biological necessities. In the case of something as fundamental as ensuring clean and abundant water supplies, this requires familiarity with the relevant scientific knowledge about sanitation, the flow of fluid in pipes, metallurgy, etc. A preference for science-based technological practices rather than magic thinking will therefore be essential. But beyond this, it requires subscription to certain ethical principles; for example, a sense of personal responsibility in those individuals whose job it is to maintain and protect the water supply. Minimisation of corruption and dishonesty, appointments to senior posts on the basis of skill and experience rather than kinship networks, some minimal level of state investment and regulation, and a degree of commitment to distributive justice, will also be necessary. These take on the status of secondary value universals. Thus, from the simple principle of respecting an individual's right to make a choice about living or dying, about being or not being, we can derive many ethical consequences that are consistent with Enlightenment thought. We may discover that much of the implicit ethos of the Enlightenment is not so narrowly Eurocentric after all.

Let us now look at the second principle: giving preference to those values that still leave individuals who do not subscribe to them free to express their own, quite different values over values that do not leave those options open. I have already noticed that this is more likely to minimise those culture wars that are marked by an intensified loathing of everyone's values by everyone else with other values. But there is more to this principle than its potential to keep the peace. It is intended to address the criticism that the Enlightenment did not appreciate that there are many different ways of understanding the purpose and meaning of human life; that, for example, some human beings may prefer unhappiness to happiness; or prefer the deferred happiness which they are expecting in the next world to happiness in this world. We should, of course, respect these latter individuals;

after all, it is possible that they may be right to focus on the After-Life (though I personally doubt it). And because we cannot pre-judge the seekers of unhappiness or early immolation as wrong, we cannot help ourselves to off-the-shelf, 'free' or 'spontaneous' universal values such as 'Nobody wants to live in permanent, unrelieved pain', 'Nobody wants to die young', of the kind that the utilitarians assumed and for which they were execrated by their Romantic critics. We have, instead, to consider how we can reconcile such absolute differences in the valuation of the things of life as between an ascetic who wishes for Hell on Earth and a hedonist who seeks happiness. We would, after all, like to give both of them the respect that is due to their deeply different values.

We could argue that the problem does not have to arise (and can therefore be overlooked) because we are all – hedonists and ascetics – tucked away in the private spaces between the meshes of society, and here we can imagine that we are solitary and practise our own way of life. But this is not an adequate response; for how individuals order their affairs is not entirely separated from how society as a whole orders its affairs; questions of self-government and government *tout court* are not clearly distinct and they have many points of inter-action – a truth which would be conceded even by those who do not agree that 'the personal is political' and vice versa. Private ascetics will have their views about how public life should be conducted, what should be permitted to others, what goals society should aim for. And private hedonists will also have views on these matters. And both will expect their views to be reflected in what actually takes place in the public sphere they share. Another-worldly ascetic, for whom flagel-lation is the proper activity of man, is unlikely to be a good team-player in institutions devoted to limiting human suffering. So how, where there are strongly opposed ascetic and hedonistic views, should society as a whole order its affairs? Where some of its members pre-fer suffering to happiness (and strongly believe that others should share their conviction), should society be ordered on Hell-on-Earth/ascetic lines; or should it be ordered on the lines of ordinary com-fort? We can answer this with the help of the general principle I have just advanced.

Consider scenario 1, in which society gives precedence to the wishes of the ascetic, so that all will have to live in material impoverishment and discomfort. Such a society will live closer to death: inattention to comfort and basic needs will mean that the lives of all – not just ascetics' lives – are in constant danger. Now many in such a society will not have chosen danger and discomfort and early death. Their values will thus be ignored, indeed trampled on. Amongst these dis-sidents will be immiserated infants and children, a significant number of whom will not, because of the general lack of concern for safety and comfort, survive to adult life to discover, even less to choose and live in accordance with, their own values. They will not even survive to choose *asceticism* or early death.

Now consider scenario 2, in which the values of comfort-seekers have precedence and society is so governed, administered and managed as to maximise happiness in the conventional sense. This option will reflect the wishes of the hedonists, but it will also leave the acetics free to starve, beat and even kill themselves in their frantic pursuit of the happy unhappiness that is to be found beyond the thick robe of the flesh. If you choose pain on behalf of society, this may deny others the pleasure or comfort they wish; if you choose pleasure or comfort on behalf of society, neither you nor any others are denied the freedom to choose pain for themselves. We should, therefore, support a culture that chooses comfort and still leaves its citizens free to be as uncomfortable as they choose. The Enlightenment thinkers would have found this a perfectly acceptable principle: that society should be so ordered as to free the citizens to make the happiness or unhappiness of their own making.

We therefore admit a hierarchy of values: a more basic value (for example, valuing life) is given priority over a less basic one, since the former is a necessary condition for the choice of the latter; we should give priority to those values which interfere least with the expression of other values. If you say you couldn't give a damn about life and therefore care even less about an adequate water supply, you would, if you were empowered to express your beliefs in the public sphere, increase my chances of premature death from cholera and hence prevent me from fully expressing my values – which wish for and require a long life. If, however, I, who love life, am given power to create a society in which life and longevity and comfort are valued, I still leave you free to choose otherwise – to drink as much infected water as you like in pursuit of your early demise through diarrhoea. Giving my preference for universal health over yours for universal suffering still leaves you at liberty to choose suffering and an early death. I do not need to live in fear of the consequences of your value-system, and you are not obliged to live in fear of the consequences of mine.

In summary, these two minimal criteria, which on close inspection requisition the kind of applied reason and justice that the Enlightenment thinkers worked for, enable us to separate benign from malign universalism; a malign *laissez-faire* cultural relativism – of the kind convenient for the arms salesman who doesn't mind to whom he sells his weapons, classifying his clients only into those who do and those who do not pay their bills – from a benign respect for the variety of human life.[13] After all, to permit or endorse or to look passively upon a culture which does not ensure that many of its members *live* – so that they may freely choose for or against material prosperity, for or against life itself – is to collude in an 'internal cultural colonialism' where some within a society or culture deny others within that society or culture the power even to choose for or against that culture. That seems to me to define a limit to relativism that cannot itself be relativised or dismissed merely as dominant-culture ethnocentricity, as Eurocentric illiberal liberalism.[14] This limit may provide us with the equipment to address John Rawls' challenge (as expressed by Hoffmann, see note 12), 'to reconcile a variety of reasonable conceptions of the good (for instance, different beliefs about the role of religion in society) with a single political conception of justice'. By means of the principle I have advanced, we may accommodate the fact that all values are contingent – even the desire to live: nothing could be more obviously groundless than my desire to continue being this utterly contingent being (me) – with the frequently pressing need to adjudicate between them.

THE SPECTRE OF UTOPIA

I could have made the defence of a greatly curtailed Enlightenment universalism easier for myself by taking for granted that certain human wishes are universal. Surely, I could have argued, every culture subscribes to the following:

1. Continuous unremitting pain – or fear of it – is bad.
2. People, in particular children, should not suffer avoidable death.

The reason I did not take even these seemingly basic and universal values for granted is that I am not too sure how universal they are. Yes, there must be very few cultures – even those with their gaze directed to the next world – in which normal people would dissent from those principles. But, to judge from the evidence, there seem to be many more cultures in which the conditions necessary to ensure the general implementation of these principles are not ardently pursued. In some cultures, whilst the death of the young may be regarded as a bad thing in itself, other considerations may override it. A raped daughter may be expected to prefer death to dishonour. Or the collective will may seem to place a higher value upon the esteem that comes from the trappings of wealth over the satisfaction derived from a job discharged conscientiously; and this ordering may mean that life-threatening diseases will continue to be carried by the water supply. There are circumstances in which warring tribesmen or the members of Bomber Command would not shrink from killing children. And so on. That is why, when trying to retain something from Enlightenment universalism, it is important to make as few assumptions as possible.

One of the assumptions made by Enlightenment thinkers – an assumption that, according to Isaiah Berlin, is one of the distinguishing features of Western thought since at least Plato – is that there is a commonality of human ends, howsoever different the means by which they try to achieve them. We have tried to avoid that assumption as far as is possible. Berlin connects this assumption of a unity or harmony of human ends with the notion of Utopia – the idea of a perfect society 'in which there was no misery and no greed, no danger or poverty or fear or brutalising labour or insecurity'. Western Utopias, he says, tend to contain the same elements:

> a society lives in a state of pure harmony, in which all its members live in peace, love one another, are free from physical danger, from want of any kind, from insecurity, from degrading work, from envy, from frustration, experience no injustice or violence, live in perpetual, even light, in a temperate climate, in the midst of infinitely fruitful, generous nature. The main characteristic of most, perhaps all, Utopias is the fact that they are static. Nothing in them alters, for they have reached perfection: there is no need for novelty or change; no one can wish to alter a condition in which all natural human wishes are fulfilled.
>
> (*The Crooked Timber of Humanity*, p. 20)

The connection between the idea of a perfect society and universalist assumptions about the ends of life is underlined by Berlin:

> The assumption upon which this is based is that men have as certain fixed, unaltering nature, certain universal, common, immutable goals, identical for all, at all times, everywhere. For unless this is so, Utopia cannot be Utopia, for then the most perfect society will not perfectly satisfy everyone.
>
> (ibid., pp. 20–1)

We may smile at the naivety of those who dream of Utopia, but there is a very real sense in which the hope of progress is implicitly Utopian, inasmuch as it is assumed that progress in specific areas will not be at the expense of progress in others; that, for example, the steps taken to control smallpox do not destroy the liberties which are necessary to ensure a tradition of civil government, or a just society; or, more narrowly, that the measures taken to control smallpox do not create conditions in which a yet more virulent disease may flourish. Even the less ambitious, piecemeal social engineering advocated by democratic socialists carries within it the assumption that all the small advances will point in a certain direction, will converge in the vector of overall progress. Reformers, however modest, are inescapably concerned with the health of society as a whole and even the most cautious, focused or narrow-minded of them – those for whom the branch-lines of social reality are everything – would lose their appetite for specific reforms if they did not feel that the net effect for society as a whole was positive. Speaking personally for a moment, I would not be terribly keen to continue my own work in the neurology of old age if I knew that every gain in this area would be bought at the cost of an equivalent loss in the field of the cardiology of old age, the nutrition of older people or diarrhoeal diseases affecting children in the Third World. Every reformer, every progressively inclined individual, everyone concerned to bring about net improvements, however local, is thus implicitly Utopian: it is only a slight exaggeration to say that every disinterested drive to local improvement contains an implicit drive to global improvement. Anyone who shares the hope of progress, however local or focal, must therefore take account of the fears the very notion of Utopia raises in the minds of contemporary thinkers.

Recent history has made us all too familiar with the way in which Utopian aspirations, in the hands of those who have real power rather than being ineffectual dreamers, have helped to transform backward, injust and impoverished countries into absolute Hells on Earth. The lessons learnt from the real experience of the twentieth century have been reinforced by numerous fictional dystopias. The former have shown the difficulty of bringing about massive social change without smashing what is worthwhile in the public sphere and replacing the little and inadequate good with universal evil. The catastrophe of communism, in particular, has caused a radical re-think amongst the radicals about what it is that makes a society work for the good of all its citizens, how the individual will to personal and collective advancement is best harnessed to promote the collective good, and how rapid and profound the processes of social change should be. Many fictional treatments of Utopia, however, have gone beyond anxieties about the process and doubts about the achievability of the goal and expressed a deep scepticism and disquiet about the goal itself – beyond questions about the practicability of bringing about Utopia to the very principle of Utopia itself. They have, that is to say, questioned the assumption, identified by Berlin as being central to the ideal of a perfect society, 'that men have a certain fixed, unaltering nature, certain universal, common, immutable goals' and have concluded that, since this assumption is unfounded and oppressive, the decline of Utopian ideals – at least in the West – is to be welcomed. We shall never agree on the ends of life; so a society that appears to express and embody a set of ends for all of its citizens and for all time, must be oppressive:

> Immanuel Kant, a man very remote from irrationalism, once observed that 'Out of the crooked timber of humanity no straight thing was ever made'. And for that reason no perfect solution is, not merely in practice, but in principle, possible in human affairs, and any determined attempt to produce it is likely to lead to suffering, disillusionment and failure.
>
> (*The Crooked Timber of Humanity*, p. 48)

This is a deeply disturbing conclusion, not simply because we cannot live without explicitly Utopian ideals. Many men of goodwill have done so. But because, as already indicated, modest melioristic ambitions, dreams of bringing about improvements in a small sphere of one's own expertise and concern, shade imperceptibly into the hope of leaving the world a better place than one had found it – into the hope of bringing about net overall gain. It would be an odd social conscience that was unconcerned if the improvements brought about in one sphere were exactly offset, or worse, by deterioration in other places. The socially concerned, whilst oppressively aware that there is much misery that they must leave to others to sort out, at least assume that the happiness they spread will not bring about deepening immiseration elsewhere. This assumption is not, of course, always well founded: society is not only the sum of intended actions and their intended consequences but also of the unintended consequences of actions. Nevertheless, all melioristic instincts, however narrowly expressed, have the seed of Utopianism in them; bear within them the assumption that many progressive actions will add up to overall net progress for the world, that they have a deeper meaning inasmuch as they may contribute, in howsoever small a fashion, towards the forwards and upwards movement of humanity as a whole out of want, fear, pain, impoverishment of all sorts; that the effects of these actions will converge in similar or compatible goals even if they are not strictly synergistic; and that, while Utopia will not be achieved as the outcome of a single revolutionary convulsion, it will be approximated, even if never achieved, as an asymptote approached by huge numbers of small advances. To abandon Utopian ideas, therefore, is not merely to foreswear the visionary passions of fanatics and lunatics, but to throw into question the very dimension of hope in human affairs. It is difficult to be anti-Utopian without being against progress itself; for there is no clear or sharp distinction between arrogant claims to provide universal – in both senses of 'global' and 'universally applicable' – solutions to the woes of mankind and wanting to contribute, however modestly, to making the world overall a better place.

The two major elements of the hope of progress – to alleviate the sufferings of mankind (the world of means); and to extend the capacity of human beings to realise new possibilities that men and women have invented (the world of ends) – are necessary for mankind to find any kind of meaning in its struggles beyond the brute desire for one's own survival and pleasures and the survival and pleasures of one's children and a circle of dependants and friends. This dimension of hope – making the indefinite future of humanity the secular analogue of the eternity of religious hope – is deep-rooted in our consciousness. Since we cannot, as we have agreed, become 'religious' by decree, we have to ask ourselves what secular society will become without the hope of progress and without fostering the visions of those who dream unselfishly of making the world a progressively better place? The implicitly Utopian hope of progress is an essential regulative idea for collective human morality in precisely the same Kantian sense as the notion of progress towards the truth is the regulative idea of science. Because we cannot forgo the dream of Utopia, however muted and low-key, without denying the hope of progress and hence rendering collective human action headless and amoral, we must take account of the fears that the spectre of Utopia awaken.[15]

There are two fundamental objections to Utopia, however painlessly it may be achieved. The first is that it may be boring and empty; and the second is that it will be rigid, uniformitarian and authoritarian.

NEED UTOPIA BE BORING AND EMPTY?

The western world is now engaged in constructing a fundamentally secular and deconsecrated industrial society. This is a society in which – if it is achieved – all men will live in comfort. Perhaps there will also be a large measure of formal freedom and religious and philosophical toleration. But it is a society that threatens to deprive human life of all spiritual content, a society in which the growth of freedom is likely to be accompanied by the growth of numbers of those whose inner emptiness robs them of the desire to use it, a society in which religious and philosophical toleration will be made all the easier to achieve as spiritual impoverishment makes religious and philosophical commitment constantly more rare.[16]

Berlin's deliciously mocking account of Utopia as an essentially static society whose members not only live in perfect harmony but also live in 'perpetual, even light' – making Utopia a rather staider, chillier version of the eternal afternoon of Tennyson's Lotos Eaters – captures the anorexia that many of us feel when imagining any kind of paradise, whether earthly or celestial. Perhaps we all subscribe, if only unconsciously, to the Schopenhauerian notion that humans are doomed to oscillate between suffering – from pain or from a frustrated desire for pleasure – and boredom. If we take away pain and unfulfilled desire, won't we be left only with boredom and a sense of emptiness, a life rendered meaningless by satiety? The answer to this is yes, only if the experience of meaning is exclusively rooted in pain and unfulfilled desires, if it is based upon lack. If this is inadequate as an account of the actual sources of meaning, it is even less adequate as an account of *possible* sources of meaning.

The relationship between pain (or suffering more widely construed) and meaning is that, except in those cases where pain is freely chosen, pain is anti-meaning, privation of meaning. To put this another way, its 'meaning', as I have argued elsewhere,[17] lies almost entirely in the meanings *it takes away*; in the lost meanings of the things it renders meaningless. To be in severe pain is to be rolled in the brazier of anti-meaning. The argument – developed extensively in David Morris's *The Culture of Pain*[18] – that pain is a profound source of meaning, and that life without suffering lacks a dimension of reality, fails to notice this most obvious feature of pain: that any meaning it has is stolen from elsewhere, is lost meaning. If there are new or original meanings created by pain, they are the second-order meanings relating to the ploys we use to alleviate, deal with or endure it. It is part of the Western religious inheritance, the Judaeo-Christian assumption that life is a vale of tears, to accept without question that the deepest meanings are associated with suffering; that a life relatively free of pain – a 'comfortable' life – is inescapably shallow. There is, however, no *a priori* reason for thinking that one sort of sensation, preoccupation or theme touches more closely upon our true nature, or reveals us more completely to ourselves, than another.

It may be historically true that, for humans, pain has had the upper hand over pleasure, sorrow over happiness; but this may be a contingent, not a necessary, feature of the human condition. Admittedly, as things are, it is true that to be in pain, to be acquainted with grief, is to be closer to the reality of most people's lives than to be part of a comfortable minority. But this is how things are *now*; not how they must be, even less should be, for all time. It may be argued that the revelation afforded by pain has a 'deep' permanent truth, for it reminds us of our finitude; that its negative meanings are the profoundest meanings because they underline what a small figure we cut in the universe and how what we are not vastly outsizes that

which we are. The demeaning impact of pain, its savaging of our
ordinary meanings, tears open the bubble of ordinary sense which
encloses us and normally insulates us from the objective reality of
our condition. In one sense this is true; but in another sense it is not.
We are finite, yes; but we are also the site through which the universe discloses what is not finite. In addition to being existentially
finite, finite in our being, we are infinite in our knowledge inasmuch
as we are a place of revelation of the infinite, a revelation that pain –
along with those other things, such as getting and spending, that
dominate the kingdom of means – obscures. In toothache, I am as
remote as possible from being a lens on the universe; I am a place of
toothache.

It may be argued, finally, that pain not only reminds us of our
finitude but also makes us, in a positive sense, more human. This is
not noticeable amongst sufferers in ordinary life: chronic pain isolates, degrades and closes off human beings. That is the common testimony of sufferers. However, our experience, though useless in itself
may make us more sensitive to the suffering of others. Morris quotes
Albert Schweitzer:

> Whoever among us has through personal experience learned what
> pain and anxiety really are must help to ensure that those who out
> there are in bodily need obtain the help which came to him. He
> belongs no more to himself alone; he has become the brother of all
> who suffer. On the 'Brotherhood of those who bear the mark of
> pain' lies the duty of medical work.
>
> (quoted in Morris, p. 287)

Levinas, also cited by Morris, goes further. Pain is negative, useless,
absurd, yes. But it may be transformed:

> it opens up the ethical dimension of the 'inter-human'. My own
> useless suffering, that is, takes on a changed meaning if it becomes
> the occasion for your empathetic, even suffering response.
>
> (p. 287)

There is a terrible sentimentality in this. The reality of the effect of
chronic pain – in this case, toothache – is expressed by Dostoyevsky's
anti-hero writing from under the floorboards:

> Those groans express, firstly, the degrading futility of one's com-
> plaint, a legalized tyranny of nature which one despises, but from
> which one, unlike nature, is bound to suffer. They also express a
> sense of the fact that at that moment one has no other foe than the

pain; a sense of the fact that one is utterly at the mercy of one's teeth; a sense of the fact that Providence is in a position either to will that your teeth shall cease on an instant to ache or to will that they shall go on aching another three months; and lastly the sense of the fact that if you do not agree with, but, on the contrary, protest against, the situation, your only comfort, your only resource, will be either to cut your throat or to go on beating the walls ever harder with your fists, since there is nothing else for you to do.[19]

This is life after two days of toothache. The inter-human dimension opened up by this relatively trivial agony is not quite as Levinas envisages it:

Well [by the third day] his groans will have become malicious and meanly irascible; and though he may continue them whole nights and days at a stretch, he will be aware all the time that he is doing himself no good by his utterances, but merely uselessly angering and annoying himself and others. Better than anyone else he will be aware that his family, as also the public before whom he is cutting such a figure, have for a long time been listening to him with disgust; they think him an utter rascal, and have it in mind that he might just as well have groaned in a simpler manner (that is to say without any turns or roulades), since his present style of groaning is due simply to temper and is leading him to play the fool out of sheer viciousness.

(ibid., p. 603)

There must surely be better ways of promoting mutual assistance, concern, even love, than through the anti-meanings of pain. And I am not aware of any evidence that shared sorrow is superior to shared pleasure and shared joy as a means to opening up and maintaining the 'inter-human'. Unalloyed happiness, the continual pursuit of pleasure, seem hard, empty, shallow, only in a world – such as the world today – in which pain has the upper hand. In Utopia, where pain has been conquered, there may be other means of being profoundly inter-human than through the mediation of pain. 'The great object in life', Byron said, 'is sensation – to feel we exist, even though in pain.' Yes, but preferably not in pain; or not in involuntary pain.[20]

So much for the anxiety generated by the prospect of a pain-free life in Utopia. What about the consequences of the satisfaction of all desires? Surely, desires are a source of positive, rather than merely privative, meaning and their universal and rapid satisfaction must lead to an emptying of life. Ironically, one of the powerful rebuttals of this assumption comes from the most vicious critic of the

Enlightenment dream of progress: Joseph de Maistre. Man, he said, 'is infinite in his desires and, always discontented with what he has, loves only what he has not'.[21] This was in support of his claim that humans are insatiable for power. But it could equally well be used to demonstrate the inexhaustible possibility for development, elaboration and refinement within desire, reflecting a boundless human capacity for discovering new sources of meaning. Development may occur in one of at least two directions: the folding and unfolding of existing desires – as witnessed by humankind's ability to transform and deepen and enrich the experiences associated with physiological needs;[22] and the discovery of new desires, new modes of rapture, fascination and preoccupation. Let us look at examples of these two possibilities.

First, the elaboration of appetites. Consider human sexuality. It is rooted in physical appetite: this is the *donnée*. But it is also deeply implicated in a specifically human wish to be *recognised* by others, a wish that, as Hegel and many others after him appreciated, reaches into the fundamentally metaphysical nature of the human animal. Sexuality has many dimensions. Here are a few at random: sensual delight; care and mutual support; awe and adoration; responsibility; the many narratives of growing love, of consent and mutual understanding; the sense of privilege; the profound vision of the otherness of the Other. These intersect and interact in an infinite number of ways. And within the sphere of sensual delight, the delightful and not so delightful tensions between orthogonal (and sometimes incompatible) dimensions: maximising sensation; enjoying privileged knowledge; savouring power; asserting companionship; expressing love. There are equally complex tensions between the long narrative – of conquest, of uncovering, of companionship – and the instant of sensation. And between the institutional and the sensational, the discursive and the tactile, the social and the individual, person-as-subject and the person-as-object, the general object of desire and the particular desired person, the physical and the metaphysical. And sexuality has, of course, other less attractive elements: the exertion of power over another; the desire to hurt, to smash, to revenge oneself, to humiliate, to destroy.

Human sexuality is susceptible, therefore, of endless elaboration. This is to be understood not only, or not especially, in terms of exploring the factorial X number of combinations of bodily surfaces and acts involving them, but in terms of an infinitely folded and unfolding consciousness of the other person and through him or her of oneself. This complexity, this foldedness, is, or may be, evident even at the 'purely' physical level. Whereas in animals, the sexual act is an operation performed by one animal upon another, by a subject upon

an object – the 'ten second jump and shriek' that Young referred to –
in human beings, there is a layered awareness of one's own sensations, of the other's sensations, of the other's awareness of one's awareness of his or her sensations; and so on. A vertigo of mutually reflected
consciousnesses can open up at every point. And there are other, richer
possibilities at the level of words, symbols, gestures. They are infinite
because humans are explicit animals and there is no intrinsic limit to
the extent to which explicitness unfolds.

This ever-enhanced mutual awareness was represented in the existentialist literature derived from Hegel's analysis of human relations
as a kind of Hell. Sexuality in Sartre's philosophical treatises, books
and plays was a battle for domination, and the sexual relationship
was an interaction between two roles: that of Master-Sadist and that
of the Slave-Masochist. The thirst for recognition – for an affirmation
of one's own reality by another's acknowledgement of that reality,
through one's consciousness of one's self – underwent malignant change
in Sartre's vision into a contest between two beings locked in a life-
and-death struggle each trying to impose their own meaning upon
the relationship and to deny the meaning the other would wish to
impose. This, in turn, was part of a larger project to enclose the other's
existence within one's own world and deny the reciprocal fact that
one is part of the other's world.

Sartre's pessimistic view – echoed by many writers in the twentieth century – is at least in part the result of a gross simplification, a
reduction of the richness and complexity of sexual relations to a naked battle between categories. Few would accept that the asymmetry
in human relationships between Master and Slave is a permanent feature
of the metaphysical condition of humanity. It may characterise many
relationships – for example, the asymmetrical relationships between
most men and women over history. It is not impossible that in Utopia
sexual relationships would be between equals and the joyful exploration of bodily and social possibility would not unfold into destructive madness. Sadism would not be necessary, for example, in order
to enhance sensation; this would come from mutual awareness of
awareness; from increasingly intense acknowledgement of the other's
unique reality.[23]

The possibilities inherent in sexuality indicate the wider possibilities for Utopians: a metaphysical journey through an ever-deepening
sensuality, a contemplative physical pleasure, towards delight in the
mystery of the world, in particular of consciousness and of human
consciousness, as revealed in the life of one's companion of choice, to
a deeper fascination with the mystery of the perceived world – an
enraptured delight in the appearance of light, in the feeling of warmth
on one's arm, in the quiet sound of the evening breeze in the trees.

Truly understood, all of these are inexhaustible. In Utopia, we may imagine individuals engaging at will in a lucid ecstasy whose object is the ordinary sensations of the body and the sensibilia of the common day. It may be that, at first, the majority of people will need the help of a gifted minority, who will serve a role similar to that of artists in the present, decidedly pre-Utopian, times; ultimately, when the education of the senses – the ability to enjoy experience for its own sake, to experience one's experiences[24] – occupies a central place in the curriculum, all will be artists, not in the sense of producing art, but in the sense of experiencing the world as artists do in their best moments.

If one still wanted to have a use for the term, one could think of the sensibility of the Utopians as being essentially religious. Their religion would not, however, be encrusted with institutions (it would have no institutions, being purely a matter of actual consciousness) or suborned to the service of power; nor would it spawn or be spawned from creeds and dogmas. Its cognitive content would be the indisputed obvious: things no one is going to argue about; meanings beyond differentiated meaning or disputable interpretation. A religion of wonder – about light, about perception, about the fact that we, inexplicably, are – which will be perhaps closer to the true spirit of religion than a thousand child-slaughtering battles over the *filioque* clause, or the persecution of an unmarried mother, or even a twenty-page proof of the existence of God.[25]

Some prophets of such Utopias – most notoriously Aldous Huxley – saw in drugs such as mescaline a rapid route to dissemination of artistic experiences and sensibilities. There are at least two important reasons for doubting the power of drugs to hasten the birth of Utopia. First, drugs are available now and yet Utopia hasn't arrived: they don't fit into the Kingdom of Means in which we most of us have to spend our lives. Not only do they fail to help bring about the Utopian future liberated from the exigencies of need: through their induction of dependency, they create new and terrible needs that demand to be met again and again and again. Secondly, drugs would undermine the conditions necessary to maintain Utopia. The lucid trances, the ecstatic enjoyment of the real, that I see as possible for Utopia, would not break up society or damage the bonds of shared delighted and mutual concern. They would be part of, integral with, one's sociality. Drugs, with their isolating and confused delights that are totally unrelated to external reality (including the reality of others' happiness and unhappiness and their needs), threaten to erode everything that makes people supportive of one another.[26] If it is possible, as I have argued, to have a pain-free life, a hunger-free life, which is not lacklustre, boring or empty, if it is possible to delight in satiety, it will not be necessary to import artificial significance into life through drugs to replace the lost meanings that were associated with struggle for survival, with suffering and with unsatisfied desire.

This may not convince those who, like Dostoyevsky's Underground Man, believe that if they were transported to Utopia they would want to smash it simply for the sake of asserting their own unique existence and demonstrating that they were not merely insigificant cogs in a vast machine. Such sentiments, echoed by a thousand existentialist and pseudo-existentialist writers and a hundred thousand sympathetic commentators since Dostoyevsky, do not take into account the fact that we are already able to cope with the fact that the universe does not perpetually acknowledge our status as its centre and irreplaceable only child. Inside or outside Utopia, in a well-organised Heaven or in an anarchic Hell, we are equally a minute part of a huge crowd. Nor do these anti-Utopian sentiments recognise that the inhabitants of Utopia will have had different formative experiences from themselves; that, unlike Dostoyevsky's Underground Man, they will not have known privation, injustice, marginalisation and humiliation, in sum, the damaging experiences that fill them with their impulses to cause damage. Likewise, the Underground Man's impatience with the unremitting rationalism of Utopia would not make sense to those who actually had the privilege of living in Utopia:

> See here: reason is an excellent thing – I do not deny that for a moment; but reason is reason and no more, and satisfies only the reasoning faculty in man, whereas volition is a manifestation of all life. . . . It is true that, in this particular manifestation of it, human life is all too often a sorry failure; yet it nevertheless is life, and not the mere working out of a square root. For my own part, I naturally wish to satisfy *all* my faculties, and not my reasoning faculty alone (that is to say, a mere twentieth portion of my capacity for living).
>
> (p. 613)

This would simply not make sense to a Utopian whose happiness was rooted in a unity of consiousness in which reason and sensation and will were all convergent in experiences of wholeness. There is no *a priori* reason why this should not be possible. Nor why the impulse to evil should remain as powerful in those whose lives have not been damaged as they manifestly are in those like the Underground Man who have been damaged:

> Does reason never err in estimating what is advantageous? May it not be that man occasionally loves something besides prosperity? May it not be that he also loves *adversity*?
>
> (ibid., p. 618)

Not necessarily in Utopia. The assumption that in Utopia we shall be bored, our lives will be empty and sooner or later we shall try to smash something – each other, society – is the result of reading the future of man from one part of his past. And that is a mistake. After all, we are able to live at a great remove from all the things – instincts, impulses, etc. – that are supposed to define us and limit our capacity for change.[27, 28]

NEED UTOPIA BE RIGID OR AUTHORITARIAN?

Utopian writers tend to assume that their Utopias will be highly ordered. The dystopian consequence of this seems to be that order will be maintained by an intensely authoritarian and obtrusive state, often underwritten by a potentially brutal, quasi-militaristic police force. The social framework which sustains the exquisite sensitivities of the privileged of Utopia will be upheld by rubber truncheons applied to the heads and electrodes applied to the genitals of dissidents.

It is easy to see why this may be considered to be an inevitable feature of Utopias. First, the kind of order Utopias exhibit cannot be relied upon to emerge as a 'spontaneous order' in the Hayekian sense. It is therefore assumed to be intrinsically unstable and vulnerable to the slightest tremors of dissent, unrest or rebellion. Secondly, a society which is well off and comfortable has much to lose collectively. There is a lot invested in not rocking the boat: it is unlikely that unplanned change will bring about further improvements; the results will most likely be deleterious rather than otherwise. Rebellions, or other sources of unplanned change, will have to be dealt with severely. For the same reason, a comfortable society will feel threatened from without and have a high level of anxiety about external enemies and their collaborators from within. 'Fortress Utopia', the result of inevitably uneven development of societies through the world, will be at risk from becoming paranoid – with predictable consequences for civil liberties. This will seem likely even in the best kinds of Utopias; those, for example, which are *not* built on the slave labour of an underclass.

More careful consideration of the kind of Utopia we are envisaging – a Utopia of Ends rather than of Means, of consciousness rather than consumption – may lead us to question the received idea that Utopia will be automatically associated with police batons and a Draconian rule of law parodying Singapore. Are we sure that the citizens will *want* to dissent in a violent and explicit and destabilising way? It is quite likely that they will have better things to do. After all, it is only in the world of tyrannies, want, greed, etc. that dissent seems natural, even noble. Nor will their failure to dissent necessarily be evidence that the citizens have been drugged, bribed ('bread and circuses'), brainwashed or cowed into conformity. For the latter assumes the division of society into a ruling class – of Guardians, Commissars, or whatever – and a ruled class who have to be kept under thumb of, or in thrall to, the rulers. But it is not impossible that in Utopia we may have moved beyond this. The exercise of power will be diffused into civic duty and the latter will not be considered to be an end in itself; it will simply be what it is – a duty, an interruption to the cultivation of the Kingdom of Ends. Power will not be concentrated and, as such, will be no more likely to lead to corruption than the power of the milk monitor. By such a diffusion of power, Utopia will have stability built into it and the impulse to dissent will be focused on particular technical questions; the adversarial habit will atrophy while individuals pursue their own ends in harmony with others. The standard dystopian image of infantilised citizens free to pursue empty private ecstacies so long as they do not seek power, their lives monitored and patrolled by brutal shock troops ready to pounce if they show any signs of independence or rebellion, assumes that power remains in the hands of a few ruthless Guardians who foreswear the Kingdom of Ends for power; or foreswear delight for power, which has become and end in itself. This is a possible, but not a necessary, scenario.

There remains, however, a further serious question. If the population in Utopia is pacific rather than pacified, then the Utopians are very vulnerable to attack from others outside of Utopia. In order not to be a sitting target, one of two things is necessary: either Utopia should be well armed; or it should be universal, so that it has no outside threat because there is nothing outside of it apart from empty space. There are difficulties with either of these solutions. 'Fortress Utopia' will need a standing army, a tradition of military organisation and discipline, of command and obedience that will be at odds with the ethos that we have envisaged, of diffused power and private citizens living largely in a kingdom of private or privately chosen ends. The spectre of highly disciplined, CS gas-wielding shock troops patrolling outside the dreams of the Lotos-eating citizenry returns. Once the concentration of power and authority necessary to requisition and maintain and run a standing army is granted, then the possibility of corruption and tyranny is raised once more. A citizenry whose power aspirations do not extend beyond that of being a milk monitor seems distinctly vulnerable. If the army is recruited from within the citizens on the Swiss model, so that the soldiers revert to being citizens after a period of duty, there emerges the new problem of transforming successive intakes of Lotos-Eaters into a fighting force. This can be solved only by having a permanent officer class. A much more attractive option is to extend Utopia to the point where it ceases to have an outside. And I suspect that this is the only viable option:

in the globally networked world of the twenty-first or twenty-second century, no localised Utopia – even less a localised Fortress Utopia – would be possible. And this must depend upon even development world-wide and upon human beings feeling themselves to be citizens of no particular country, owing an allegiance only to the human race. As we reach the end of the twentieth century, with demagogue-inspired reassertions of ethnic allegiances, this, alas, seems even less likely than at the beginning of the century. However, an economy based upon infinitely renewable bits (of information) rather than exhaustible supplies of available energy required to move atoms (of matter) raises the hope of cooperation rather than brutal and bloody competition between countries. Nicholas Negroponte[29] has sketched an account of some of the possibilities arising out of the 'irrevocable and unstoppable' change from atoms to bits. His optimistic account of the digital future is based upon the fact that digital technology is 'decentralizing, globalizing, harmonizing, and empowering' (p. 229). Nothing could be more remote from the ethnocentric passions of the demagogues.

Utopia, then, may not necessarily be authoritarian, forbidding and prone to dealing brutally with dissent; for, with power diffused and fewer grounds for dissent, the relationship between the individual and authorities will not be an adversarial one; at any rate, the adversarial moments will not be organised into factions and rebellion. The idea that there might be nothing to dissent from may be distressingly unromantic, undramatic to some and may invoke the image of the state (the world-state, one assumes), as a massive herd of shepherdless sheep. It is the uniformity that may be most deeply offensive: the prospect of all the citizens obediently chewing the grass of identical experience is not inspiring. (We have already quoted de Tocqueville on this in note 16.) The assumption of empty, mindless uniformity would, however, be mistaken. There is no reason why what Emerson called 'the infiniteness of the private man' should be eliminated by comfort and contentment. It is pain, rather than lack of it, that rivets us to our finitude. Even rebellion is narrowing: it takes its themes, its agenda from its object. Rebellion 'against the system', if it is anything other than rhetoric and gesture, if it is in any sense constructive, is in danger of converging to single issue narrowness. Utopia, on the other hand, should permit, in the interstices of uniformity, an infinite depth and variety of experience.

The model here is sexual behaviour. From a certain distance its stereotyped nature is almost comic. The picture from within is different. Inside the relationship, inside every act, there is the possibility of an inexhaustible variety of experiences, awarenesses, understandings and misunderstandings. Even between the same two people, no two successive acts of intercourse are the same. In Utopia, where individuals are committed to experiencing and reflecting upon their experiences, such variety would be understood and experienced. The richness and fathomlessness and variety of life in Utopia would come from the content of consciousness, not from its descriptive silhouette or from the remote, third-person viewpoint of comparative ignorance. The actual content of consciousness would provide the essential inner distances within Utopia.

In *Two Concepts of Liberty* Berlin asserted, against an oppressive paternalism that would deny us the right to define ourselves, that 'we must preserve a minimum core of personal freedom, if we are not degrade or deny our nature'. This is incontrovertible. But it is only from the point of view of material necessity that we may be captured by the generalising gaze, only from the utilitarian point of view that experiences are classified into successive and identical instances of a finite number of types. It is war and hunger that classify weathers into wet and dry, foggy and clear; in peacetime and satiety, the thousand different cloud formations and the thousand modes of shadowiness and sunniness are there to be appreciated. And it is the control exerted by societies dominated by material want that reduces sexual behaviour to recognisable forms that can be allocated to the categories of transgressions or permitted modes.

This last point can be usefully connected with wider doubts that have been expressed about Utopias of all kinds, and about the Enlightenment project in particular: that it is at once too prescriptive and too shallow. The specific target of this criticism is the notion of a society based upon reason. Let us return to Dostoyevsky's anti-hero:

> See here: reason is an excellent thing – I do not deny that for a moment; but reason is reason and no more, and satisfies only the reasoning faculty in man, whereas volition is a manifestation of all life. . . . It is true that, in this particular manifestation of it, human life is all too often a sorry failure; yet it nevertheless *is* life, and not the mere working out of a square root. For my own part, I naturally wish to satisfy *all* my faculties, and not my reasoning faculty alone (that is to say, a mere twentieth portion of my capacity for living).
>
> (p. 613)

Yes, of course, reason and utility and the Kingdom of Means is not sufficient in itself; the reasoning faculty is not the whole of what we are. Man doth not live by reason, even less *in* reason, alone. Reason is only a shell. It creates the framework within which we are able to live whole lives. But only the framework. This is not, however, a case for irrationalism. The reason-based technology that liberates us from toothache leaves us free to choose, or to find, our own meanings and values and pleasures and depths and revelations. Reason is the beginning, not the end, of the full life. There is thus no enmity between reason and creativity: reason creates the conditions in which creativity is possible. The true enemies of creativity are toothache, unreasonable oppression, unchallenged cruelty, unaccountable government, hunger, fear, etc.

The two complaints that first, reason does not satisfy the whole man and secondly, political theory based upon Enlightenment thought was too shallow because it did not address the fundamental wishes of mankind are really two aspects of the same complaint and can be answered in the same way. Indeed, they answer one another. Yes, a politics based upon reason is shallow and thank God that it is. The social order should not be, or prescribe, the ends of life: politics should fall short of final ends. A successful political system *should* be 'shallow' in this sense, so that it can free people to discover and make their own depths.[30] May we be protected from political leaders who want to legislate for our depths, who think they can answer to our deepest needs rather than address the things that prevent us from discovering, inventing and meeting our deepest needs. The history of the world has so often been disfigured by religious theocrats and secular dictators (of Right and Left) who busied themselves too much with the people's depths instead of permitting them to be free to find their own depths. Three cheers for the beneficent power of reason. And three more cheers for its shallowness.

UTOPIA AND THE REPRESSION OF TRAGEDY

Utopian thinkers have often been condemned for failing to take account of, or for actively repressing, the essentially tragic nature of the human condition. In his *Sketch*, Condorcet envisaged an indefinite extension of human life as part of the indefinite perfectibility of the human race:

> Organic perfectibility or deterioration among various strains in the vegetable and animal kingdoms can be regarded as one of the general laws of nature. This law also applies to the human race. No one can doubt that, as preventive medicine improves and food and housing becomes healthier, as a way of life is established that develops our physical powers by exercise without ruining them by excess, as the two most virulent causes of deterioration, misery and excess wealth, are eliminated, the average length of human life will be increased and better health and a stronger physical condition

will be ensured. The improvement of medical practice, which will become more efficacious with the progress of reason and of social order, will mean the end of infectious and hereditary diseases and illnesses brought on by climate, food or working conditions. It is reasonable to hope that all other diseases may likewise disappear as their distant causes are discovered. Would it be absurd to suppose that this perfection of the human species might be capable of indefinite progress; that the day will come when death will be due only to extraordinary accidents or the decay of vital forces, and that ultimately the average span between life and decay will have no assignable value? Certainly man will not become immortal, but will not the interval between the first breath that he draws and the time when in the natural course of events, without disease or accident, he expires, increase indefinitely?[31]

As Condorcet admitted, no amount of progress will alter the fact that we live a life that underlines our finitude in an infinite world, and that we die after a brief period of time that falls infinitely short of eternal life. Our potentially limitless longings will still receive a limited answer in any conceivable Utopia. Material progress will not palliate the sadness that comes from the transience of everything – including those whom we love. In short, Utopian thought does not take account of the inevitability of death and, for some, it is therefore invalidated or dismissed as incorrigibly shallow.

How shall we answer this criticism, which may, with equal justice, be directed against any melioristic philosophy? Perhaps by pointing out that the human condition has two aspects: an irremediable transience; and remediable woes that disfigure the (comparatively) brief stay on earth our transience affords us. Progressive thinkers, concerned with the Kingdom of Means, address the quality, and to some degree the duration, of that stay. The earthly paradise of the *philosophes* does not offer eternal life. 'No-one is so old that he does not think he could live another day' as Cicero noted in *De Senectute*, but immortality will never be achieved. For all forms of life, and in particular human life, are highly complex and entrain high levels of negative entropy;[32] in this sense, human beings persist in the teeth of their own intrinsic improbability. The Second Law of Thermodynamics ('the most metaphysical of all laws of physics', as Boltzmann proclaimed it) predicts that all systems tend towards increased disorder. Immortality, therefore, seems about as likely as a perpetual motion machine, though not as precisely forbidden because the energy to maintain order can be requisitioned from elsewhere. Repair mechanisms retard the expression of the Second Law in living organisms and, as Kirkwood[33] has pointed out, the allocation of resource between reproduction and repair is an evolutionary choice. However, not even the quasi-immortal sea anenome would survive the heat death of the universe or the Big Crunch.

At any rate, we may anticipate that future progress in medical science will make possible only finite additions to lifespan, palliating rather than curing our transience. This raises the question of what (finite) additions to a finite lifespan are worthwhile? The answer will change as our perceptions of the curve of life are altered by medical and social advance; in particular the definition of premature, 'tragically early', death will be revised upwards. Nietzsche's Zarathustra recommended that, since we cannot live for ever, we should at least die at the right time. When is the right time to die? According to Paul Valéry's *M. Teste*:[34]

> It is said that there are two kinds of death, the *natural* (complete) and the *ordinary – giving back* to the world nothing but a corpse *empty of its possible consciousness.*
>
> The *ordinary* is the ordinary dead man (and on his features, the expression of a man surprised and slightly shocked, impolitely interrupted by some trifle in an interesting conversation).
>
> The *natural* or true death would be the total exhaustion of the possibilities of the system of an individual man. All the inner combinations of his capacities, incomplete in themselves, would be exhausted. He has told himself everything he knew.

This seems an unlikely prospect and we may assume that all human beings will leave much unfinished business behind when they die and death will remain as poignant. Does not the Utopian dream of progress, therefore, distract from this fundamental certainty and so render us spiritually more shallow? I don't think so; indeed, I would argue the reverse: life and death in Utopia will be more, not less, metaphysical.

With more effective ways of retarding the onset of diseases and limiting their adverse effects, it seems likely that 'old age' will come to play a bigger role in limiting the quality and duration of life.[35] The distinction between disease and ageing is not as clear-cut as has been suggested by those who have been appropriately anxious that woes in older people should not be dismissed as (untreatable) 'ageing' and opportunities for improving (treatable) illnesses lost. Even, however, supposing ageing and disease were clearly separable, they would still interact and converge, having a common ultimate outcome – death – and a common pathway to that outcome – homoeostatic failure. The question that then concerns us is whether death purely or predominantly by ageing would be an advance over death by clearly defined disease. Death *in* old age will, of course, seem more appropriate (or less inappropriate) than death in youth; but, beyond this, death *from* old age may be less unpleasant, not being associated with intrusive symptoms such as pain, nausea, shortness of breath and gross disability. Instead, we may envisage a subtle and progressive reduction in life-space associated with an increased probability of a demise that is more easily achieved – as if the distance to be traversed between life and death had been abbreviated. The image of death by ageing as the end-result of gradual but harmonious failure of all organs is attractive. It is compatible with current conceptions of ageing in the absence of clearly defined disease, which suggest a picture of progressive, roughly synchronous decline in function of many different organs.[36] Such a death would seem to be likely to be more conscious, more metaphysical, than death typically is at present. 'Do not go gentle into that good night.' No; but do not go kicking and screaming, either. Instead, proceed by a series of grey-scale gradations of evening to oblivion. The tragedy is not blunted, but purified of the kind of distractions that dominate decline and death at present.[37] Physical suffering is not necessarily a more translucent metaphysical window than painless decline; quite the reverse: to suffer is to be nailed to the particular, to endure an involuntary narrowing of an attention made almost absolute.

Utopia and Utopian medicine will not, therefore, cure transience, but may permit a death that is more in keeping with the possibilities of man the metaphysical animal. It is absurd, therefore, to see progress towards Utopia as being a means by which humankind is made shallower; on the contrary, it may be the means by which human beings come nearer to fulfilling the mysterious potential within them to become ever more richly and complexly aware of themselves and of the world around them.[38]

THE ENLIGHTENMENT, NATURE, REASON
AND HUMAN NATURE

Enlightenment thinkers were inspired by the example of science and by the dream of applying the methods of science to the social and political sphere:

> What science had achieved in the sphere of the material world, it could surely achieve also in the sphere of the mind; and further, in the realm of social and political relations. The rational scheme on which Newton had so conclusively demonstrated the physical world to be constructed, and with which Locke and Hume and their French

disciples seemed well on to the way to explaining the inner worlds
of thought and emotion, could be applied to the social sphere as
well. Men were objects in nature no less than trees and stones; their
interaction could be studied as that of atoms and plants. . . . Nature
was a cosmos: in it there could be no disharmonies; and since such
questions as to what to do, how to live, what would make men just
or rational or happy, were all factual questions, the true answers to
any one of them could not be incompatible with true answers to
any of the others. The idea of creating a wholly just, wholly virtuous, wholly satisfied society, was therefore no longer utopian.[39]

Unfortunately, as Berlin points out, 'the central dream' – 'the demonstration that everything in the world moved by mechanical means,
that all evils could be cured by appropriate technological steps, that
there could exist engineers both of human souls and of human bodies'
– proved to be a delusion. And it could be argued that both the successes *and* the failures of the Enlightenment programme for the progressive improvement of the lot of mankind and the indefinite
perfectibility of man himself, turned on the conception of man as a
piece of nature. For this conception is both true and not true. It is
certainly true in so far as humans are bodies, which may be fruitfully
seen as complex physico-chemical systems whose properties are covered by the same physical and chemical laws observed in the rest of
the natural world. They are living organisms which share much with
other living organisms. It is equally certainly untrue in so far as humans are thinking, reflecting agents utilising those laws in order to
achieve goals that they themselves have set. Man is uniquely that
piece of nature in virtue of which nature is made explicit as the scene
of his unfolding life.

The Enlightenment thinkers correctly saw that, in a sense, man was
an animal; but they crucially failed to see the extent to which he was
an *explicit* animal able, amongst other things, to see his status as an
animal and, for example, to draw certain conclusions from this about
the way to reduce the dependency and limitation originating from
the animal condition. If human beings were merely collections of atoms
pushed around like other collections of atoms in the universe (as
the more hard-line reductionist faction amongst the Enlightenment
thinkers thought), was it not strange that they should be also, uniquely,
collections of atoms that developed (collective) theories about the nature
of atoms and about the ways in which collections of atoms might be
expected to behave? In common with many other thinkers (including
contemporary physicalists) some of the *philosophes* and their materialist successors overlooked the very peculiar and mysterious nature of
a piece of nature that, alone amongst all the pieces of nature, was in

possession of the abstract concept 'Nature' so that it could, among other things, assert its own status as a piece of nature.

I won't here reiterate the inexhaustibly long list of consequences of the fact that man is, uniquely, the 'Explicit Animal', and that, as the early Marx pointed out, free, conscious activity is his 'species-characteristic'. I have done so at length elsewhere.[40] But it will be helpful to relate this overridingly important fact about us to the dream of Utopia.

It makes Utopia more difficult for a start. As Popper has pointed out, changing social and political systems is not the same as changing physical systems, for the members of the former are aware of what is going on and may have their own views about it. Even descriptive laws of society are difficult to derive, not merely because those who would derive them are themselves part of society – and so do not have the disinterested transcendent relation to their object of study that scientists studying bacterial colonies do – but also because the very fact of making predictions alters social behaviour and so confounds the laws. But beyond this obvious negative point, which was obvious only after Popper had pointed it out in his classic critiques of historicism, there is a more positive one: that the explicitness of human beings opens up a limitless proliferation of possibilities and is an inexhaustible source of depth. This was touched upon when we addressed the question of whether Utopia would be boring and/or empty. It is equally relevant when we consider whether or not a life enacted within a framework of reason would be regimented, dull, oppressive and threatening.

The common assumption that such a life would have all these unpleasant negative qualities is based on a misunderstanding as to the nature of reason, in turn rooted in a failure to see the extent to which subordinating oneself to reason involves active participation. Reasonable behaviour requires the deployment of the imagination – at the very least to ensure an understanding of the basis and scope of the rules to which one is submitting. It does not require blind submission to an algorithm, passive acquiescence in a felicific calculus. Being reasonable makes complex demands upon a perceptiveness that goes beyond any kind of mechanical, syllogistic calculation. Reasonableness is not a just a matter of mobilising a blueprint rationalism or logic machinery – of mere passive submission.[41] It is, therefore, necessary to dispose of the idea that rationalism and active creativity are somehow opposed.

At the most superficial level, it is obvious that reasoning on the basis of knowledge is not alone sufficient to dictate a course of action. This is evident in medicine, where despite the availability of very detailed knowledge of diagnostic procedures and treatment protocols, there is a good deal of room for personal style, and where decisions have often to be made on the basis of incomplete knowledge and imagination has to inform even the most casual encounter with a patient. If it is true of medicine, which is comparatively unusual among the humane arts in the degree to which its practice is based upon carefully sifted evidence gathered under conditions designed to minimise bias, how much more true must it be in the wider conduct of everyday life, where the printed directions are even more scanty. There will always be a gap between the general rules and the actual choice of action in a particular situation. There is, therefore, no danger that life in Utopia will be conducted by algorithm. At a deeper level, it is even more obvious that creativity and reason are not mutually exclusive. The human animal follows rules consciously, rather than being manipulated by them. In order to do so, it is necessary for her to understand the rule, to assent to it, to see that it applies to a particular situation – in other words, to see that a rule can be invoked – and to see how to apply it. At every level, and in every sense of the word, this is consciousness at its most productive and creative.

The Enlightenment notion of man as a piece of nature, therefore, requires cautious interpretation, if we are to avoid falling into *l'homme-machinisme* and are to ensure that the notion of a rule-governed rational society is not seen to be a straitjacket and inimical to the natural creativity of man. For the idea of man as a part of nature is ultimately a liberating one: it loosens the hold of religious authorities, prevents us from seeing humanity as composed of pieces of spirit opaque to all except the priests. The positive aspect of the Enlightenment naturalistic anthropology may be retained, therefore, by thinkers, without their falling into the traps I have discussed at length in Part I, if they acknowledge the unique status of man as an explicit animal, distanced from physical and biological nature, not by a transcendent angelic spirit but by the capacity to make things explicit and to live explicitly. It is in virtue of this capacity *l'homme surpasse infiniment l'homme*.[42]

CONCLUSION: THE ENLIGHTENMENT DREAM AND THE HOPE OF PROGRESS

The deficiencies in the Enlightenment notion of human nature (in so far as a common picture can be extracted from the writings of so many disparate thinkers) and in the programmes of the *philosophes* for improving the conditions of human life are easily stated. The simplistic assumption that there would be a universal solution to human ills – based upon the methods of and guided by the findings of physical science – is foremost among these. Arrogant scientistic

universalism has not, of course, been the sole preserve of Enlightenment thinkers, but some of them were perhaps the first to make it
programmatic.[43] And like many other thinkers before and since, they
failed to recognise the extent to which they were children of their
time. It is, however, a historicist exaggeration to see Enlightenment thought, as Goldmann does, as merely a stage in Western bourgeois thought and rooted in a society that is based on exchange value
(whence come its ideas of equality, the social contract, toleration,
universalism); but it is equally mistaken to see it as Man's first
and definitive discovery of his own unchanging essence. We recognise that many of the values that the *philosophes* saw as self-evident,
eternal truths are neither self-evident nor eternal; and we also recognise that the assumption that all values can be brought into harmony
requires careful interpretation. When Helvetius asserted that moral
laws come from the individual's pursuit of his own happiness, and
that it is in the individual's interest to promote the general welfare,
since his own happiness depends on other people, he not only failed
to take account of the people having conflicting interests in the pursuit of the same ends, but also failed to recognise that people might
have fundamentally different ultimate ends; that, in other words,
it is not only intermediate means but final goals that may be the subject of dispute. The belief that everyone could be paid-up members of
the Party of Mankind – patriots for humanity rather than for their
country or simply for themselves – that there were values that tran-
scended one's self, one's faction, the interests of one's social group,
brings to mind too many memories of doomed ventures into universal brotherhood – malign ones such as International Communism and
benign but ineffective ones such as the League of Nations or The
Committee for World Government. The scientistic version of universalism carried additional dangers of demeaning those whose only role
was to be the substrate of the beneficent, improving intentions of the
legislators. We have discovered how important it is to recognise and
curtail

> the tendency – difficult to avoid but disastrous – to assimilate all
> men's primary needs to those that are capable of being met by these
> methods: the reduction of all aspirations and questions to disloca-
> tions which the expert can set right. Some believe in coercion, others
> in gentler methods; but the conception of human needs in their
> entirety as those of the inmates of a prison or reformatory or a
> school or a hospital, however sincerely it may be held, is a gloomy,
> false, and ultimately degraded view, resting on the denial of the
> rational and productive nature of all, or even the majority of, men.[44]

This fault is not, of course, peculiar to the *philosophes* and their heirs; indeed, it has been most strikingly practised by anti-Enlightenment (Marxist, Fascist) legislators; but the Enlightenment programme always, on account of its confident universalist scientism, carried the possibility of becoming coercive, patriarchal and physicianly, on the grounds that, as Heracleitus expressed it, 'the beast has to be driven to the pasture with blows'. The Committee of Public Safety was not entirely a travesty of Enlightenment paternalism. The journey to ruthlessness in Utopian dreamers possessed of universal truths is not always a long one, and does not always have to pass first through corruption. Moreover scientism – and the belief or pretence that the pursuit of life, liberty and happiness can be reduced to a matter of utilitarian calculation – harbours other dangers. A morally neutral, purely technical and managerial approach to social problems, runs the risk of creating a moral vacuum which may predispose to political instability. Once this is admitted, it has also to be acknowledged that moral passions infused with a desire to control others, and reinforced by appeal to a transcendental authority and supported by power structures, are no less dangerous.[45]

I hope I have shown that progressive thought does not necessarily entail subscribing to an oppressive universalism. It is possible to identify a core of human values that it is not unreasonable to assume would be invariant across cultures; and from these, much follows that would be consistent with the Enlightenment approach to the woes of the human condition. This would include individual accountability and the deployment of reason. As has already been pointed out, reason does not necessarily strangle creativity and freedom – not only because its intelligent deployment requires a creative imagination, but also because its meshes are not tightly drawn. Likewise, politics based upon certain principles such as reason, accountability of the governors to the governed, a commitment to respecting the lives of all citizens, while it may influence the content of consciousness and the meanings that are available to people, are not themselves those meanings and the whole story of that consciousness. It is important that they are not, that they do not pretend to be. The personal is political only in so far as the political obtrudes beyond its proper sphere – and people are either incompetently or oppressively governed.

Faith in the application of reason to human affairs should be hard to challenge now that we have ample experience of the consequences of irrationalism translated from the private to the public sphere. The 'shipwreck of rationalism' described in John Stuart Mill's *Autobiography*, where he discovers that the fulfilment of utilitarian ends is simply not sufficient an aim to nourish the spirit, has been overshadowed by the greater and more appalling shipwrecks wrought by irrationalism.

Likewise, it will be difficult to go back to the charismatic authority of priests with their transcendental warrant to guide human affairs now that we can see what is possible when practice is evidence-based and evidence derived from cooperation in uncertainty rather than fabricated out of autocratic certainty. The occasional outbreak of charismatic leadership – Branch Davidian-style omniscience and other manifestations of the horse-shit of the Apocalypse – does not inspire confidence in the benefits of regression to earlier modes of establishing authority. And while, as Barthes argued,[46] the Enlightenment-universalist notions of the Great Family of Man may be sentimental masks to conceal the reality of exploitation, they are infinitely preferable to the fevered *volkisch* ideology that had its roots in Counter-Enlightenment thought and culminated in the *Volksgemeinschaft* concept in National Socialist ideology, where it provided the rationale for the Nazis' *Judenpolitik* and the concentration camps.[47] And although outrage at inequality and suffering, and a desire *ecraser l'infame*, has sometimes led to outrages of its own and added to human suffering, it is still more likely to improve the lot of mankind than a passionate concern to preserve the privileges of the privileged, concealed beneath some patrician rhetoric about the intrinsic superiority of the upper classes and their God-given right to rule.[48] The fundamental vision of the Enlightenment – that we are all of us, above all, members of the human race – not only elevates the oppressed, but unmasks the oppressors: it strips away the aura behind which they hide from critical evaluation. And the belief in the indefinite perfectibility of man is greatly to be preferred to an inert or paralysed pessimism, 'a drooping despondency that offers no remedy for the abuses it bewails'.[49]

There is much to criticise in dreams and fantasies of the Enlightenment *philosophes*; but it is difficult to quarrel with their wish to make life better and their rage that unnecessary suffering is permitted and that superstition, intolerance and illegitimate authority are allowed to work so effectively to maintain a world in which such suffering continues. Nor can one quarrel with their belief in the essential goodness of man – at least as a 'regulative idea' in the Kantian sense. The fact that Condorcet's dream of the indefinite perfectibility of man received its finishing touches while he was being hunted to his death by his fellow heirs of the Enlightenment is jaw-achingly funny only to the malicious and shallow, those who, from positions of comfort, may bear the sufferings of the world with considerable ease.

In the Preface, I have described this book as a 'Yes–but' to Berlin's 'Yes–but to the Enlightenment'. In particular, I have focused on those more recent thinkers who believe that there are fundamental reasons why what we call, in shorthand, 'the Enlightenment project' – something between a programme and a dream – of progressive improvement of the lot of mankind is misconceived. These contemporary prophets of the Counter-Enlightenment seemed to be vindicated by the man-made catastrophes of recent history; by the depressing fact that, in Adorno and Horkheimer's words, 'the fully enlightened earth radiates disaster triumphant'. And, indeed, there are some reasons for believing that this age, 200 years on from the *philosophes*, is the worst of ages: wars on an unprecedented scale; monstrous institutions and regimes marked by mass torture, concentration camps and genocide; greed-driven ecological disasters; and so on. However, the scale of wars may reflect technological advance (including better methods of transport and communication permitting mass mobilisation, as well as more destructive weapons) rather than moral decadence. Increased technological capability may also account for the greater expression of evil in evil regimes; after all, mass murder and unremitting persecution is not new and there is no horror in the twentieth century that has not been perpetrated on a huge scale in previous centuries. The treatment of the slaves by the Egyptians, the Mayan sacrifice of the prisoners they captured in wars launched specifically for that purpose, the Athenians' savage destruction of Samos match what has been achieved in the twentieth century, once one allows for the constraints imposed upon earlier barbarism by technological limitations. The history of the world is, as Nietzsche said, the refutation by experiment of the idea of a moral world-order; it has always been a story of injustice, cruelty, chaos, oppression and of privations suffered by ordinary people cheek-by-jowl with over-provision for the privileged few. Indeed, in view of the enormously enhanced power of individual human beings through technological advance, it is a tribute to progress that it has *not* led to universal oppression and universal warfare, along the lines predicted in many dystopic fictions.

For perhaps the first time in history there are sizeable enclaves of comfort and justice in the world and, even more surprisingly, these enclaves include many ordinary people. This is the first century in which, in some countries at least, the rhetoric of the Rights of Man has been taken seriously and translated into the accountability of officials, politicians, the establishment, to the mass of the people. Abuses of power are still present even in these countries, but they are visible and challenged. In the United Kingdom, the experience of work, of illness, of going out into the street, of being a school-child, have all changed for the better – beyond all recognition. To take some aspects at random: work is no longer, or rarely, excruciating labour supervised by a self-indulgent bully against whose reign of terror there is no appeal; the care of ill people is not only technically more competent, but is administered by individuals whose natural tactlessness (we are all born congenitally tactless) is modified by an awareness of

how things look from the patient's point of view; a walk down a
busy street is not a journey through offal; and school-children are no
longer required to tolerate the capricious cruelty of ignorant teachers
and malignant peers without redress. This is not to say that cruelty
of all sorts does not occur in the United Kingdom: child abuse, wife
battering, gang terrorism are daily news. But what has changed is
the scale of the cruelty and the disappearance of the assumption that
it is normal, acceptable or natural: it is more visible precisely because
it is no longer acceptable. In short, in every sphere, individuals treat
each other better and with more respect. This may be because, on
account of modern technology, they have more spare capacity to do
so. If so, this cannot be a case against modern technology – or against
this century.[50]

The idea of the twentieth century as a time when people treated
each other with greater respect is not one that will receive ready assent: respect for other people was not the notable feature of the 'great'
wars, of the concentration camps, of the Gulags, or of the colonial
wars any more than it has been particularly in evidence in recent
years in Rwanda, the former Yugoslavia, Somalia, South Africa, and
so on. The list of times and places where respect has been wiped out
by atrocity is dismayingly long. But that is not the point I am making. I am arguing that, perhaps for the first time in history, there are
now some societies, where most people, most of the time, not merely
the well-off or the powerful, are treated by most other people most
of the time, in a manner that is consistent with the assumption that
they have equal entitlement to respect. The gross abuse of 'inferiors'
is not universal and unchallenged.

This statement may prompt two other questions: Is this improvement more than offset by deterioration elsewhere? Is the deterioration elsewhere actually *caused* by the improvement in those countries
such as the United Kingdom where ordinary people are better off?
Should we regret the precipitous decline in infant mortality amongst
working-class Britons because of its cost to other countries? These are
unanswerable questions: answering them would oblige us to carry
out a 'global feliufic calculus' which is simply impossible; and oblige
us to trace causal relations beyond the point at which one can do so
with any kind of confidence. Whether, for example, the improved
sensitivity of the nurses caring for terminally ill patients in hospices
in the United Kingdom has been bought at the cost of more brutal
treatment of child workers in sweat-shops in Bombay is not easy to
answer. Certainly, it could be argued that the labour- and revenue-
intensive care of dying cancer patients is dependent upon the investment of a greater proportion of Gross Domestic Product in health
care, and that this depends directly or indirectly upon the importa-

tion of cheap goods, and that the cheapness of these goods depends upon a brutalised child labour force. One could equally well say that a country such as India, which has a long and terrible tradition of classifying some of its citizens as 'untouchables', has always condoned the exploitation of the lower orders by the wealthy middle and upper classes and that child labour is eternal business-as-usual.

All that we need to do is to put into question the myth of 'The Myth of Progress'.[51] Yes, it *is* possible that, overall, progress has not been made; that the sum of human suffering is no less now than it was 100 years ago. But this is far from self-evident; so it cannot be taken for granted, as it is by so many humanist intellectuals. There is, in fact, a considerable body of evidence to support the contrary claim that there *has* been overall global improvement in standards of living, and even, possibly, in quality of life. John Tierney[52] has pointed out that:

1. Over the last few decades, infant mortality has decreased and life expectancy has increased, most dramatically in the Third World.
2. Although many people in the Third World have been affected by war, persecution, drought and disastrous agricultural policies, the number of people affected by famine has been declining over the last three decades. The number is lower than it was during the same decades of the last century, even though the world's population is much larger.
3. The average person in the Third World is better nourished now than in the late 1960s. Food production has increased faster than the population.
4. The average worker world-wide can buy more coal, metals and food with an hour's pay than he could a century ago.

But even if life were not overall better for the average human being than it was a century ago, we still have to ask ourselves what message we would draw from this about the way forward. Is there any alternative to that advocated by the Enlightenment thinkers?

Let me put the question more bluntly. If you don't believe in reason – on the grounds that it is an instrument of domination (e.g. of females by males or of non-Europeans by Europeans) – what will you put in its place? Unreason?

As Gellner has said:

In our human self-image and self-assessment, the claims of unreason appear to be overwhelmingly strong, though it is not very clear just how we should live by the contrary, irrationalist vision. Its oracles speak in nebulous and murky language, and their pronouncements are allergic to clarity.[53]

If you throw away reason (and its despised cousins, logic and common sense), will there be any restraints on choice of actions, on methods, etc. of achieving human ends? And if you relativise all ends and all values, will you be happy to support a global policy that aimed, let us say, for the greatest unhappiness of the greatest number? (In a sense de Maistre takes this path; though he does not make clear whether he regards the eternal bloodbath of life as positively desirable or merely inevitable.)[54] And if you don't believe in science and technology, what alternatives do you have to offer when it comes to meeting human needs – for shelter, warmth, food, drink, etc.? How shall these be provided except by better understanding of how safe, clean food, etc. is produced, and what the body needs, by the kind of understanding exemplified by the discovery of vitamins and the progressively more subtle understanding of their actions? And what do you have to offer when it comes to solving the secondary problems that arise as an unlooked for consequence of scientific solutions to human needs? (As Medawar pointed out, 'the deterioration of the environment produced by technology is a technological problem for which technology has found, is finding, and will continue to find solutions'.)[55] After all, as I have noted on another occasion, it was high science and not The Children of the Celtic Dawn who discovered the hole in the ozone layer (as well as the ozone layer in the first place – and ozone – and its role in filtering out harmful radiant energy – and radiant energy – and the harm it sometimes does). We need as much science in the application of science as in science itself, as much technological tact and subtlety in regulating the application of technology to solving problems as in developing the technology in the first place.[56] And if you don't like modern democracy – which, in principle at least, regards each individual as equal to every other, an equality that is most directly expressed in the principle of one person one vote – what will you put in its place? Is there any other form of government – autocracy, oligarchy, ochlocracy, the varieties of dictatorship of Left and Right – that have proved to be better safeguards of the welfare and aspirations and dignity of ordinary people? Democratic accountability at all levels of public life may be imperfect, but are there other ways in which accountability may be better secured? Or is the unaccountability of a charismatic leader with transcendent authority that comes from either spiritual or temporal power better? History tells us not.

And if, finally, you are disaffected with contemporary 'Western' civilisation, what alternative do you have to offer? If you don't like the comforts it brings, will you seek out discomfort for yourself and your children? If you don't like its emphasis on material progress, will you seek to extend destitution? All the endless talk about 'aliena-

tion' and moaning over the modern division of labour that has supposedly reduced individuals to mere functions has not contributed
one quantum of light to understanding the precise nature of the
discontents humanist intellectuals routinely express with respect to
civilisation. The grumblers have never, for example, addressed the
question of whether hoeing a turnip field by arthritic hand hour after
hour in a biting wind is more or less alienating than typing memo-
randa on a word processor in a warm office; or whether the rational
(and hence, we are to understand, the dehumanising) approach to
disease results in more or less alienation – whether a course of antibiotics is more dehumanising than the unremitting savagery of untreated cystitis.

It seems that, if we are really to will progress, we have no alternative but to work within the framework of Enlightenment thought,
though our approach will be greatly modified in the light of lessons
learnt. These lessons would include:

1. Entertaining more modest ambitions than the *philosophes* did for
the scope of interventions and the rate of change. Modest interventions, whose effects are carefully monitored (so that, at the very least,
lessons can be learned from the unexpected outcomes) and infinite
patience – these are crucial. Humility and admitted uncertainty should
be the guiding emotions of social planners.[57]

2. Avoiding scientism – that is to say, using a cargo-cultists parody
of what is believed to be '*the* scientific method' in areas where even
the approaches of successful science would be inappropriate – and
the related assumption that a single method can be used to understand, even less to bring about, desirable social change. This does
not, however, mean failing to recognise that there are transferable
virtues in science: models of cooperation; a commitment to testable
hypotheses; recognising the limits of certainty to which the existing
evidence commits us.

3. Thinking through as clearly as possible the nature of universal
or species wishes as a guide to identifying culturally invariant values
and at the same time learning how to respect differences more actively. This will require all of us being aware of 'repressive tolerance', whereby dissenting world pictures are accommodated by not
being taken seriously.

4. Being wary of excessive centralisation of power: men and women
of apparently good intention must be as closely regulated and as
answerable and accountable as people of less good intentions. Being
accountable to invisible forces, to transcendent auditors – God, the
National Interest or the Future of Man – is an insufficient safeguard.

5. Respecting the anti-civic heart of man; understanding, as J.S. Mill
came to understand in the anguish of his mental breakdown, that

reason and duty do not reach to the very bottom of our souls. Thinking how to reconcile this recognition with the need to ensure that all contribute to the process by which the civic order necessary for private depths to be developed is maintained.

6. Recognising that, even in a pre-Utopian world of scarcity, utility cannot be the only criterion of value.[58] This acknowledgement of the validity of non-utilitarian values, recognises that many human activities are expressive – self-exteriorising, self-exhibiting, 'that I am' – rather than usefully directed towards some practical end and cannot be captured by a reforming rationalism that would make them more practical, efficient, etc. Language, for example, would probably not be improved as a means of communication – and certainly not enriched – by being transformed into a *characteristica universalis*. Where non-utilitarian values become anti-utilitarian – even if only by virtue of diverting scarce resources from basic needs to less basic ones – is a difficult and delicate question and will require much careful thought. If a Welshman still insists on having road signs in Wales in Welsh, because of their cultural significance, even if the result is an increase in road accidents; or if he insists on having them in both English and Welsh even though the resource to do this will be indirectly taken from the Special Care Baby Unit budget – how shall we, Welsh and non-Welsh, respond? In setting a legislative course that apportions respect to both the expressive and utilitarian aspects of life, reason in the narrow sense and reason in the wider sense of a faculty that incorporates an empathetic understanding of human wishes are both essential.

7. Acknowledging that, ultimately, the Kingdom of Ultimate Ends must be left to itself.

This would probably be accepted by moderate and reasoned critics of the Enlightenment, such as Berlin. His 'Yes-But', informed by an intense awareness of the horrors unleashed upon the world by intemperate and arrogant social reformers of all kinds, still leaves most of the framework in place:

> Yet what solutions have we found, with all our new technological and psychological knowledge and great new powers, save the ancient prescriptions advocated by the creators of humanism – Erasmus, Spinoza, Locke, Montesquieu, Lessing and Diderot – reason, education, responsibility – above all, self-knowledge? What hope is there for men, or has there ever been?[59]

There may be some who feel that the 'ancient prescriptions' ask too much of mankind; or that Berlin's list of the Great and the Good is suspiciously Eurocentric; or that, even if these 'prescriptions' are acceptable, genuine human universals, they will never bring about net progress because sooner or later they run into contradictions – if only because, they argue, the sum of even well-intended actions will inevitably bring about unintended consequences most of which will be unpleasant. To such we can only say that there is no *a priori* reason why progress is not possible. Yes, we still have a long way to go; there is much that is sickening, angering, horrifying about the condition in which a significant part of humanity still lives. But it is nevertheless still possible to admit that there has been progress (in some places) without being complacent or cruelly indifferent to the continuing avoidable suffering of many of our fellow-men and without denying that progress for some has been bought at the cost of further immiseration of others. This, surely, cannot be too difficult to understand – so long as one is not a humanist intellectual with a vested interest in crying universal woe. And acknowledging the serious things that are amiss with the modern world, some of them resulting from misapplied technology, does not justify crying 'failure!'

If we deny or rubbish the progress that mankind has already made, and at the same time are aware of the huge efforts mankind have made to ameliorate the human condition, we shall inevitably conclude that no progress is possible. Such 'principled' despair will be a thousand times worse in terms of quietism than the most arrant care-nothing, do-nothing conservatism. An attitude wavering between fatalism, cynicism and moral superiority may suit the purposes of humanist intellectuals who prefer the comfort of the seminar room to the relative discomfort of the places where the real work of bringing about a better future must take place. It lets them morally off the hook – just as does the idea that there is no truth (only the dominant rhetoric of particular interpretive communities) and no genuine agency (only passivity in the seas of history, discourse, the unconscious, or whatever). 'Drooping despondency' makes very little demands on one's free time.[60] But we must refute those for whom (to parody Keats) 'the miseries are the world are misery and let them rest'. For, as Medawar has pointed out,[61] although humans have been around for 500 000 years, it is only during the past 5000 years that they have won any kind of reward for their special capabilities and only during the past 500 years have they begun to be, in the biological sense, a success. 'Only during the past 10 to 15 minutes of the human day has life on earth been anything but precarious.' Technology has been really effective – because driven by science and a fundamental understanding of natural laws – only in the last 50 years. Reason is a comparative newcomer in human affairs and a neonate in the history of living things.[62]

Opposition in principle to the idea of progress, based upon assumptions about the nature of mankind – Original Sin, aggressive animal nature (ethology, Social-Darwinism), incurable irrationality (anthropology) – or about society (it is too deep to be understood, a collection of opaque forces rather than the summed activity of human agents) – simply fails to see the whole story. None of the theoretical reasons for denying the hope of progress is decisive. Nor, it must be admitted, are there irrefutable reasons for assuming that progress is guaranteed or inevitable. One would have to be a Hegelian or a Marxist to be stupid enough to believe that progress will come about of its own accord. If we believe, as I believe, that it has to be brought about by human effort, human beings mobilising the abstract intelligence and universalising goodwill that they uniquely possess, there is no certainty that the future will be better than the past. So we are left with a secular equivalent of Pascal's wager, which I commend to the reader.

As Pascal pointed out, nobody can be absolutely certain that God exists. We are in the position of best-guessing gamblers, making absolute and irreversible decisions in the context of uncertainty. What, then, should we do? Pascal recommends believing in God, for this will place the believer in a no-lose situation. If he is right, then he will be appropriately rewarded when he meets his Maker face to face. If he is wrong, he will will not suffer for his credulity in the after-life of total oblivion. If, on the other hand, he wagers on the non-existence of God, his reward, if there is no God after all, is to enjoy the same oblivion as the believer. But if God really does exist, then he will be condemned to Eternal Damnation as punishment for his error. Pascal's wager is not an entirely full or fair statement of the case, if only because there is quite a range of gods to choose from and the result of choosing the wrong one could be persecution on earth and damnation in the after-life. Nor does it take account of the psychology of religious belief: the true experience of God should be (as Nijinsky proclaimed) 'a fire in the head' rather than the outcome of a prudent calculation of probabilities. We can, however, usefully transpose Pascal's wager to the secular sphere and use it to think about the hope of progress. If we believe in the possibility of progress, we may or may not be successful in bringing it about. But if we deny the possibility of progress, then, since it will not happen of its own accord, we shall ensure that progress shall most definitely not come about. For the sake of the hungry child in the dust, we should not allow those who prophesy doom and gloom to speak unopposed; otherwise their prophecies will help to bring about their own hideous fulfilment. And more hungry children will die in the dust, while the prophets of gloom, of course, continue to enjoy life in the library and the seminar room.

And perhaps for our own sake as well. Once you throw away belief in progress and the desire to make progress – the passion to alleviate human suffering here and now and in the future, on a small scale and on a large scale, locally and globally (and, as we denizens of the global village are aware, the distinction between these categories is not absolute) – then you have thrown away one of the deepest and most noble and fertile sources of goodness in human beings and, effectively, much of the underpinning of civilisation. For a truly human culture is always – though never exclusively – preoccupied with improving the lot of mankind and in modern times this has taken the form of concern about justice for all, about the rights of the many, about enrichment of the poor and empowerment of the powerless. Great, rich cultures have a generosity that is implicitly on the side of progress (even if it is not Utopian or explicitly progressive). The only question for such cultures is whether progress is pursued well or badly, effectively or ineffectively. As Medawar has said, 'The idea of improvement must be pretty well coeval with human speculative thought. In one form or another it embodies almost the whole spiritual history of mankind.'[63]

The enemies of hope have found their own reasons for dismissing the Enlightenment dream, without, perhaps fully realising what they are doing – or what they would be doing if the world took them seriously. Hitherto, those who have rejected earthly happiness have had alternative, next-worldly, futures to look forward to. In the absence of such alternatives, to dispense with the hope of progress, to mock 'the grand narratives of emancipation and enlightenment', is to lead humanity towards a collective despair perhaps unprecedented in articulate cultures. Or, more likely, since even the most articulate pessimists are not notably lacking in personal ambition and concern for self-advancement, to set an example to the well-heeled sections of the race that will encourage them to pursue their own happiness and forget that of humanity as a whole.

For the sake of our humanity, then, as well as for the welfare of those whose lives would otherwise be Hell on earth, we must believe in, and strive for, progress, as did those noble philosophers of the Enlightenment.[64] 'To deride the hope of progress', as Medawar says, 'is the ultimate fatuity, the last word in poverty of spirit and meanness of mind.' This book has been written in the hope that such poverty of spirit and meanness of mind will not have the last word.

Notes and References

1. Lucien Goldmann, *The Philosophy of the Enlightenment: The Christian Burgess and the Enlightenment*, trans. Henry Maas (London: Routledge & Kegan Paul, 1973), p. 1.
2. Isaiah Berlin, 'The Counter-Enlightenment', in *Against the Current*, ed. Henry Hardy (London: Hogarth, 1979), p. 1. To say that this Prologue is indebted to Isaiah Berlin's writings would be a grotesque understatement. Berlin's map of the Counter-Enlightenment – reflecting his profound insight into the passions that moved its major players and the imaginative universes they inhabited, his understanding of the connections between various strands of Counter-Enlightenment thought and his wide knowledge and grasp of the pervasive influence on the thought and history of the two succeeding centuries of those who opposed the *philosophes* – is definitive. Anyone wishing to defend the Enlightenment finds himself in constant dialogue with Berlin. The present author is no exception. Wherever I went, I found that Berlin had been there before me, usually clarifying the issues, only very occasionally muddying them. Eventually I had to accept that it was not sufficient to acknowledge him in the text of this chapter. That is why he appears in its title.
3. See Ernest Gellner, *Reason and Culture* (Oxford: Blackwell, 1992), to which the present discussion, and much else in this book, is greatly indebted.
4. See Peter Gay, *The Enlightenment: An Interpretation.* Vol. I *The Rise of Modern Paganism* (London: Weidenfeld & Nicolson, 1967), p. 311. The entire discussion of the respective influences of Bacon and Descartes on the Enlightenment thought (pp. 310–3) is illuminating. Gay points out that the Enlightenment was in fact the culmination of an intellectual movement that began with the Renaissance.
5. What has been described as 'the autocritique of the Enlightenment' was not confined to Rousseau. Goldmann (op. cit., pp. 44–9) reminds us how, in works such as *Rameau's Nephew* and *Jacques the Fataliste* – and even in the essay form of many of his works – Diderot questioned not only the values of bourgeois society and its ideology but also many of the fundamental categories of the Enlightenment itself. He was, Goldmann tells us, 'the only *philosophe* to understand that, while men's behaviour may be determined by their social circumstances, these circumstances themselves result from the actions of men' and, although this understanding fell short of a dialectical or historical understanding, 'he was more aware than any of the other *philosophes* how complex the social world is' (p. 49). We shall discuss presently how this autocritique was present even before the Enlightenment itself. For example, Bayle, one of the grandfathers of the Enlightenment, had shown how reason could, if pursued to the end, lead to the view that all fundamental ideas were 'all equally cogent, all equally probable and all mutually destructive' (Paul Hazard, *The European Mind 1680–1715*, trans. J. Lewis May, Harmondsworth: Penguin Books, 1964).

411

6. Isaiah Berlin, *The Crooked Timber of Humanity* (London: Fontana Press, 1990), p. 175.
7. René Descartes, *Discourse on Method*, Part IV, quoted in Gellner, op. cit., p. 1.
8. Pierre Bayle, *Dictionnnaire Historique et Critique*, art. 'Takiddin', quoted in Paul Hazard, *The European Mind 1680–1715*, trans. J. Lewis May (Harmondsworth: Penguin Books, 1964), p. 134.
9. J.G. Hamann, Letters to Kant, extracted and translated by Ronald Gregor Smith in *J.G. Hamann, 1730-1788. A Study in Christian Existence* (London: Collins, 1960), p. 241.
10. For a subtle examination of the complementary roles of rationalism and empiricism in advancing reliable, transcultural, transcendent knowledge, a discussion in which the history of ideas is seamlessly united to epistemological analysis, see Gellner's *Reason and Culture* (op. cit.), especially the section 'Rationalism and empiricism in partnership'. 'In early rationalism', Gellner notes,

   > a double transcendence was involved: transcendence of nature and the senses on the one hand, and independence from accumulation of social ideas, of culture, on the other. The aspiration for unprecarious, sound knowledge seemed to require the overcoming both of the unreliable, ephemeral, context-bound nature of sense perception, and of social prejudice alike. Each seemed to provide manifestly unsuitable building bricks for a reliable cognitive edifice.
   >
   > (p. 166)

   This idea of knowledge of a realm beyond the world of the senses and nature is, however, a 'a snare and a delusion'. As Gay (op. cit., p. 311) points out, Descartes also appreciated the importance of observation:

   > For his part, Descartes the rationalist knew that he could not do without a 'store of experience to serve as matter for my reasonings' and he laid down as a rule that 'observations' become 'the more necessary the further we advance in knowledge'.

11. Though not exclusively by induction, as Gay (p. 311), again, points out:

    > In one of his striking images, Bacon compared reasoners to spiders 'who make cobwebs out of their own substance', and experimenters to ants who 'only collect and use'. His ideal natural philosopher was the bee, gathering 'its material from the flowers of the garden and of the field' and digesting it 'by a power of its own'.

12. Pierre Bayle, *Commentaire philosophique*, Part One, i, quoted in Hazard, *The European Mind 1680–1715*, p. 130.
13. 'What is the Enlightenment?' In the same essay, Kant observed ruefully that, although he was living in an Age of Enlightenment, he was not living in an enlightened age. Enlightenment was a process and it was far from complete.
14. Quoted in Isaiah Berlin, *The Age of Enlightenment* (New York: New American Library, 1956), p. 270.
15. Alfred, Lord Tennyson, *In Memoriam*, LIV, first stanza.

16. See R. Dawkins, and J.R. Krebs, 'Arms races between and within species', *Proceedings of the Royal Society of London* (1979), B, 205: 489–511.

17. Raymond Tallis, *The Explicit Animal* (London: Macmillan, 1991). This double status of humankind and human reason – as both part of nature and as revealer of nature – lies at the heart of the problems of consciousness and of freedom, and is also the basis of the hope of progress. Yes, we are caused to have reasons but reasons are not mere effects in us, like the beating of our heart or the reflexes that are mobilised when we stumble. There is a fundamental difference between choosing to do something on the basis of explicit reasons and making unconscious adjustments to soften our landing when we fall down stairs.

18. Isaiah Berlin, *The Hedgehog and the Fox* (London: Phoenix Paperback, 1978), p. 76. In this essay, Berlin traces the deep connections between Maistre's explicit doctrines and Tolstoy's more usually implicit world-picture. Maistre and Tolstoy share 'the same sardonic, almost cynical, disbelief in the improvement of society by rational means, by the enactment of good laws or the propagation of scientific knowledge'; they both loathe the progressive, liberal dream 'of the ordering and planning of society in accordance with some man-made formula'; and they are both acutely aware of the deep, infra-rational forces that move men and societies.

Other writers upon whom Maistre has been influential bring out his sinister, inverted carnality, his obsession with the bloodthirsty, self-immolatory craving he believed – or perhaps hoped – had been implanted by God in all living creatures, and in human beings most of all. Baudelaire, for example, was entranced by Maistre's emphasis on Original Sin and the sanction of the executioner. His admiration for Maistre does not seem unconnected with his sexuality, which was deeply stained by sado-masochistic fantasies of humiliation and degradation. Indeed, it could be argued that his dark theologicised sexuality was a projection of Maistre's savage vision of society on to sexual relations. For Baudelaire sexual pleasure was rooted in a knowledge of doing evil:

> Moi, je dis: la volupté unique et suprème de l'amour gît dans la certitude de faire le *mal*. Et l'homme et la femme savent, de naissance, que dans le mal se trouve toute la volupté. (*Journal Intime*)

> [What I say is this: the unique and supreme sensual pleasure of love comes from the certainty of doing *wrong*. Man and woman know from the day they are born that in doing wrong they will find sensual pleasure.]

The dark mystery of sex had nothing to do with love or companionship. He dreamt of appearing to his lover in the guise of executioner ('A une Madonne'); and of a definitive sexual act which consisted not of giving or receiving orgasmic pleasure, but of infecting an innocent with venereal disease through wounds torn open in her flesh ('A celle qui est trop gaie'). Thus are *Soirées de Saint-Petersbourg* transposed into *Nuits de Paris*.

19. Not all Enlightenment thinkers believed in the natural goodness of human beings. Helvetius' *De l'esprit* combined Locke's sensationalism and La Mettrie's materialism in a deeply unflattering picture of man that, as Peter Gay notes, was repulsive even to the more tough-minded of his contemporaries: they found his 'gloating insistence on universal, unrelieved egoism a little repulsive' (*The Enlightenment: An Interpretation*. Volume II

The Science of Freedom p. 513). According to Helvetius, men are the recipients of sensation and the centres of passion and, thus constituted, each acts to realise his desires by following his self-interests with unremitting consistency. However, this pessimistic view of man's nature was offset by an optimistic view of man's possibilities:

> What man thinks, believes, even what he feels, is open to the most extensive modifications through the social environment – man, in other words, can be educated to be almost anything, even a good citizen. (Gay, p. 513)

> it is solely through good laws that one can form virtuous men. Thus the whole art of the legislator consists of forcing men, by the sentiment of self-love, always to be just to one another ... to make such laws one must know the human heart, and to know first of all that men, responsive to themselves, indifferent to others, are born neither good nor bad, but ready to be one or the other
>
> (ibid., pp. 514–15)

20. Albert Camus, *The Rebel*, trans. Anthony Bower (Harmondsworth: Penguin Books, 1971), p. 12.
21. Isaiah Berlin, *The Magus of the North: J.G. Hamann and the Origins of Modern Irrationalism* (London: Fontana, 1994), p. 4.
22. Cited in Isaiah Berlin, *The Age of the Enlightenment*, 'J.G. Hamann', p. 247.
23. *Kant's Critique and Metacritique*, extracted and translated by Ronald Gregor Smith, in *J.G. Hamann, 1730–1788. A Study in Christian Existence* (London: Collins, 1960), p. 216.
24. Letter to Herder in ibid., p. 246.
25. Letter to F.H. Jacobi in ibid., p. 252–3 and p. 257.
26. *Kant's Critique and Metacritique*, in ibid., p. 216. Hamann's view of language was profoundly ambivalent or perhaps deeply confused. On the one hand, he seemed to be a nominalist, for whom there was no simple correspondence between words and extra-linguistic reality. As such, he seemed to regard language as a mere human convention. On the other hand, he believed that language was a divine fabric unfolding in articulation of itself so that D'Alembert's views were an unnatural abstraction from the divinity of the world.
27. Hamann has other reasons for limiting the power of reason. Not only is it not universal, as is claimed; it is also of limited scope: Hume (whose works Hamann had translated) 'needs faith if he is to eat an egg and drink a glass of water'. Indeed, the knowledge with which reason has to operate hardly reaches outside of our own experience. As Vico had claimed, without divine revelation we can know only what we have made; factual truths are *facta*, things that we have made. The natural sciences do not give us secure knowledge because God, not we, made nature and we can never grasp his designs; we can know only human creations and intentions. History – the acccount of what men have made – is therefore the only true science.
28. Marquis de Condorcet, *Sketch for a Historical Picture of the Progress of the Human Mind*, 'The Tenth Stage: The Future Progress of the Human Mind', trans. June Barraclough (New York: Noonday Press, 1955).
29. Martha Nussbaum, 'Feminists and Philosophy', *New York Review of Books*, XLI, No. 17, 20 October 1994, pp. 59–63. This informed, balanced and

persuasive essay is strongly recommended reading for all those who think, or fear, that abandoning the fundamental principles of rational thought would further the cause of women's liberation or do anything to diminish the injustices and oppression from which they suffer.

30. As Devaney has set out in her detailed study (M. J. Devaney, Macmillan, Forthcoming), *At Least since Plato' and other Post-structuralist Myths*, there is even a widespread argument to the effect that 'not p and not-p' is yet another oppressive constraint on thought. Devaney has convincingly identified this myth as central to post-modern thought.

31. On second thoughts, why stop at the whole individual? Is it not a kind of tyranny to insist that an individual should be held over time to being or having a single essence? Is it not a kind of despotism if legislation extends beyond the single moment of the single individual?

32. It is worth recalling, when Enlightenment thinkers are criticised for oppressive universalism – for what Lovecraft called 'uniformitarianism' – that the extent and density of their prescriptions for human life and the scope of their universalism is as nothing compared with that of the religions against which they directed their brilliant critiques. This is evident if one compares the universalism of the *Encyclopédistes* with that of the Catholic Church – which offered salvation only to the believers and varying degrees of damnation to those who had incurable doubts.

As Julien Benda had pointed out (in *La Trahison des clercs*), the universalism of the intellectuals is an immemorial tradition, deriving from the time when the intellectuals were clerics: then, as ambassadors of the Church, they upheld what they believed to be eternal values, transnational and trans-ethnic. For Benda, universalism was a moral obligation of intellectuals; not to serve universal ideas and principles but to identify with local ethnic, class or national interests, was to betray their high calling. 'To be a thinker' as someone once said, 'is to have no intellectual fatherland'. How remote was this idealistic conception of the role of the intellectual from the post-war behaviour of the intellectual *engagés* for whom the betrayal of truth and justice for the sake of the anti-bourgeois revolution was the norm, expressed most vividly in Sartre's refusal to admit the existence of the Gulags not because they did not exist or because he did not believe they did but because it would have demoralised the French proletariat to have done so.

33. David Hume, *A Treatise of Human Nature*, Book 2, 'Of the Passions', 3.3.4.

34. This and the issues addressed in the paragraphs that follow are dealt with in much greater detail in my *Newton's Sleep* (London: Macmillan, 1995).

35. See 'The Uselessness of Art', Part II, of *Newton's Sleep*. Camus's observation that 'The future is the only kind of property that the masters willingly concede to the slaves' (*The Rebel*, op. cit., p. 162) seems apposite. A sense of the inadequacy of the Kingdom of Means as the dwelling place of the meaning of life lay behind the famous crisis of J.S. Mill's early twenties, inaccurately characterised as 'the shipwreck of rationalism'.

36. *Notes from the Underground*, trans. C.J. Hogarth (London: Great Chancellor Press, 1994), p. 613.

37. Theodor Adorno and Max Horkheimer, *Dialectic of Enlightenment*, trans. John Cumming (London: Verso, New Left Books, 1979).

38. Cf. Aldous Huxley's complaint: 'Modern man no longer regards nature as being in any sense divine, and feels perfectly free to behave towards her as an overweening conquerer and tyrant.' Quoted in Walter Moore, *A Life of Erwin Schrodinger* (Cambridge: Canto, 1994), p. 251.

39. Herbert Marcuse, *One-Dimensional Man* (London: Sphere, 1968), p. 23.
40. Edmund Burke, *Reflections on the Revolution in France* (London: T. Nelson and Sons, n.d.), p. 246.
41. 'Essay on the Generative Principle of Political Constitutions', in *The Works of Joseph de Maistre*, selected, translated and introduced by Jack Lively (London: George Allen & Unwin, 1965), p. 147.
42. In *Rationalism in Politics* (London: Methuen, 1962).
43. My account of Oakeshott's thought is greatly indebted to W.H. Greeleaf, *Oakeshott's Philosophical Politics* (London: Longmans, 1966).
44. Laurent Dobuzinskis, 'The Complexities of Spontaneous Order', *Critical Review* 3(2) (1989): 241–66.
45. Karl Popper, *The Poverty of Historicism* (London: Routledge, 1957), p. 65.
46. *Thinking Green! Essays on Environmentalism, Feminism and Non-Violence* (London: Parallax, 1994).
47. T.S. Eliot, *Notes towards the Definition of Culture* (London: Faber, 2nd edition, 1962), p. 94.
48. See, for example, 'Anti-Science and Organic Daydreams', in *Newton's Sleep*, op. cit., and 'Is Reality no Longer Realistic?' in Raymond Tallis, *In Defence of Realism* (London: Edward Arnold, 1988).
49. Bryan Appleyard, *Understanding the Present* (London: Picador, 1992), p. 112.
50. David Lehman, *Signs of the Times* (London: André Deutsch, 1991) p. 185. A different kind of thread connecting 'organicist' intellectuals (quite different from agrarian ones such as Heidegger and Hamsun) and Fascism has also been been noted by Ernest Gellner:

> Fascism, or Nazism, was based on the idea that industrial society could best be run on values which had characterised much of the agrarian past: aggression, hierarchy, territoriality, loyalty to community rather than universalism.
>
> (*TLS*, 16 July 1993, p. 3)

It has also been suggested that the Taylorisation of labour and industrialisation of society, which organicists rejected, themselves prepared the way for the concentration camps on the grounds that extreme division of labour compartmentalises wider moral and social awareness. This is a dubious argument: the myth of the inorganic, atomistic society is examined in my *In Defence of Realism* (London: Edward Arnold, 1988), ch. 1; and this volume, ch. 4.
51. The brevity of the path from Herder to Hitler is captured in the closeness of the terms *das Volk* and *volkisch*. For Herder, Berlin tells us, 'every action, every form of life, has a pattern which differs from that of every other. The natural unit for him is what he calls *das Volk*, the people, the chief constituents of which are soil and language, not race or colour or religion' (*The Crooked Timber of Humanity*, p. 40). This was a benign notion of 'the peaceful coexistence of a rich multiplicity and variety of national forms, the more diverse the better' (ibid., p. 245). However, 'under the impact of the French revolutionary and Napoleonic invasions, cultural or spiritual autonomy, for which Herder had originally pleaded, turned into embittered and aggressive nationalist self-assertion'. And this determined the path that *volkisch* thought took. It has been characterised by Avraham Barkai as follows:

Through all its variations *volkisch* thought accorded a transcendental 'essence' to a group of people, which transformed it from a solely in-strumental or contractual union of individuals into a mythical, self-perpetuating organism. Whereas nations or classes are compared with a heap of stones, the *Volk* appears as an indissoluble rock. In it every individual is bound by his or her inborn nature and emotions. It is the source of well-being or creativeness. Only in union with other mem-bers of the *Volk* can the individual find full self-expression of his indi-viduality.

> (Avraham Barkai, '*Volksgemeinschaft*, "Aryanization" and the Holocaust', in David Cesarini (ed.), *The Final Solution: Origins and Implementation*, Routledge, 1994, p. 34)

The linking of the *Volksgemeinschaft* ideal with anti-Semitism (easy since every pure essence requires impurity against which to define and defend itself) inevitably led to the classification of the Jews as the *Volksfeind* and the hideous *Judenpolitik* that resulted from this.

52. Robert Marshall, *The Storm from the East* (London: BBC Publications 1993), p. 54.
53. These facts are drawn from the monograph on Pre-Columbian civilisa-tions in Volume 26 of the Macropedia of the *Encyclopedia Britannica*, 15th edition (Chicago: University of Chicago Press, 1993).
54. The most plausible summary is that wars were no less horrible in the past – they were probably in many respects more horrible – but they were closer in horribleness to the horror of what was accepted in the past as ordinary life. This is argued by C.V. Wedgewood in her classic account of the Thirty Years' War:

> The impact of war was at first less overwhelming than in the nicely balanced civilization of today. Bloodshed, rape, robbery, torture, and famine were less revolting to a people whose ordinary life was encom-passed by them in milder forms. Robbery with violence was common enough in peace-time, torture was inflicted at most criminal trials, horrible and prolonged executions were performed before great audiences; plague and famine effected their repeated and indiscriminate devastations.

55. 'The Significance of the Final Solution', in David Cesarini (ed.), *The Final Solution: Origins and Implementation* (London: Routledge, 1994), p. 304. However, Bauer's position is more complex than I have suggested. In *The Holocaust in Historical Perspective*, he at once asserts the uniqueness of the terrible events, inasmuch as 'the Holocaust was the policy of the total, sacral act of mass murder of all Jews they could lay hands on', and then questions whether it is right to see the events in this way:

> Not to see the difference between the concepts [Holocaust and geno-cide], not to realise that the Jewish situation was unique, is to mystify history. On the other hand, to declare that there are no parallels, and that the whole phenomenon is inexplicable, is equally a mystification.
>
> (p. 36)

The terrible uniqueness of the Holocaust was that it was the result not of revenge, nor of territorial conflict, nor of insecurity – as previous huge massacres have been; it was the cold enactment of a policy, arising out of an ideology, to try to drive the Jewish people from the face of the earth.

This uniqueness has been argued passionately in Steven Katz's massive *The Holocaust in Historical Context. Vol I: The Holocaust and Mass Death before the Modern Age*, discussed by Malcolm Bull ('One and Only', *London Review of Books*, 23 February, 1995, pp. 11–12). In fact Katz redefines 'genocide' (as the killing of entire 'national, ethnic, racial, religious, political, social, gender or economic groups' rather than definable sub-groups) in such a way as to exclude all previous events that have been considered as genocidal. The reason for this is understandable and honourable: the casual way in which the word has come to be used to describe any large-scale outrage threatens to trivialise the terrible events experienced by the Jews in the 1930s and 1940s (though Raphael Lemkin, who introduced the terms in 1943, had a much wider definition, which would have encompassed the fate of quite a number of peoples in Nazi Europe, as Bauer (op. cit., p. 35) points out). And Katz also wants (in Bull's words) 'to bridge the gap between those who claim that the Holocaust is morally, metaphysically or theologically unique, and those who claim that it must be placed in historical, and thus comparative, perspective, by arguing that it has a "phenomenological uniqueness" as a historical instance of genocide'.

The debate as to whether the Holocaust was qualitatively different from every previous act of destruction visited by one people upon another carries the danger of covering these terrible events with academic dust. It should be conducted by individuals who, for a variety of reason, will have greater moral authority in these matters than myself. Anything that threatens in any way to diminish the scale of the Nazi atrocities should be treated with disgust. However, it is important not to render it a-historical and, as has been suggested by critics of some Holocaust historians, to 'sacralise' it and so render it inexplicable in human terms.

In pursuit of the latter aim, we need to relate it to the other terrible things that collectives of human beings have carried out – for example, the Athenians' slaughter of all the adult males at Melos, Genghis Khan's 'trays of blood', the systematic slaughter and/or forced conversion of the native Americans in fifteenth- and sixteenth-century 'Latin' America (the very name states the cultural rape), the equally systematic slaughter and dispossession of native Americans in North America in the nineteenth century, the murder between 1941 and 1945 of 487 000 Serbs by the Nazi-backed Pavelic regime in the newly independent state of Croatia. All these outrages were facilitated by demonising the victims or reducing them to the status of sub-human vermin, or both. The fact that these hideous events cannot be plausibly laid at the door of Enlightenment thought (or the hyper-rationalist, bureaucratic instrumentalism that it is sometimes accused of importing into political, social and anthropological thinking) must undermine any causal relationship between Enlightenment universalism and genocide and argue a more universal human propensity to aggression to which Enlightenment scepticism, tolerance and universalism, must surely act as a corrective rather than a stimulus.

56. See my *In Defence of Realism*, chapter 1, op. cit.
57. The final sentence of Mark Lilla's penetrating review of Berlin's *The Magus of the North: J.G. Hamann and the Origins of Modern Irrationalism*, in *London Review of Books*, 6 January 1994, pp. 12–13.

58. Isaiah Berlin, 'The Apotheosis of the Romantic Will', in *The Crooked Timber of Humanity*, ed. Henry Hardy (London: HarperCollins, Fontana, 1991), p. 237.
59. This is argued at length in my *The Explicit Animal* (London: Macmillan, 1991).
60. Mark Rosenau, quoted in Joel Handler's 1992 Presidential Address to the Law and Society Association (*Law and Society Review* 26 (4) (1992): p. 726).
61. 'The doctrine of madness', in *Sincerity and Authenticity* (Cambridge, MA: Harvard University Press, 1973).
62. Frank Lentricchia, *After the New Criticism* (London: Methuen, 1980), p. 301. I quote this in *In Defence of Realism* (op. cit.) where, in the section 'The Radical Critic Against Himself', I deal with this 'strange discrepancy' in some detail.

   The appalling ethical consequences of the deeply muddled thinking promoted by de Man, his colleagues and his disciples were dramatically illustrated by the response of Derrida and some of his fellow deconstructionists to the posthumous discovery of de Man's collaborationist, anti-Semitic war-time articles. This response has been brilliantly documented in David Lehman, *Signs of the Times* (op. cit.), which should be required reading for all those who feel that deconstruction is harmless fun.

   Derrida had the moral blindness and temerity to suggest that a condemnation of de Man would 'reproduce the exterminating gesture' (Lehman, p. 243). It is difficult to know which is more sickening: the suggestion that those who criticised de Man were no better than the Nazis engaged in extermination; or the use of the word 'gesture' in this context – as if the murder of six million people were a kind of gesture. What many people are simply unable to grasp is (to quote Lehman) 'the dangers that ensue when metaphors substitute for facts, when words lose their meaning, and when signifiers and signifieds part company, with the deconstructionist's blessing'.

   Once you start declaring lucidity to be a dangerous rhetorical ploy and assert that 'meaning is Fascist' (to be discussed later), you have played into the hands of those, such as de Man, for whom a refusal to recognise the extra-textual referent of discourse was of such strategic importance.
63. David Macey, *The Lives of Michel Foucault* (London: Vintage, 1994), p. 388.
64. Michel Foucault, *Surveiller et Punir*, quoted and translated by Macey, ibid., pp. 329–30.
65. He had the same profound insincerity in his relation to medicine. According to the books that made him famous, there had been no real advances over the last 200–300 years in the understanding, classification and treatment of diseases – only a change in discursive formations. And yet Foucault would not have tolerated either himself or his friends being treated in eighteenth-century fashion in an eighteenth-century hospital with its standards of care and hygiene.

   There are internal contradictions in his writings as well. At one point he suggested that 'the needs of the revolutionary armies, and regular decimation of physicians in their ranks, made the *improved* and accelerated training of doctors a military and political necessity' (italics mine). This was meant to show that medical advances were driven by military necessity rather than beneficence. But it also indicates a desire to direct two incompatible sneers at the history of medicine: there have been no objective or real advances in treatments; and these advances were merely driven by political expediency.

His published views on psychiatry were equally at odds with his personal practice. Although his *œuvre* was dedicated, as the historian Lawrence Stone has pointed out, to challenging the 'humanitarian values and achievements of the eighteenth-century Enlightenment' (Macey, p. 433), he would have objected if the conditions that prevailed in pre-Enlightenment lunatic asylums were replicated in modern psychiatric hospitals. And he did not protest when his great friend and teacher, Louis Althusser, was admitted to a psychiatric hospital following the murder of his wife rather than being hung, drawn and quartered.

INTRODUCTION TO PART I

1. Dudley Young, *Origins of the Sacred: the Ecstasies of Love and War* (London: Little, Brown and Company, 1992).
2. Despite its awe-inspiring and wonderfully empathetic erudition, *Origins of the Sacred* is not difficult to read. Young has an easy, engaging style and his reassertion of down-to-earth common sense and studied lapses into colloquialism sound the right note rather than being queasily blokey. His occasional crudity is refeshing and a signal that, unlike Freud for example, he is aware that sex even on the page should not be treated as an abstract category and that it is about the interaction of bodies as well as the interlocking of complexes, about the physical as well as the metaphysical. *En route* to its big conclusions, *Origins of the Sacred* is rich in incidental insights. I would not wish my critical account of the book, and my use of it as a springboard to a wider critique of the contemporary critic of culture, to discourage others from reading it. Quite the reverse.
3. That, at least, is what he says in his Introduction. However, Young, in common with many hostile critics of culture, is wildly inconsistent as to when the trouble began: in pre-human times when the apes got out of hand; in prehistory, when, after the fatal parricidal dance, the alpha and shaman roles were separated; in the Mycenaean culture, where the violent origins of the sacred were suppressed; or a mere 400 years ago when, according to Young, a rift appeared in the Western soul, 'as science and religion went their separate ways' (p. xvii).

1. FROM APES TO PLATO

1. Young seems uncertain just how widely distributed love is. In the next sentence, he asserts that 'Love is something so subtly insane that only humans are clever enough to fall into it' (p. 79). So much for the Bonobo apes.
2. This is a dubious argument, as the connection between sexual intercourse and pregnancy was probably not established until neolithic times. Anyway, the time-interval between the penetration of the partner and her death would make the gratification somewhat delayed; perhaps it appealed only to middle-class Bonobo apes.
3. The chronology of 'where it all went wrong' is again confusing. Clearly, this is somewhat at odds with his assumption that the trouble really began with the Renaissance and his claim, elsewhere, that Homer's poetry 'represents ... Western man's decisive break with the primitive wisdom we have been trying in some measure to remember' (p. 282).

4. There is a third ecstasy, separate from the violations of love and war: the ecstatic love of beauty, which, according to Plato, awakens the earth-bound soul to remembrance of eternal perfection and prompts the desire to become united with it. From this point of view the trajectory to the sexual climax is a diversion that does not do justice to the beautiful object. The question posed by the lover of beauty is: 'How to possess and be possessed by the beautiful without despoiling and being despoiled by it.' Young's discussion here leads to the conclusion that homosexuality is crucial to the development of romantic love. For the most central question affecting the relationship between warriors is how to penetrate another body without degrading it. The urge to penetrate and explore is the urge to know and command. One way out is in Platonic love. Here the energies for love and war are reconciled in the 'generative perceptions of beauty'; after both have been renounced for the deeper perception that comes from non-possessive appreciation of beauty.

5. In some cultures, notably the Minoan in Crete, this evolved to a different ritual: the marriage of a female human figure with a male animal one – usually a bull. The union symbolised the marriage of Nature and Culture, the absorbing of Nature's violence by cultural complexity, and represented the faith that Culture may absorb and transform the energies of Nature without killing them off.

6. To imagine that *Origins of the Sacred* is, on this account, a 'dangerous' book is, of course, pure fantasy: it is one small signal emitted among the noise of a ten million more powerful signals. *Pace* Nietzsche, it is not on dove's feet but on electromagnetic waves that come the thoughts that change the world. And the structures of power are such that they do not tumble when rational thought questions their legitimacy: it has ever been thus.

2. FALLING INTO THE HA-HA

1. Raymond Tallis, *The Explicit Animal* (London: Macmillan, 1991). The distances between man and the other animals are set out at length (perhaps at too great length) in chapter 6, 'Man, the Explicit Animal'.

2. E.R. Dodds, *The Greeks and the Irrational* (Berkeley: University of California Press, 1951), p. 273.

3. Mary Midgley, *Beast and Man: The Roots of Human Nature* (London: Methuen, 1980). Although I disagree with this book in many respects, there is much with which I am in profound agreement. It is a lucid, engaging, rich and persuasive account of its central thesis that 'We are not just like animals; we *are* animals' (p. xii).

4. This error is addressed in more detail in my *Psycho-Electronics* (London: Ferrington, 1994). See especially the introduction ('Thinking by Transferred Epithet') and the entries on 'Information' and 'Misplaced Explicitness'.

5. For example:

> Our clinical adjective for such a breakage [when music gives up its 'physiological presupposition'] is 'schizoid', and I believe one can hear it foreshadowed in late Beethoven; in the Ninth Symphony clamourously, in the pastiche of the *Missa Solemnis* anguishingly, and in the discarnating last quartets sublimely.
>
> (p. xxxi)

My rough sketch of [the Romantic soul of bourgeois Europe] may be
concluded by briefly considering Beethoven's last piano sonata (Opus
111). This piece is a wonder of sanity regained, also a prophecy: after
some initial argument melody returns gracefully to its home in the
human body, and the syncopated rhythms point unmistakably forward
to ragtime and the jazz age. As he leaves the stage, the last of the
European grandmasters quietly announces that he has concluded (consummated and killed) a musical life that had begun in medieval plain-
song.

(p. xxxii)

Etc. Some chimp!
6. The reader will recall that Winnie the Pooh's poems began as 'hums'.
7. The relationship between 'instinctive', 'primitive' knowledge and hostile
   prejudice is connected with the dangerous so-called 'wisdom of myths',
   discussed in chapter 3.
8. Jane Goodall, *In the Shadow of Man* (Boston: Houghton Mifflin, 1971), pp.
   250–1. Quoted in Tallis, *The Explicit Animal*, p. 166. What is unique about
   man, of course, is not that he has a propensity to degenerate into barbarism, but that he rose out of it in the first place. As Philip Hensher points
   out in a recent review (*Guardian*, 16 June 1995), this is something that
   novelists such as Golding are simply not up to thinking about. Which is
   particularly remiss given the vast quantity of human time that has been
   spent on the admittedly dull but non-barbarous activity of setting, sitting and marking examination papers whose topic is the essential barbarity of man as illustrated by *The Lord of the Flies*.
9. This is set out in the discussion of the *pour-soi* in *Being and Nothingness:
   An Essay on Phenomenological Ontology*, trans. Hazel Barnes (London:
   Methuen, 1957).
10. *A Treatise of Human Nature*, book I, part III, sec. XIV.
11. This is discussed in the section 'Exhibiting Consciousness' on the final
    chapter ('Recovering Consciousness') of *The Explicit Animal*, pp. 217–23.
12. This journey is described in the entry 'Memory' in Tallis, *Psycho-Electronics*.
13. Edmund Leach, *Lévi-Strauss* (London: Fontana, 1970), p. 84.
14. One the great modern masters of Disneyfication is the poet Ted Hughes.
    When he is simply describing the appearance and presence of animals,
    he is a genius. When he uses the personae he has projected in them to
    draw portentous lessons about human nature and The Predicament of
    Modern Man (alienated from his animal nature, etc.), he is a bore. *Crow*
    must be the worst volume of poems written by any great poet, not only
    because the verse is written with an unbelievable lack of inventiveness
    and technical skill, but because of the endless implicit sermonising based
    upon the presumed nature of Crow who is, of course, also a mythologi-
    cal as well as an anthropomorphised carrion bird.
15. We could extend the metaphor a little further by suggesting that the
    hominids who formed the link between human and non-human primates
    are like troglodytes or garden gnomes squatting in the Ha-Ha enabling
    the well-trained to step on them as they leap over the Ha-Ha!
16. Appealing to our essentially animal nature is a two-edged sword, as we
    have seen with respect to both the arguments within Enlightenment thought
    and in the arguments between Enlightenment and Counter-Enlightenment
    thinkers.

For most Enlightenment thinkers, the fact that man was a piece of nature was a source of hope. It meant that his needs, both individually and collectively, could be addressed using the methods that had been so powerful and successful in the exploration of physical nature: a science of man and a science of society that would bring to an end the strife that grew out/man's ignorance of his real needs and of how to meet them was on the cards. But it also implied that man was no more intrinsically moral than any other piece of nature. Nevertheless, *L'homme machine*, actuated by physiological needs and the blind, selfish will to survive, could still be educated to be good because this was in its best interest. But it was difficult to see where the educators would emerge from the community of biological machines and equally difficult to be sure that the animal will to survival would be sufficiently enlightened, or enlightenable, to ensure a commitment to the civic virtues and the common good.

We have already cited the famous passage from Joseph de Maistre, which argues that man's status as a piece of nature, in particular his animality, was the clue to his essential nastiness.

> In the whole vast dome of living nature there reigns an open violence, a kind of prescriptive fury which arms all the creatures to their common doom: as soon as you leave the inanimate kingdom you find the decree of violent death inscribed on the very frontiers of life. You feel it already in the vegetable kingdom ... but from the moment you enter the animal kingdom, this law is suddenly in the most dreadful evidence. ... There is no instant of time when one creature is not being devoured by another. Over all these numerous races of animals man is placed, and his destructive hand spares nothing that lives.
>
> (quoted in Isaiah Berlin, *The Crooked Timber of Humanity*, London: Fontana, 1991, p. 111)

Seeing man as a piece of nature is clearly not wholly reassuring; and it does raise questions, as well, not only about the origin of values that go beyond appetites, about the source of civic virtue, but also about the nature of the reason upon which the Enlightenment thinkers placed so much store. How, ultimately, in a world of causes, does reason-directed agency arise? We shall return to this question in chapter 11. In the meantime, we note the strangeness of Young's confidence that our salvation lies in recognising our animal nature. If this were wholly true, it would more likely be a recipe for despair; for if we are lousy as human beings, we would be even worse as animals.

And the appeal to man's animality is also an alibi for scoundrels. Of course, there is a difference between seeing everyone (including oneself) as an animal versus seeing other human groups as animals in the sense of being subhuman. But there is the also a danger in viewing oneself as an animal. It encourages the belief that one cannot help what one does and that it is not only acceptable but necessary to live by the law of the jungle: eat or be eaten. Moreover, it rather too easily licenses the Social Darwinism that has disfigured so much of contemporary history. It has often been pointed out that there is a world of difference between acknowledging that there are potentially powerful instinctual forces operating within one that may undermine the civic virtues to which one is encouraged to subscribe and condoning the behaviour attributed to those

forces. But I am not entirely sure that positing the existence of such forces
– and arguing that they are part of our inescapable essence – is itself
morally neutral, particularly if, as I believe, to do so is to misread human nature.

Does it really matter if people, on theoretical grounds, see themselves
and others in animal terms? Perhaps, if I am honest, I have to say I am
not sure. The world is propelled towards its futures by technological advance (itself driven by ideas that fall well short of the metaphysical or
even the metabiological), by unargued large- and small-scale prejudices,
by the accidents of climate and geography, by the inheritance from previous generations and by the inherent instability of societies – huge dynamical systems open to the four winds of chance. The epidemiology of
kindness and nastiness towards others cannot be correlated with world-
pictures rigorously derived. Large, carefully argued ideas lack the strength
that comes from malice, anger, hatred and power of prejudice. For an
idea to make its way in the world, it should have armies, not arguments
behind it. Its proponents are required to marshall not syllogisms but
batallions of the brutishly ignorant and the brutishly well-trained. The
animal nature of man is not one of those ideas. Behind the actual evil
attributable to Social-Darwinism are not the careful arguments of *The Origin
of Species* but racism (an immemorial characteristic of mankind), the lust
for power, the motives of evil politicians: like many ideas that seem to
have had a harmful effect on the course of history, Darwinism is merely
a septic tank in which the pus of the collective soul may collect, rather
than a primary source of, or force for, evil.

3. THE WISDOM OF MYTHS AND THE MYTHS OF WISDOM

1. As already noted, this chronology is at odds with Young's own assertion
   that it all went wrong in pre-Hellenic times. We shall discuss his unstable
   chronology in due course.
2. Isaiah Berlin, 'The Counter-Enlightenment', in *Against the Current: Essays
   in the History of Ideas*, ed. Henry Hardy (London: Hogarth, 1979) p. 5.
   I am not too sure how much this restores the validity of myths and how
   much it patronises them, as I shall discuss. Or how sincere it is. I suspect the reception of Herder's ideas has confused a benign mythophilia
   – which recognises the wholeness, completeness and incommensurability
   of different imaginative universes in different eras – with a less benign
   mythomania which asserts that myths express eternal, transcendent wisdom that we have lost – and to which we must return by an aggressive
   re-assertion of cultural identity.
3. Margaret Anne Doody, 'Fear of Rabid Dogs', review of Marina Warner's
   Managing Monsters (*London Review of Books*, 18 August 1994, pp. 14–15).
   Precisely as autocrats, dictators and their supporters knew. This was probably why Maistre deeply distrusted the empirical scientific tradition. As
   Jack Lively says (in the Introduction to his translation of selections from
   his work, *The Works of Joseph de Maistre*, London: Unwin, 1965), 'he saw
   the natural science of his own time as being wholly inferior to the knowledge of ultimate causes enjoyed by primitive men, and inferior by this
   fact to the wisdom embodied in national tradition and prejudice' (p. 17).
   Doody also points out the myth of the ancientness of myths: there is
   no sacred original of a myth, quoting Warner on this point: 'Every tell-

ing of a myth is a part of that myth; there is no Ur-version, no authentic prototype, no true account'. 'The hearthside crone who passes on the wisdom of the tribe has always been a polyglot cosmopolitan.' (So much for Herder!) It is, as Warner points out, 'a piece of social engineering to pretend that the myth-story is always the same, and the same from time immemorial, because it is true'.

4. The 'peasant' is often doubly trapped by the wisdom of myths. Not only does he have to swallow the myth from the myth-teller, but he is imprisoned in myth-swallowing mode by a meta-myth. His openness to the wisdom of myths – his very simplicity, his child-like credulity – is itself taken as a form of wisdom. The awakening of critical intelligence – a contraction of the toothless omega to pursed omicron – is consequently read as a sign of corruption. Those who purport to believe in peasant wisdom – from the early churchmen to Tolstoy ('except ye be as little children') – don't really believe in it, of course. They know that the toothless omega will swallow anything – not only the prescribed myths but also other groundless but unauthorised beliefs called 'superstitions', not to speak of a hogshead or two of low-grade vodka or cloudy cider. Praising the wisdom of the peasants is essentially a way of berating their more sophisticated but less wise colleagues, in order, that is, to assert and exhibit their own wisdom. The horny-handed wisdom of the field serves predominantly as a means of criticising the manicured cleverness of the Salon.

5. It would not be entirely unfair to describe the conscious revival of myths in the late nineteenth and the twentieth centuries – self-aggrandising myths of the *volk*, of race, of national destiny, demonising myths fomenting hatred and loathing of others – as one of the most effective means for organising diffuse hatreds into mass bloodshed.

6. G.M. Carstairs, *The Twice-Born* (Bloomington, Indiana: Indiana University Press, 1967), p. 67. This appalling racism is often forgotten when First World countries are (rightly) condemned for their treatment of immigrants. Structural racism is less visible than episodic racism. Both are unacceptable and yet those who fight against the latter would often hesitate to condemn the former, being blinded by sentimental ideas about 'cultural pluralism' and respect for 'indigenous cultures' even if they are indigenous tyrannies. For myself, I am as angry with a thousand children dying in the dust in the Third World because they are regarded as belonging to an inferior caste as I am with the intermittent brutalities of racially prejudiced individuals in the First World. This may be because my liberalism does not extend to patronising individuals in other cultures by assuming that they cannot be judged as rigorously as those in my own because they do not know any better.

7. In one of his more irresponsible passages, Young tries to add to the confusion by merging modern concepts of hygiene – rooted in microbiology – with ancient magical notions of pollution control:

> The medical metaphors are apposite: primitive pollution operates very much as we understand the modern virus to do, a contagious and effectively invisible life that feeds upon health. And just as today we say that hygiene keeps you healthy whereas dirt spreads disease, so did the primitive: if pollution is defilement that calls for ritual cleansing or purification, the most obvious material 'signature' for such defilement is dirt; and so the primitive adopted it as his major 'symbol' of pollution. And just as today we purify the dirt of a wounded body by cleansing

with water, so did the primitive magician. Water was the primitive agent of *karthasis*, whether a ritual sprinkling of a body recently exposed to noxious influence, or, more importantly, the tears of lamentation through which we sought to purge ourselves of the grief instilled by some calamity.

(p. 182)

To call this delirious association of ideas 'thinking' is too kind and to call it 'crap' would be perhaps over-determined. Perhaps we should simply thank *pneuma* that Dr Young is not in charge of Central Sterile Supplies or has no control over public health policy. He would certainly increase the flow of tears of lamentation. (Incidentally, we shall deal with the assimilation of science to magic – of which this is a prize example – in the next section.)

8. It may be apposite to say a little about the two major ways in which anthropologists have baled out myths that seem factually wrong: functionalism and structuralism.

The functionalist approach saw superficially maladaptive and counterfactual myths as having a deeper adaptive function (and, consequently, a deeper, 'pragmatic'): that of maintaining the social system of which it was a part. The myth (or ritual) might of itself be useless or even worse; but it earned its keep by supporting the belief system of the society and by this means helping to ensure the continued effective working of the society as a whole. (Small consolation for the 'polluted' little girl with the hole in her bladder.)

The structuralist approach claims that myths don't have meanings in isolation; their meanings are to be found in the system of oppositions that is expressed through the sum total of all myths. This system of oppositions is an unconscious (at least until the birth of M. Lévi-Strauss) expression of the human mind and, specifically, of its way of so ordering the world as to reflect its own structure. By this means, does the mind turn the world into its own thing and make itself at home in the world.

Once you've adopted a structuralist view of myths, it is difficult to believe any myth except, perhaps, the myth of the wisdom of the Parisian mythographer. Lévi-Strauss's mythography was, of course, his way of bringing modern thought, and in particular modern science, down a peg or two: he found taxonomies implicit in 'primitive' myths at least as complex as anything in contemporary science. Unfortunately, their complexity was known only to Lévi-Strauss and certainly not to the savages whose thought he was decoding.

9. The transition from 'science' to 'secular scientism' will not have escaped the attention of the alert reader. The ability to merge the two concepts is an essential part of the tool-kit of the scientophobe.

10. See *Newton's Sleep* (London: Macmillan, 1995), in particular the first part ('The Usefulness of Science'), where I also point out that the attitude of the Romantics to science was considerably more complex than the simpleminded hostility of those who have taken some of their more memorable sayings as the last word on the matter.

11. See *Newton's Sleep*, 'The Eunuch at the Orgy', where I address the contradictions of the Strong Sociology of Scientific Knowledge using my own arguments and others requisitioned from John Durant and Lewis Wolpert. It is interesting to note that where the SSSK writers have tried to support their claims with actual sociology, they have been proved wrong.

For example, the common SSSK assertion that scientists command belief for their claims in so far as they have a social warrant would lead to the prediction that accepted 'great scientists' should come from gentlemanly backgrounds. The examples of Newton, Faraday and the large number of major Jewish figures in twentieth-century science indicates that entry to the Scientific Pantheon does not depend on the same criteria as membership of a golf club.

12. That, and the fact that science has been almost from its inception, transnational, the least ethnocentric of all human activities. The scientific community is not impressed by hierarchies because it is not confined to the boundaries within which hierarchies hold sway. A British scientist whose idea of truth was answerable only to the authority of British scientists would not cut much ice with Japanese scientists, and vice versa. Lysenko-science is a transient aberration and carries no international clout.

    Young, characteristically, refers to 'Western' science. Of course, science is not confined to the 'West': both the production and consumption of scientific theories and data and science-based technology take place outside the West – unless he takes China and Japan to be part of the West – or, patronisingly, assumes scientific outside the West as either unimportant or imposed. His Occidentalism – politically correct, but historically inaccurate – makes him seem to forget that *Black Athena*, which he cites, tries to demonstrate that 'Western' science did not originate with the Greeks, but that it came from Africa via Egypt. Whatever one thinks about the factual basis of the Black Athena argument, it does seem to emphasise that 'Western' science is part of a universal human culture – that it is not a European aberration. The assumption that science and reason are not part of a universal human heritage rather reminds me of the assumption in much politically correct health care planning that the hospital is not part of the community.

13. Young also tries to assimilate science to magic by asserting that 'there is common ground between sympathetic magic and what, since Plato and Aristotle, we have called rational discourse: both believe in the magic of representation' (p. 170). The reasons he gives for this are so muddled as not to warrant detailed examination but sufficiently typical to justify at least a glance:

> For sympathetic magic to 'work', the representation must be sufficiently like the thing expressed to call its soul before us; i.e., the whole thing turns on a relation of *sameness*. In rational discourse, similarly, for any proposition to be deemed true, the words spoken must be sufficiently like the things they represent for them to grasp or capture the situation – sameness again.
>
> (p. 171)

This is nonsense, of course, as anyone not brought up in the Academy of Lagado would know. Mimetic theories of language – even crypto-mimetic theories like Wittgenstein's Picture Theory based upon logical isomorphism – have long since been discredited, even among those who have not read Saussure on the arbitrariness of the linguistic sign. (For a detailed discussion, see my *Not Saussure*, London: Macmillan, 2nd edition, 1995.) Young uses worries about mimetic theories to argue that, since the grounds of 'adequate sameness' are difficult to specify (and the problem of universals is anyway unsolved), both magical and scientific discourse are groundless!

And since, anyway, mathematical equations not only represent the world but get rockets to the moon, the magical activity of representation is powerful, after all! Unfortunately, Young is not alone in having to be told that equations themselves don't do anything. They may guide the calculations that instruct the technologists who instruct the craftsman; but repeating the equations over and over, intoning them like a mantra, would not produce a moon rocket.

There is little evidence of the power of mythic. Of course, there is plenty of evidence of the wrong sort that magical remedies kept primitive man in close touch with nature: no one could accuse anyone who rotted in the earth at 30 years of age, after a life peopled by many dead children and enjoyed in a body infested with malaise-producing worms, of having a stand-offish attitude to Mother Nature.

Those who emphasise how science grew out of magic and chemistry out of alchemy in order to support the conclusion that science and magic, chemistry and alchemy, are essentially the same have a good test case in Newton who was, in his day, both a scientist and alchemist of European renown. His impotent alchemy has been forgotten; and his powerful science still remains the great cornerstone of physics and of our understanding of the material world.

14. The idea of a culture that is mad is also a deeply suspect one – as is any diagnosis of society, for reasons I shall address in chapter 4. It is one of which Young, as are so many others, is inordinately fond. On p. 26, he tells us that 'We are paranoid because culture is paranoid and vice versa; a liberating perception'. I am not too sure how liberating that is, but it is very familiar from the writings of Laing and other discredited gurus as well as from the Frankfurt School's critique of the Enlightenment and their assertion that 'paranoia is the dark side of cognition' (see the Prologue).

15. In Richard P. Feynman, *Surely You're Joking Mr. Feynman!* ed. Ralph Leighton (London: Unwin, 1986).

16. He offers no evidence in support of the supposed causal relationship between science-induced stupefaction and 'lurid fantasies of salvation'. Such fantasies have been an occupational hazard of humanity since time immemorial and certainly do not follow from the the rise of science. If anything, the reverse is the case. Of course, one can see why he wanted to believe this and why, in the absence of fact, he said it: he is in the grip of a fashionable theory: the advent of science has extended the realm of reason and this has led to a build-up of pressure of unreason which finds its expression in 'lurid fantasies of salvation'.

17. Bryan Appleyard, *Understanding the Present: Science and the Soul of Modern Man* (London: Picador, 1992).

18. Quoted by John D. Barrow, *Theories of Everything* (Oxford, Oxford University Press, 1991), p. 9.

So much, incidentally, for the 'arrogance' of science, which its critics – in particular, the practitioners of *kulturkritik* – make so much of. In fact, science is considerably less arrogant than most culture criticism and, compared with magicians, scientists are almost pathologically lacking in confidence. The very assumption that we have to find out what we do not know – the empirical approach – and the related assumption that there is much that we cannot know would be totally incomprehensible to the magician, who knows everything already and who could not admit to ignorance for fear of loss of face. In magic, the magician himself is always to the fore, as one who authorises knowledge, who transmits the expertise based upon revealed truths that only he has access to. The knowledge the scientist uses or adds to does not uniquely belong to him, and its authentication lies with anyone who is able to follow the method he has used; he does not regard knowledge as his own property, as attaching to his own person; and if others are not able to replicate his findings, he cannot appeal to his personal authority to defend his claims. The essential humility of science is expressed in the way that so much of its discourse is tentative, taking the form of hypotheses which are available for testing – verification or refutation – by others, or, indeed, oneself:

> In general we look for a new law by the following process. We guess it. Then we compute the consequences of the guess to see what would be implied if this law that we guessed is right. . . . If it disagrees with experiment it is wrong. In that simple statement is the key to science . . . it does not make any difference how smart you are, who made the guess, or what his name is – if it disagrees with experiment it is wrong.
> (Richard Feynman quoted in Lewis Wolpert, *The Unnatural Nature of Science*, London: Faber, 1992, p. 60)

This humility is dramatically expressed in the double-blind controlled clinical trial of new medication, the design of which incorporates the clinical scientists' own willingness to suspect himself of potential bias. How remote this is from the witch doctor's groundless confidence in his arcane knowledge! And it is less dramatically, though no less tellingly, expressed in the fact that, apart from a handful of exceptions, the anniversaries of scientists are rarely noticed. While a minor poet has a good chance of a centenary celebration, one has to be a very great scientist indeed to attract any kind of posthumous recall. The contributions of individuals tend to be lost in the collective advance of knowledge: the grain added to the ant-heap does not carry a flag indicating credit and possession.

The professed scope of scientific knowledge still, despite millenia of accumulated expertise, falls considerably short of that of the magician or the 'mythic' healer:

> It is necessary to recognise that with respect to unity and coherence, mythical explanation carries one much further than scientific explanation. For science does not, as its primary objective, seek a complete and definitive explanation of the Universe. . . . It satisfies itself with partial and conditional responses. Whether they be magical, mythical or religious, the other systems of explanation include everything. They are applied to all domains. They answer all questions. They account for the origin, for the present and even for the evolution of the universe.
> (François Jacob, quoted in Barrow, *Theories of Everything*, p. 8)

There is another sense in which science is self-effacing: in trying to identify the principles that enable us to determine how best to serve our needs, the scientist goes beyond, or outside, those needs. The scientist recognises that nature is other than man and is not centred on his needs; that the world does not share, is not even a mirror of, our hungers; we must consequently approach the problem of meeting our needs by indirection. By getting ourselves and our immediate, clamant, hungers – for food, fame or whatever – out of the way, only then shall we (to modify Matthew Arnold) 'see the object as it is', in itself. This distinctive approach of science is underlined in the passage from Frances Yates quoted earlier.

Precisely because science doesn't fill the heavens with human needs and man's most immediate preoccupations – unlike 'sympathetic magic' which simply projects these things into the entire universe – it says more about the what than the why of things. It is astrology, not astronomy, that finds human destiny in the stars. That is why it doesn't presume to tell us how to live – or not, at any rate, the ultimate ends of human life. For some, of course, this is a serious defect of science. I regard it as yet another saving grace (as I shall discuss presently).

If there is anything that deserves the charge of arrogance, it is cultural criticism, with its general statements of huge scope, based upon opinion informed by minimal data and not amenable to formal evaluation by others. But of this more later.

19. Though science, and reason in general, are given a major share of the blame for the military catastrophes of recent times. To quote the passage from Appleyard (*Understanding the Present*) I cited in the Prologue:

> The innocence of the easy, progressive Enlightenment myth, struck down in the trenches, finally died at Hiroshima and Auschwitz. Scientific reason was as capable of producing monsters as unreason. It is no good arguing that Auschwitz was unreasonable – in its way, it was, and secular society cannot be sure that it can offer any higher, purer rationality – nor is there any point in claiming that Hiroshima was a necessary evil designed to prevent more deaths from a protracted conventional war. That may have been true, but the atomic genie had been let out of the bottle and new, evil rationalisms would find other justifications for its use. Above all, nuclear weapons seemed to confirm our sense that there was something unprecedentedly and uniquely corrupted about our age.
>
> (p. 122)

And Appleyard gives passing notice to the famous report from the Club of Rome (Donella H. Meadows, Dennis L. Meadows, Jorgen Randers and William Behrens III, *The Limits to Growth: A report for the Club of Rome's project on the predicament of mankind* (Potomac Associates, 1972, p. 141):

> The application of technological solutions alone has prolonged the period of population and industrial growth, but it has not removed the ultimate limits to that growth.

Just so. But this is an argument for limits to population (where scientists can help) and limits to greed (scientists probably can't help), not evidence that 'the dream of omnipotent, problem-solving technology was over' (p. 131). On the contrary, defining the problem on a global scale and communicating and implementing the solutions will require extremely high-tech support – for example, electronic communication systems and databases. For more on this, see the first part of my *Newton's Sleep*.

20. For most of us, the question of status is personal rather than collective and comes more from social than from metaphysical considerations, from the hierarchies that the religious vision has done so much to uphold, to underwrite and condone. Being personally at the bottom of the heap has been a greater preoccupation with most people than not being species-wise at the centre of the universe. Besides, as Diderot said:

No one was ever frightened by the thought that there is no God; terror comes from the thought that there may be one, a God of the kind that is depicted (*Pensées Philosophiques*, quoted in Lucien Goldmann, *The Philosophy of the Enlightenment*
(London: Routledge & Kegan Paul, 1968), p. 61)

21. As I have argued at length in *The Explicit Animal* (London: Macmillan, 1991).
22. Pain is worse than meaningless. For not only does it largely lack meaning itself, but it also drains meaning from everything else by commanding rapt and exclusive attention to its meaninglessness. Pain stands to pain-free consciousness not as silence to sound but as Smetana's tinnitus to his music. Consequently, it does not simply fail to enrich, it actively impoverishes, does not elevate but degrades, does not bring human beings together but isolates them. It may seem unnecessary to point this out, but it is often overlooked by those who seem comfortable with the idea of flesh-tearing rituals or the daily discomforts and agonies of the pre-scientific past. There are even some who feel that there are deep intrinsic meanings in pain which, as a result of our more scientific approach to the relief of suffering, have been lost. This perverse sentimentality is criticised in my article on David Morris's *The Culture of Pain* (TLS, 1 May 1992, pp. 3–4). I shall return to this theme in the final chapter.

It may be apposite here to question Young's bald assertion that 'the signal achievement of modern science was ... to "deconsecrate" both space and time' (p. 7). We have to ask how sacred space and time were to the cold, aching, itching bodies that passed through it. This question might not readily occur to a culture critic whose image of the past is largely mediated through libraries. The ecstatic ritual moments have been memorialised, preserved in artefacts, and so retained in the collective memory; the long centuries of ordinary discomfort, on the other hand, have vanished without trace.

23. And some are even reassured by the Newtonian world-picture: 'As a result of Newton's discovery' [of universal gravitation] the cosmos itself became unfrightening, regulated, unhedged by chaos' (Young, p. 11). And they still had the hubbub in the corner as a back-up calmative.
24. Perhaps we may see in medicine – an art with manifestly human purposes rooted in biological science – a model of how the gap between the two cultures may be bridged. The gap is largely illusory, arising out of the mistake of contrasting science and 'the humanities', as if science were the 'inhumanities'. We must never forget that the sciences are human achievements and that they are driven by human dreams of knowledge and clamant human needs.
25. We ought to think of her in particular when we come across statements like this one which Appleyard quotes (on p. 238) with such approval from Allen Bloom's *The Closing of the American Mind* (Harmondsworth: Penguin Books, 1989, p. 26): 'Science in freeing men, destroys the natural condition that makes them human. Hence, for the first time in history, there is the possibility of a tyranny grounded not on ignorance, but on science.'
26. As Heidegger put it, *Dasein* is that being whose being is an issue for itself. Or that form of matter which most matters to itself.

Value is presupposed in any human activity, including science. This does not however mean that, as the Sociologists of Knowledge would have us believe, the findings of science are totally determined by individual or collective social values, that they are an internal affair of historicised humanity, having nothing to do with extra-human nature.

27. Dostoyevsky's existentialist anti-hero, whom we discussed in the Prologue, would certainly be foremost among the complainers:

> But if you were to tell me that all this could be set down in tables – I mean the chaos, and the confusion, and the curses, and all the rest of it – so that the possibility of computing everything might remain, and reason continue to rule the roost – well, in that case, I believe, man would *purposely* become a lunatic, in order to become devoid of reason and therefore able to insist upon himself.
>
> (*Notes from the Underground*, p. 615)

28. Of course, if Young's views were right – and there were no essential difference between primitive magic and science (only the inessential one that the former doesn't work and the latter does) – then there would be no point to his call for a return to a '*neo*primitivism, a retrieval of archaic (and childlike) attitudes': you can't regain what you haven't lost, indeed are enmired in. Equally difficult would be the rather self-conscious determination to make 'the retrieval of innocence as our project for the next millenium'.

29. In contrast, these questions are addressed head-on by Ernest Gellner in *Reason and Culture* (Oxford: Blackwell, 1992), which should be required reading for all anti-scientific, anti-rational cognitive relativists. The following passage is of particular relevance:

> If indeed it were one further case of deluded self-glorification [to claim a special status for rationalism and science], it would be naive vainglory of a rather unusual kind. Science has conquered the world without encountering much resistance, and virtually no *effective* resistance. It is true that it is subject to a great deal of abuse and denigration, both in its home areas and in the regions it has colonized; but for practical purposes, and in the pursuit of the serious business of life (production, coercion), few spurn its help. Most men and societies are eager, often indecently eager, to avail themselves of it. It makes surprisingly few claims for itself nowadays, but then, it does not need to do so: practice speaks louder than words.
>
> (p. 165)

Gellner is deeply aware, also, of the miracle, the mystery, of the way in which the universe surrenders its secrets to the combination of reason and observation. 'Prometheus Perplexed', the chapter from which the above passage is drawn, is not only a defence of appropriate rationalism but also a moving exploration of the mystery of the 'partial intelligibility of the world' and the progress of human understanding through science.

30. Quoted in Walter Moore, *A Life of Erwin Schrodinger* (Cambridge: Canto, 1994), p. 251.

31. Diderot, *Pensées Philosophiques*, quoted in Lucien Goldmann, *The Philosophy of the Enlightenment*, trans. Henry Maas (London: Routledge & Kegan Paul, 1968), p. 72.

32. Quoted in Lewis Wolpert, 'Science and anti-science', *J. Royal College of Physicians*, 21(2) (1987): 159–65.

33. C.V. Wedgewood, *The Thirty Years War* (London: Penguin Books, 1956), pp. 14–15.

   Historians and social commentators have often been subject to Arcadian fantasies. Since thinkers first consciously contrasted the condition of life in the present age with that of life in the past, they have reported a sense of being latecomers, of living in a secondary, or second-order, age, in an age that is a pale reflection of the primordial time when creative giants stalked the earth. In this fabulous age, the world of man and nature was experienced first hand, the great, foundational, genre-setting works were written and the world-pictures were formed and their parts given names. What followed was reaction, responses to responses, criticism rather than creation. Each age has, in other words, to some extent been, and characterised itself as, a critical age.

34. T. Luhrmann, *Persuasions of the Witch's Craft: Ritual Magic and Witchcraft in Present-day England* (Oxford: Blackwell, 1989), gives ample evidence of widespread belief in the crudest of supersitions.

35. Max Weber, *The Sociology of Religion*, trans. Ephraim Fischoff (Beacon Press, Boston, 1964) p. 137. Appleyard also quotes the Weberian historian R.H. Tawney's *Religion and the Rise of Capitalism*:

   > From a spiritual being who, in order to survive, must devote a reasonable attention to economic interests, man sometimes seems to have become an economic animal, who will be prudent, nevertheless, if he takes due precaution to ensure his spiritual well-being.
   >
   > (p. 56)

   Again, he doesn't fully realise how damaging this is to his own spiritual prescriptions. Of course, if God definitely existed, *and* He needed our belief and worship *and* we lived in material Paradise *and* science stopped us from worshipping God, then a case might be made for the iniquity of the scientist. As none of these seems to be true or certain, the case against science *per se* (as opposed to certain uses to which it is sometimes put) is infirm indeed.

36. Goldmann quotes Diderot:

   > What voices! What cries! What groans! Who has enclosed these piteous corpses in this dungeon? What crimes have all these wretched ones committed? Some beat their breasts with stones, some tear their bodies with nails of iron. The eyes of all are filled with remorse, sorrow and death. Who condemns them to these torments? The God whom they have offended. . . . But what kind of God is this? Could a God of goodness take delight in bathing himself in these tears? Are not these terrors an offence to his mercy? If these were criminals assuaging a tyrant's wrath, what more could they do?
   >
   > (*Pensées Philosophiques* VII)

   Such a God would certainly not be good enough for the 'me generation', who would be concerned less with what they should do for God, and his awareness of this, than with what God should do for them. When they pray, they expect him to have the courtesy to reply.

37. John Updike, in his Introduction to *Soundings in Satanism*, ed. F.J. Sheed (Mowbrays, 1972), p. vii, in Richard Webster, *Why Freud was Wrong: Sin,*

Science and Psychoanalysis (London: HarperCollins, 1995) p. 6. This is particularly applicable once we appreciate the rather debased origins of our religious rituals:

> thus through the ages we see the identity of the totem feast with the animal sacrifice, the anthropic human sacrifice, and the Christian eucharist, and in all these solemn occasions we recognise the after-effects of that crime which so oppressed men, but of which they must have been proud. At bottom, however, the Christian communion is a new setting aside of the father, a repetition of the crime that must be expiated.
>
> (S. Freud, *Totem and Taboo*, London: Hogarth Press, pp. 153–4)

Young seems to accept this interpretation of the aetiology/significance of the Eucharist.

Moreover, we have no way of predicting the effect of a return to Dionysus, the primitive and the ritualistic, what, if we threw away all that distanced us from myths, the result in a moral, legalistic society would be. In an advanced industrial civilisation accoutred with, amongst other things, nuclear weapons, the result of destabilisation might not be a universal outbreak of sweetness and light.

38. The starkest self-contradiction relates not to religion but to the project that he proposes for the next millenium – 'the retrieval of innocence' – which, at the very least, would mean the conscious shedding of a certain order of consciousness! After such knowingness as humanist intellectuals such as Young displays, what innocence is possible? Can innocence anyway be consciously sought after?

39. Religious tourists who urge on others beliefs they themselves seem from internal evidence not to believe are common amongst the *kulturkritiks*. I am inclined to be a little more gentle on them than some others are. Aletheia Jackson is a bit more blunt when she talks about Robert Nozick's *The Examined Life*:

> Perhaps the most offensive aspect of *The Examined Life* is Nozick's cavalier, voyeuristic treatment of religion. Either one is a believer or one isn't. If one is a Christian, for example, one believes in the historical reality and transcendent meaning of Christ's crucifixion and resurrection. One writes about them as historical facts. If one is not a Christian, then one cannot meaningfully repeat the Christian understanding of Christ's crucifixion. This fact simply cannot have the same meaning for a non-believer. But Nozick, who is not a Christian – or a believer of any kind – nevertheless writes in a matter-of-fact way about the sundry beliefs of various religions. In short, he wants to have his cake and eat it too. How does he manage it? *He simply isn't serious.* He doesn't take his readers seriously. Hence the pretentious shallow eclecticism of his whirlwind tour through Christianity, Zen, the kababalah, Hinduism, etc. Hence the glibness and utter frivolity of his treatment of solemn matters of ultimate concern.
>
> ('For the Love of Whizdom', *Critical Review*, 1990, pp. 345–64)

40. Dennis Diderot, *Addition aux Pensées Philosophique* (1770), quoted in Goldmann, op. cit.

41. Grevel Lindop, 'Newton, Raymond Tallis and the Three Cultures', *PN Review* 18(1) (1991): 36–42.
42. I explore an analogous problem to this in relation to mind-altering drugs and other ways of divorcing experience from external reality in 'The Work of Art in an Age of Electronic Revolution', in *Theorrhoea and After* (forthcoming).
43. Isaiah Berlin, *The Magus of the North: J.G. Hamann and the Origins of Modern Irrationalism* (London: Fontana/Collins, 1993). I cannot resist quoting John Stuart Mill here:

> In this age the quiet surface of routine is as often ruffled by attempts to resuscitate past evils as to introduce new benefits. What is boasted of at the present time as the revival of religion, is always, in narrow and uncultivated minds, at least as much a revival of bigotry; and where there is the strong permanent leaven of intolerance in the feelings of a people, which at all times abides in the middle classes in this country, it needs but little to provoke them into actively persecuting those whom they have never ceased to think proper objects of persecution.
>
> (*On Liberty*, chapter II)

Once they have found a god to love, the next thing people usually do is seek is a fellow human to hate for not loving that god as passionately as they do.
44. In the *Mantle of the Prophet: Learning and Power in Modern Iran* (1985), discussed by Robert Irwin, *Times Literary Supplement*, 3 February 1995, p. 10.
45. These facts are drawn from an article by Margaret Coles, 'Women of Iran "treated as subhumans"', in *The Observer*, 4 December 1994.
46. Lear's cry seems apt here:

> Thou rascal beadle, hold thy bloody hand!
> Why dost thou lash that whore? Strip thine own back;
> Thou hotly lusts to use her in that kind
> For which thou whipp'st her
>
> (IV, vi, ll, 158–60)

Except that the lustful beadle is real, while the 'whore' is a virgin.
47. It is interesting to consider, in the light of these facts, what Michel Foucault had to say about the Khomeini's revolution following a lightning visit in which he saw enough to satisfy himself that he was sufficiently well-informed to make a pronouncement. In Iran, he concluded, there existed 'a spiritual politics, a model for all the world'. In an interview published in Claire Brière and Pierre Blanchet's *Iran: la révolution au nom du Dieu* (1979), Foucault praised the Iranian revolution as the 'spirit of a spiritless world' – consciously echoing what Marx said about the revolutionary potential of religion. What was emerging in Iran showed that Shi'ite Islam was 'a religion which has, throughout the ages, constantly given an irreducible strength to everything within the people that can oppose state power' (quoted in David Macey, *The Lives of Michel Foucault* (London: Vintage, 1994), p. 409). When a letter to a newspaper, written by a woman who had for her own safety to remain anonymous, protested about what the Islamic revolution had meant for women, he haughtily

dismissed her protestations as simply a mobilisation of the 'age-old reproach of fanaticism'.

This was at a time when Foucault was deeply preoccupied with issues of civil liberties and involved in protest after protest against the abuses of power that he saw in France. These were real abuses, but on a rather different scale from those of the regimes – Cambodian, Maoist China, Iranian – that he supported. A classic case of not being able to see the beam in another nation's eye for preoccupation with the mote in one's own nation's eye.

48. The beadle's obsession with women's pelvises – what goes into them and what comes out of them – has led to unspeakable cruelties. Genital mutilation – ranging from clitoridectomy to complete infibulation – is still a widespread practice. Its appalling consequences – going far beyond its intended purpose of totally denying sexual pleasure to women – are set out in recent article in the *British Medical Journal* ('Female genital mutilation in Britain', J.A. Black and G.D. Debelle, *BMJ*, 310 (1995): 1590–2). The article, which describes the difficulty of eliminating the practice amongst certain ethnic minorities in Britain, where it is against the law, makes sickening reading. And genital mutilation is only one small part of the terrible impact of the beadle–mullah–priest's desire to control the pelvises of women.

49. Much of what follows is derived from the following articles: Douwe A.A. Verkuyl, 'Two World Religions and Family Planning', *Lancet*, 342 (1993): 473–4; Editorial, 'Unholy struggle with the third world genie', *Lancet*, 342 (1993): 447–8; Luc Bonneux, 'Rwanda: a case of demographic entrapment', *Lancet* 344 (1994): 1689–90; Editorial, 'Abortion: one Roumania is enough', *Lancet* 345 (1995): 137–8.

50. This story illustrates, *en passant*, the real relationship between knowledge and power – more precisely, between exclusion from knowledge and powerlessness – that Foucault, for all his obscure ramblings about Power/Knowledge, failed to grasp. As Verkuyl points out, leaders of Islam and the Roman Catholic Church are united in the feeling that women should not take important decisions: their duty is to suffer the consequences of those decisions in their bodies and, if necessary, to give their lives up to them.

51. The bitterly ironic force of 'natural' here is underlined by Verkuyl, who points out that the argument that family planning is unnatural 'should not be used by someone [the Pope] who flies in an airplane all over the globe and has a natural tumour removed by unnatural surgery under unnatural general anaesthesia'.

52. The crucial role of the Church in promoting the spread of AIDS must not, of course, be forgotten. If one wanted to ensure maximum spread of this catastrophic disease, one could not do better than block information, condemn attempts at education and forbid protection.

53. M. King, 'Health is a sustainable state', *Lancet*, 1991 (ii): 664–7.

54. Luc Bonneux 'Rwanda: a case of demographic entrapment', *Lancet*, 344 (1994): 1689–90.

55. This, too, seems to add a questioning footnote to the standard Marxist claim that the economic life of capitalism simultaneously created the first non-believing, deconsecrated society and the most acute forms of human alienation. A *believing* society may, also, be deeply alienated – particularly if the beliefs, as in the case of Roman Catholicism in Rwanda – have been imported from elsewhere and could not possibly have any

kind of adaptive value so far from the lands where they were first cooked up.

56. This story is told in Walter Moore, *A Life of Erwin Schrodinger* (Cambridge: Canto, 1994), p. 287.

57. J.J. Rousseau, *The Social Contract*, chapter VIII, 'Of Civil Religion'.

The challenge, which I shall address in the final chapter of this book, will be to work out how the richness and depth of a sense of 'meanings beyond ordinary meaning' can be made available to human beings without this creating pretexts for Churches, factions and bloodshed. After all (as I have discussed in *Newton's Sleep* (London: Macmillan, 1995)) religious ritual was the earliest expression of human desire to complete the sense opened up by consciousness, to answer the question 'To what (ultimate) end' all our various means – useful activities – tend. A question that has hardly lost its importance. (Of course, early rituals were quasi-technological as well: rain-making, placating the gods, making the world safe, etc.) Art – in particular dance, self-transformation through disguise, etc. – evolved as an adjunct to such rituals. It gradually acquired an autonomous existence outside the actual rituals, though it typically existed to support religious rites or to affirm religious beliefs. Increasingly, it distanced itself from the religion it served and it developed values of its own; an aesthetic emerged. The glorification of God gradually passed into glorification of the world, of the work of art and, even, of the artist. The latter lost his anonymity and retained a personal association with a body of work, became a mini-god ('present everywhere, visible nowhere', as Flaubert said), the author of a world of his own. Finally, a largely humanist, and individualistic as opposed to collective, art emerged, separate from religion. An art that served religion has been replaced by an art that may supplant religion.

This is the background to the view, developed in 'The Difficulty of Arrival', in *Newton's Sleep*, that art may now be thought of as having inherited the mantle of religion in being the vehicle through which ultimate meanings are sought and realised. Like religion, art is concerned with making humanity aware of its difference, its essential nature; of distancing humanity from the world and then making it at home in the world; of opening up the sense of the world and completing that sense, if only momentarily, in perfected experience. Art shares the property of religion, of being an instrument by which humanity (to borrow Lévi-Strauss's phrase) 'makes the world its thing'.

This brief excursus will make clear that my own assumptions are secular. My distancing human consciousness from organic life (in particular in *The Explicit Animal*) is not intended to justify re-connecting it with the domain of the angels and spirits. My position is as remote from revealed religion as from scientific materialism. The failure to provide an evolutionary explanation of consciousness is not proof that earlier, theological views are right after all: the gaps in our understanding are not very satisfactory portals for admitting gods into our world systems. My purpose is to try to free consciousness from the coffin of large theories that conceal its essential nature; to exhume from the carapace of such theories the rich and mysterious consciousness Valéry explored and celebrated in the magic hours of the foredawn – the consciousness whose great hymn is *La Jeune Parque* and whose broken mirror is the *Cahiers*. A true understanding of the unfathomed nature of consciousness should awaken atheistic religious sensibilities to a tautologous wonder at what is the case (rather

than fear at what might be), a faith that worships the obvious, the everyday, the things no one can deny and to which everyone has equal access.
58. Francis Bacon, 'Of Unity in Religion', in *Essays or Counsels, Civil and Moral*.
59. Anyone who may feel that the anxieties expressed here at the prospect
of a universal religious revival are an over-reaction should read Christopher
Hitchens' case history of one of the prize specimens of the contemporary
religious scene. Although it is a mere 98 pages long, *The Missionary Position: Mother Teresa in Theory and in Practice* (London: Verso, 1995) confirms Hitchens' position as a leading commentator of our age.
   Mother Teresa emerges as a calculating power-broker, one who has
supported pretty well every evil cause going, for the sake of her mission. Her mission which has nothing to do with relieving suffering and
everything to with fighting secular humanism, has some singular achievements; foremost among these must be the creation, with donations large
enough to found a dozen first-rate teaching hospitals, of a chain of Poor
Law Hells-on-Earth where the dying and the destitute receive no comfort
for their material woes. Ordinary medical care, based on ordinary medical science, is eschewed because 'Mother Teresa prefers providence to
planning; her rules are designed to prevent any drift towards materialism'
(quoted from Robin Fox's *Lancet* article, 8925 (September 1994), pp. 807–8).
As Hitchens comments:

> As ever, the true address of the missionary is to the self-satisfaction of
> the sponsor and the donor, not to the needs of the downtrodden. Helpless
> infants, downtrodden derelicts and the terminally ill are the raw material
> for demonstrations of compassion. They are in no position to complain
> and their passivity and abjection is considered a sterling trait.
>
> (p. 50)

In continuation of a venerable tradition, Mother Teresa has always placed
her sentiments behind the poor and her power at the disposal of the
rich. She has provided moral support for the heirs of Enver Hoxha, for
the Contras, for Jean-Claude Duvalier (while claiming to be apolitical),
in exchange for support on the issues of controlling women's reproduc-
tive freedom; and for crooked industrialists such as Robert Maxwell and
Charles Keating (while claiming to be above material consideration) in
exchange for hard cash. Meanwhile she adds to the sufferings of those
who come to die in her hell-houses in the tragically mistaken belief that
they will there receive comfort. She has no compunction about this: the
pains of the dying are 'the kisses of Jesus' and as such should not be
alleviated by analgesics. (Her own illnesses have been treated in the world's
best hospitals.)
   Greater even than these two sources of damage are the effects of her
powerful opposition to abortion and contraception. In this one 'simple, humble'
woman (who has managed to cut quite a figure on a world stage for a long
time), most of the dangers of a religious revival are spelt out large.
60. Much lip-service, however, is still paid to it. In many cases, attributing
'wisdom' to the old is a way of patronising them, and an indirect way of
acknowledging their lack of practical function and ordinary competence.
It is analogous to the treatment D.H. Lawrence complained of when he
said that the aristocrats who admired him 'told me I had genius, as if
this were a way of consoling me for not having their incomparable privilege of birth'.

The wisdom of the old is under siege from many directions. The discrediting of the notion of wisdom in general in a world of technical know-how and competence particularly impinges on them, because they are prone to be out of touch with the latest technology. They therefore rely on others to guide them through life – if only to show them how to get the video to work. (The child explaining to the octogenarian how to use a video-cassette recorder – which the former can understand because he approaches it with an absolute attention undiluted by preoccupation or by preconceptions brought from earlier technology – is an updated version of the myth of Anchises carried by the son, Aeneas, he once dandled on his knee.)

And then there is the demographic fact that many old people survive to have prolonged periods of ill health that may undermine any claim to wisdom by their constant assault on good humour and their undermining of independence. Worse still, the high prevalence of dementing illnesses among the very old creates an association between old age and extremes of foolishness as least as strong as that between old age and wisdom, so that to be 90 and not to have 'lost one's marbles' is to become an object of amazement.

The greater frankness in biographies has also contributed to destroying the association between old age and the wisdom that come from no longer having ambitions, hopes, vested interests and illusions. The much-biographied old age of Tolstoy – a pathetic, selfish, Lear-like figure squabbling to the end with his wife over trivia – undermined the image of a world-sage seeing deeper and further as he got older and increasingly detached from the things of this world. And many other Great Men have been similarly exposed as whinging domestic tyrants wallowing in self-pity and moral indignation over trifles. This damages the idea of serenity in an old age in which all passion is spent and replaces it with an old age of bickering in which all *grand* passion is spent and trivial passion remains to be diffused over a multitude of minor irritations. None of which stops the old laying claim to a kind of meta-wisdom that prides itself on being able to see through claims to wisdom – and the related claim that old age is 'happy' because the lives of the old are 'rounded off'.

61. The assumption that wisdom is an affair of *men* will raise a few eyebrows, but this only reflects the demographic facts of the myth. Of course, wisdom has also been attributed to priestesses, earth mothers, etc. but it has usually been, most explicitly in the case of nuns, a wisdom that has a deeper understanding of men's wisdom – the wisdom to see into the depths of men's understanding, to see that they are right. Independent female wisdom has usually been treated with suspicion, as a source of anarchy and danger – most spectacularly in the case of Kali. Or, where it is deeper, it is subordinated to some more important male purpose – the wisdom of the White Goddess and other, lesser, Muses whose function was to inspire men to the essential act of creation. In any case, its status as wisdom has always depended upon the male imprimatur. The female wisdom that comes from closer contact with the earth, with organic reality, is severely circumscribed: after all, the earth is but one of the planets and only one of the fallen sublunary realm. 'Earth' wisdom is thus wisdom about things that pass, not about, for example, the eternal thoughts of God.

Recently there have been two sorts of endeavours to move female wisdom centre stage. The first is to see that it has a special place in private life and the intimate conduct of human relations, because women are thought to be more in touch with their emotions and more able to hearken to those intuitions which are a surer guide than male reason as to what is really going on between and within people. Female empathy is superior to male abstraction – to his instrumental cleverness. And the second is to suggest that female yieldingness and passivity – reflected in everything from a disinclination for war, a non-territoriality and a tolerance of multi-valued logics (A _and_ not-A) – is what we need in the public sphere if the technological, acquisitive, competitive world created by males is to be saved from self-destruction.

This kind of thing – encapsulated in Graves' 'Man does, woman is' and Young's talk of 'female centre and male-margin, stillness and movement' – is as often resented by women as it is welcomed. It is seen as a way of confining women to her sphere: women are too good, wise, pure, etc. to get mixed up in the real, corrupt world of power and control, the world where the real (if wrong) decisions are made. Ascribing menstrual wisdom to women is as marginalising as ascribing barbic wisdom to old men. And it is, of course, no less mythical in itself, having a precarious existence in a world of know-how, competence and answerability.

62. It has often pointed out that only about 20 per cent of medical procedures have been properly evaluated. I have pointed this out myself. But then I add two things: (a) it does not follow from this that the other 80 per cent are all useless; and (b) the 20 per cent of properly evaluated procedures sets medicine above and apart from pretty well every other human activity. 0 per cent of educational, legal and political activities have been subjected to the rigours of a double-blind controlled trial, which is the only way of testing efficacy. (Moreover, a recent study which examined the proportion of medical interventions on an in-patient population that were based upon evidence of efficacy, came up with a higher proportion than the conventional figure: about 70 per cent.) (J. Ellis et al., _Lancet_, Vol. 346 (1995), 407–10.)

63. A typical example is Richard Selzer's _Mortal Lessons: Notes on the art of surgery_ (London: Chatto and Windus, 1981), where he advocates that a surgeon should be seen as a priest, 'a celebrant standing close to the truth curtained in the Ark of the body' and where he dismisses 'mere' technical competence. The absurdity of this book and its attitude to medicine is discussed in Raymond Tallis, 'First Love, Last Rites', _Quarto_, September 1981, p. 5.

64. We shall return to Freud in chapter 8. Suffice it to say for the present that Freud's incompetence as a doctor, his unrepentant attitude to his mistakes, his cocaine addiction, his unscrupulous way with his patients when he was bent on finding his own ideas in their souls, his cavalier approach to scientific truth and the sophisticated nature of the PR (or p.r.) machine that he mobilised in order to conceal all of these faults and to present himself as a courgeous discoverer of unpleasant truths, a martyr to science, are all now well documented – notably in Richard Webster's _Why Freud Was Wrong: Sin, science and psychoanalyis_ (London: HarperCollins, 1995). Key references are given in chapter 8.

65. Freud and all those doctors who come enveloped in a nimbus of 'wisdom' that prevents their performance from being evaluated by the usual outcome measures typically present themselves as 'healers'. This is a usefully ambiguous terms because it suggests an alternative, deeper competence than that of the doctor who makes you better in the usual sense – so that you actually feel better and are alleviated of the problem that

brought you to the doctor in the first place: it hints at a competence that cares for the pilgrim soul in you and which, as with the competence of the priest, may actually first make you feel worse and may never make you feel better in the conventional sense. The sense that 'he cares for me', 'he (alone) understood me', is a potent draw for deeply vulnerable people.

66. As the aphorism goes: a physician makes the same mistake a thousand times and calls it experience – and then expects you to defer to his superior wisdom.

An experience with a long-retired colleague was in part responsible for prompting me to put together my scattered thoughts about wisdom. He was a reasonably competent doctor, though his anxieties about his own competence made him avoid as far as possible acute medicine or medical meetings where his practice would be exposed to discussion. His posture, however, was always that of a 'wise' doctor who was cynical about the benefits of change. Whenever a reform was proposed, he used to urge 'caution' (always apologising for being 'anecdotal' but 'we mustn't get hooked on numbers'). In a fit of exasperation, we looked back on his record and it suddenly became clear that every contribution he had made to planning had been directed to obstructing progress and to keeping things just as they always had been when they were more carefully arranged around his own convenience, inclinations and fears.

67. Isaiah Berlin, 'Joseph de Maistre and the Origins of Fascism', in *The Crooked Timber of Humanity* (London: Fontana, 1991), p. 167.

68. Christopher Hitchens' observations, apropos Mother Teresa, are very much to the point:

The naive and simple are seldom as naive and simple as they seem, and this suspicion is reinforced by those who proclaim their own naivety and simplicity. There is no conceit equal to false modesty, and there is no politics like antipolitics, just as there is no worldliness to compare with ostentatious antimaterialism.

(op. cit., p. 86)

Likewise the 'humility' of the spectacularly humble.

69. Karl Popper, *The Open Society and its Enemies*, Preface to the 1st edition (London: Routledge & Kegan Paul, 1945). Wise men, like great men, are not, of course, very good team players. Why should they listen to others when they have access to superior sources of kowledge and expertise? This passage from Burke, which Popper uses as an epigraph, seems to me to be particularly apposite:

In my course I have known and, according to my measure, have co-operated with great men; and I have never yet seen any plan which has not been mended by the observations of those who were much inferior in understanding to the person who took the lead in the business.

70. Perhaps my sceptical and jaundiced attitude to wisdom is influenced by the practical, detailed responsibilities of my professional life. I have noticed that 'wisdom' tends to be preached most of all by those who, either by accident or by design, have limited responsibilities in the real world or limited technical expertise. The aura of wisdom is the last resort of the incompetent and is apt to be preached by those, such as literary critics, whose competence is not of great urgency or easily testable.

71. T.S. Eliot, *The Dry Salvages*, V.ll. 7–8.
72. I have quoted this from an interview with Gould, but I can no longer find the source.
73. Perception, matter, etc. are certainly not explained. Even if they were, it would not invalidate my point: why should the explained be less interesting than the unexplained?
74. The surrealists, too, tried to re-enchant the world by resurrecting a sense of the magic of everyday life. This was fine so long as it was directed at awakening our sense of the mysteriousness of the ordinary, less so when it became violent and sentimental in turn, as it tried to abolish reality by jeering ordinary life out of existence. For a rather mean-spirited but remarkably accurate diagnosis of surrealism, see Raymond Tallis, In Defence of Realism, 7.2, 'Among the Whimlings'.

4. A CRITIQUE OF CULTURAL CRITICISM

1. Octavio Paz, *On Poets and Others* (London: Paladin and Manchester: Carcanet, 1987), pp. 39–40.
2. Lionel Trilling, 'On the Teaching of Modern Literature' in *Beyond Culture. Essays on literature and learning* (Harmondsworth: Penguin Books, 1967).
3. Francis Fukuyama, 'In the Zone of Peace', *Times Literary Supplement*, 26 November 1993, p. 9.
4. Richard Rorty, 'Professionalised Philosophy and Transcendental Culture', *Georgia Review* 30 (1976): 763–4.
5. 'Jacques Derrida in Discussion with Christopher Norris', in *Deconstruction*, ed. Catherine Cooke and Andrew Benyon (London: Academy Editions, 1989), p. 7.
6. It is easy to see why someone should want to be thought wise. This competence-beyond-and-above-competence requires less effort to maintain than ordinary competences – such as getting one's facts right, making something work or discovering a general law that stands up to testing and is a sound basis for prediction and practical activity. And the rewards – material and spiritual, financial and sexual – are enormous as the careers of the great modern exemplars, the Parisian *maîtres à penser*, illustrate so dramatically. The difficult question is why anyone should want to grant them the title.

    The case of Jacques Lacan throws some light on this. I have written elsewhere on Lacan. The nullity of his ideas is discussed in *Not Saussure* ('The Mirror Stage: A Critical Reflection', London: Macmillan, 2nd edition, 1995). The extraordinary adulation that protected his ideas from critical appraisal is discussed in 'The Strange Case of Jacques L.', in *Theorrhoea and After* (forthcoming). However, this is worth briefly addressing again here.

    There is a deep paradox in intellectual life, where what Kant called 'misology' or hatred of reason, seems to live side by side with an extreme rationalism. This misology can take many forms: a worship of peasant wisdom; a love of irrationality for its own sake; admiration of naked uncontrolled power; a hunger for religious conversion based on blind faith; an uncritical acceptance of authoritarian political regimes that imprison and murder intellectuals. In Paris, the self-betrayal of *les clercs* is manifested in all these ways; but in addition, one also sees a pattern where the radical young, who purport to reject everything that has gone before, seem to long to prostrate themselves at the feet of a dogmatic *maître à penser*, and those who despise all hierarchies and would question everything grant the Chosen One absolute intellectual power and unquestioning belief. This is connected with the central importance of the personal lecture which makes an air of omniscience and charisma essential equipment for the upwardly mobile academic.

Lacan was certainly a charismatic figure, able to evoke this kind of response from the members of his breakaway Ecole Freudienne de Paris:

> Yes, I loved him, and like most of my generation I was in love with thought – this fascination with thought irritates those who do not participate in it. When you are always on time for an appointment and nothing can make you miss it, when you leave disappointed but charmed, what else is it but love?.
>
> (Catherine Clement, *Vie et légendes de Jacques Lacan*, quoted in Bice Benvenuto and Roger Kennedy, *The Works of Jacques Lacan*, London: Free Association Books, 1986)

Bonjour tristesse, adieu critical faculties. Webster's penetrating and lucid critique cites the reminiscences of Lacan's own disciples. Elizabeth Roudinesco, for example, has noted how

> Lacan's texts were sacralised, his person was imitated; he was made into the sole founder of the French psychoanalytical movement. Subdued, an army of barons spoke like Lacan, taught like Lacan, smoked Lacan's cigars.... If that army had been able, it would, like Lacan, have carried its head inclined to the left or had the cartilage of its ears stretched in order to have them, like his, stand out.
>
> ('The Cult of Lacan')

And she reports how, in his famous seminars, which eventually attracted *la toute Paris*,

> Lacan did not analyse; he associated. Nor did he expatiate; he produced resonances. At every session of that exercise in collective therapy, his students had the impression that the master was speaking of them and for them, in a coded message secretly addressed to them alone.
>
> (quoted Webster, p. 306)

He aroused the Sacred Terror of love in disciples who treasured his words, however harsh, however incomprehensible. As Webster points out, 'the urge towards self-humiliation in front of an ineffable wisdom is one of the most significant elements in our religious tradition' (p. 309). Webster's conclusion could not be bettered:

> Driven by his own fierce ambition and his simultaneous need to be loved, celebrated and feared, Lacan eventually stumbled upon a way of exploiting this aspect of our cultural psychology by inventing an explanatory system which does not explain and a version of psychoanalysis which renders human nature infinitely obscure.

The unconscious aim of his ritual obscurity appears to be domination. The obscurity is a means of creating anxiety about not understanding, while at the same time preserving the mystery of Lacan's thought and personality.

Lacan is perhaps the supreme contemporary example of the Wise Man as Intellectual – of the Intellectual Guru who is able to disarm not the Common People but the intellectual elite of their resistance to belief, indeed of their critical faculties. Paris, for reasons we have alluded to earlier, seems to specialise in producing and admiring such individuals. But she does not have the monopoly of such trade.

7. J.G. Merquior, *From Prague to Paris* (London: Verso, 1986), p. 260. For a review essay on this book, see Raymond Tallis, 'A Cure for Theorrhoea', *Critical Review* 3(1) (1989): 7–40

There is perhaps a less cynical explanation of *kulturpessimismus*: the feeling that modern man is a debased creature is a secularisation of the sense of Original Sin, the feeling that we are rotten. That the idea of Original Sin should outlive the religious beliefs of which it is an integral part may have an existentialist explanation: that, by contingently being, we are superfluous, unoccasioned, gratuitous and so are in debt in virtue of existing, a debt we can discharge only by our death.

8. Anyone who wishes to see what is happening in humanities departments dominated by theory and to understand why it has happened should read at least the last chapter of *From Prague to Paris*. It is a classic of intellectual history. Merquior's scepticism about the humanist intellectuals who 'have every interest in being perceived as soul doctors to a sick civilisation' reminds me of Saul Bellow's *Herzog* who complained wryly of those existentialist philosophers 'who toted the void like so much real estate'.

There is also a distinctly unfunny side of the hegemony of Theory, as illustrated all too clearly by the sorry tale published recently in *PN Review*. It is a miserable story of the persecution and ostracisation of two individuals who had the temerity to question the uncritical retailing of critical theory to first year students in English literature. It is significant that it had to be published anonymously ('An Open Letter', *PN Review* 103 (May–June 1995): 30–3).

9. A diagnosis that Nietzsche himself would not have approved of. As Merquior points out (p. 237), the *kulturpessimismus* of Foucault or Adorno is precisely the kind of decadent nihilism that Nietzsche so deeply loathed in Schopenhauer.

10. To put Eliot's claims about Donne's sensibility in some kind of context, we need to think about actual history, real life, in Donne's time. The passage from C.V. Wedgewood, cited in the previous chapter (see note 33), should do the trick. Re-reading this passage may suggest that having an undissociated sensibility was not a great comfort to the general population. It might be argued that a few privileged individuals experienced life with a wholeness and intensity unavailable to anyone today but that is speculative and not, anyway, a sound basis for the implicit felicific calculus necessary to determine whether things were better then than they are now. All the objective evidence seems to be against it.

11. These usually have a specific trigger – a moral panic occasioned by some folk devil (to use the sociologist's scornful phrase about lay prejudices

that are only slightly less ill-informed than their own diagnoses). The recent shocking murder of James Bulger, a two-year-old child, by two children aged 10 and 11, released a torrent of speculation in the media to the effect that children in 1990s Britain are being brutalised by parental neglect, consequent upon the breakup of the traditional family and the need for mothers to go out to work.

Rosalind Coward ('Are children becoming more corrupt and violent or is the real problem middle-class bigotry?', *Guardian*) put this 'brutalisation' of the children of the 1990s in perspective by noting that:

1. In the fifteenth century children were sold by their parents in marriage to the highest bidder and were often severely beaten if they refused.
2. In 1720 (according to Defoe), of children at work in the textile industries, 'there was not one child in the town or the villages round it of above five years old, but, if it was not neglected by its parents and untaught, could earn its bread'.
3. The Society for the Prevention of Cruelty to Children dealt with over 5 000 000 cases in 1884, the year of its foundation. The customary brutal flogging of children continued in public schools.

Anyone who still feels that modern children are uniquely neglected by modern parents should read the horrifying pages of *Capital*, where the large-scale sexual and physical abuse of brutalised, neglected children often worked to their death in factories, is set out in hideous detail.

12. J.G. Merquior, 'In Quest of Modern Culture: Hysterical or Historical Humanism', *Critical Review* 5(1) (1991): 399–420.
13. Richard Feynman has emphasised the centrality to true science – as opposed to 'cargo cult' science – of a special kind of integrity:

> a principle of scientific thought that corresponds to a kind of utter honesty – a kind of leaning over backwards. For example, if you're doing an experiment, you should report everything that you think might make it invalid – not only what you think is right about it: other causes that could possibly explain your results; and things you thought of you've eliminated by some other experiment, and how they worked – to make sure the other fellow can tell they have been eliminated.
>
> ('Cargo Cult Science', in *Surely You're Joking, Mr. Feynman!*, London: Unwin, 1986, p. 341)

This is both a methodological and a moral injunction. The diagnosticians of society may sometimes seem to get as far as establishing apparent causal relations between some social phenomena and others by showing crude correlations but being unable to correct for other, uncontrolled variables, cannot validate these relations. Their methodological impotence is matched by the lack of scientific integrity that would recognise or acknowledge these confounding factors. Social diagnosticians – marshalling all the evidence for their hypotheses and downplaying the probability of confounding factors – sometimes seem closer to lawyers than scientists, advocates for their own viewpoints than for the truth.

14. The contrast between the merciless treatment of scientists and technicians who get the diagnosis wrong and the indulgent handling of the

errors of diagnosticians of society is striking. Doctors usually get the diagnosis and treatment of, say, appendicitis right and yet the few occasions in which they get it wrong are regarded as evidence of the failure of medicine. Science and technology are marked by episodic disasters and routine, continuous success and yet they are treated as a constant source of scandals (even when it has, as is often the case, required science of a high order to uncover the problems brought about by science-based technology).

Unlike a scientist, a cultural critic can make a career out of foolish prognostications. (At least, 'they stimulate discussion'). The failure of every single one of the historians of Eastern Europe to predict the momentous events that took place towards the end of the 1980s – the bringing down of the Berlin Wall, the collapse of one communist regime after another, the dissolution of the Soviet Union – has not, it appears, shaken anyone's faith in their wisdom. Professor Norman Stone, the most prominent among the historian pundits, is still approached for advice by everyone from the makers of television programmes to prime ministers. His failure, after 20 years as an established expert on contemporary Eastern European history, to anticipate any of the really important events or to see any of the most significant trends, seems to have left his reputation as an expert untarnished.

15. See 'Anti-Science and Organic Daydreams', in *Newton's Sleep* (London: Macmillan, 1995).
16. J.G. Merquior, 'For the Sake of the Whole', *Critical Review* 4(3) (1990): 301–25.
17. Quoted in Aaron Wildavsky, 'Can norms rescue self-interest or macro explanation be joined to micro explanation?', *Critical Review* 5(3) (1991): 301–24.
18. We are not, according to Barthes, safe from the Fascism of language even in solitary communion: 'Once uttered, even in the subject's deepest privacy, speech enters the service of power' (Roland Barthes, Inaugural Lecture, Collège de France).
19. The quotation from Goethe:

> He who cannot give to himself
> An account of the past three thousand years
> Must remain unacquainted with the things of darkness,
> Able to live only from day to day

is sufficiently important to Young to figure as one of the epigraphs at the head of the book.
20. His repeated scathing references to 'over-educated' readers suggest another order of ingratitude – and self-contradiction. Without such readers, he would have had no audience: his book would not therefore have been published, for publishers are not in the business of publishing books that will remain unread. There is an element of projected self-hatred in Young's dislike of the 'over-educated': it is Academe raging against itself. The over-education he bemoans is not a great problem outside the Arts Faculties of universities. I am not surrounded by individuals overburdened with erudition. But then I do not spend my days surrounded by arts graduates.

The insincerity of Young's hostility to over-education is betrayed early: in the Preface to his book, he bemoans the lack of the teaching of Greek in schools (p. xii).

21. The anti-abstraction argument modulates into a slightly tricky pro-feminist position reflected in Young's expressed wish to put more of 'the feminine touch' into the affairs of humankind and the world at large. There is, he tells us, a wisdom in females that has been suppressed by the male emphasis on abstraction and symbols; suppression of this has led to the uncontrolled violence of our present era. In order to avoid the accusation of being patronising to women, he advocates talking about yin and yang elements in humanity – although he then quickly reverts to talking about 'men' and 'women' and he seems to tread an uneasy course between a politically correct feminism and a nervous chauvinism. The latter is interesting because it is based on, and so makes particularly visible and open to question, the animal essentialism that is the fundamental presupposition of his book. Let us deal with this first.

For Young, the male is still a hunter even if he hunts with a briefcase and his prey is a pay cheque. The invention of love and, indeed, the role of the female, are to be understood in relation to this insuperable male characteristic. Throughout creation, Young tells us, 'the male needs to be coerced into child-rearing' (p. 72). The male may be uncertain of his role in the child's life: after all, he may not be sure which child is his; whereas there can be no question as to who is the mother. The male's commitment 'is primarily to his territory and only secondarily to his mate' – and her children. However, infants need male nurture: it is the male who will bring the swag back from the hunt. A strategy is needed to ensure that this will happen. The female, with the help of ventral intercourse, where glances are exchanged as well as secretions, invents love. This is a rival attraction to hunting; without love, 'the feckless males, high on hunting and wanderlust, would otherwise stay out all night and bring home no bacon' (p. 64).

And this connects the case against abstraction with the case for 'the feminine touch': the male is concerned with ritual, with rank, with symbols; as for the female, 'inheriting her social position from her mother, all she needs in order to acquire full citizenship is a fruitful womb'. She has no need for complex rites of passage: she lives in the real world: 'while her attention is principally focused on things as they are, his is absorbed in a theatre full of signs and codes where an invisible thing called status is being constantly negotiated'. This contrast between male and female is enhanced by other contrasts: male abstractness with female empathy; the female-centre with the male-margin; her stillness with his movement. We are in the territory of 'Man does, woman is'.

These distinctions have little or no relevance to daily life in modern society. One does not have to point out that there are female mathematics dons or even that in Young's own specialty of English literature (where in many cases abstraction proliferates without bounds) females are well represented. One just has to think how many hours individuals of both sexes spend caught up in abstractions (the early morning rush to get the children to school involves building up a complex card castle of particulars mediated through abstractions of a very high order) or images (the many hours spent watching, thinking about and discussing television). As for gradations of rank, females are for the most part as sensitive as males to markers of social difference and distinction.

Animal essentialism takes one as far in understanding the behaviour and needs of the Explicit Animal as a briefcase would in a bison hunt. Young's prescription – the introduction of the female touch into the affairs of the world – is therefore unhelpful as well as being somewhat creepily patronising. It invites a response analogous to Leavis's famous 'What shall I do now, ejaculate?' – perhaps 'What shall I do now, menstruate?' Moreover, Young's compliments are definitely two-edged; at least they have a pronounced back hand. For there is an assumption that religion is very much an affair of males; it solves their needs and problems. He cites, jestingly, the argument about culture: 'the god-haunted male insisting on ritual ablutions to stave off the devil while his frivolous wife is content simply to sit down and have an apple for breakfast' (p. 65). And yet his own view is not far from that: the home-making home-body practical wife versus the husband high on the ecstasy of the hunt, feeling that there is more to life than this.

It is difficult to believe that religion – which after all is driven by the idea of death, and the sense of one's own mortality – is sex-specific. Surely women are as concerned as men are with the fact that they die? And does it not follow from this that they, too, have religious and metaphysical yearnings and that their needs extend beyond keeping Monday, Tuesday and Wednesday going? Of course it does; but you would not think this from reading Young. Historically, religion has been discovered, developed and dominated by men – at least its rituals and power structures have – while its main substrate – human sacrifices, the preached to, the flower arrangers, the faithful attenders – have been women. But this simply reflects the politics of this world, not different attitudes to the next world.

That Young lets this absurdity slip though is revealing of something else that we have adverted to earlier: his uncertainty as to how to take religion. We have already noted his ambivalence as to whether the religious ecstasy cuts any metaphysical ice. In his discussion of male–female differences and relations, the male sense of the sacred, of something 'beyond' the family circle, beyond the tribe, is mocked as male silliness, rooted in his obsession with rank and ritual:

> One might say he [the male] can go anywhere but belongs nowhere, and if he belongs to anything, he belongs to the 'tribe'. 'What's that?' the female might ask him one day, 'I've never seen it'. 'It's the whole thing' he replies, 'invisible certainly, everything and nothing. I've just discovered it while standing in the centre of the circle'; and soon he will tell her that it is collected in the sacred body of an animal, which will one day be called a totem. She will scratch her head in skeptical wonderment and return to realistic matters, reminding us of the origin of this conversation in the male baboon's attachment to a symbolic theatre of dominance rituals.

If religion is not mere silliness, but a response to a hunger, common to both sexes, for life more abundant, for a life that can overcome the terror of death, for a hope that can compensate for the sense of finitude, then Mrs Flintstone's attitude to Mr Flintstone as outlined above must surely be inadequate and the female cannot be flawed in a way that chauvinists from Moses onwards have suggested: Man thinks, woman gets on with things. From what one can understand of Young's advocacy of the feminine touch, it would appear that its here-and-now-centred practical, down-to-earth common sense would be inimical to restoring our sense of the sacred.

Young's congratulating women on being more down-to-earth is reminiscent of other, earlier modes of patronisation, Michelet's, for example:

> History goes forth, ever far-reaching, and continually crying to him: 'Forward!' A woman, on the contrary, follows the noble and serene epic that nature chants in her harmonious cycles, repeating herself with a touching grace of constancy and fidelity.
>
> (in Ruth Butler, *Rodin* (New Haven: Yale University Press, 1993), p. 179).

22. See, for example, 'Reality is no longer realistic', in Raymond Tallis, *In Defence of Realism* (London: Ferrington, 2nd edition, 1995), and the Prologue and final chapter of the present book.
23. David B. Morris, *The Culture of Pain* (Berkeley: The University of California Press, 1991), p. 124.
24. The quotations from MacKinnon and Dworkin are taken from Lynne Segal's excellent *Straight Sex: The Politics of Pleasure* (London: Virago, 1994).
25. Within the women's movement, there is deep resentment of the appropriation of victimhood by well-off sisters exhibiting their symbolic wounds. Segal (ibid.) documents this resentment very clearly. She quotes one feminist who reminds certain angry gender feminists that 'working-class and Third World women can be seen actively engaged in sex-related issues that *directly* affect life-and-death concerns of women (abortion, sterilization, abuse, health care, welfare, etc.)' (p. 62). Another points out that 'it is because white feminists already have most of their basic needs met, that they can afford to highlight their sexual experiences as the most restraining, if not sole, source of women's oppression' (pp. 62–3).
26. One might expect that a successful, well-off white, heterosexual, employed, middle-class male would find this particularly challenging as he would seem, by virtue of the categories to which he belongs, to be a meeting place of every kind of oppressor. This is to underestimate the ingenuity of those for whom victimhood is crucial to their moral superiority. If you are Terry Eagleton, for example, you can mobilise your working-class background, your Irish-Catholic roots and the fact that you were born in the North and make a fair bid to be considered an underprivileged outsider, even when you have been appointed as the Warton Professor of English at the University of Oxford.

    In a famous toe-curling interview in the *Guardian* shortly after he had been appointed to this Chair (brilliantly reported by Catherine Bennett, 31 August 1991), Eagleton reconciled his cushy insiderdom with his claim to victim status by describing himself as 'the Barbarian inside the Gates' and referring to his Irish ancestry. 'Wear your roots as your leaves' and you may join the ranks of the underprivileged. You won't even have to resort to tricks such as claiming that, despite the appearance of privilege, you are a victim after all because you suffer dreadfully from false consciousness or are marginalised because you are excluded from the ranks of the marginalised.
27. I have addressed the theme of moral superiority in 'The Survival of Theory (3): Moral Overstanding or Halo-tosis', *PN Review*, collected in *Theorrhoea and After* (forthcoming).
28. Bryan Appleyard *Understanding the Present* (London: Picador, 1993), p. 102.
29. See Paul Lawrence Rose, *Wagner: Race and Revolution* (London: Faber, 1992).
30. See Raymond Tallis, *Newton's Sleep* (London: Macmillan, 1995), Part 1, 'The Usefulness of Science'.
31. The habit of omniscientism is addictive indeed. Young, like many other

cultural critics, makes huge general statements, even when they are not required for his arguments. He cannot, for example, quote from T.S. Eliot's *East Coker* without telling us that it is 'one of the few great poems of the century' (p. 38). Consider what one would need to know and how confident one would have to be of one's judgement, in order to make and sustain that statement. It presupposes (a) that one has read all the poems of the century, (b) one is able to judge them with sufficient objectivity to determine whether they are great or not.

32. ISIS–4, 'A Randomised Factorial Trial Assessing Early Oral Captoril, Oral Mononitrate, and Intraenous Magnesium Sulphate in 58050 Patients with Suspected Acute Myocardial Infarction', *Lancet*, Vol. 345, March 1995.

33. This is discussed in *In Defence of Realism* (London: Ferrington, 2nd edition, 1995); see section 4.1 'Reality as Artefact', and the section on Helmholtz in 'Marginalising Consciousness', the second part of this book.

INTRODUCTION TO PART II

1. Michael Foucault, *The Archaeology of Knowledge* (London: Tavistock, 1972) p. 22.

2. Lucien Goldmann, *The Philosophy of the Enlightenment*, trans. Henry Maas (London: Routledge and Kegan Paul, 1973), p. 32.

3. See, for example, the chapter on the rise of science in Bertrand Russell, *The History of Western Philosophy* (London: Unwin University Books, 1961). Newton himself was not happy with the mechanistic world-picture of his own laws. His nostalgia for an animistic view of the material world was given full expression in the massive corpus of his alchemical writings. For a full discussion, see Richard Westfall's biography of Newton (*Never at Rest*, Cambridge, 1980). See also Raymond Tallis, *Newton's Sleep: Two Cultures and Two Kingdoms*, (London: Macmillan, 1995).

4. Of course, consciousness continues to obtrude upon the thoughts of even the most hard-line physicalists. I have elsewhere (Raymond Tallis, *The Explicit Animal*, London: Macmillan, 1991) discussed some of the ways in which the embarrassment posed by consciousness has been dealt with by philosophers and scientists. Consciousness may be biologised: its origin is 'explained' by its adaptive value for the organism, and its content identified with events in the nervous system. It may be computerised: mind (or consciousness) is 'information processing' and can be modelled on artefactual information processing systems. Or, as a combined result of both of these moves, it may be reduced to a set of input–output relations, connected by chains that are at once biological-causal and mathematical-algorithmical. In short, it may be mechanised or mathematicised to the point where it is emptied of intrinsic content. Functionalism, the logical outcome of the biologisation and computerisation of consciousness, simply disposes of it, dissolving it without remainder into the functions it is supposed to serve, ignoring the process (our consciously *being there*) in favour of some kind of product – behaviour or energy exchange.

The elimination of mind, to the point of denying specific mental contents such as qualia (see Daniel Dennett, 'Quining qualia', in A.J. Marcel and E. Bisiach, eds, *Consciousness in Contemporary Science*, Oxford: Clarendon Press, 1988), is now almost orthodox in the English-speaking world. Notwithstanding its recent repudiation by one of its founding fathers (Hilary

Putnam, *Representation and Reality*, Cambridge, MA: MIT Press, 1988), functionalism remains the dominant philosophy of mind among philosophers and cognitive scientists such as neurophysiologists, psychologists and the artificial intelligentsia.

5. This period has not entirely passed with the replacement of Behaviourism by Cognitive Psychology. Many cognitive psychologists applaud the functionalists' dismissal of 'folk psychology', with its belief in the reality of such things as 'fears', 'sensations', 'mental images', etc. See *The Explicit Animal*, esp. pp. 116–18.

6. Emile Durkheim, review of A. Labriola, Essais sur la conception matérialiste de l'histoire', in *Revue Philosophique*, December 1897, trans. and quoted in Peter Winch, *The Idea of a Social Science* (London: Routledge and Kegan Paul, 1958), pp. 23–4.

7. Emile Durkheim, *The Rules of Sociological Method*, trans. S.A. Solovay and J.H. Mueller, ed. G.E.G. Catlin (New York: The Free Press, 1938), pp. xliii–xliv.

J.G. Merquior in *From Prague to Paris* (London: Verso, 1986, pp. 66–7) has commented interestingly on the relationship between the denial of the subjective viewpoint (and the relevance of the testimony of individual) and the aspiration to scientific status with reference to Claude Lévi-Strauss (who, as we shall see, is one of Durkheim's most influential intellectual descendants):

> Six years later, with *Elementary Structures* already behind him, Lévi-Strauss wrote that the linguistic model could free social science from its old feelings of inferiority before the science of nature ... the linguist is able to study his subject without any significant interference [from his subject matter] because language is structured by unconscious rules unaffected by observation.

This could be seen as turning Vico's *Scienza Nuova* on its head. Vico, it will be recalled, argued that we could not really know nature because it had not been made by ourselves, but by God. We can truly know only that which we have made ourselves: and hence the only true sciences are the social, human sciences – in particular history – because history has been made by men. We know history from the inside; or can do if, as the Viconian Herder thought, we are willing to mobilise empathy, a kind of projected introspection.

5 MARX AND THE HISTORICAL UNCONSCIOUS

1. Karl Marx, *Capital* Volume 1, trans. Samuel Moore and Edward Aveling, Afterword to the second German edition (London: Lawrence and Wishart, 1970), p. 19.

It is important not to accept Marx's exaggerated account of the distance between himself and Hegel. G.A. Cohen (*Karl Marx's Theory of History: A Defence*, Princeton: Princeton University Press, 1978), quoted by Allen Wood ('Hegel and Marxism', in Frederick C. Beiser, (ed.), *The Cambridge Companion to Hegel*, Cambridge, 1993) has expressed the affinity between Hegelian and Marxian theories of history, as follows: 'Marx's conception of history preserves the structure of Hegel's but endows it with fresh content' (p. 433). They both regard history as the history of human activity and identify human history with the development of objective social practice. From the point of view of the present argument, the crucial sense in which Hegel is a precursor of Marx, is that he, like Marx, mounts a radical 'critique of *individualist* models of agency, especially self-conscious, rational agency' (R.B. Pippin, 'You Can't Get There From Here: Transition Problems in Hegel's *Phenomenology of Spirit*, in Beiser, ibid., p. 53).

2. Karl Marx, *A Contribution to the Critique of Political Economy*, trans. N:I. Stone (Chicago: Charles H. Kerr, 1904), p. 30.

3. Raymond Aron, *Main Currents in Sociological Thought*, vol. 1 trans. Richard Howard and Helen Weaver (London: Penguin Books, 1968), p. 121.

4. K. Marx and F. Engels, *Manifesto of the Communist Party*, trans. Samuel Moore (Moscow: Progress Publishers, 1952), p. 72.

5. Isaiah Berlin, 'Political Ideas in the Twentieth Century' in *Four Essays on Liberty* (Oxford: Oxford University Press, 1969), pp. 13–14.

The connection between Marxist and right-wing anti-rationalistic political and social thought is set out very clearly in this essay. What they have in common is a belief that

There is one and only one direction in which a given aggregate of individuals is conceived to be travelling, driven thither by quasi-occult impersonal forces, such as their class structure, or their collective unconscious, or their rational origin, or the 'real' social or physical roots of this or that 'popular' or 'group' 'mythology'.

6. This paradox has been expressed succinctly by Berlin (ibid., p. 14), who pointed out that Marxism, 'in the main a highly rationalistic system', denied 'the primacy of the individual's reason in the choice of ends and effective government alike'.

7. This unresolved contradiction had major ideological, political and tactical consequences. If the proletariat were the sole repositories of undistorted consciousness, they would clearly not require to be led or driven towards the revolution by bourgeois intellectuals. Kautsky grappled with this and came to the conclusion that knowledge of society and the practical application of that knowledge were not the same thing:

socialism was a theory that could only be the result of scientific observation and not of the spontaneous evolution of the proletariat. Socialist theory was bound to be the creation of scholars, not of the working class, and must be introduced from the outside into the workers' movement as a weapon in the struggle for liberation.
(Leszek Kolakowski, *Main Currents of Marxism II: The Golden Age*, trans. P.S. Falla, Oxford: Oxford University Press, 1978, p. 42).

The theory 'of socialist consciousness implanted in the spontaneous working-class movement from the outside' had fateful consequences. As Kolakowski points out, it became, for Lenin, a political instrument by supplying the theoretical basis for the 'new idea of a proletarian party directed by intellectuals versed in theory – a party expressing the authentic, scientific consciousness of the proletariat, which the working class was unable to evolve for itself'. And for Kautsky, as for Lenin, it was the theoretical justification 'of a socialist party transforming itself into a party of professional politicians and manipulators'. The seed for a million party-inspired deaths, for the terrorisation of the People by those acting on their behalf, was sown.

8. Albert Camus, *The Rebel*, trans. Anthony Bower (London: Penguin Books, 1971), p. 168.

9. The account that follows is an abbreviated version of my discussion of Althusser and the notion of 'ideology', in *In Defence of Realism* (London: Edward Arnold 1988). See chapter 4, 'Realism and the Idea of Objective Reality'.

10. Louis Althusser, 'Ideology and Ideological State Apparatuses', in *Essays on Ideology* (London: Verso, 1984), p. 11.

11. As I have argued in *The Explicit Animal* (see note 4).

12. I owe this way of expressing Althusser's view to Geoffrey Hawthorn, *Enlightenment and Despair* (Cambridge: Cambridge University Press, 1976), p. 228. The privileged status of science would not, of course, be accepted by the Sociologists of Knowledge but, as we have seen, the sociologisation of knowledge comes up against precisely the difficulty that Marxists acknowledge: science-as-ideology does not explain why science is so successful in explaining and predicting events, nor why science-based technology *works*.

13. This contradiction in Marxist thought is clearly expressed in Isaiah Berlin's account of the thinking behind Lenin's contempt for the democratic process and his emphasis upon the need for the exercise of absolute authority by the revolutionary nucleus of the party:

> Democratic methods, and the attempts to persuade and preach used by earlier reformers and rebels, were ineffective . . . due to the fact that they rested on a false psychology, sociology and theory of history – namely, the assumption that men acted as they did because of conscious beliefs which could be changed by argument. For . . . Marx . . . had shown that such beliefs were mere 'reflections' of the condition of the socially and economically determined classes of men. . . . A man's beliefs . . . could not alter . . . without a change in that situation. The proper task of a revolutionary therefore was to change the 'objective' situation, i.e. to prepare the class for its historical task in the overthrow of the hitherto dominant classes.
>
> ('Political Ideas in the Twentieth Century', in *Four Essays on Liberty*, Oxford: Oxford University Press, 1969, p. 18)

The very claim that no one can be trusted to have disinterested ideas or to be amenable to persuasion by people of goodwill undermines the assumption that Lenin et al. were acting disinterestedly in the light of objective truth. And it is a powerful argument for the severe curtailment, rather than the infinite extension, of their powers. (The blindness here is reminiscent of that discussed earlier of the proto-Fascist Maistre, who concluded from the intrinsic evil of men that power should be concentrated in the hands of a few – presumably evil – men! As we noted in the Prologue, the evil of mankind argues for the dispersion of power, not for its concentration.)

6. DURKHEIM AND THE SOCIAL UNCONSCIOUS

1. Introduction to Ferdinand de Saussure, *Course in General Linguistics*, trans. Wade Baskin (London: Fontana, 1974).

2. Jean-Paul Sartre, *Being and Nothingness*, trans. Hazel Barnes (London: Methuen, 1957), p. 458.

3. Sartre's objection to Freud is not directed against the latter's implicit structuralism. It is on the grounds that 'the dimension of the future does not exist for psycho-analysis' (ibid., p. 458). 'Human reality loses one of its ekstases and must be interpreted solely by a regression towards the past from the standpoint of the present'. The meaning of acts is not available to the actor but only to the informed observer: 'his acts are only a result of the past, which is in principle out of reach, instead of seeking to inscribe their goal in the future.' In short, the actor is somehow absent from his acts. We shall return to this point in due course.

4. Jean-Jacques Rousseau, *The Confessions*, trans. J.M. Cohen (Harmondsworth: Penguin Classics, 1954), p. 17.

5. These two are not, of course, equivalent, as was pointed out to me by an anonymous referee of an earlier version of this chapter. Heidegger, for instance, denied the first but not the second. Structuralists cannot even allow that the individual invests the world with meaning-content whilst the structure shapes that meaning, for the structuralist vision does not allow content distinct from form.

6. Emile Durkheim, *The Elementary Forms of Religious Life*, quoted in Anthony Giddens (ed.), *Emile Durkheim: Selected Writings* (Cambridge: Cambridge University Press, 1972), p. 232. This does not prove the impossibility of transcending socialisation – it shows only that socialisation is characteristic of humanity – but it has been taken to reinforce the inescapability of the social dimension.

7. Emile Durkheim, *The Rules of Sociological Method*, 8th edition, G.E.G. Catlin, trans. S.A. Solovay and J.H. Mueller (New York: Free Press, 1938), p. 106.

8. Quoted in Anthony Giddens, *Durkheim* (London: Fontana, 1978).

9. Durkheim, quoted in J.H. Abrahams, *Origins and Growth of Sociology* (London: Penguin, 1973), p. 165.

10. This hidden purpose could not be made generally known without being defeated – a point that seems to have escaped writers like Dudley Young, as we have discussed in Part I (especially Chapter 3, (pp. 141ff).

11. Emile Durkheim, *Selected Writings*, ed. Anthony Giddens (Cambridge: Cambridge University Press, 1972), p. 263.

12. Durkheim's interpretation of religion is not, of course, without precursors, most notably Feuerbach. The assertion in his *Essence of Christianity* that 'the mystery of God is only the mystery of the love of man for himself' is half-way to Durkheim's more functionalist social interpretation.

13. Jeffrey Friedman has pointed out to me that there is a significant difference between Durkheim on suicide on the one hand, and Durkheim on religion (and his structuralist descendants) on the other. This difference may be obscured by exaggerating the dichotomy between individual choice of, and social causation of, suicide. Social and individual factors may interact or dovetail: adverse social conditions (which may apply across entire populations) such as the collapse of the economy may bring individuals closer to the threshold at which suicide seems rational. In contrast, if religion is the worship not of a transcendent being but of society itself, the individual is totally deceived and there is a true opacity, a deep misconception, an impenetrable unconsciousness at the heart of his belief. The difference is illuminated by the fact that Durkheim's analysis might give one good reason for not participating in religious activity,

but is unlikely to dissuade an individual from committing suicide; it 'unmasks' religion, but not the rationale of suicide. It is Durkheim the unmasker of religion rather than Durkheim the suicidologist who is the precursor of Lévi-Strauss.

14. Ernest Gellner, *Reason and Culture* (Oxford: Blackwell, 1992), p. 35.
15. Durkheim, *Selected Writings*, p. 265.
16. Durkheim, *The Rules of Sociological Method*, p. xlvii.
17. I owe this way of putting it to Jeffrey Friedman.
18. Claude Lévi-Strauss, 'Overture to *Le Cru et le Cuit*', in Jacques Ehrmann, ed., *Structuralism* (Garden City, New York: Anchor-Doubleday, 1970), pp. 31–55.
19. Jeffrey Friedman (personal communication) has questioned whether Lévi-Strauss's structuralism is any more radically undermining of the individual as conscious agent than Durkheimian functionalism as reflected in his theory of religion. To be reduced to a plaything of society (Durkheimian functionalism) may be no less demeaning than to be reduced to the plaything of a system of symbols.
20. A sceptical account is available in J.G. Merquior, *From Prague to Paris: A Critique of Structuralist and Post-Structuralist Thought* (London: Verso, 1986), to which I am indebted for much clarification. Merquior's brilliant book (written with all the insight of an ex-pupil) contains a sharp critique of Lévi-Strauss. A more detailed criticism of the supposed philosophical basis of structuralism and post-structuralism is my own *Not Saussure* (London: Macmillan, 1988). Many of the positions of that book, and of Merquior's critique, are summarised in Raymond Tallis, 'A Cure for Theorrhoea', *Critical Review* 3, no. 1 (Winter 1989): 7–39.
21. Merquior, p. 38. Merquior wittily re-reads an article that Lévi-Strauss wrote in the 1940s on 'French Sociology' which implicitly gives an account of his relationship to Durkheim. Act One: Durkheim's emphasis on social facts as things clear the way for a scientific sociology. Act Two: Mauss avoids the cumbersome 'collective unconscious' by stressing that social forces act as *structures*, 'total facts' that are internalised as rules operating at the unconscious level. Act Three: unconscious structure, spelled out by psychoanalysis and linguistics, is installed as the master concept in modern social science, which takes the form of social anthropology (p. 45).
22. This is argued in detail in *The Explicit Animal*. See, in particular, chapter 7, 'Man, the Explicit Animal'.
23. Sartre, *Being and Nothingness*, p. 459.
24. In the spirit of Gellner, we might ask, for example: Who organises the dances? Do they take place at a certain time? Does everyone have to turn up and if they don't, are they condemned to focal anomie or mental retardation? Are all the concepts covered systematically? Do children have to go to a series of dances? Are there different rites for beginners? etc. Enough of that.
25. 'How We Got Here', *Critical Review* 2, no. 1 (Spring 1988): 111–44.
26. I have discussed this in *The Explicit Animal* and also in Raymond Tallis, *Psycho-Electronics* (London: Ferrington, 1994).

7. FREUD AND THE INSTINCTUAL UNCONSCIOUS

1. Adolf Grunbaum, *The Foundations of Psychoanalysis. A Philosophical Critique* (Berkeley: University of California Press, 1984).

2. Jeffrey Masson, *The Assault on Truth: Freud's Suppression of the Seduction Theory* (New York: Viking-Penguin, 1985). See also Janet Malcolm *In the Freud Archives* (New York: Vintage Books, 1985), for further documentation. The most damning evidence comes from Freud's own correspondence which had been suppressed by the 'keepers of the flame' in the Freudian archives. Freud's brutally inquisitorial methods are also documented in the published works:

> The work [of therapy] keeps coming to a stop and they keep maintaining that this time nothing has occurred to them. We must not believe what they say, we must always assume, and tell them too, that they have kept something back. . . . We must insist on this, we must repeat the pressure and represent ourselves as infallible, till at last we are really told something. . . . There are cases, too, in which the patient tries to disown [the memory] even after its return. 'Something has occurred to me now, but you obviously put it into my head'. . . . In all such cases, I remain unshakably firm. I . . . explain to the patient that [these distinctions] are only forms of his resistance and pretexts raised by it against reproducing this particular memory, which we must recognise in spite of all this.
> (*Standard Edition*, 2: 279–80, quoted in Frederick Crews, 'The Revenge of the Repressed', *New York Review of Books*, 17 November 1994, pp. 54–60)

3. Allen Esterson, *Seductive Mirage: An Exploration of the Work of Sigmund Freud* (Chicago and La Salle, Illinois: Open Court Books, 1993).
4. E.M. Thornton, *Freud and Cocaine: The Freudian Fallacy* (London: Blond and Briggs, 1983).
5. Richard Webster, *Why Freud was Wrong: Sex, Sin and Psychoanalysis* (London: HarperCollins 1995). My essay on Webster's book ('Burying Freud', *Lancet* 1996, 347: 669–71) is a brief summary of the case against Freud as a clinician and scientist.
6. Robert Wilcocks, *Maelzel's Chess Player: Sigmund Freud & the Rhetoric of Deceit* (Maryland: Rowman & Littlefield, 1994). Wilcocks' witty and subtle book draws attention to, amongst other things, Freud's use of the middle mode of discourse in which 'apparent referentiality' is combined with 'the absence of assessable reference'. The consequent 'semantic slippage'

> is harnessed to promote rhetorical persuasion of the validity of his inventions within which context the metaphor does duty of a whole network of thought patterns, each of which subverts the primary referential trope.
> This semantic slippage is precisely a quality of the Freudian oeuvre whereby an analogy taken metaphorically is quietly subsumed into the text of of the psychological inquiry so that it reemerges into the reader's consciousness as a concrete representation of a psychic event – itself now understood by means of the analogy offered, but *no longer in terms of analogy.*
> (p. 45)

7. Ernest Gellner, *The Psychoanalytic Movement* (London: Paladin, 1985). Gellner compellingly accounts for the attractiveness of Freud to the contempo-

rary mind. Freud's theory, by seeming to be constructed from the elements upon which clinical medicine is based, was acceptable to modern sensibility. At the same time, however, 'Freud's system also provided pastoral services comparable to those previously offered by religion, but in a manner suited to the ethos and customs of the new age'. Moreover, the ideas were 'manned by a well-groomed clerisy who promised a new kind of salvation' and who were incorporated into closely regulated guild.

8. Frank J. Sulloway, *Freud: Biologist of the Mind* (New York: Basic Books, 1979).

9. Elizabeth Roudinesco, *Jacques Lacan, Esquisse d'une vie, histoire d'un système de pensée* (Paris: Fayard, 1993). The baselessness of Lacan's ideas is discussed in 'Jacques Lacan: A Critical Reflection', in Raymond Tallis, *Not Saussure* (London: Macmillan, 2nd edition, 1995) and Raymond Tallis 'The Strange Case of Jacques Lacan', *PN Review* 14(4) (1987), 60: 23–6.

10. Frederick Crews, 'The Revenge of the Repressed', *The New York Review of Books*, Part I, 17 November 1994 pp. 54–60; Part II, 1 December 1994, pp. 49–58.

Not the least harmful effect of the 'recovered memory' merchants is that they discredit the testimony of those many who really have been sexually abused. The dismissive attitude to the testimony of ordinary individuals has been a long-standing tradition of psychoanalysis. After all, it was Freud who first overrode the protestations of some of his patients that they had not been sexually abused as children and then dismissed the claims of his female clients that they had been abused, by asserting that every infant fantasises sexual relationships with its parents. The overt abuses of the Recovered Memory Movement provide the ideal cover under which the guilty may escape justice. This point is made very clear by Richard Webster (op. cit.) in the final chapter ('Freud's False Memories') of his mighty demolition of Freud.

Crews' articles, incidentally, show how orthodox Freudians cannot be allowed to distance themselves from the behaviour of unscrupulous Recovered Memory therapists, which has led to so much suffering. For what the card-carrying members of the Guild and the charlatans have in common is more important than anything that separates them. What they have in common is a belief in the repression of sexual memories and in the privileged access the therapist has to them, and the habit of discrediting the ordinary testimony of conscious individuals appealing to the usual sources of evidence. The road to Hell is paved with denials of ordinary intentions, ordinary beliefs, ordinary memories in favour of the preconceptions and diagnoses of experts. Once you assert – as Renée Fredrickson, a 'recovered memory expert' who has contributed to the epidemic of children who are persuaded that they have been abused by their parents, asserts – that 'the existence of profound disbelief is an indication that memories are real' (Crews, p. 54), then everything – every kind of suggestion, false accusation, manipulation, etc. – is possible. According to Blume (also cited by Crews), having 'no awareness at all' of having been sexually molested is also a symptom of having been abused. This, too, is a logical consequence of Freud's theory of the mind. ·

11. I have discussed the happy after-life of discredited theories among the uncritical minds of theorists in my chapter on Lacan in *Not Saussure*. The other reason why Freud is so indestructible is that, quite simply, he was a genius: he was a great prose artist; he was a peerless connector of

ideas (who else would think of linking – to use Richard Webster's example – 'the sexual anatomy of pre-historic birds to the obstinacy of two-year-old children, and the evolution of crocodiles to the meanness of Viennese aristrocrats'?); he was a great leader and persuader of men; and he was willing to go wherever his ideas took him.

12. Quoted in Richard Wollheim, *Freud* (London: Fontana, 1971), p. 174. In writing this section about the Freudian unconscious, I have been greatly assisted by Wollheim's exemplary guide, although he would not share my evaluation of Freud.

13. As we have already noted, Sartre arrives at the same conclusion from a different direction. He emphasises the active role of consciousness (or the for-itself) in sustaining the world into which the unconscious has irrupted. This consciousness is not merely the plaything of the past, driven by a *vis a tergo* deriving from the unresolved conflicts of childhood and infancy; it is also oriented towards a future into which it projects itself. The absence of the dimension of the future, and so of the active meaning- and value-conferring role of consciousness, in Freud's interpretation of neurotic enactment is crucial to Sartre's penetrating critique of Freud:

> The dimension of the future appears not to exist for psychoanalysis insofar as acts seem to be constituted by the past.... Human reality loses one of its ekstases and must be interpreted solely by a regression toward the past from the standpoint of the present. At the same time the fundamental structures of the subject, which are signified by its acts, are not so signified *for him* but for an objective witness who uses discursive methods to make these meanings explicit.... His acts are only a result of the past, which is on principle out of reach, instead of seeking to inscribe their goal on the future.
>
> (*Being and Nothingness*, New York: Simon & Schuster, 1966, pp. 458–9)

Sartre also criticises the very concept of unconscious repression (see *Being and Nothingness*, pp. 50–4). The main thrust of his criticism is that the unconscious has to know what it is that has to be repressed in order (actively) to repress it. It has also to know that it is shameful material appropriate for repression. If, however, it knows both of these things, it is difficult to understand how it can repress it without first being conscious of it. The only way round this would be to reduce repression to forgetfulness, which would undermine the fundamental Freudian principle that repression is quite unlike forgetfulness, inasmuch as it is active and targeted.

14. A.H. Chapman and M. Chapman-Santana, 'Is it possible to have an unconscious thought?' *Lancet* 344 (1994): 1752–3. The examples the authors give in this paper also show how, when Freudians talk about the Unconscious, they are often simply talking about things we are conscious of but are not yet conscious of reflectively. In accordance with their own theories, they shouldn't, of course: the Unconscious is supposed to be composed of psychic things that have been actively repressed rather than simply not yet brought into full consciousness (or, as discussed in note 13, simply forgotten).

15. Theodor Adorno and Max Horkheimer, *Dialectic of Enlightenment* trans. John Cumming (London: Verso, New Left Books, 1979), p. 192.

16. As Young's book repeatedly emphasises, the sacrificial ritual is often associated with, or is in permanent danger of degenerating into, human sacrifice. The natural human sacrifice is the woman, who traditionally

has provided the human tribute, after the first, mythical parricidal or theophagous orgies. The probabilities are that opening up the orgiastic dimension of the past would lead to increased instability, mismanagement and violence. The return of the irrational that Young so longs for is more likely to blow a hole across the universe than seal the hole in the ozone layer – a task that requires a long process of technological change, of legislation and of reform of human expectation and of the means by which they seek life more abundant remote from the immediacy of the dance and its orgiastic consequences.

17. W.H. Auden 'Vespers', from *Horae Canonicae, Selected Poems* (London: Faber & Faber, 1968), p. 114.

18. De-centring seems to have been a peculiarly French pursuit: Lacan, Foucault, Barthes, Lévi-Strauss, Derrida were all committed to denying the centrality of the rational self and even the reality of self-presence. There is a particular irony in this predilection; for the writers in question were noted for their massive egos, their boundless self-confidence and personal ambition, and their remorseless deployment of something that looked superficially like reason.

19. Louis Althusser, 'On Marx and Freud', trans. Warren Montag, *Rethinking MARXISM*, 4(1) (Spring 1991). I am grateful to Jeffrey Friedman for drawing my attention to this paper.

20. This is a variant of Althusser's Lacanian idea that it is the Ideological State Apparatuses, designed to reproduce the conditions of production, that transform individuals into (unified) subjects. I have examined this in *In Defence of Realism* (London: Edward Arnold, 1988).

21. It is difficult in practice to know how much Freud really believed about the hegemony of the Unconscious and his own rhetoric about Copernican revolutions. But he certainly gave plenty of evidence to suggest that he sometimes swallowed himself hook, line and sinker. His reference to the 'symptom' of consciousness (quoted earlier) is itself symptomatic. And the way that, in the analyst's consulting room, the entire world changes and behaviour is reduced to a symbolic or symptomatic expression of the Unconscious forces, is decisive. Suddenly, a woman fiddling with her handbag is no longer just fiddling with her handbag – hesitating whether or not to take out a handkerchief or worrying whether she has enough money to pay the analyst – but is symbolically masturbating in response to memories evoked by the permissive atmosphere of analysis. Suddenly ordinary life, even outside the consulting room, is a place where psychopathology is acted out rather than explicit goals pursued; or the latter are mere rationalisations of the former. In the all-encompassing vision of depth psychology, the difference between ordinary and neurotic behaviour disappears. If Freud has been misunderstood – if he did not after all believe that most of us were in the grip of dark forces for most of the time and never more so than when we were protesting the rationality of our behaviour – then he deserves to have been misunderstood.

22. See Raymond Tallis, *The Explicit Animal*, for a reflection on the mysterious property of consciousness to make things progressively more explicit, to be increasingly conscious of itself. Interestingly, even those who support Freud's 'revolution' in psychology have difficulty dealing with this. Thomas Nagel, for example, in a recent laudatory review of Richard Wollheim's *The Mind and its Depth* ('Freud's Permanent Revolution', *New York Review of Books*, 12 May 1994, pp. 34–7), takes issue with Wolheim's Kleinian account of the development of the moral sense.

Wollheim rejects the standard Freudian story which bases moral development on the internalisation of the disapproving father, in favour of a story which emphasises the reconciliation of ambivalent feelings towards the mother whom it both loves and hates. Moral development is constituted by the integration of the good self and the bad self, the one that loves and the one that hates. Morality, far from transcending infantile needs and wishes and the mental structures established before the age of two, needs to be rooted in them, otherwise it will be very superficial and, consequently, fragile. Nagel's dissent from this view is very much to the point:

> I do not see how a theory of this kind could by itself explain more than the very beginnings of the complex system of restraints on aggression, acknowledgment of formal obligations and of the rights and claims of others, that make up a fully developed morality.... [M]orality in the strict sense requires forms of thought that are much more impersonal than fear of, love for, or identification with particular external or internal 'objects', whether fathers or mothers. It aims to supply objective standards in the realm of conduct which will allow us to justify ourselves to one another and agree on what should be done.

Unfortunately, Nagel fails fully to see the implications of this point he himself is making; or the larger reminder that it should trigger. That the Kleinian (or the Freudian) vision of the motors of human behaviour do not take account of the most striking thing about us: that we do things for quite complex, abstract reasons, and that what we actually do is frequently remote from our appetitive goals, though it may be directed towards them. Even if our reasons were mere rationalisations (which they are not), the fact that we have to deploy rationalisations cannot itself be explained in terms of the vision that the psychoanalysts have of the structure and function of the human mind. If we root our adult behaviour in the bodily, instinctual, world of the infant, it is difficult to see how that behaviour could have acquired such abstract and abstractly driven forms as ordinary behaviour does. Stamp collecting may be a substitute for witholding one's faeces and frustrating one's mother, construed as a hate object; but this does not begin to account for the plans I form to save up to attend the Stamp Collectors' Congress next year and the pangs of conscience I feel when I realise that the money I save could have gone to a worthier cause, such as a new pair of football boots for my nephew. Nagel, in short, misses the fundamental point that humans are explicit animals, able to elaborate and morally and imaginatively inhabit enormously complex distances between themselves and their quasi-instinctual goals.

Wollheim denounces the 'phantasy that morality marks the spot where human beings part from animal nature'. There is a much more pervasive, and damaging, fantasy that human nature is a form of animal nature. Morality arises out of the distances between human nature and animal nature, between reflexes and instincts on the one hand and explicitly formulated and negotiated intentions, goals and purposes on the other.

8. THE LINGUISTIC UNCONSCIOUS

1. *The Explicit Animal*, 'Consciousness Restored'.
2. This claim has to be treated with a good deal of non-literal-minded care. See Raymond Tallis, *Psycho-Electronics* (London: Ferrington, 1994), especially the entry 'Information'.
3. George Steiner, *After Babel* (London: Faber, 1974).
4. I owe this way of putting the matter to an anonymous referee of an earlier draft of this section. Her/his criticisms have been invaluable. It is important to note the *two-way* relationship: the self constitutes language as well as being constituted in/by language. I would not wish it to be thought that I subscribe to the notion that 'language speaks us' – the very viewpoint that I criticise later in this section.
5. This claim is subjected to critical examination in the chapter on Lacan in *Not Saussure*.
6. See *Not Saussure*. Shorter versions of my critique are available in *Theorrhoea and After* (forthcoming) and Raymond Tallis *Critical Review*.
7. Ferdinand de Saussure, *Course in General Linguistics*, trans. Wade Baskin (London: Fontana; Glasgow: Collins, 1974), pp. 117–18.
8. See Stephen Cox, 'Devices of Deconstruction', *Critical Review* (Winter 1989), pp. 56–76 for a clear exposition and critique of the Derridan notion of the General Text. This notion, according to Derridans, ruins any project of interpretation. We shall encounter in due course the argument that, because language is intrinsically undecidable, definite meanings can be derived from statements only through coercively imposing such meanings upon them. All (definite) meaning is therefore 'Fascist'.
9. This belief is most famously expounded in Roland Barthes, 'The Death of the Author', in *Image – Music – Text*, selected and translated by Stephen Heath (London: Fontana, 1977).
10. Jonathan Culler, *Structuralist Poetics* (London: Routledge and Kegan Paul, 1975), p. 28. Culler is, of course, a secondary source, but the formulations in his extremely (and deservedly) popular expositions have had enormous influences. The lucid (if deeply uncritical) texts of Hawkes and Eagleton have also been crucial mediators between the gurus and the congregation.
11. 'The statement that everything in language is negative is true only if the signified and the signifier are considered separately; when we consider the sign in its totality, we have something that is positive in its own class. ... Although both the signifier and the signified are purely differential and negative, when considered separately, their combination is a positive fact: it is even the sole type of facts that language has' (*Course*, pp. 120–1).
12. The crucial role of deixis in ensuring reference to particulars – or in ensuring that (general) meanings have particular referents – is significantly overlooked by post-Saussureans. Without the explicit deixis of the present speaker or the implicit deixis (anaphora), parasitic upon the notion of explicit deixis, of the absent narrator in a written story, reference – the cash value of language – would fail. Deixis is not just a matter of the body being available as the 0,0,0 of a set of coordinates, as the basis of 'here' and 'now' and so of the determination of specific, unique (i.e. actual) referents. Deixis has to be mobilised, exploited, *meant*. This shows how wide of the mark are characteristically post-Saussurean claims such as Dragan Milovanovic's assertion that 'Language speaks the subject,

providing it with meaning at the cost of being' (*Postmodern Law and Disorder*, Liverpool: Deborah Charles Publications, 1992, p. 7). This is something to which we shall return. (See also note 15.)
13. Terence Hawkes, *Structuralism and Semiotics* (London: Methuen, 1977), pp. 16–17.
14. I have discussed this passage in more detail in *Not Saussure*, pp. 70–9.
15. It is worthwhile spelling this out a little bit further, with reference to Derrida's notion of the transcendental signified. The term is introduced in a seemingly crucial but very obscure passage in his *Of Grammatology*, trans. Gayatri Chakravorty Spivak (Baltimore: The Johns Hopkins University Press, 1976), p. 50.

> Since one sign leads to another *ad infinitum*, from the moment there is meaning, there are nothing but signs. *We think only in signs.* Which amounts to ruining the notion of the sign at the very moment when, as Nietzsche says, its exigency is recognised in the absoluteness of its right.
>
> (ibid., p. 50)

One is left only with 'play' – the free play of the signifier –

> the absence of the transcendental signified as the limitlessness of play, that is to say as the destruction of onto-theology and the metaphysics of presence.
>
> (ibid, p. 50)

The chain of signs never terminates at anything that is simply present; it always points to the next sign, so that it is reduced to signs of itself – to traces. More generally, we never touch presence unmediated by signs – immediate presence, presence itself. Immediacy is an impossible, elusive dream. Thus Derrida.

It is, of course, untrue that the emergence of 'meaning' in signs results in the evaporation of presence to traces of traces. The paw marks are a sign to me of a lion. But, over and above their character as signs of a *general* meaning, they have *particular* existence as depressions in the dust. They are that which means 'lion', they carry the meaning 'lion' on this occasion; but that is not all that they are. They continue to exist when they are not meaning and they have features that are quite independent of their meaning, or that are not involved in the specification of the meaning 'lion'. Their location two inches rather than two feet from a particular bush, their being dampened by rain, their being seven in number rather than six, etc. are not features relevant to the discrimination of their general meaning. So, existing and signifying, being present and signifying something that is absent, are not alternative states. On the contrary, being a sign is predicated upon being an existent that is present.

We are no more entitled to infer from the fact that one sign may lead to another *ad infinitum* that the signified is never reached than to conclude from the fact that since every effect is itself a cause and the causal chain is interminable that there are no effects – that the chain of causes never 'arrives at' effects. Of course, there is no 'transcendental effect' which would bring the causal chain to an end; but this does not mean that there are no effects at all. *Omne causa de causis* – all causes themselves arise from causes – does not imply that there are 'no effects/ things/

events'. The fact that the signifier does not reach a 'transcendental signi-fied' should be cause for concern only if the signifier were the sign itself and the 'transcendental signified' were a referent. And manifestly they are not.

The 'transcendental signified' is a useful smokescreen. It is used, vari-ously, to mean: the signified; the meaning of a sign; the referent of a chain of signs in use; or the ultimate termination of the chain of signs – in plenitude or closure of meaning, in absolute presence or in God. So those who believe in the reality of the signified and do not believe that language is an endless chain of signifiers apparently also believe in the transcendental signified; and to believe in the transcendental signified is to believe that the chain of signs comes to an end, that a final meaning can be reached and that the place where the latter is reached is identical to the place where signs give way to absolute presence – to believe, in other words, in God, Who is both absolute presence and final meaning. Since most contemporary readers are liable to be atheists and since, too, Husserlian 'absolute presence' is so elusive, merging the notion of the signified with that of the *transcendental* signified is certain to discredit it and to give plausibility to the idea that discourse is an endless chain of signified-less signifiers. The concept of the 'transcendental signified' enables Derrida to move almost imperceptibly from the position that no sign opens directly on to a plenitude of meaning/presence, i.e. is underwritten by God, to the claim that there is no signified at all, or none, anyway that the linguistic signifier can reach out to.

Incidentally, Derrida, and his epigones, endlessly repeat the unfounded assertion that the notion of (unmediated) presence is a specifically West-ern illusion. This is factually incorrect, as this passage indicates:

> In the [Eastern] mystical consciousness, Reality is apprehended directly and immediately, meaning without any mediation, any symbolic elabo-ration, any conceptualization, or any abstractions; subject and object become one in a timeless and spaceless act that is beyond any and all forms of mediation. Mystics universally speak of contacting reality in its 'suchness', its 'isness', its 'thatness', without any intermediaries; beyond words, symbols, names, thought images.
> (*Quantum Questions*, ed. Ken Wilber (New Science Library, Shambhala, Boulder and London, 1984), p. 7)

16. See 'The Mirror Stage, A Critical Reflection', in *Not Saussure*.
17. See Emile Benveniste, 'Subjectivity in Language', in *Problems in General Linguistics*, Miami Linguistics Series No. 8, trans. Mary Elizabeth Meek (Coral Gables, FLA: University of Miami Press, 1971). See also, 'Changes in Linguistics' and 'Man and Language', in the same collection.
18. Catherine Belsey, *Critical Practice* (London: Methuen, 1980), p. 59.
19. Benveniste, 'Subjectivity in Language', p. 224.
20. Milovanovic, *Postmodern Law and Disorder*, p. 7.

It was, of course, Heidegger, rather than the post-Saussureans, who first asserted that language speaks us – 'Die Sprache spricht', as he said in his essay on Georg Trakl. He has been a major influence on post-modernist thought, as are the hermeneutic philosphers such as Gadamer and Ricoeur, who were themselves influenced by Heidegger. For the lat-ter, language constitutes us and we are unable to break out of the hermeneutic circle within which understanding moves. This notion has been an all-pervasive one, linked with Derrida's infamous assertion that 'there is nothing outside of the text' and Rorty's popular relativisation of truth to 'interpretive communities'. Moreover, it has a venerable ancestry – being pre-modern, as well as post-modern. The most relevant ancestors are, of course, the Counter-Enlightenment thinkers Hamann and Herder. It was Hamann who (as we have already noted in the Prologue) asserted that 'language was the organ and criterion of truth'. And for Herder, thought was inseparable from language, and hence from the culture expressing and expressed in the language. Truth resided in the world-picture, the spirit of the *volk*.

The political expressions of organicist, anti-universalist, *volkisch* thought should not need spelling out, but it is surprising how few post-Saussurean thinkers seem to be aware of the potential consequences of their liberation from the notions of transcendent truth and from 'the grand narratives of emancipation and enlightenment'. Paul de Man's intellectual affiliations are regarded as an anomaly and Heidegger's politics quite separate from the philosophical notions that have been so powerfully influential on post-Saussurean thinkers. The relationship between post-modernism and far right-wing thought has been set out with exceptional clarity by Rainer Friedrich, in 'The Deconstructed Self in Artaud and Brecht', *Forum for Modern Language Studies* xxvi (1990): 282–95. What is not often appreciated is that the excited post-modernist gigglers are too shallow even to be right-wing; or to notice that they are not on the side of the (progressive) angels.

21. Grice's theory of discourse – with its emphasis on the production as well as the reception of meaning and the role of the recipient's understanding of the producer's intentions, of what the producer intends to mean, in ensuring communication – is crucial to an understanding of the errors in post-Saussurean discourse theory. I have discussed Grice's theories and their metaphysical implications in *The Explicit Animal*, chapter 8, 'Recovering Consciousness'.

22. See the chapter on Derrida ('Walking and Difference') in *Not Saussure*.

23. This notion of an 'enabling constraint' (a term I owe to a generous and painstaking, anonymous referee of an earlier version of this section) is a cousin of the Chomskeian notion of 'rule-governed creativity'. Chomsky's achievement was to recognise a problem that no one else had fully appreciated – which is that of explaining how it is that we are all inducted, through language, into a common world organised in roughly the same way for each of us. It is, as Gellner points out, the same problem as Durkheim's, except that Chomsky considers not only major categories, such as 'God', 'Man', etc., but also 'the rank and file of our conceptual army'. There must be constraints upon the ways in which we organise experience, and those constraints come from language. This raises the further problem of how we acquire the syntactic and semantic rules of language. Associationist psychology is simply unable to account for the fantastic number of rules regulating the use of individual words and the generation of verbal-strings that a five-year-old child has acquired in its short life. Associationism is the road not to the shared world but private delirium. 'All our concepts', Gellner points out, 'are compulsively disciplined' (*Reason and Culture*, Oxford: Blackwell, 1992, p. 126). According to Chomsky, we are endowed with a Language Acquisition Device built into the brain that is primed to extract from the sloppy phonetic and

semantic material served up to us a set of general principles, so that we are able to combine a restricted set of phonetic elements in a phonetically, syntactically and semantically disciplined, and yet creative, way. Within these constraints, we are able to generate an infinity of novel sentences that meet our individual communicative needs.

Gellner worries that this account of the constraints built into language mean that 'Language has reasons of which the mind knows not' (ibid., p. 124):

> We mainly, or exclusively, think through language; but if our speech is bound by deep rules of which we know nothing (and the unravelling of which in linguists' inquiries which are arduous and highly contentious), then it would seem that we are not and cannot be in control of our own thought.
>
> (ibid., p. 127)

The examples we have already given indicate that, in practice, things are not as bad as Gellner fears. Theoretical considerations cannot abolish the distinction between ordinary, seemingly deliberate, chosen talk and a fever of words shouted out in a delirium. Nor can they undermine the fact (discussed in the next paragraph of the main text) that I can, and frequently do, use even off-the-shelf 'empty' phrases such as 'Hello' to express very complicated and calculated communicative intents. What Chomsky has done is not to produce evidence that we do not express ourselves through language but to show that our self-expression is even more mysterious than we thought it was. Likewise a physiologist's demonstration of the existence of reflexes does not prove that there is no such a thing as deliberately walking to one place rather than another, only that the interaction between mechanisms and free action (or the use of mechanisms to bring about free action) is yet more puzzling.

24. 'Language [*langue*] is not a function of the speaker; it is a product that is passively assimilated by the individual.... Speech [*parole*], on the contrary is an individual act. It is wilful and intellectual' (*Course*, p. 14).

25. The displacement of the self by language – summarised in the assertion that language speaks us rather than vice versa – has recently lost some of its fashionability, since it has been recognised that making language a 'concrete entity' and an agent is no more convincing than ascribing concreteness and agency to the self. As Geoffrey Galt Harpham has expressed it:

> It is now easier than it once was to see in the new discourse on language not a severely literal description but a dense mesh of metaphors of intentional agency that were necessarily drawn from the only possible model for such agency, the old-fashioned subject. Without anyone's remarking on the fact, the 'concrete entity' of language had implicitly been described ... as a 'who' making choices, guiding decisions, encouraging values, giving and taking orders.
>
> ('Who's Who', *London Review of Books*, 20 April 1995, pp. 12–14)

In short, behaving like a 'who' rather than a 'what'. This is an example of what I have elsewhere characterised as the Fallacy of Misplaced Explicitness, whereby consciousness, agency, deliberation, etc., which are denied in the place where they are intuitively to be found, are ascribed to material objects or automatic processes (see the relevant entry in *Psycho-Electronics*, op. cit.).

What, of course, goes missing from the self or subject when it is replaced by discourse is the personal coherence that characterises a self; for example, the link between my being a witness to an accident (because of being in a particular place) and feeling obliged to call for help; or between my undertaking to do something at time t and my actually doing it at time t'.

26. See *The Explicit Animal*, 'Recovering Conciousness', where the relationship between the arbitrariness of linguistic signs and the limitless human capacity for developing self-consciousness, for making things explicit, is set out in greater detail.

9. UNCONSCIOUS CONSCIOUSNESS

1. This aping is prevalent even in ethics. Mary Midgely (*Beast and Man*, London: Methuen, 1979) has pointed out that the title of G. E. Moore's *Principia Ethica* echoes Newton and symptomatises an explicitly Newtonian ambition amongst moral philosophers, which goes back at least as far as Hume. See also Karl Lashley, 'The Problem of Serial Order in Behaviour', in L.P. Jeffress, ed., *Cerebral Mechanisms in Behaviour: The Hixon Symposium*, 1951: pp. 112–36, where the common aim of the sciences is seen to be to explain all phenomena in terms of the differential equations that capture the properties of the physical world.

2. J.B. Watson, *Psychology from the Standpoint of a Behaviorist* (Philadelphia, 1919) advanced the view that 'consciousness' was a fruitless object of study – especially as the only means of access to it was via the unreliable method of introspection – and that it could be safely ignored.

3. Disguised behaviourism, in the form of functionalism, however, is flourishing and currently dominates thought about the mind and consciousness. See Raymond Tallis, *The Explicit Animal*, chapter 5, 'Emptying Consciousness: Functionalism', pp. 141–60.

4. See Karl Lashley, 'The Problem of Serial Order in Behaviour', op. cit.

5. N. Chomsky's review of B.F. Skinner, 'Verbal Behavior', *Language* 35 (1959): 26–58.

6. G.A. Miller, E. Galanter and K. Pribram, *Plans and the Structure of Behavior* (New York: Holt, Rinehart and Winston, 1960).

7. I have throughout relied on the translations by R.M. and R.P. Warren in their *Helmholtz on Perception: Its Physiology and Development* (New York: John Wiley and Sons, Inc, 1968).

8. In the preface to his standard text (*Sensation and Perception in the History of Experimental Psychology*), Boring wrote:

> No reader of this book will need to ask why I have dedicated it to Helmholtz. There is no-one else to whom one can owe so completely the capacity to write a book about sensation and perception. If it be objected that books should not be dedicated to the dead, the answer is that Helmholtz is not dead. The organism can predecease its intellect, and conversely. My dedication asserts Helmholtz's immortality – the kind of immortality that remains the unachievable aspiration of so many of us.
>
> (quoted in Warren and Warren, Preface, p. i)

9. P.N. Johnson-Laird, *The Computer and the Mind* (London: Fontana, 1988), p. 60. I have examined this claim in more detail in *The Explicit Animal* – see chapter 4, 'Computerising Consciousness', and *Psycho-Electronics* (op. cit.), see especially the entries on 'Calculations' and 'Logic'.

10. David Marr, *Vision, a Computational Investigation into the Human Representation and Processing of Visual Information* (San Francisco: W.H. Freeman, 1980). My critique of Marr is heavily indebted to Peter Hacker's article, 'Seeing, Representing and Describing', in J. Hyman, ed., *Investigating Psychology: Sciences of the Mind after Wittgenstein* (London: Routledge, 1991).

11. P.M.S. Hacker, 'Experimental Methods and Conceptual Confusion: An Investigation into R.L. Gregory's Theory of Perception', *Iyyun, The Jerusalem Philosophical Quarterly* 40 (July 1991): 289–314.

12. The draining of both utility and conscious content from perception in contemporary cognitive psychology – whose characteristic philosophical framework is qualia-free functionalism – is discussed in *The Explicit Animal*. See especially chapter 5.

13. R.L. Gregory, *Eye and Brain* (London: Weidenfeld & Nicolson, 1966). A similar argument can be applied to unconscious mechanisms in action. It is obvious that we have little or no direct control over, say, the spinal mechanisms necessary for our voluntary (and so moral or immoral) behaviour. Deliberate action is clearly predicated on more or less intact physiological mechanisms of movement. (Just as, to take the argument deeper into the springs of action, the formulation of goals and intentions depends upon being possessed of a certain level of consciousness, upon not being in coma or in a confusional state, requires an orientation in space and time that lies beyond our will, etc.). At a higher level, conscious doing is rooted in unconscious mechanisms; for example, we can perform learned actions when other aspects of memory have been lost. We can still remember how to do a certain action even when the memory of learning it, and the circumstances under which one learnt it, have been forgotten. One can 'know how', even when one has forgotten all the relevant 'knowing that'; one can retain procedural or habit memory when one has lost the relevant occurrent memories. Automatism seems to penetrate to the very heart of even the most clearly voluntary action.

    Even so, this does not invalidate the concept of voluntary action. Indeed, it sharpens our sense of its extraordinariness and of the extraordinariness of those rather privileged voluntary actions that consist of investigating the mechanisms that underlie voluntary action.

14. For a more detailed discussion, see *The Explicit Animal*, chapters 6 and 7, and *Psycho-Electronics*, especially the entry on 'Calculations'.

15. There is an interesting discussion of 'unconscious inference' by S.G. Shanker ('Computer Vision or Mechanist Myopia?') in S.G. Shanker, ed., *Philosophy in Britain Today* (London: Croom Helm, 1986).

16. See 'Misplaced Explicitness', in *Psycho-Electronics*.

17. Daniel C. Dennett, *Consciousness Explained* (London: Penguin Books, 1993), pp. 253–4.

18. Daniel C. Dennett *The Intentional Stance* (Cambridge, Mass.: Bradford Books, 1987), p. 5.

10. RECOVERING THE CONSCIOUS AGENT

1. This phrase derives, I think, from Paul Ricoeur. I am here borrowing it from David Lehman's brilliant *Signs of the Times: Deconstruction and the Fall of Paul de Man* (London: André Deutsch, 1991). I cannot resist quoting his comment about literary criticism in the 'age of suspicion':

> Acquiescing in the notion that disinterested inquiry is an impossibility and that every value judgement is necessarily a power play before it is anything else, they make their decisions by ideological litmus tests and determining the sexist and racist quotient in any piece of writing, from Plato to the present. This is, at bottom, a conception of the literary critic as an agent of the thought police, single-minded, obsessively concerned to enforce the party line, willing to subject chosen works to a violent form of interrogation, and more than happy to eliminate literature altogether in favor of pure theory.
>
> (p. 263)

2. Isaiah Berlin, 'Political Ideas in the Twentieth Century', in *Four Essays on Liberty* (Oxford: Oxford University Press, 1969), p. 18.

   Later in the same essay, he tries to distance Marx and Freud from the bad influence they had on twentieth-century thought:

> By giving currency to exaggerated versions of the view that the true reasons for men's beliefs were most often very different from what they themselves thought them to be, being frequently caused by events and processes of which they were neither aware nor in the least anxious to be aware, these eminent thinkers helped, however, unwittingly, to discredit the rational foundations from which their own doctrines derived their logical force.
>
> (p. 21)

   This is characteristically generous but not, alas, sustained by all the facts.

3. They are not, of course, alone in this. I have already noted how thinking by transferred epithet and misplaced consciousness is not unique to psychologists and philosophers of a cognitive persuasion. How often have we been told that 'It is language that speaks' and it will be recalled that it was Lévi-Strauss who stated that 'myths think themselves out in men and without men's knowledge'. The reader is also referred to my *Psycho-Electronics*, which is devoted almost entirely to examining such dodges.

4. *Times Literary Supplement*, 3 February 1995, p. 6. The contradictions in the post-modern critique of the narratives of emancipation and enlightenment have been uncovered by many critics; see, in particular, note 10 below.

5. Of course, Marx saw the Enlightenment through the eyes of Hegel's critique. So, although he inherited the Enlightenment spirit of criticism and healthy scepticism and was fired by a Voltairean anger at the permanent conditions of oppression sponsored by those in power and the mystifications of their ideologues such as priests and cultural functionaries, he also inherited Hegel's dialectical obfuscations, his belief in historical necessity and his irrationalism – or, rather, his sidelining of ordinary reason in favour of the dialectical unfolding of Great Reason. And all that was necessary for Marx's writings to cause maximum damage was to ensure

they had their most enduring effects in countries that had been relatively untouched by the Enlightenment. His writings, therefore, became a means by which feudal autocracies continued their tradition of oppression with a different rationale. The unaccountable czars, bureaucrats and priests were replaced by unaccountable dictators, commissars and ideologues – illustrating not the Hegelo-Marxist laws of history but Pareto's circulation of elites, in which, however, the main motor is not ability but brutality and ruthlessness.

6. Joel Handler, Presidential Address to the Law and Society Association in *Law and Society Review* 26(4) (1992): 697–731, at pp. 697–8.

7. Just in case it might be thought that Barthes was joking, here is an earlier passage from the same lecture:

> power is the parasite of a trans-social organism, linked to the whole of man's history and not only to his political, historical history. This object in which power is inscribed, for all of human eternity, is language, or, to be more precise, its necessary expression: the language we speak and write.
>
> Language is legislation, speech is its code. We do not see the power which is in speech *because we forget that all speech is a classification, and that all classifications are oppressive.* Jakobson has shown that a speech-system is defined less by what it permits us to say than by what it compels us to say ... Thus, by its very structure my language implies an inevitable relation of alienation. To speak, and with even greater reason, to utter a discourse, is not, as is too often repeated, to communicate; it is to subjugate.
>
> (ibid., p. 460, emphasis added)

In common with many of the *maîtres à penser*, Barthes does not pause to differentiate between the legitimate and beneficial use of 'the power which is in speech'. We shall return to this.

8. *After the New Criticism* (Chicago: the University of Chicago Press, 1980, pp. 283–4). For the followers of Paul de Man (blissfully unaware that they were following a real Fascist), it was not only language but meaning itself that was 'Fascist'. Lehman (ibid.) gives an account of how this was explained to him by one of de Man's disciples:

> We inhabit [Lehman's informant told him] an indeterminate universe. Everything is mediated entirely through language – the only way we can know anything is by using words. And the words of any discourse constantly shift their meaning. Everything depends on interpretation and no interpretation is more correct than any other. The proper attitude is to regard all interpretations as equally 'not true and not false'. To insist that a given piece of discourse means something specific and decided is to elevate one meaning at the expense of others. It is to uphold a hierarchy of values and that renders one guilty of a dictatorial urge. Fascism, in short.

The argument is, of course, self-refuting in two ways. First, it assumes that it has a definite meaning itself and this means that meaning is determinate – at least to some degree. And secondly, it assumes that those who force determinate meaning upon statements must be capable of doing so, even though their words, too, are subject to unlimited indeterminacy.

If they are not able to do *that* then there can be no way that meaning can become Fascist.

9. He is reminiscent of many other preachers of disorder who expect their own lives to be cocooned in a high level order. I am reminded of John Weightman's observation about the Marquis de Sade:

> There is something unconvincing, perhaps even comic, in the frantic excess of his Gothic destructiveness, when, in real life, he expected society to function efficiently, at least with regard to his privileges as an aristocrat. In the intervals of preaching chaos in the name of unbridled lust, he wrote indignant letters demanding immediate payment of the feudal dues from his estates.
>
> (*TLS*, 1 May 1992, p. 5)

10. This is a recurrent theme in my *Not Saussure* (London: Macmillan, 2nd edition, 1995). The way in which post-modern thinkers 'saw off the branch upon which they are sitting' (to use Jonathan Culler's metaphor) is also addressed with exemplary clarity in the special issue of *Critical Review* on post-modernism (Volume 5, Issue 1, Spring 1991).

Carl Rapp's essay on Lyotard ('The Crisis of Reason') deals a fatal blow to the latter's reputation as a thinker. According to Lyotard, it is the achievement of post-modern thought to recognise that the narratives of progress towards the truth are without foundation. Lyotard's *The Postmodern Condition* asserts that there are two different kinds of knowledge – narrative knowledge and scientific knowledge. The 'metanarrative of knowledge' is that the shift from the former to the latter represents true progress. Lyotard asserts that post-modernism now appreciates that this itself is a myth without justification: no form of knowledge gives us access to the truth; and narrative knowledge and scientific knowledge cannot be ranked in the way that intellectual historians usually have done. The problem, which Lyotard overlooks, is that this 'metanarrative of metanarratives' claims itself to be true and to transcend the pre-post-modern narrative, by explaining where the latter stands in the history of thought. It represents exactly that kind of knowledge which Lyotard and other standard bearers of post-modernism considers to be meaningless.

Jeffrey Friedman, in 'Postmodernism v Postlibertarianism', put his finger on the post-modern 'conceit'

> that we can have it both ways: that we can gain enough critical perspective on our thought to call it false, at the same time that we deny our ability to transcend our own context. This, of course, merely reproduces the hubris that postmodernists claim was invented by the Enlightenment and that, they contend, they somehow managed to escape.
>
> (p. 152)

Post-modernism, as Friedman phrases it, 'propounds a metanarrative of the end of metanarrative' (p. 151), and claims to have achieved 'a transcendence of transcendence'.

This is particularly notable in those who follow Richard Rorty in deconstructing any hint of universalist pretensions in thought, or context-transcending notions of truth and reality. Rorty's neo-pragmatist attack on philosophy, his assertion that truth is relative to interpretive communities and his rejection of 'the universal, timeless and necessary' is triply self-contradictory: it is itself a philosophical position; it is not itself presented as being relative to a specific interpretive community; and it seems to have the status of a recently discovered universal, timeless and necessary truth. (The notion of interpretive communities as the final arbiters of truth is, of course, deeply flawed. It does not, for example, account for the difference between an outbreak of racist 'thought' and a law of physics – both being sustained by 'interpretive communities' in Rorty's sense; between groundless prejudices and useful truths; or between scientific laws that lead to effective technologies and magic that doesn't. Nor does it explain how interpretive communities become established or what determines their boundaries, nor even what we should think of as such a community – 'me on a bad day', 'our household', 'everyone in our street', 'all the people in British medicine', 'Europeans since the pre-Socratics'. Connected with all of these is a failure to explain why the interpretive community which accepts one truth – for example, Newton's laws of motion – should be so much larger than the interpretive community that accepts another, such as the prophecies of the Reverend Jones of Jonesville.

11. When Foucault was admitted to hospital after a road traffic accident, Simone Signoret was the first to be informed of the event. 'All are agreed on her reaction; she was startled and horrified that neither the police nor the hospital staff had recognised Foucault' (Macey, op. cit., p. 370). This response, which Macey allows to pass without comment, is multiply revealing. It shows, first of all, that the ordinary notion of identity seemed to be applicable to the man who denied it (and who, indeed, had erased the concept of 'man' from the intellectual map). At least his close friends believed so. Secondly, it exposed a deeply anti-egalitarian expectation amongst his friends. Presumably, Foucault would have been treated differently had he been recognised and 'differently' would not, of course, have meant 'worse'. Finally, there is the comical assumption that policemen, doctors and nurses are derelict in not knowing the identity of a certain intellectual who loomed so large in certain intellectual circles. To loom large in those circles is presumably to loom large in the world, and those who cannot identify the large-loomers – doctors, nurses and policemen culpably busy with other preoccupations – quite naturally evoke the 'horrified' reactions of other large-loomers.

12. The reader is referred to chapter 3, 'Magic and Science', and to my *Newton's Sleep* (London: Macmillan, 1995), where the Strong Sociology of Knowledge is criticised. This critique was in part based upon Lewis Wolpert's excellent *The Unnatural Nature of Science* (London: Faber, 1992).

Of course, Faraday was not working in isolation – he will have been influenced by ideas 'in the air' at the time (though his notions of field-forces were revolutionary); but not every scientist of his time who was equally exposed to such influences made so much of them, rose so far above them, and had such an enduring impact on the way we see and operate in the world. Likewise, his notion of forces and fields was greatly influenced by the world-picture he derived from his Sandemanian background, which had encouraged him to seek a unity of forces. But the proof of his theory was its basis in observation, its ability to be generalised and to predict novel observations and its practical applications. The Clerk Maxwell–Hertz–Einstein revolution in theoretical science, the technological transformation of the world would not have come out of Faraday's work if he had simply been acting out childhood influences. After all,

Einstein was no Sandemanian, and one does not have to be a Sandemanian to get a dynamo to work.

13. See especially chapter 6, 'Man, the Explicit Animal'.

14. Or at least the pre-Marxist Sartre of *Being and Nothingness*.

15. The post-Saussurean marginalisation of consciousness is, of course, at least in part an over-reaction against Sartre's failure to recognise the bounds of freedom and the historical influences on the self. Foucault's empty cipher, vanishing into the historical transformations of discursive forces, is a response to Sartre's all-powerful, anhistoric, abiological, purely metaphysical *pour-soi*. See the Appendix.

16. That 'Men make their own history, but they do not make it just as they please; they do not make it under circumstances chosen by themselves, but under circumstances directly encountered, given and transmitted from the past' (Karl Marx, *The Eighteenth Brumaire of Louis Bonaparte*) is unlikely to be dissented from by many.

17. For a further examination of this problem of specification-by-rules (with an incomplete understanding of its devastating implications for his own theories), see Daniel Dennett, 'Cognitive Wheels: The Frame Problem of AI', in Christopher Hookway, ed., *Minds, Machines, Evolution* (Cambridge University Press, 1984). See also my discussion in *The Explicit Animal*, pp. 223–6. The experience of the imaginary anthropologist, incidentally, should cast doubt on what Althusser (op. cit., p. 24) describes as 'the golden rule of materialism': 'never judge a being by its consciousness of itself'. I am afraid that a statistical, objective analysis of my greeting behaviour and that of a large group of individuals considered to belong to the same class as myself would yield no insights into why, with what purpose, or what feelings, I greeted, or failed to greet, anyone on a particular occasion. You would have to know a fantastic amount about me really to know why.

18. Man (and woman) is, very importantly, The Piss-taking Animal – and this is a mark of his/her distinctive consciousness – a consciousness of the signs that he/she not only responds to but consciously deploys – a consciousness of signs *as signs*.

19. Ironically, this distance between language and experienced reality has been used as an argument to support the idea that language does not make contact with the world and that discourse is, therefore, a sealed system that refers only to itself. At the same time, discourse theorists overlook actual experience as that which informs, subverts, acts as a critical check upon, the world of discourse. See Raymond Tallis, *In Defence of Realism* (London: Edward Arnold, 1988).

20. The role of understanding – individual understanding – and assent in rule-following has been grossly underestimated. The cloudy notion of 'internalisation' of rules has contributed to this failure. According to internalisation theorists, rules become transformed into something more mechanical and automatic – into unconscious pressures. However, rule-following in many cases is like responding to an appeal to reason. This combines individuality – I have to see and understand and assent to its conclusions – and collectivity – I submit to general, universal rules. (This is also an answer to romantics and those others who see the application of reason in human affairs as being anti-individual and anti-creative.)

After Grice's detailed and brilliant analyses of the production of meaning in ordinary conversation, there should be no excuse for anyone to forget the extent to which rule-following is explicit and based upon a complex understanding not only of the rules but of your understanding of the rules. This is one of the points implicit in The Frame Problem (see note 17).

21. This is well illustrated by the Critical Legal Studies movement which, for example, suggests that the state should be held responsible in cases of child murder, since 'contemporary modes of social thought ... recognise the pervasive relationship between observer and observed and deny the primitive notion that subjects act upon a background of distinct, fixed objects rather existing in a reciprocal and ever-changing subject–object tension' (cited in Lehman, op. cit., pp. 39–40).

22. And herein lies the deep meaning of the Paul de Man affair which Lehman described and analysed with such brilliance. The question of Paul de Man was not a question about a piece of the endless and undecidable text of society; it was a question about the actual behaviour of an extra-textual individual – as was accepted, without question, by his defenders, such as Derrida, as well as his attackers. De Man had always believed, or pretended to believe, that writing should be approached as examples of the text-system playing with itself. 'Considerations of the actual and historical existence of writers are a waste of time from a critical viewpoint' (op. cit., p. 137) As Lehman says, 'how poetically just it would be if so antibiographical a theory of literature should be vanquished by the discovery of a ruinous biographical fact' (p. 140).

23. Perhaps we should be more intensely and more continuously astonished than we are at the way automaticity and responsibility interact, at how physiological and social reflexes interact with our agency.

APPENDIX: PHILOSOPHIES OF CONSCIOUSNESS AND CONCEPT

1. Ludwig Wittgenstein, *Philosophical Investigations* trans. G.E.M. Anscombe (Oxford: Blackwell, 1963), p. 104e, paragraph 314.
2. Wittgenstein's position as a philosopher of the concept as opposed to a philosopher of consciousness is not as secure as that of Frege, as we shall see presently.
3. This was spelt out in Michael Dummett's great book *Frege: Philosophy of Language* (London: Duckworth, 1973). The final chapter assesses 'Frege's Place in the History of Philosophy' and concludes not only that 'he achieved a revolution as overwhelming as that of Descartes' (p. 665) but also that he reversed the hierarchical ordering established by Descartes between various areas of philosophy. Where Descartes had placed consciousness at the centre of philosophy and made the theory of knowledge fundamental, Frege placed language at the centre of philosophy and made the theory of meaning – the analysis of linguistic meanings, creating 'a model for what the understanding of an expression consists in' – the first, the fundamental, the central, task of philosophy. Moreover, for Frege, as we shall see, psychological processes and mental images are irrelevant to the theory of meaning.
4. G. Frege, 'Review of E.G. Husserl, *Philosophie der Arithmetik I*', available in *Collected Papers* ed. B. McGuinness, trans. M. Black and others (Oxford: Blackwell, 1984).
5. Antony Flew, *A Dictionary of Philosophy* (London: Macmillan Press, 1979), p. 272.
6. David Bell, *Husserl* (Routledge: London and New York, 1990), p. 80. My discussion of the clash between Frege and Husserl has also been greatly

influenced by Edo Pivcevic's admirably lucid *Husserl and Phenomenology* (London: Hutchinson, 1970).

Bell points out that Husserl has been unfairly dealt with by scholars and historians and that his psychologist programme was only one part of two complementary investigations, the epistemological one that dealt with the origin and nature of *our grasp of* arithmetical concepts. Nevertheless, Husserl did not make this clear, and perhaps it did not become clear to himself until he was licking his wounds after his savaging by Frege.

7. The imagist-associationist theory of meaning is not only an atomic theory of language, an item-centred theory as linguists would say, but also a name-based theory. It assumes that all words are *names* of individuals or classes. It cannot accommodate words that do not stand for anything – function words such as conjunctions and articles. This, of course, is connected with the inability of the theory to deal with the fact that words have meaning only in the context of sentences – that words have to work together to generate meanings, senses or references.

8. Dummett wonders why the imagist-associationist theories of meaning 'could have retained their grip on the minds of philosophers for so long' and suggests that part of the explanation may lie in 'the ambiguity in the word "idea" as applying to mental images and the meanings of expressions' (p. 640).

9. 'Frege: The Twentieth Century Descartes', *Times Literary Supplement*, (30 November 1973): 1461–2.

10. Ernest Gellner, *Word and Object* (London: Penguin Books, 1959). It is full of memorable epigrams; for example: 'When a priest loses his faith, he is unfrocked; when a philosopher loses his, he re-defines his subject.' Gellner, in particular, shared Popper's contempt for the nominalist flavour of much of Oxford philosophy, its tendency to shift the focus of discussion from the nature of things to the meaning of words.

11. In the lively account of 'The Rise of Structuralism' that opens his effervescent and wonderfully perceptive critique of structuralist and post-structuralist thought, *From Prague to Paris* (London: Verso, 1986).

12. The rationale of the anti-historicism of the structuralists was ambiguous. It was not clear to what extent it was rooted in a belief in the primacy of the synchronic over the diachronic – of unchanging structure over temporalised event – and to what extent it was a methodological constraint, with the preference for synchronic analysis being merely a heuristic tactic.

13. Or, more precisely, demonstrated to their own satisfaction.

14. Lévi-Strauss also disapproved (in *Tristes Tropiques*) of the tendency that Existentialists had 'To promote private preoccupations to the rank of philosophical problems' since it was likely to 'end in a kind of shopgirl's philosophy'.

15. See *Not Saussure* (London: Macmillan, 2nd edition, 1995), chapter 6, 'Walking and Différance'.

16. In Gerd Brand, *The Central Texts of Ludwig Wittgenstein*, translated and with an introduction by Robert E. Innis (Oxford: Blackwell, 1979).

17. Ludwig Wittgenstein, *Tractatus Logico-Philosophicus*, trans. D.F. Pears and B.F. McGinness (London: Routledge and Kegan Paul, 1961), pp. 117–18.

18. See my discussion of Derrida's reading of Husserl's *Expression and Meaning* (the first part of *Logical Investigations*), in *Not Saussure* (London: Macmillan, 2nd edition, 1995), pp. 189–202.

19. There are important similarities in the two philosophers' final positions, as Bell (op. cit., pp. 228–31) points out. The Wittgenstein of *On Certainty* who pointed out that even scepticism could get a foothold only against the background of a presupposed common, real world would have approved of the holistic Husserl for whom the *lebenswelt*, the given intersubjective world presupposed in all practical activities, was 'the ultimate foundation of all objective knowledge'.

20. A tradition that continued, of course, into the post-Fregean era of Russell, the Vienna Circle and philosophers such as Ayer who were influenced by them. It was as if only half of Frege's message had been assimilated by these thinkers. In consequence the later Russell and Ayer were regarded as something of an anachronism by the second-wave post-Fregeans, especially those influenced by the later Wittgenstein.

21. For a more detailed discussion of the interpenetration of language and experience, see *In Defence of Realism* (2nd edition, London: Ferrington, 1995), 'Realism and the Idea of Objective Reality' and *Not Saussure*, section 3.1, 'The Articulation of Reality'.

22. David Hume, *A Treatise of Human Nature*, Book I, Part 6, Section 6, 'Of Personal Identity'.

23. Thomas Nagel, *The View from Nowhere* (Oxford: Oxford University Press, 1986).

24. The investigation of this unique agent that is the site of (and in some sense composed of) classifiable events that are therefore not unique could begin with the inadequacies of Hume's critique of the notion of personal identity.

    His main argument, set out in the passage already cited, that the self is a fiction, depends upon his observation that, when we introspect we find specific impressions, or their echo in recall, but nothing corresponding to the self. There is no distinct substance to underpin our identity. Our error, according to Hume, is to assume that the succession of sensations, each of which modifies the whole only slightly, amounts to a fixed, stable object. What he overlooks is that, if there *is* such a succession of sensations (which he does not deny), there must be some basis for relating those sensations to the same series; only in this way could they amount to a *succession*. It is this to which the notion of the self has to answer. The self could not, therefore, be a distinct sensation – part of the series. Nor could it be a distinct tone that would brandmark all the impressions as belonging to the same series – for, unless we had access to impressions arising from other series (i.e. belonging to other selves), this would not count as a distinguishing feature: the brand-mark would factor out. No, the very notion of the self requires that the sensations are assumed to belong to the same series. So my fear of going to the dentist today and the experience of being at the dentist tomorrow have the same referent – myself in the dentist's chair.

    In other words, to look to the series of sensations for marks of the self is to put the cart before the horse. That these are *my* sensations, that they are of *me*, that they concern *me*, is assumed in their being assigned to a single series (and being first-person accessible to memory); the self is presupposed in the series and cannot therefore be found in it. It is in this presupposition that the reality of the self as something transcending the succession of 'fugitive impressions' resides. And there, too, lies its mystery.

25. There are two main versions of this principle. The so-called weak form

is that 'what we can expect to observe must be restricted by the conditions necessary for our presence as observers'. The 'strong' form is that 'the Universe (and hence the fundamental parameters on which it depends) must be such as to admit the creation of observers within it at some stage'. Neither is very satisfactory and neither seems to say much more than that what is the case must have the conditions necessary for it to be the case.

26. For a critique of the application of neo-Darwinian assumptions to conciousness, see *The Explicit Animal*, passim, especially chapter 2 'Biologising Consciousness: I Evolutionary Theories'

27. Francis Steegmuller, in his edition of Flaubert's letters (*The Letters of Gustave Flaubert 1830–1857*, London: Faber, 1981), quotes George Eliot's execration of 'that dead anatomy of culture which turns the universe into a mere ceaseless answer to questions'.

EPILOGUE: THE HOPE OF PROGRESS

1. Isaiah Berlin, 'The Apotheosis of the Romantic Will', in *The Crooked Timber of Humanity*, ed. Henry Hardy (London: HarperCollins, Fontana, 1991), p. 237.

2. Isaiah Berlin, Introduction to *The Age of Enlightenment* (New York: New American Library, 1956), p. 29.

3. This has, of course, led to deep conflicts between diagnosis and treatment; we have seen how Dudley Young, typical of many cultural critics, advocates the revival of the sacred while, at the same time, demystifying and naturalising – desacralising – the sacred.

4. Francis Fukuyama, 'In the Zone of Peace', *Times Literary Supplement* (26 November 1993): 9.

5. Behind this is the same kind of Counter-Enlightenment *schadenfreude* that generated and circulated the false rumours that Voltaire (whose end was peaceful) died in hideous pain, eating his own excrement.

6. Jeffrey Friedman, 'Postmodernism v Postlibertarianism', *Critical Review* 5(2) (Spring 1991): 145–58.

7. Albert Camus, *The Rebel*, trans. Anthony Bower (London: Penguin Books, 1953), p. 146. This complements Berlin's melancholy observation that 'the disciples of those who first exposed the idolatry of ideas frozen into oppressive institutions – Fourier, Feurbach and Marx – should be the most ferocious supporters of the new forms of "reification" and "dehumanization" is indeed an irony of history' (*Four Essays on Liberty*, Oxford: Oxford University Press, 1969, p. 34).

8. Isaiah Berlin, 'The Counter-Enlightenment', in *Against the Current*, ed. Henry Hardy (London: The Hogarth Press, 1979), p. 3.

9. Isaiah Berlin, 'Joseph de Maistre and the Origins of Happiness', in *The Crooked Timber of Humanity*, p. 121.

10. This seems to be Maistre's answer, too. It is because of the intrinsic evil of men that he opposes the Enlightenment. Non-hierarchical societies of equals do not have the means to constrain that evil: men need to be hemmed in by the terror of authority to be saved from themselves. His remedy, of course, is at odds with his analysis, as we pointed out in the Prologue: if men are evil, it is at least arguable that power should be dispersed rather than concentrated in a few hands.

11. I am reminded by what Gellner (and after him Merquior) pointed out, 'that all the romantic binges proferred by the counter-culture (and soon

commercialized) . . . depend on the rational basis of the self-same culture they profess to scorn' (J.G. Merquior, 'In Quest of Modern Culture: Hysterical or Historical Humanism', *Critical Review* 5(3) (Summer 1991): 399–420).

12. The supposed Eurocentricity of liberal values, in particular the liberal conception of justice, must, however, not be accepted 'on the nod'. As Stanley Hoffmann points out ('Dreams of a Just World', *New York Review of Books*, 2 November 1995, pp. 52–6), this notion (made in their defence by many repressive regimes) doesn't take into account the fact that one can 'find believers in liberal values, believers who are often repressed, everywhere, and many anti-liberals in the West' (p. 55). And he cites a refutation, by the Asian Yash Ghai, of the so-called 'Asian', anti-liberal point of view. Hoffmann succinctly defines the central philosophical and ethical problems in a non-ideal world:

> how to produce both order and justice in a world of different "corporate bodies" and regimes, which reflect different conceptions not only of the social good but of political justice and are based at least as often on coercion and repression as on consent.
>
> (ibid., p. 55)

13. I may be thought of as having smuggled in not only rationalism but also western science into those things that are not merely cultural options. My own view (which I have expressed at length elsewhere, e.g. *Newton's Sleep*, London: Macmillan, 1995) is that many human problems *are* technological and that to these a technical solution is required. The background for the most successful technical solutions is so-called 'Western' science. (It is perhaps worth noting that it is *not* 'Western', though it may have begun in the West – if Egypt and Greece may be categorised as parts of the 'West'. It is now part of a universal human heritage and to suggest otherwise is to insult all of those non-Europeans who have made major contributions.) This is recognised even in those countries where much is made of indigenous cultures and crafts and the superiority of native over imported solutions. When the rich fall ill, seek to travel great distances, need to communicate or want to have a good time, they do not hesitate to utilise 'Western' science and technology.

14. It may seem that I am making too much of a meal of defending core Enlightenment values. Surely, no one would oppose – at least in principle – a reign of tolerance, justice, democratic rule by accountable leaders fired by a sense of justice and directed by reason rather than irrationality or expediency. Am I not making a rather dragged out case for Motherhood and Apple Pie? Anyone who asks this question has probably forgotten how these values – which do not seem to be sufficiently valued by humanist intellectuals within the academy, though they are beneficiaries of them – are routinely trampled on in the real world. The Second World War and the collapse of the Berlin Wall did not bring to an end theocratic tyranny, secular despotism and religious intolerance. The enemies of the Enlightenment are at large. In many cases, in common with a significant number of humanist intellectuals, these enemies believe they have good reasons for rejecting tolerance, accountability, etc. Toleration, for example, is seen in many places of the world to be a lesser value than inculcating the One True Faith that will save the community from Armaggedon and its members from eternal damnation. And (to move

inwards from the real world to the world of the library) there is a massive literature – the work of a nexus of writers whom we may capture in
the portmanteau term 'Boas-deconstructors' – which not only denies that
there is such a thing as cultural evolution and progress but that we should
do everything possible to protect so-called primitive peoples from development. Their views are echoed by certain romanticising professional
travellers such as Wilfred Thesiger who positively regretted that the no-
mads in the Empty Quarter should share the values of the West. As Michael
Asher notes in his biography (*Thesiger*, Viking, London, 1994), Thesiger
assumes 'that the primitive only wishes to be left alone and has no desire to see his children grow up healthy, to emerge from poverty and to
be able to read and write'.

15. In what follows, I am going to address anxieties about an achieved Utopia, not anxieties about the process by which it is achieved or deeper
anxieties about whether the world is in fact moving forward or backwards. I shall deal with the latter question towards the end of this chapter.

What about the process? Most of the anxieties are relevant to explicit
programmes for bringing about Utopia. The latter can provide the immemorial excuse for large-scale iniquities: you cannot make an omelette
without breaking eggs. Alas, it is easier to break eggs than to make an
omelette – which is why so many revolutions have broken millions of
eggs and made no omelettes.

The alibi of the future has also provided justification for generic injustices directed against entire categories of individuals – those who are
identified as being opposed to the coming Utopia, either explicitly in
their words and actions or implicitly in virtue of their 'class' or 'historical' situation. Such people – or, as they are often called, 'elements' –
may be destroyed, given the overriding importance of realising the Utopian dream. The alibi of the future also justifies the (temporary) further
oppression of the oppressed by their liberators: the beast has to be driven
to the pasture with blows; things inevitably get worse before they get
better; change always hurts even those who benefit from it; etc.

In short, the Utopian imperative justifies the way in which (to quote
Marx, cited in Camus, op. cit., p. 172) 'Progress resembles that horrible
pagan god who only wished to drink nectar from the skulls of his fallen
enemies.' Explicit Utopianism allows the leaders to subordinate the present
experience of the people in the name of the indefinite future, sacrificing
several generations to the needs of an infinite series of future generations, giving the commissars and the *nomenklatura* free access to the riches
of the earth while they plan the next massacre or prepare the ground for
the next unplanned famine. Utopia, in short, makes unacceptable demands,
requiring the majority of one generation to abjure the hope of happiness
in the only life they will have, to pave the way for the possible happiness of future generations.

The unquestionable goodness and historical rightness of the explicit
Utopian dream and its prophets and guardians license the untrammelled
power of those whom the alibi of the future – of being on the side of
generations yet unborn – lifts above moral or legal judgement. Since the
prophets of the revolution and the midwives of the coming Utopia cannot be contradicted, the failure of Utopia to materialise must be due to
saboteurs. The Utopian dream reduces all discourse to brutalising
simplifications, as Camus captured in his incomparable analysis in *The
Rebel*:

To the extent to which Marx predicted the inevitable establishment of the classless city and to the extent to which he established the good-will of history, every check to the advance toward freedom must be imputed to the ill-will of mankind.

(p. 207)

Or, more precisely, certain 'reactionary elements' within it – capitalist lackeys and other members of the evil classes. These will be singled out for special attention, with the familiar, dispiriting consequences:

The principles which men give to themselves end by overwhelming their noblest intentions. By dint of argument, incessant struggle, pol-emics, excommunications, persecutions conducted and suffered, the universal city of free and fraternal man is slowly diverted and gives way to the only universe in which history and expediency can, in fact, be elevated to the position of supreme judges: the universe of the trial.

(ibid.)

16. Lucian Goldmann, *The Philosophy of the Enlightenment*, trans. Henry Maas (London: Methuen, 1973), p. 95. This is an indirect echo of de Tocqueville's famous critique of democracy:

I see an innumerable multitude of men, alike and equal, constantly circling around in pursuit of the petty and banal pleasure with which they glut their souls.... Over this kind of men stands an immense, protective power which is alone responsible for procuring their enjoy-ment and watching over their fate. It provides for their security, fore-sees and supplies their pleasure, manages their principal concerns, directs their industry, makes rules for their testament, and divides their in-heritance.

17. Raymond Tallis, 'Terrors of the Body', *Times Literary Supplement* (1 May 1992): 3–4.
18. David B. Morris, *The Culture of Pain* (Berkeley: The University of Califor-nia Press, 1991).
19. Fyodor Dostoyevsky, *Notes from the Underground*, Part One, chapter 4.
20. Morris looks forward to an era of 'post-modern' pain. This is pain of which the patient takes charge. Taking charge of pain includes 'assum-ing personal responsibility for its meaning' (ibid., p. 289). This may seem a distinct possibility in the first hour of toothache or vomiting; by the third month, the pain will be calling the shots, especially if it is a symp-tom of a fatal disease. Assuming personal responsibility for the meaning of a tunnel of barbed wire leading out of the light into endless darkness is not a realistic option and it is an inhuman expectation to suggest that it is. Give me diamorphine and I'll forgo the post-modernism.
　　The positive valuation of pain is, of course, intimately tied up with the religious vision of the world as a vale of soul-making. In accordance with this vision, pain is of spiritual benefit both to those who suffer and to those who grow through their self-sacrificing service in tending those who suffer. Even when this vision does not tilt into the overt sado-maso-chism explored in books as diverse as Nietzsche's *Anti-Christ* and Huysmans' *A Rebours*, it is far from benign. Christopher Hitchens' ac-count of the damage caused by Mother Teresa's mission to the wretched

of the earth, the destitute and terminally ill, is an illuminating and terrible example, sufficient by itself to persuade the uncertain of the horrors that may result when a religious agenda hijacks human need (*The Missionary Position*, London: Verso, 1995).

21. Joseph de Maistre, 'Study on Sovereignty', in *The Works of Joseph de Maistre*, selected, translated and introduced by Jack Lively (London: George Allen and Unwin, 1964), p. 118.

22. See Raymond Tallis, *The Explicit Animal* (London: Macmillan, 1991), especially chapter 6, 'The Explicit Animal'.

23. Utopians have often worried over the major barriers to happiness that would come from sexual jealousy. They have suggested two solutions.

    The first is that sexual relations should be free and untrammelled: everyone should screw everyone else. This rather assumes that individuals will no longer want to choose with whom they make love – that no one will find anyone else, or their sexual proclivities, unattractive. This seems rather unlikely and would anyway diminish the 'specialness' of the sexual relationship. Moreover, even amongst small groups, where there has been some pre-selection, free love invariably leads to the exploitation of those with a less intense sexual drives by those with more intense drives and, to a lesser extent, vice versa; and of the less powerful by the more powerful.

    The second is to forgo sexual relations altogether (with the continuation of the human race being assured by artificial insemination). The history of abstinence is not encouraging: the imperatives of sexual desire are an inescapable reality.

    There is a third way, where long-term sexual relationships are so highly developed that their metaphysical possibilities are fully realised and the physical specifics of the partners become less important. The more metaphysical sexual relationships become, the less pronounced will become the differences between values of sexual partners. Each will become to the other the means of exploring, understanding, possessing the otherness of the world, a singular who is at once unique and stands for an entire class, a concrete universal, an archetype. This is the path that I am suggesting here; where perfectly developed sexual relationships open on to something wider and even deeper than themselves.

    Incidentally, the lucrative assumption, common to much quasi-literary fiction exploring the 'outer limits' of human behaviour, that the ever more frantic search for enhanced sexual sensation will inevitably modulate into a murderous frenzy reminiscent of Young's hunting pack, is just that – a lucrative assumption. The psychopathic hero of Bret Easton Ellis's *American Psycho*, for example, is utterly empty of emotion and has his murderous sexual experiences with strangers. This is not where most people are starting from. We have no reason to assume that the inhabitants of Utopia will be affectless serial killers.

24. The difficulty of experiencing our experiences and, in the wider sense, of 'arriving' (in the Kingdom of Ends) is explored in Raymond Tallis, *Newton's Sleep* (London: Macmillan, 1995).

25. Utilitarianism, and in particular John Stuart Mill's version of it, was dismissed by right-wing romantics such as Carlyle and Nietzsche as 'the philosophy of pigs'. This is utterly unjust. Anyway, my own utilitarianism may escape this charge because (like Mill) I see material comfort as the beginning, not the end, of human development. The pigs fall asleep next to the trough once they have supped their fill. For human beings, satiety is the beginning of wakefulness, of an ever-unfolding conscious-

ness undistracted by the struggle for survival, by the tyranny of hunger, by preoccupation with getting and spending.

26. For an extended discussion of this, see 'The Work of Art in an Age of Electronic Reproduction', in *Theorrhoea and After* (forthcoming).

27. The assumption that Utopia is boring may also be based upon the idea that it is static. This is a misunderstanding: it will always be an asymptote towards which humanity will struggle for complete achievement. The future struggle, however, will not be for adequate nutrition or for decent treatment by others or for a cure for toothache but for the perfection of mutual understanding or the realisation of delight inherent in consciousness of the sunlight and for establishing the conditions in which that becomes increasingly possible. Nor will it be the fulfilment of a pre-established blueprint: when it is approached, it will emerge as something unprecedented. And therein lie potential dangers – as well as unforeseen sources of hope.

28. Since the romantics inaugurated the tradition of rebellion against the social order, it has been natural to sympathise with the rebel. After all, most social orders have been absurd, unjust, obtrusive, etc. – just the kind of thing that should be rebelled against. It would be less easy to sympathise on existentialist grounds with rebels who brought down a genuinely beneficent and harmonious society in order to assert that they were not mere cogs in the machine. Such a collapse would increase suffering. One would not readily forgive a rebel who, for the sake of self-expression, participated in destructive acts that increased infant mortality and, in consequence, denied infants the right to participate in or dissent from the social order. In many ways, Bakhunin's psychopathic visions of setting fire to Paris, of the great anarchistic act of destruction, are the purest expression of the 'Apotheosis of the Romantic Will'. The hatred of reason and order, in other words, leads in one direction to the burning of children. In the other direction, it leads to the endearing, impotent, Petrushka-like gestures of the Idiot, or the Clown, or the Pooterish clerk who writes 'I do not like my work' on an official form.

29. Nicholas Negroponte, *Being Digital* (London: Hodder and Stoughton, 1995). The major barrier to the switch from atoms to bits, however, is the uncontrolled increase in the population of the world. As Negroponte points out, bits are not edible, they cannot stop hunger. Bodies survive by utilising available energy from the environment in order to maintain the low levels of entropy that characterise physiological life. The more bodies there are, the more energy is required and the more the planet will be polluted. While the shift from an energy-based to an information-based economy will reduce per capita consumption of energy, there is an irreducible minimum for each body. Population over-growth will prevent any kind of Utopia emerging; indeed, it is the recipe for Hell on earth. This is why those who believe in the next world rather than the present one are so adamantly opposed to contraception and are unconcerned about the consequences of this. The earth is *meant* to be a vale of suffering – a doctrine from which the most articulate amongst the believers do not noticeably suffer.

30. It may be questioned whether the private depths individuals cultivate inside the meshes of the public domain are somehow invalidated by lacking public acknowledgement. (We earlier discussed this in relation to the notion of a private religion – see 'Religion and the Re-Enchantment of the World' above, p. 157ff). If the question is posed in this way, the answer seems

fore-ordained. However, the analogy of a 'mesh' may not itself be a valid one. For we are talking about the actual content of consciousness: it is in the very fact that consciousness has content – and not just abstract form, constraints and conditions – that liberation from the public sphere lies. The image of private experience being contained in small boxes drawn by public (political, social) constraints is misleading. It overlooks the asocial core in all sensation, in all experience.

31. *Sketch for an Historical Picture of the Progess of the Human Mind. The Tenth Stage: The Future Progress of the Human Mind*, trans. June Barraclough, introduction by Stuart Hampshire (New York: Noonday Press, 1955; reprint Westport, Conn.: Hyperion Press, 1979).
32. This was first appreciated by Erwin Schrodinger who discussed it in *What is Life?* (Cambridge: Canto, reprinted, 1993).
33. T.B.L. Kirkwood, 'Biological Origins of Ageing', in *Oxford Textbook of Geriatric Medicine*, eds J.G. Evans and T.F. Williams (Oxford: Oxford University Press, 1992), pp. 41–8.
34. Paul Valéry, *M. Teste*, trans. Jackson Matthews (London: Routledge and Kegan Paul, 1973), p. 99.
35. This is a more complex and controversial claim than may seem at first sight and has been widely discussed – see, for example, Raymond Tallis, 'Medical Advances and the Future of Old Age', in *Controversies in Health Care Policies* ed. Marshall Marinker (London: BMJ Books, 1994), pp. 76–88.

  The notion that old age will take over where disease leaves off assumes that the two are distinct and this, as I have suggested, is not self-evident. Could not 'ageing' merely be the sum of sub-clinical disease processes that have not advanced far enough to assume the distinctive features of decay categorisable according to the International Classification of Diseases? This suggestion is also more complex than it might initially appear. First, subclinical disease likely to be mistaken for ageing would need to be multiple to have sufficient cumulative impact and one could argue that the vulnerability necessary to fall victim to a *multiplicity* of diseases is due to ageing: it is precisely what ageing is. Alternatively, low levels of disease might cause visible deterioration or death only in an organism that had been brought near to the threshold by other changes – presumably those of ageing. For such an organism, less disturbance would be required for a pathological process to produce dysfunction and, because of 'homoeostenosis' (narrowing of the range of physiological disturbances which can be accommodated by adaptive mechanisms), displacements from the normal range, are more likely to be irreversible. Thirdly, a disease is subclinical only so long as it has not been recognised by a clinician, or at least reported to a clinician by a patient. Clinician – and patient – recognition of disease depends upon *observation* of characteristic features: subclinicality may be maintained to a high level of damage by age-related failure of the body to produce characteristic (usually adaptive) responses, modifying the manifestations of disease, so that it may present as non-specific decline. If ageing is characterised by *de-differentiation of disease* – and the narrowing repertoire of disease manifestations expressed in the predominance of the 'geriatric giants' (falls, immobility, confusion and incontinence) – we could postulate a point of convergence of ageing and disease, where disease elicits no specific features or, alternatively, argue that ageing is a form of disease that, due to the failure of adaptive responses, elicits no specific features.

  These conceptual difficulties in demarcating ageing from disease are compounded by the empirical ones. At present, there are few symptoms and signs that fully meet the criteria for a 'true' ageing process: one that occurs universally in old age and only in old age. Conditions that seem to meet that criterion are often trivial (and scarcely life-threatening), such as wrinkling of the skin; or developmental, such as the menopause which has only incidental dysfunctional consequences. Some significant and serious pathology comes close to universality – for example, senile deterioration of the retina of eye, osteoarthritis, benign enlargement of the prostate – but no one is going to suggest that these can be dismissed as 'mere' ageing with the implication of therapeutic inactivity.

The uncertainty surrounding the characteristics of 'the ageing process *per se*' – if it exists – in part reflects the methodological problem of defining study populations in order to identify 'pure' or 'physiological' ageing. It also reflects the fact that there is often no sharp demarcation between changes that are typically attributed to age and those that are given a specific diagnostic label; for example, the pathological changes seen in Alzheimer dementia are also seen in normal brains, the difference being only one of distribution and quantity (though this may change when more robust genetic markers are identified).

As I have already indicated, there are obvious clinical, didactic and (perhaps) political reasons for saying, with perhaps more confidence than the facts and concepts justify, that ageing should be distinguished from disease. For disease is traditionally something that requires a therapeutic response; whereas, so far, ageing is something that patients, carers, clinicians and politicians feel no obligation to do anything about. Older people are at risk from being treated not as individuals but as points on a negative regression curve mapping 'inexorable' decline in function.

36. It is not inconceivable, therefore, that we may be able to slow down the ageing process at some time in the future, though the effects of this in the long term may be difficult to predict; after all, the human organism is a very complex dynamical system and for such systems predicting the outcome of interventions is notoriously difficult.

37. There is evidence to suggest that the period of disability before death is shrinking despite an increased lifespan; that there is, to use Fries and Crapo's phrase, a 'compression of morbidity'. For a review of some of the evidence on this, see Raymond Tallis, 'Medical Advances and the Future of Old Age', op. cit.

38. I have not addressed the ethical problems of increased longevity, which will almost certainly be achieved to a different extent in different societies. As each generation lives longer, it consumes more of the world's goods. If immortality were possible, we would be faced with the ultimate example of inequity, in which coming generations had freehold on time, where all previous generations had enjoyed only leasehold. This would be unlikely to be achieved world-wide, or not at least at the same time, so that such an advance would probably be at first an exacerbation of current inequities, whereby a minority is able to requisition a major share of the goods of the world, including medical care, while about a quarter of the world's population is destitute. Hitherto, some have been fortunate enough to live in the leeward side of history but all have lived in the windward side of time. The future prospect of a select few living in the leeward side of time while all others suffer the common condition of mortality seems an inequity yet more morally repulsive than those we see already. It is, however, only a distant prospect.

39. Isaiah Berlin, Introduction to *The Age of Enlightenment* (New York: New American Library, 1956), pp. 27–8.
40. Raymond Tallis, *The Explicit Animal*. It is worth noting that the more materialist of the Enlightenment philosophers were not alone in failing fully to understand that man is an explicit animal. Although early Marx accepted that self-consciousness was the 'species-characteristic' of mankind, he forget this when he turned Hegel on his head:

> My dialectic method is not only different from the Hegelian, but is its direct opposite. To Hegel, the life-process of the human brain, i.e. the process of thinking, which under the name of 'the Idea', he even transforms into an independent subject, is the demiurgos of the real world, and the real world is only the external phenomenal form of 'the Idea'. With me, on the contrary, the ideal is nothing less than the material world reflected by the human mind, and translated into forms of thought.
> (*Capital*, Volume 1, Afterword to the second German edition, trans. Samuel Moore and Edward Aveling, p. 19)

He gives priority to material nature without thinking how it is that nature becomes explicit in human beings – how, that is to say, it is (nonoptically) reflected in their brains and therein translated into 'forms of thought'.

I have said nothing here about the complexity of human consciousness; its million dimensions; its countless stories of reflection, modification, internal allusion, of conscious and unconscious inner distancings. My discussion in chapter 10 and elsewhere of the use of the greeting 'Hello' hardly begins to tap into the rich fabric of implicature sustaining even this linguistic tic.

41. Even operating with logic, with the abstract skeleton of reason, involves the imagination. Whitehead and Russell's *Principia Mathematica* did not write itself. And logic, of course, is not the whole of reason; reason is not the whole of reasonableness; reasonableness is not the whole of good sense. None the less each is the *sine qua non*, the necessary if not the sufficient condition, of the other: there is no reason without the constraint of logic; no reasonableness without the constraint of reason; no good sense without reasonableness.

The role of understanding in the deployment of reason is seriously underestimated by many critics of the Enlightenment; for example, Lucien Goldmann (*The Philosophy of the Enlightenment*):

> We have to choose between morally neutral technical knowledge and the synthesis of knowledge with immanent faith in a human community to be created by men; between understanding (*Verstand*) and reason (*Vernuft*); between capitalism and socialism. It is for us to determine which of these is to be the future vision of mankind.

Of course, we do not have to choose between *Verstand* and *Vernuft* in practical everyday life: on the contrary, the one cannot operate without the other.

42. It is interesting that the two antecedents of Enlightenment thought – rationalism and empiricism – did not see themselves as requiring a passive (rule-bound or impingement-shaped) individual consciousness:

The answer seems to be that these two philosophies [rationalism and empiricism] share the same fundamental concept: the treatment of the individual consciousness as the *absolute origin* of knowledge and action. Pure rationalism finds this origin in clear innate ideas existing independently of experience; pure empiricism, rejecting entirely the notion of innate ideas, finds the origins in sense-perceptions more or less mechanically organized into conscious thought.

(Goldmann, op. cit., p. 19)

43. Indeed, those who oppose universalism seem to be unable to do so without themselves being trapped into making universalistic statements. For example, the assertion that all truths are relative to interpretive communities is itself a truth that claims to encompass all interpretive communities, or to transcend particular communities. And the anti-Enlightenment Parisian *maîtres à pensers* were particularly fond of pronouncements laying claim to a breath-taking omniscience.
44. Isaiah Berlin, *Four Essays on Liberty*, p. 35.
45. Goldmann's observation is apposite here: that capitalist society 'splits . . . the individual bourgeois into two fundamentally opposed forms: the "economic man", amoral, unfeeling and irreligious when he is earning a living, and the kind father, affectionate friend and good Christian in the rest of his life.'
46. Roland Barthes, 'The Great Family of Man', in *Mythologies*, selected and trans. Annette Lavers (London: Jonathan Cape, 1972).
47. See Avraham Barkai, '*Volksgemeinschaft*, "Aryanization" and the Holocaust', in *The Final Solution: Origins and Implementation*, ed. David Cesarini (London: Routledge, 1994).
48. Marx cites a prize specimen in *Capital*. He quotes a certain Townsend, a Church of England parson, who glorified the misery of the masses as a necessary condition of the overall wealth of the nation:

'Legal constraint [to labour] is attended with too much trouble, violence and noise. . . . whereas hunger is not only a peacable, silent, unremitted pressure, but as the most natural motive to industry and labour, it calls forth the most powerful exertions'. Everything therefore depends upon making hunger permanent among the working class, and for this, according to Townsend, the principle of population, especially active among the poor, provides. 'It seems to be a law of Nature that the poor should be to a certain degree improvident' [i.e. so improvident as to be born *without* a silver spoon in the mouth] 'that there may always be some to fulfil the most servile, the most sordid, and the most ignoble offices in the community. The stock of human happiness is thereby much increased, while the more delicate are not only relieved from drudgery . . . but are left at liberty without interruption to pursue those callings that are suited to their various dispositions . . . it [the Poor Law] tends to destroy the harmony and beauty, the symmetry and order of that system which God and Nature have established in the world.'

(pp. 646–7)

49. Peter Medawar, 'On "The Effecting of All Things Possible"', in *The Hope of Progress* (London: Methuen, 1972), p. 125. This essay is one of the great masterpieces of English prose.

50. Anyone who doubts the reality of progress in some places should read
the horrific account of English working-class life in *Capital*. I found it
particularly illuminating to read about infant mortality rates in my own
town of Stockport; and about life in Doveholes, an ordinary village some
15 miles from where I am writing this. According to the report of the
(Poor Law) Relieving Officer of the Chapel-en-le-Frith Union (Chapel-
en-le-Frith is another delightful little village), at Doveholes,

> a number of small excavations have been made into a large hillock of
> lime ashes [the refuse of lime-kilns], and which are used as dwellings,
> and occupied by labourers and others employed in the construction of
> a railway ... through that neighbourhood. The excavations are small
> and damp, and have no drains or privies about them, and not the slightest
> means of ventilation except up a hole pulled through the top, and used
> for a chimney. In consequence of this defect, small-pox has been rag-
> ing for some time and some deaths amongst the troglodytes have been
> caused by them.
>
> (*Capital*, p. 665)

Some progress has been made since then. (One of the problems of assessing progress, is that one tends to forget where one has come from.)
51. A myth that is bolstered by two other myths: the myth of the organic
communities of the past; and the myth of the inorganic non-communities
of the present and the future. I have dealt with the former in *Newton's
Sleep* (London: Macmillan, 1995) and will not repeat here what I have
said there. Let me, however, say a few words about the supposed inorganic non-communities of the present.

The image that this invokes is of atomised individualistic lives passed
in consuming whatever is thrown at them by a television screen or a
computer terminal. This image is incomplete. In our household the younger
members spend much time interacting with electronic screens of various
sorts (though this is punctuated by frequent forays to play football and
basketball in the drive and elsewhere). During this time, they are no less
in communication with one another than they would be if one were playing
football for one team and the other for another. There is much passionate discussion of the game being played on the screen and the supposed
inorganic, post-modern, traditionless experience is frequently punctuated
by distinctly organic and immemorial traditions such as rolling on the
floor, fighting. This is simply an emphatic manifestation of the fact that
even the participants in a computer game are not dissolved into an ethernet
of elsewhere: they are here and now, rooted in the present tense and the
present location. Their experience of events on the screen is clearly situ-
ated in their experience of the house, each other, the changing lights and
sounds of their immediate environment.

The myth of the organic unity of the past was that individuals were
not separated by their fears and pains; and the myth of the inorganic
community of the present is that individuals are not related to their environment through the bodily experiences of warmth, shared physical
environments, etc.
52. Data taken from John Tierney's article 'Betting the Planet', published in
The Guardian, 28 December 1990. Of course, these facts, while undermining the case for pessimism, do not provide grounds for complacency.

War, famine, hunger, premature death, oppression, persecution and torture are still all too prevalent on the planet. Even so, it is astonishing how much things have improved, despite the continuation of disasters – natural and man-made – and an explosively rising population. How much better things would be if the population expansion had been curbed and men had behaved better towards men. At any rate, it is a tribute to the power of technology and to human ingenuity that it can still bring about an improvement in the lot of human beings despite the tendency of some of mankind to shoot itself through the foot with an AK-47.

53. Ernest Gellner, *Reason and Culture* (Oxford: Blackwell, 1993), p. 181. Gellner also points out that 'the claims of unreason are not equally persuasive in all spheres. They are not very persuasive in cognition, notwithstanding the fact that the absence of a warranty for rational procedures is undeniable. Cognition continues to function admirably, even given the absence of any such guarantee' (ibid., p. 181).

54. The more intelligent question is not whether we should accept reason or reject it – whether we should on the one hand, see the world in entirely rational terms or, on the other, assume a ferocious irrationality, whether we should be 'dry' or 'wild' – but how much wildness we can allow ourselves without forgetting the thirsty child screaming in the dust. But we are a long way from even beginning to ask this sort of question.

55. Medawar, op. cit., p. 125. The entire passage is worthy of citation:

> Many different elements enter into the movement to depreciate the services to mankind of science and technology. . . . We wring our hands over the miscarriages of science and technology and take its benefactions for granted. We are dismayed by air pollution but not proportionately cheered up by, say, the virtual abolition of poliomyelitis. . . . There is a tendency, even a perverse willingness to suppose that the despoliation sometimes produced by technology is an inevitable and irremediable process, a trampling down of Nature by the big machine. Of course it is nothing of the kind. The deterioration of the environment by technology is a technological problem for which technology has found, is finding, and will continue to find solutions. . . . I am all in favour of a vigorously critical attitude towards technological innovation: we should scrutinize all attempts to improve our condition and make sure that they do not in reality do us harm; but there is all the difference in the world between informed and energetic criticism and a drooping despondency that offers no remedy for the abuses it bewails.

56. This is particularly obvious in medicine, where enormous distances separate the well-validated idea or treatment from its routine good use in everyday practice. One has only to think of the poignant contrast between the care with which drugs are developed and tested and how they are used in the real world; between pre- and post-marketing.

57. Medicine may provide some useful models here. The incremental improvement in the treatment of illnesses, the recognition of uncertainty as an inescapable aspect of interventions and the caution with which changes are introduced (at least by non-charlatans) has much to teach planners and reformers. The recent ISIS studies alluded to earlier have led to the recommendation of a series of treatments which, if universally implemented would significantly reduce mortality from heart attacks. Several

hundreds of thousands of patients had to be recruited in many hundreds of centres in scores of countries in order to produce robust answers. The next step is to look at ways of implementing this newly established good practice and to investigate how services are organised to ensure this. This brings the challenges of medicine closer to the wider challenges of social reform. The Cochrane collaboration, which is keeping a world-wide database of properly validated double-blind controlled trials for universal use, is another model for social reformers to examine.

58. And this applies to future as much as present utility. Yes, we must work for the future but also live in the present because this is the only life we have; besides, there is a danger that tyrants may use the alibi of the future to justify present suffering, present iniquity. As Camus said, the future is the only estate that the masters freely make over to the slaves.

59. 'John Stuart Mill and the Ends of Life', in *Four Essays on Liberty*, op. cit.

60. I am perhaps being too generous, here, in following Merquior's diagnosis of the diagnosticians of society:

> That a deep cultural crisis is endemic to historical modernity seems to have been much more eagerly assumed than properly demonstrated, no doubt because, more often than not, those who generally do the assuming – humanist intellectuals – have every interest in being perceived as soul doctors to a sick civilisation.

There may be baser instincts at work. There is a delighted child in all of us that rejoices when things go wrong. And many of those who have loathed the modern world and detested progress have been themselves seriously damaged individuals: Baudelaire, Eliot, Benn – the list speaks for itself.

61. 'On "The Effecting of All Things Possible"', op. cit.

62. Just how new it is, and how extraordinary, is captured in Gellner's 'Prometheus Perplexed', the penultimate chapter of his *Reason and Culture*. A few passages will have to suffice:

> Reason is a foundling, not an heir of an old line, and its identity or justification, such as it is, is forged without the benefit of ancient lineage. A bastard of nature cannot be vindicated by ancestry but only, at best, by achievement. . . .
> The Cosmic Exile [reason], opting out of culture, is impractical. But it constitutes the noble and wholly appropriate charter of myth of a new kind of culture, a new system of a distinctively *Cartesian* kind of Custom and Example. Custom was not transcended: *but a new kind of custom altogether was initiated*. The separation of referential cognition from other activities, the systematic submission of cognitive claims to a severely extra-social centralized court of appeal (under the slogan of 'clear and distinct ideas', or of 'experience'), and the establishment of a single currency of reference, had burst open the limits of knowledge. It initiated and made possible an age of totally unprecedented, fabulous cognitive and economic growth. Through its associated technology, it brought the Malthusian age to an end. Henceforth resources would, and generally did, grow faster than population. Coercive political systems were no longer imposed on mankind by the need to enforce an inevitably unjust distribution on members of society endowed with inherently limited resources. Oppression, from now on, was to be our option, but no longer our destiny.

... in the one great and irreversible transition or *coupure* between the traditional and the rational spirit, pragmatic considerations overwhelmingly and decisively favour one of the two contestants. At one particular crossroads, the verdict of history is categorical, unambiguous, decisive and irreversible.

(pp. 160, 165)

This should be read in conjunction with the following passage from Hermann Hesse:

Since the end of the Middle Ages, intellectual life in Europe seems to have evolved along two major lines. The first of these is the liberation of thought and belief from the sway of all authority. In practice this meant the struggle of Reason, which at last felt it had come of age and won its independence, against the domination of the Roman Church. The second trend, on the other hand, was the covert but passionate search for a means to confer legitimacy on this freedom, for a new and sufficient authority arising out of Reason itself. We can probably generalise and say that Mind has by and large won this often strangely contradictory battle for two aims basically at odds with each other.

(Hermann Hesse, *The Glass Bead Game*, trans. Richard and Clara Winston (London: Picador, 1987), p. 19).

63. 'The Genetic Improvement of Man', in *The Hope of Progress* (London: Methuen, 1972), p. 69.

This essay usefully distinguishes three main kinds of vision of the future of man: the Olympian, the Arcadian and the Utopian. In the Olympian version, 'men can become like gods; can achieve complete virtue, understanding and peace of mind, but through spiritual insight, not by mastery of the physical world'. Arcadian visions of the future are bound up with the ancient legend of a Golden Age, 'it is directed backwards': 'men remain human but in a state of natural innocence. They retreat into a tranquil pastoral world where peace of mind is not threatened, intellectual aspiration is not called for, and virtue is not at risk.' Authority is replaced by fraternity in this 'world without strife, without ambition, and without material accomplishment'. The Utopian vision assumes that man improves the world through his own exertions: 'he begins as a tenant or lodger in the world, but ends up as its landlord; and as his environment improves, so, it is alleged, will he': 'Virtue can be learned and will eventually become second nature, understanding can be aspired to, but complete peace of mind can never be achieved because there will always be something more to do. Men look forwards, never backwards, and seldom upwards.'

My own hope for the future combines elements of all three visions. From the Olympian vision, I would take the emphasis on spirituality. This, however, would be possible only on the basis of universal, or near-universal, liberation from material want. From the Arcadian vision, I would borrow the emphasis upon the light-handededness of authority, itself possible because the emphasis upon the 'metaphysicalisation' of experience (or the discovery of the mystery inherent in experience) would limit the appetite for material gain and the strife that follows from this. The logic of increasing consumption would be challenged by a shift of emphasis towards experiencing more deeply and more thoroughly the things that we have – learning how to possess our possessions. (I have discussed this at greater length in 'The Work of Art in an Age of Electronic Reproduction', in my forthcoming *Theorrhoea and After*.) And from the Utopian vision, I would take the belief that material progress is a necessary condition of spiritual liberation – to paraphrase Brecht, 'Grub first and then metaphysics' – and the assumption that individuals would behave better towards one another when they were themselves treated better – 'Grub first, then ethics'.

And I would reject elements of each. The Olympian vision seems to suggest that man can live by the spirit alone and that each can survive in a solipsist bubble of ecstatic contemplation. On the contrary, we need the material and spiritual succour of the companionage of others. The Arcadian vision is based upon a sentimental idea about human nature in the remote past, in particular about organic, agrarian communities. We have little evidence that such communities are as fraternal as Rousseau and others would have us believe: spite is not an invention of the drawing room. As Jon Elster has pointed out, pre-industrial societies are riddled with envy:

> A depressing fact about many peasant societies is that people who do better than others are often accused of witchcraft and thus pulled down to, or indeed below, the level of others. Against this background, ruthless selfishness can have a liberating effect.
> (quoted in Aaron Wildavsky, 'Can Norms Rescue Self-Interest?' *Critical Review*, (1991) 5(3): 305–25, at p. 315)

And, finally, most Utopian visions focus too exclusively on material advancement.

64. I have expressed the view that lack of belief in progress takes away an important part of the meaning of life. Others have suggested that, on the contrary, belief in progress undermines the meaning of the past and, indeed, of the present in so far as it is regarded as the past of the future. For example, Bryan Appleyard talks of the reduction of 'history to an insignificant landscape of ages that were trying and failing to become our age', adding that 'in time, our age will be reduced to the same condition' (*Understanding the Present*, London: Picador, 1992, p. 237). I think that the hope of progress and respect for the past can be reconciled: we have simply to recognise what those who brought about the better future (which is our present) achieved; to remember, as Newton did, those giants upon whose shoulders we stand.

Index